Undergraduate Texts in Physics

Undergraduate Texts in Physics (UTP) publishes authoritative texts covering topics encountered in a physics undergraduate syllabus. Each title in the series is suitable as an adopted text for undergraduate courses, typically containing practice problems, worked examples, chapter summaries, and suggestions for further reading. UTP titles should provide an exceptionally clear and concise treatment of a subject at undergraduate level, usually based on a successful lecture course. Core and elective subjects are considered for inclusion in UTP.

UTP books will be ideal candidates for course adoption, providing lecturers with a firm basis for development of lecture series, and students with an essential reference for their studies and beyond.

More information about this series at http://www.springer.com/series/15593

Arnt Inge Vistnes

Physics of Oscillations and Waves

With use of Matlab and Python

 Springer

Arnt Inge Vistnes
Department of Physics
University of Oslo
Oslo, Norway

Translated by Razi Naqvi

ISSN 2510-411X   ISSN 2510-4128   (electronic)
Undergraduate Texts in Physics
ISBN 978-3-319-72313-6   ISBN 978-3-319-72314-3   (eBook)
https://doi.org/10.1007/978-3-319-72314-3

Library of Congress Control Number: 2018950787

Translation from the Norwegian language edition: *SVINGNINGER OG BØLGERS FYSIKK* by Arnt Inge
Vistnes, © CreateSpace Independent Publishing Platform, 2016. All Rights Reserved.

This Springer imprint is published by the registered company Springer Nature Switzerland AG
The registered company address is: Gewerbestrasse 11, 6330 Cham, Switzerland

To
Kirsten
Ingunn, Torunn, Maria
with families

Preface

Origin

The University of Oslo in Norway is one of the first universities to introduce numerical methods as an integral part of almost all mathematically oriented courses for science students (first attempts started in 1997). This created the need for textbooks in physics covering all the topics included in the syllabus. There were many textbooks on oscillations and waves on the market, but none adhered well with the learning objectives we adopted.

The Norwegian version of this book was originally written in 2008 for use in the course *"FYS2130 Svingninger og bølger"* (Oscillations and Waves) and has undergone many revisions and expansions since then. The course is given in the fourth semester to students enrolled in the Department of Physics at the University of Oslo. These students have taken courses in Python programming, classical mechanics and electromagnetism, but have had limited education in oscillations and wave phenomena.

Scope

In the present book, I have mostly adhered to traditional descriptions of the phenomena; however, I have also tried to point towards potential limitations of such descriptions. When appropriate, analogies between different phenomena are drawn.

The formalism and phenomena are treated quite differently from section to section. Some sections provide only qualitative descriptions and thus only a superficial or introductory understanding of the topics while other sections are more mathematical and demanding. Occasionally, the mathematical derivations are not essential to understand the material, but are included to show the connection between basic physical laws and the phenomena discussed in the text.

Principles from numerical methods are employed as they permit us to handle more realistic problems than pure analytical mathematics alone, and they facilitate to obtain a deeper understanding of some phenomena.

Program codes are given, ready to use, and is a tool for further exploration of the phenomena that are covered. Our experience from teaching this topic to students over years is that, numerical methods based on "hands-on computer code development" expand the experimental attitude and facilitate the learning process.

We try in this book to emphasize how so-called algorithmic thinking can improve understanding. As a personal example, the algorithm for calculating how a wave evolves over time has given me a much deeper understanding of the wave phenomena than by working with analytical mathematics over years. Another example is the realization that all variants of classical interference and diffraction can be calculated using a single computer program, demonstrating not only that numerical methods are powerful, but also that the underlying physical mechanism is identical in all these cases.

We have made an effort to ensure a logical and reader-friendly structure of the book. Especially important parts of the core material in the text are marked by coloured background, and various examples show how the core material can be used in different contexts. Supplementary information and comments are given in small print. Learning objectives point to the most important sections of each chapter. Most of the chapters include suggestions to further reading.

There are three types of exercises in the book. The first type of exercise consists of a list of concepts in each chapter that can be used by students in various ways for active learning. Thereafter follow comprehension/discussion questions and more regular problems often including calculations. Best learning outcome is achieved by trying all the three types of tasks, including oral discussions when working with understanding concepts and the comprehension/discussion questions. The problems used in the exercises are taken from daily life experiences, in order to demonstrate how physics is relevant in many aspects of our everyday life.

For the more regular problems, the aim is to encourage the reader to learn how to *devise* a strategy for solving the problem at hand and to select the appropriate laws. A "correct answer" without an adequate justification and reasoning is worthless. In many tasks, not all the relevant quantities are supplied, and in these cases, the reader must search for the necessary information in other books or the Internet. This is a natural part of working with physics today. A list of answers for the problems is not worked out yet. Some problems require particular data files to be analyzed that will be available from a web page advertised by the publisher.

Content

In our daily life, oscillations and waves play an important role. The book covers sound phenomena, our sense of hearing, and the two sets of measurements of sound and units that are in use: one for physical purposes solely and the other related to the sense of hearing. Similarly, the book treats light phenomena and our sense of vision, as well as the two sets of measurements and units that are in use for these purposes. In addition, we also discuss colour mixing and important differences between our senses of hearing and vision.

By introducing Fourier transform, Fourier series and fast Fourier transform, we introduce important tools for analysis of oscillatory/wave phenomena. Our aim is to give the reader all necessary details so that she/he can utilize this numeric method to its full potential. We also point out a common misconception we often find in connection with Fourier analysis.

We introduce continuous wavelet transform with Morlet wavelets as a kind of time-resolved Fourier transform and explain why we have chosen this method instead of a short-term Fourier transform. Much emphasis is put on optimizing the analysis and how this is closely related to the time-bandwidth product; a classical analogue to Heisenberg's uncertainty principle. A computer program is provided for this topic as well as for many other parts of the book.

One chapter is devoted to numerical method, mainly in how to solve ordinary and partial differential equations of first or second order. Other topics covered in the book are geometric optics, interference, diffraction, dispersion and coherence. We also briefly cover skin effect, waveguides and lasers.

Intended Audience

The reader of the book should have some basic programming experience, preferably in Matlab or Python, and know basic mechanics and electromagnetism. The principal ingredients of the book encompassing physical phenomena and formalism, analytical mathematics, numerical methods, focus on everyday phenomena and state-of-the-art examples are likely to be of interest to a broader group of readers. For instance, we have experienced that established physicists who want to look up details within the themes like colour vision, geometrical optics and polarization also appreciate the book.

Computer Programs

In this book all computer programs are given in Matlab code. However, all the these programs are available as separate files both in Matlab and in Python code at the "additional resources" Web page at https://urldefense.proofpoint.com/v2/url?u=http-3A__www.physics.uio.no_pow_&d=DwIFAw&c=vh6FgFnduejNhPPD0fl_yRaSfZy 8CWbWnIf4XJhSqx8&r=XEZMlNM4R7ifyRi4yqEXrX4PijpiPm_iDxB1eUZS6p7 kGbAxcoXv9UvW2zchY1tK&m=mR4-YUO9Vkp_XIQj5cqH36JdicZTu7wQVqj 1FXexQYg&s=sGWNtpJ6F9-V_b46f9GYoA2AFPJayGu6bE4AdSa-Yts&e=.

Some introduction is given to programming style, reproducibility and documentation, but not at a level as is expected for a course fully devoted to programming. We do not provide an introduction to "dimensionless variables".

Acknowledgements

I want to take this opportunity to thank everyone who contributed to this book, particularly Borys Jagielski, Knut Kvaal, Jan Henrik Wold, Karl A. Maaseide, Irina Pettersson, Maria Vistnes, Henrik Sveinsson, Cecilie Glittum and colleagues and students at the Department of Physics; I owe special gratitude to Anders Johnsson for offering valuable hints and comments. I am also indebted to K. Razi Naqvi, who translated the book from Norwegian to English and contributed to many substantial improvements of the material presented in the original version. Many parts of the book are modified after the translation, so do not blame Prof. Razi Naqvi if you find bad English sentences here and there.

Morten Hjorth-Jensen is thanked for his perennial support and interest in issues related to teaching. Thanks are also offered to Hans Petter Langtangen for inspiration and hints regarding programming. I must also thank my former teachers, among them Svenn Lilledal Andersen and Kristoffer Gjøtterud, for creating an environment in which my physics understanding grew and developed, and to Gunnar Handal, who challenged me in a constructive manner as regards university pedagogy.

A generous grant from *The Norwegian Non-fiction Writers and Translators Association* allowed me to be free from teaching obligations for two fall semesters, during which the first version of the book and some of the illustrations were prepared. A warm "thank you" to Anders Malthe-Sørenssen for providing inspiration for teaching in general and for securing financial support for the translation of the book from Norwegian to English.

Most of all, I thank my dear Kirsten and our children for their loving forbearance during the periods when I have been busy working with this book. I now look forward to take more part in family life.

Kurland, Norway   Arnt Inge Vistnes
June 2018

Contents

Chapter 1
Introduction

Abstract Initially, the introductory chapter deals with different ways people comprehend physics. It might provide a better understanding of the structure of the book and choices of the topics covered. It continues with a description and discussion on how the introduction of computers and numerical methods has influenced the way physicists work and think during the last few decades. It is indicated that the development of physics is multifaceted and built on close contact with physical phenomena, development of concepts, mathematical formalism and computer modelling. The chapter is very short and may be worth reading!

1.1 The Multifaceted Physics

Phenomena associated with oscillations and waves encompass some of the most beautiful things we can experience in physics. Imagine a world without light and sound, and then you will appreciate how fundamental oscillations and waves are for our lives, for our civilization! Oscillations and waves have therefore been a central part of any physics curriculum, but there is no uniform way of presenting this material.

"Mathematics is the language of physics" is a claim made by many. To some extent, I agree with them. Physical laws are formulated as mathematical equations, and we use these formulas to calculate the expected outcomes of experiments. But, in order to be able to compare the results of our calculations with actual observations, more than sheer mathematics is needed. Physics is also an edifice founded on concepts, and the concepts are entwined as much with our world of experience as with mathematics. Divorced from everyday language, notions and experiences, the profession would bear little resemblance to what we today call physics. Then we would just have pure mathematics! The Greek word $\phi\upsilon\sigma\iota\varsigma$ ("physis") means *the nature* and physics is a part of *natural* science.

People are different. My experience is that some are fascinated primarily by mathematics and the laws of physics, while others are thrilled by the phenomena in themselves. Some others are equally intrigued by both these facets. In this book, I will try to present formalism as well as phenomena, because—as stated above—it is the combination that creates physics (Fig. 1.1)! A good physicist should be in close

© Springer Nature Switzerland AG 2018

A. I. Vistnes, *Physics of Oscillations and Waves*, Undergraduate Texts in Physics,

https://doi.org/10.1007/978-3-319-72314-3_1

Fig. 1.1 Oscillations and waves are woven into a host of phenomena we experience every single day. Based on fairly general principles, we can explain why the most common rainbow has invariably a radius of $40-42°$ and is red outward, and the sky just outside the rainbow is slightly darker than that just inside. You already knew this, but did you know that you can extinguish the light from a rainbow almost completely (but not for the full rainbow simultaneously), as in the right part of the figure, by using a linear polarization filter? The physics behind this is one of the many themes covered in this textbook

contact with phenomena as well as formalism. For practical reasons and with an eye on the size of the book, I have chosen to place a lot of emphasis on mathematics for some of the phenomena presented here, while other parts are almost without mathematics.

Mathematics comes in two different ways. The movement of, for example, a guitar string can be described mathematically as a function of position and time. The function is a solution of a differential equation. Such a description is fine enough but has an ad hoc role. If we know the amplitude at a certain time, we can predict the amplitude at a later instant. Such a description is a necessity for further analysis, but really has little interest beyond this. In the mechanics, this is called a *kinematic* description.

It is often said that *in physics we try to understand how nature works*. We are therefore not satisfied by a mere mathematical description of the movement of the guitar string. We want to go a little deeper than this level of description. How can we "explain" that a thin steel string under such-and-such tension actually gives the tone C when it is plucked? The fascinating fact is that with the help of relatively few and simple physical laws we are able to explain many and seemingly diverse phenomena. That gives an added satisfaction. We will call this a *mechanical* or *dynamic* description.

Mathematics has traditionally been accorded, in my opinion, overmuch space, compared with the challenge of understanding mechanisms. This is due in part to the fact that we have been using, by and large, analytical mathematical methods for solving the differential equations that emerge. To be sure, when we use analytical methods, we must penetrate the underlying mechanisms for the sake of deducing the equations that portray the phenomena. However, the focus is quickly shifted to the challenges of solving the differential equation and discussing the analytical solution we deduce.

This approach has several limitations. First of all, the attention is diverted from the content of the governing equations, wherein lie the crucial mechanisms responsible for the formation of a wave. Secondly, there are only a handful of simplified cases we are able to cope with, and most of the other equations are intractable by analytical means. We often have to settle for solutions satisfying simplified boundary conditions and/or solutions that only apply after the transient phase has expired.

This means that a worrying fraction of many generations of physicists are left with simplified images of oscillations and waves and believe that these images are valid in general. For example, according to my experience, many physicists seem to think that electromagnetic waves are generally synonymous with plane electromagnetic waves. They assume that this simplified solution is a general formula that can be used everywhere. Focusing on numerical methods of solution makes it easier to understand why this is incorrect.

1.2   Numerical Methods

Since about the year 2000, a dramatic transformation of physical education in the world has taken place. Students are now used to using computers and just about everyone has their own or have easy access to a computer. Computer programs and programming tools have become much better than they were a few decades ago, and advanced and systematic numerical methods are now widely available. This means that bachelor students early in their study can apply methods as advanced as those previously used only in narrow research areas at master's and Ph.D. level. That means they can work on physics in a different and more exciting way than before.

Admittedly, we also need to set up and solve differential equations, but numerical solution methods greatly simplify the work. The consequence is that we can play around, describing different mechanisms in different ways and studying how the solutions depend on the models we start with. Furthermore, numerical solution methods open the door to many more real-life issues than was possible before, because an "ugly" differential equation is not significantly harder to solve numerically than a simple one. For example, we could write down a nonlinear description of friction and get the results almost as easily as without friction, whereas the problem is not amenable to a purely analytical method of solution.

This means that we can now place less emphasis on different solution strategies for differential equations and spend the time so saved for dealing with more real-life issues. I myself belong to a generation which learned to find the square root of a number by direct calculation. After electronic calculators came on the market, I have had no need for this knowledge. We are now in a similar phase in physics and mathematics. For example, if we use the Maple or Mathematica computer programs, we get analytical expressions for a wealth of differential equations, and if a differential equation does not have a straightforward analytical solution, the problem can be solved numerically. Some skills from previous years therefore have less value today, while other skills have become more valuable.

This book was written during the upheaval period, during which we switched from using exclusively analytical methods in bachelor courses to a situation where computers are included as a natural aid both educationally and professionally. We will benefit directly from this not only for building up a competence that everyone will be happy to employ in professional life, but also by using it as an educational tool for enhancing our understanding of the subject matter. With numerical calculations, we can focus more easily on the algorithms, basic equations, than with analytical methods. In addition, we can address a wealth of interesting issues we could not study just by analytical methods, which contributes to increased understanding. Numerical methods also allow us to analyse functions/signals in an elegant way, so that we can now get much more relevant information than we could with the methods available earlier.

Using numerical methods is also more interesting, because it enables us to provide "research-based teaching" more easily. Students will be able to make calculations similar to those actually done in research today. There are plenty of themes to address because a huge development in different wave-based phenomena is underway. For example, we can use multiple transducers located in an array for ultrasound diagnostics, oil leakage, sonar and radar technology. In all these examples, well-defined phase differences are used to produce spatial variations in elegant ways. Furthermore, in so-called photonic crystals and other hi-tech structures at the nanoscale, we can achieve better resolution in measurements than before, even better than the theoretical limits we believed to be unreachable just a few years ago. Furthermore, today we utilize nonlinear processes that were not known a few decades ago. A lot of exciting things are happening in physics now, and many of you will meet the topics and methods treated in this book, even after graduation.

1.2.1 Supporting Material

A "Supplementary material" web page at http://www.physics.uio.no/pow is available for the readers of this book. The page will offer the code of the computer programs (both Matlab and Python versions), data files you need for some problems, a few videos, and we plan to post reported errors and give information on how to report errors and suggestions for improvements.

1.2.2 Supporting Literature

Many books have been written about oscillations and waves, but none of the previous texts covers the same combination of subjects as the present book. It is often useful to read how other authors have treated a particular topic, and for this reason, we recommend that you consult, while reading this book, a few other books and check, for example, Wikipedia and other relatively serious material on the Web. Here are some books that may be of interest:

- Richard Fitzpatrick: "Oscillations and Waves: An introduction". CRC Press, 2013.
- H. J. Pain: "The Physics of Vibrations and Waves". 6th Ed. Wiley, 2005.
- A. P. French: "Vibrations and Waves". W. W. Norton & Company, 1971.
- Daniel Fleisch: "A Student's Guide to Maxwell's Equations". Cambridge University Press, 2008.
- Sir James Jeans: "Science and Music". Dover, 1968 (first published 1937).
- Eugene Hecht: "Optics", 5th Ed. Addison Wesley, 2016.
- Geoffrey Brooker: "Modern Classical Optics". Oxford University Press, 2003.
- Grant R. Fowles: "Introduction to Modern Optics". 2nd Ed. Dover Publications, 1975.
- Ian Kenyon: "The Light Fantastic". 2nd Ed. Oxford University Press, 2010.
- Ajoy Ghatak: Optics, 6th Ed., McGraw Hill Education, New Delhi, 2017.
- Karl Dieter Möller: "Optics. Learning by Computing, with Model Examples Using MathCad, Matlab, Mathematica, and Maple". 2nd Ed. Springer 2007.
- Peter Coles: "From Cosmos to Chaos". Oxford University Press, 2010.
- Jens Jørgen Dammerud: "Elektroakustikk, romakustikk, design og evaluering av lydsystemer". http://ac4music.wordpress.com, 2014.
- Jonas Persson: "Vågrörelselära, akustik och optik". Studentlitteratur, 2007.

Chapter 2
Free and Damped Oscillations

Abstract This chapter introduces several equivalent mathematical expressions for the oscillation of a physical system and shows how one expression can be transformed into another. The expressions involve the following concepts: amplitude, frequency and phase. The motion of a mass attached to one end of a spring is described by Newton's laws. The resulting second-order homogeneous differential equation has three solutions, depending on the extent of energy loss (damping). The difference between a general and a particular solution is discussed, as well as superposition of solutions for linear and nonlinear equations. Oscillation in an electrical RCL circuit is discussed, and energy conservation in an oscillating system which has no energy dissipation is examined.

2.1 Introductory Remarks

Oscillations and vibrations are a more central part of physics than many people realize. The regular movement of a pendulum is the best-known example of this kind of motion. However, oscillations also permeate all wave phenomena. Our vision, our hearing, even nerve conduction in the body are closely related to oscillations, not to mention almost all communication via technological aids. In this chapter, we will look at the simplest mathematical descriptions of oscillations. Their simplicity should not tempt you into underestimating them. Small details, even if they appear to be insignificant, are important for understanding the more complex phenomena we will encounter later in the book.

2.2 Kinematics

In mechanics, we distinguish between kinematics and dynamics, and the distinction remains relevant when we consider oscillations. Within kinematics, the focus is primarily on *describing* motion. The description is usually the *solution* of differential

© Springer Nature Switzerland AG 2018

A. I. Vistnes, *Physics of Oscillations and Waves*, Undergraduate Texts in Physics,

https://doi.org/10.1007/978-3-319-72314-3_2

equations or experimental measurements. The underlying physical laws are not taken into consideration.

In dynamics, on the other hand, we set up the differential equations of motion based on known physical laws. The equations are solved either by analytical or numerical methods, and we study how the solutions depend on the physical models we started with. If we seek physical understanding, dynamic considerations are of greater interest, but the kinematics can also be useful for acquiring familiarity with the relevant mathematical description and the quantities that are included.

How do we describe an oscillation? Let us take an example: A mass attached to one end of a spring oscillates vertically up and down. The top of the spring is affixed to a stationary point.

> The kinematic description may go like this: The mass oscillates uniformly about an equilibrium point with a definite frequency. The maximum displacement A relative to the equilibrium point is called the *amplitude* of oscillation. The time taken by the mass to complete one oscillation is called *time period* T. The *oscillation frequency* f is the inverse of the time period, i.e. $f \equiv 1/T$, and is measured in reciprocal seconds or hertz (Hz).

Suppose we use a suitably chosen mass and a limited amplitude of displacement for the spring. By that we mean that the amplitude is such that the spring is always stretched, and never so much as to suffer deformation. We will be able to observe that the position of the mass in the vertical direction $z(t)$ will almost follow a mathematical sine/cosine function:

$$z(t) = A \cos(2\pi t/T) .$$

However, such a description is not complete. There is no absolute position or absolute time in physics. Therefore, when we specify a position z (along a line), we must also specify the point with respect to which the measurement is made. In our case, this reference point is the position of the mass when it is at rest.

Similarly, we must specify the reference point relative to which the progress of time is measured. In our case, the origin of time is chosen so that the position has a maximum value at the reference time $t = 0$. If there is a mismatch, we must compensate by introducing an *initial phase* ϕ, and use the expression

$$z(t) = A \cos(2\pi t/T + \phi) .$$

Since the quantity $2\pi/T$ occurs in many descriptions of oscillatory movements, it proves advantageous to define an *angular frequency* of ω as follows:

$$\omega \equiv 2\pi/T = 2\pi f$$

Fig. 2.1 A harmonic
oscillation is characterized
by amplitude, frequency and
phase; see the text

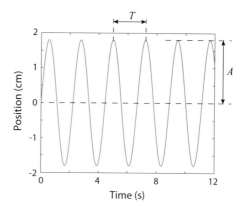

where f is the frequency of oscillation. This is a fairly common way to describe an oscillation (Fig. 2.1).

However, a "simple harmonic oscillation" can be described in many ways. The most common mathematically equivalent ways are:

$$z(t) = A \cos \omega t + B \sin \omega t \tag{2.1}$$
$$= C \cos(\omega t + \phi) \tag{2.2}$$
$$= \Re\left\{ \mathscr{D} e^{i\omega t} \right\} \tag{2.3}$$
$$= \Re\left\{ E e^{i(\omega t + \phi)} \right\} \tag{2.4}$$

$\Re\{\}$ indicates that we take the the real part of the complex expression within the braces, and $\mathscr{D}$ is a complex number.

Euler's formula for the exponential function (complex form) has been used in the last two expressions. According to Euler's formula:

$$e^{i\alpha} = \cos \alpha + i \sin \alpha \ .$$

This formula forms the basis for a graphical representation of a harmonic motion: First, imagine that we draw a vector of unit length in a plane. The starting point of the vector is placed at the origin and the vector forms an angle α with the x-axis. The vector can then be written as follows:

$$\hat{x} \cos \alpha + \hat{y} \sin \alpha$$

Fig. 2.2 A phasor is a vector of a given length. The phasor rotates at a given *angular frequency* and with a definite initial phase. The figure shows the position of the phasor at one point in time. See the text

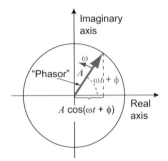

where $\hat{x}$ and $\hat{y}$ are unit vectors along the x- and y- direction, respectively. The similarity to the previous expression is striking, assuming that the real part of the expression is taken to be the component along the x-direction and the imaginary part as the y-component.

This graphical vector representation can be extended immediately to represent a harmonic oscillation. We then use a vector with a length corresponding to the amplitude of the harmonic motion. The vector rotates with a fixed angular frequency of ω about the origin. The angle between the vector and the x axis is always $\omega t + \phi$. Then the x-component of the vector at any given time indicates the instantaneous amplitude of the harmonic oscillation. Such a graphical description is illustrated in Fig. 2.2 and is called an *phasor* description of the motion.

Phasors are very useful when multiple contributions to a motion or signal of the same frequency are to be summed up. The sum of all contributions can be found by vector addition. Especially in AC power, when voltages over different circuit components are summed, phasors are of great help. We will come back to their uses later. Phasors are useful also in other contexts, but mostly when all contributions in a sum have the same angular frequency.

It is important to learn all the mathematical expressions (2.1)–(2.4) for simple oscillatory motion so that they can be instantly recognized when they appear. It is also important to be able to convert quickly from one form to another. This book is full of such expressions!

2.3   Going from One Expression to Another

Phasors are of immense aid. As mentioned, a phasor is a vector that rotates in the complex plane as time passes (see Fig. 2.3). The vector rotates at an angular velocity equal to ω. The component of this vector along the real axis represents the physical value of our interest, and it is this component that can be expressed in more than four equivalent ways.

Fig. 2.3 Sketch of a phasor that rotates around the origin with an angular velocity ω.

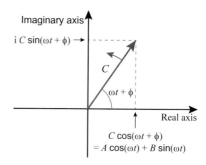

2.3.1 First Conversion

Let us first show the transition from Eqs. (2.2) to (2.1). We use Rottmann's compilation of mathematical formula (an important tool when working with this book!), and use the trigonometric addition formula for cosines to get:

$$z(t) = C \cos(\omega t + \phi)$$
$$= C \{\cos \omega t \cos \phi - \sin \omega t \sin \phi\}$$
$$= [C \cos \phi] \cos \omega t + [-C \sin \phi] \sin \omega t.$$

This expression is formally identical to Eq. (2.1), from which it follows that:

$$C \cos(\omega t + \phi) = A \cos \omega t + B \sin \omega t \quad \text{if we set} \quad A = C \cos \phi \quad \text{and} \quad B = -C \sin \phi.$$
$$(2.5)$$

2.3.2 Second Conversion

We can go the opposite way by utilizing the details given in Eq. (2.5):

$$A^2 + B^2 = (C \cos \phi)^2 + (C \sin \phi)^2 = C^2 (\sin^2 \phi + \cos^2 \phi) = C^2$$

$$C = \pm \sqrt{A^2 + B^2} .$$

And, by dividing the last two relations in Eq. (2.5), we get:

$$\frac{B}{A} = \frac{-C \sin \phi}{C \cos \phi} = -\tan \phi$$

This is a fraction whose numerator is the y-component and the denominator the x-component of the phasor at $t = 0$. Then, it follows that

$$\phi = -\arctan\frac{B}{A} \ .$$

It should be noted here that both the tan and arctan have a periodicity of π, and one has to be careful about which of the two possible solutions one chooses. What quadrant ϕ is in depends on the sign of A and B separately. We must keep this in mind to make sure we choose the correct ϕ!

If a computer is used for calculating arctan, the $atan2(B, A)$ variant is recommended for both Matlab and Python. Then the angle comes out in the correct quadrant.

With these reservations, we have shown:

$$A\cos(\omega t) + B\sin(\omega t) = C\cos(\omega t + \phi) \quad \text{where} \quad C = \sqrt{A^2 + B^2} \quad \text{and} \quad \phi = -\arctan\frac{B}{A} \ .$$
$$(2.6)$$

2.3.3 Third Conversion

The transition from Eqs. (2.4) to (2.2) is very simple if we use Euler's formula:

$$e^{i\alpha} = \cos\alpha + i\sin\alpha \ .$$

From this, it follows that:

$$\Re\left\{Ee^{i(\omega t + \phi)}\right\} = \Re\left\{E\left[\cos(\omega t + \phi) + i\sin(\omega t + \phi)\right]\right\} = E\cos(\omega t + \phi) \ .$$

If this is equal to $C\cos(\omega t + \phi)$ one must have:

$$\Re\left\{Ee^{i(\omega t + \phi)}\right\} = C\cos(\omega t + \phi) \quad \text{if} \quad C = E \ . \qquad (2.7)$$

This simple relation holds equally well both ways (from Eqs. (2.4) to (2.2) or the opposite way).

2.3.4 Fourth Conversion

The last rendering to be considered here is also based on Euler's formula. It is the conversion of Eqs. (2.3) to (2.1). It is crucial to note that $\mathscr{D}$ is complex. We write this number as a sum of a real and an imaginary part:

$$\mathscr{D} = D_{\text{re}} + i D_{\text{im}}$$

where D_{re} and D_{im} are both real. This leads (once again through Euler's formula):

$$\Re\left\{\mathscr{D}e^{i\omega t}\right\} = \Re\left\{(D_{\text{re}} + i D_{\text{im}})(\cos\omega t + i\sin\omega t)\right\}$$

$$= \Re\left\{D_{\text{re}}\cos\omega t + i D_{\text{re}}\sin\omega t + i D_{\text{im}}\cos\omega t + i^2 D_{\text{im}}\sin\omega t\right\}$$

$$= D_{\text{re}}\cos\omega t - D_{\text{im}}\sin\omega t \ .$$

When this is compared with

$$A\cos\omega t + B\sin\omega t \ ,$$

one is led to the simple relation:

$$\Re\left\{\mathscr{D}e^{i\omega t}\right\} = A\cos(\omega t) + B\sin(\omega t) \quad \text{if} \quad \mathscr{D} = A - iB \ . \qquad (2.8)$$

This simple relationship also works both ways (from Eqs. (2.3) to (2.1) or the converse).

We could also look at the expression $z(t) = C\sin(\omega t + \phi)$ instead of $z(t) = C\cos(\omega t + \phi)$, but with the procedures outlined above it should be easy to navigate from one form to the next.

When we come to treat waves in later chapters, we will often start with harmonic waves. The expressions then become almost identical to those we have in Eqs. (2.1)–(2.4). It is important to be familiar with these expressions.

2.4 Dynamical Description of a Mechanical System

Let us come back now to physics. A spring often follows Hooke's law: the deviation from the equilibrium point is proportional to the restoring force exerted by the spring.

Suppose that the suspension hangs vertically without any mass at the end. It has a length of L_0. If a mass m is attached to the free end, and we wait until the system has settled, the spring will have a new length, say L_1, that satisfies the equation

Fig. 2.4 Definition of
different lengths of the
spring with and without an
attached mass; see the text

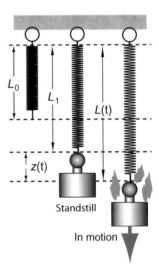

$$k(L_1 - L_0) = mg$$

where the experimentally determined k is called the spring constant, and g, the
acceleration due to gravity, is considered constant (disregarding the variation of g
with the height) (Fig. 2.4).

If the mass is pulled down slightly and released, the force acting on the mass will
always be

$$F(t) = k[L(t) - L_0] - mg$$

where $L(t)$ is the instantaneous length of the spring. Upon combining the last two
equations, one gets

$$F(t) = k[L(t) - L_0] - k(L_1 - L_0)$$
$$= k[L(t) - L_1] .$$

Important: The elongation of the spring from length L_0 to L_1 is a consequence of
the force of gravity. Therefore, in later expressions, neither L_0 nor g, the acceleration
due to gravity, will enter.

The displacement from the equilibrium point, i.e. $L(t) - L_1$ is renamed to $-z(t)$.
The force that acts on the mass will then be

$$F(t) = -kz(t) .$$

*The negative sign indicates that the restoring force is in the opposite direction with
respect to the displacement.*

According to Newton's law, the sum of the forces acting on the mass is equal to the product of the mass and the instantaneous acceleration:

$$F(t) = m\ddot{z}(t) = -kz(t) .$$

Note once more that the gravitational force is *not* directly included in this expression. This is because the restoring force due to the spring and the gravitational pull counterbalance each other when $z = 0$.

$\ddot{z}$ is the double derivative of z with respect to time, i.e. acceleration in the vertical direction:

$$\ddot{z} \equiv \frac{d^2z}{dt^2} .$$

The equation of motion can then be written as:

$$\ddot{z}(t) = -\frac{k}{m}z(t) . \tag{2.9}$$

This is a second-order homogeneous differential equation with constant coefficients, and we know its general solution to be

$$z(t) = B \sin\left(\sqrt{\frac{k}{m}}t\right) + C \cos\left(\sqrt{\frac{k}{m}}t\right)$$

where B and C are two constants (with dimensions of length). We can identify this solution as Eq. (2.1) if we set the angular frequency ω in the latter equation to

$$\omega = \sqrt{\frac{k}{m}} .$$

The constants B and C are found by imposing the initial conditions, and the particular solution for the oscillatory motion is thereby determined with one particular amplitude and one particular phase.

The angular frequency ω is convenient to use in mathematical expressions. However, when we observe an oscillating system, it is expedient to use frequency f and period T. Their interrelationship is stated below:

$$f = \frac{\omega}{2\pi} ,$$

$$T = \frac{1}{f} = \frac{2\pi}{\omega} .$$

For the mechanical mass–spring oscillator one gets:

$$f = \frac{1}{2\pi}\sqrt{\frac{k}{m}} \, ,$$

$$T = 2\pi\sqrt{\frac{m}{k}} \, .$$

What have we learned in this section? Well, we have seen that a mass, attached to a spring and experiencing the forces exerted by the spring and gravity, will oscillate up and down, executing a simple harmonic motion with a certain amplitude and time period. We have managed to "explain" the oscillatory motion by combining Hooke's law and Newton's second law.

The kinematic description gave in Sect. 2.1 is identical to the *solution* of the dynamic equation we set up in this section based on Newton's law.

2.5   Damped Oscillations

No macroscopic oscillations last ceaselessly without the addition of energy. The reason is that there are always forces that oppose the movement. We call these frictional forces.

Frictional forces are often difficult to relate to, because they arise from complicated physical phenomena occurring in the borderland between atomic and macroscopic dimensions. A basic understanding of friction has *begun* to grow during the last decades, because grappling with this part of physics requires extensive modelling by means of computers.

Air friction is complex and we need at least two terms to describe it:

$$F_f = -bv - Dv^2$$

where v is the velocity (with direction), and b and D are positive constants, which will be called friction coefficients.

An expression that also indicates the correct sign and direction is:

$$\vec{F_f} = -b\vec{v} - Dv^2\frac{\vec{v}}{v} = -b\vec{v} - D\,|v|\,\vec{v} \, . \tag{2.10}$$

In other words, the friction force $\vec{F_f}$ works in a direction opposite to that of the velocity $\vec{v}$.

If we start with a system executing harmonic motion without friction, and we add friction as given in Eq. (2.10), it is not possible to find a general solution using analytical mathematics alone. If the problem is simplified by setting the frictional force to $-bv$ only, it is possible to use analytical methods. The solution is useful for slow motion in air. For small speeds, the term Dv^2 will be less than the term bv in Eq. (2.10) so that the v^2 term can be neglected.

Remarks: $-Dv^2$ is a nonlinear term that is often associated with turbulence, one of the difficult areas of physics, often associated with chaotic systems. Friction of this type depends on a number of parameters that can be partially included into the so-called Reynolds number. In some calculations, the quantity D must be replaced by a function $D(v)$ if Eq. (2.10) is to be used. Alternatively, the Navier–Stokes equation can be used as a starting point. Reasonably accurate calculations of the friction of a ball, plane or rocket can be accomplished only by using numerical methods (Those interested will be able to find more material in Wikipedia under the headings "Reynolds number" and "Navier–Stokes equation".).

Since no great skill is needed for solving the *simplified* differential equation, we accept the challenge! The solution method will consolidate our familiarity with complex exponents and will show the elegance of the formalism. Moreover, this is standard classical physics widely covered in textbooks, and the results are useful in many contexts. The mathematical approach itself finds applications in many other parts of physics.

The starting point is, as before, Newton's second law, and we use it for a mass that oscillates up and down at the end of a spring in air. The equations can now be written:

$$\sum F = ma \equiv m\ddot{z}$$

$$-kz(t) - b\dot{z}(t) = m\ddot{z}(t)$$

$$\ddot{z}(t) + \frac{b}{m}\dot{z}(t) + \frac{k}{m}z(t) = 0 . \tag{2.11}$$

This is a homogeneous second-order differential equation, and we choose a trial solution of the type:

$$z(t) = Ae^{\alpha t} . \tag{2.12}$$

Remark: Here, both A and α are assumed to be complex numbers.

Differentiation of the exponential function (2.12), insertion into (2.11) and finally the abbreviation of exponential terms and the factor A gives the characteristic polynomial

$$\alpha^2 + \frac{b}{m}\alpha + \frac{k}{m} = 0 .$$

We rename the fractions to get a tidier expression:

$$\frac{b}{m} \equiv 2\gamma \tag{2.13}$$

$$\frac{k}{m} \equiv \omega^2 . \tag{2.14}$$

The equation now becomes:

$$\alpha^2 + 2\gamma\alpha + \omega^2 = 0 .$$

This is a quadratic equation whose roots can be written as:

$$\alpha_\pm = -\gamma \pm \sqrt{\gamma^2 - \omega^2} . \tag{2.15}$$

There arise three different types of solutions, depending on the discriminant:

- $\gamma > \omega$: **Supercritical damping, overdamping**

> If the frictional force becomes large, we get what is called overdamping. The criterion of overdamping $\gamma > \omega$ is mathematically equivalent to $b > 2\sqrt{km}$. In this case, both A and α in Eq. (2.12) are real numbers, and the general solution can be written as:
>
> $$z(t) = A_1 e^{\left(-\gamma + \sqrt{\gamma^2 - \omega^2}\right)t}$$
> $$+ A_2 e^{\left(-\gamma - \sqrt{\gamma^2 - \omega^2}\right)t} . \tag{2.16}$$
>
> where A_1 and A_2, determined by the initial conditions, involve the initial values of velocity and displacement.

- This is a sum of two exponentially decaying functions, one of which goes to zero faster than the other. There is no trace of oscillatory motion here.
  Note that, for certain initial conditions, A_1 and A_2 may have different signs, and the time course of the displacement may hold surprises!
- $\gamma = \omega$: **Critical damping**
  The frictional force and the effective spring force now match each other in such a way that the movement becomes particularly simple. Based on Eqs. (2.12) and (2.15), we find one solution: It can be described as a simple exponential function:

$$z(t) = A e^{-\gamma t} .$$

It is known from the theory of differential equations that the general solution of a second-order differential equation must have *two* arbitrary constants, so that one may satisfy two initial conditions. This means that we have yet to find the full solution. To find the missing solution, we will use a simple trial solution of the type:

$$z(t) = f(t)e^{-\gamma t} .$$

If this trial solution is substituted into our differential equation (2.11) with $\gamma = \omega$, we find easily that $\ddot{f}$ must be equal to 0. After two integrations with respect to t, we find $f(t) = A + Bt$.

Thus the general solution of Eq. (2.11) for critical damping is then:

$$z(t) = Ae^{-\gamma t} + Bte^{-\gamma t} . \tag{2.17}$$

Critical damping in many cases corresponds to the fastest damping of a system and is the one sought for, for example, in vehicle shock absorbers.

- $\gamma < \omega$: **Sub-critical damping; underdamping**
  In this case, α in Eq. (2.15) becomes complex, which means that the solution will contain both an exponential decreasing factor and an oscillating sinusoidal term. From Eq. (2.15), we get then:

$$\alpha_{\pm} = -\gamma \pm \sqrt{\gamma^2 - \omega^2} \tag{2.18}$$
$$= -\gamma \pm i\omega' . \tag{2.19}$$

where $\omega' \equiv \sqrt{\omega^2 - \gamma^2}$ is a real number. The general solution then becomes:

$$z(t) = e^{-\gamma t}\Re\left\{\mathscr{A}e^{i\omega't} + \mathscr{B}e^{-i\omega't}\right\}$$

where $\mathscr{A}$ and $\mathscr{B}$ are complex numbers, and $\Re$ means that we take the real part of the expression.

The solution for sub-critical damping can be put in a simpler form:

$$z(t) = e^{-\gamma t}A\cos(\omega't + \phi) . \tag{2.20}$$

Here the constant A and ϕ must be assigned such values as to make the particular solution conform to a given physical system. The mass will oscillate on both sides

Fig. 2.5 Examples of overcritical, critical and sub-critical damping of an oscillation that would be simple harmonic in the absence of friction. The friction is increased by a factor of four from one curve to another: sub-critical, critical and overcritical damping

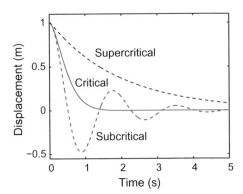

of the equilibrium point while the amplitude decreases to zero. The oscillation frequency is lower than when there is no damping (something that is to be expected since the friction acts to slow down all movement).

It is common in textbooks to present a figure that typically shows the time course for a damped harmonic motion, and Fig. 2.5 perpetuates the tradition. However, it should be noted that such figures can be very misleading, because they often assume that the initial velocity is zero (as in our figure). In a task last in this chapter, we ask you to investigate how an overdamped harmonic motion looks under some other initial conditions. If you solve that task, you will see that the solution is more diverse than the traditional figures indicate!

2.6   Superposition and Nonlinear Equations

When we tried to figure out how a damped oscillation changes with time, we assumed the validity of the differential equation:

$$\ddot{z}(t) + \frac{b}{m}\dot{z}(t) + \frac{k}{m}z(t) = 0 \tag{2.21}$$

and found a general solution that consisted of two parts. For overcritical damping, the solution looks like this:

$$z_A(t) = A_1 e^{\left(-\gamma+\sqrt{\gamma^2-\omega^2}\right)t} + A_2 e^{\left(-\gamma-\sqrt{\gamma^2-\omega^2}\right)t}$$

where γ and ω are defined in Eqs. (2.13) and (2.14) above.

In the interests of simplicity, we set:

$$f_1(t) = e^{\left(-\gamma+\sqrt{\gamma^2-\omega^2}\right)t}$$

and

$$f_2(t) = e^{\left(-\gamma - \sqrt{\gamma^2 - \omega^2}\right)t} .$$

One solution can then be written as:

$$z_A(t) = A_1 f_1(t) + A_2 f_2(t) .$$

Another solution of the differential equation could be:

$$z_B(t) = B_1 f_1(t) + B_2 f_2(t) .$$

It is easy then to see that

$$z_{AB}(t) = [A_1 f_1(t) + A_2 f_2(t)] + [B_1 f_1(t) + B_2 f_2(t)]$$

$$z_{AB}(t) = (A_1 + B_1) f_1(t) + (A_2 + B_2) f_2(t)$$

will also be a solution of the differential equation. This is due to the fact that the differential equation (2.21) is a *linear equation*.

This is called the "superposition principle". This principle pervades many parts of physics (and notably also in quantum mechanics).

Previously, many people considered superposition principles to be a fundamental property of nature, but it is not. The reason for the misunderstanding is perhaps that most physicists of those days worked only with linear systems where the superposition principle holds. Today, thanks to computers and numerical methods, we can tackle physical systems that were previously inaccessible. This means that there has been an "explosion" in physics in the last few decades, and the development is far from over.

Let us see what differences arise when nonlinear descriptions are used. By nonlinear description, for example, we mean that forces describing a system showing a nonlinear dependence on position or speed. For example, when we described damped oscillations, we found that friction must often be modelled with at least two terms:

$$F = -bv - Dv^2 .$$

The second term on the right-hand side makes a nonlinear contribution to the force.

The differential equation would then become:

$$\ddot{z}(t) + \frac{b}{m}\dot{z}(t) + \frac{D}{m}[\dot{z}(t)]^2 + \frac{k}{m}z(t) = 0 \tag{2.22}$$

In this case, we can prove the following:

If $f_A(t)$ is one solution of this equation, and $f_B(t)$ is another solution, it is in general *not* true that the function $f_A(t) + f_B(t)$ is a solution of Eq. (2.22).

In other words, when we include a second-order term to complete the friction description, we see that the superposition principle no longer applies! Even if we find a possible solution for such an oscillating system, and then another solution, the sum of these individual solutions will not necessarily be a solution of the differential equation.

The term Dv^2 is a nonlinear term, and when the physics is such that nonlinear terms play a nonnegligible role, the superposition principle does not apply.

Take a look at the "list of nonlinear partial differential equations" on the Wikipedia to get an impression of how important nonlinear processes have now become within, for example, various areas of physics. The overview indirectly shows how many more issues we can study today compared to what was possible a few decades ago. Despite this, we still have a regrettable tendency to use formalism and interpret phenomena, in both classical and quantum physics, as if the world was strictly linear. I dare say, physicists will have, within a few decades, such a rich store of experience to build on that the general attitude will change. Time will show!

2.7 Electrical Oscillations

Before we proceed with forced oscillations, we will derive the equation of oscillatory motion for an electrical circuit. The purpose is to show that the mathematics here is completely analogous to that used in mechanical system.

In electromagnetism, there are three principal circuit elements: Resistors, inductors (coils) and capacitors. Their behaviours in an electrical circuit are given by the following relationships (where Q stands for the charge, $I = dQ/dt$ is electric current, V is voltage, R is resistance, L inductance and C capacitance):

$$V_R = RI \tag{2.23}$$
$$V_C = Q/C \tag{2.24}$$
$$V_L = L\,dI/dt$$
$$= L\,d^2Q/dt^2 \, . \tag{2.25}$$

If the circuit elements are connected in a closed loop, the total voltage change will be zero when we go around the loop from any point to the same point (Kirchhoff's law). For example, we connect a (charged) capacitor to a resistor (by closing the switch in Fig. 2.6), the voltage across the capacitor will always be the opposite of the voltage across the resistor. Thus, it follows that

$$RI = -Q/C$$

$$\frac{\mathrm{d}Q}{\mathrm{d}t} = -\frac{1}{RC}Q \,.$$

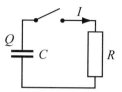

Fig. 2.6 The voltage across a charged capacitor will decrease exponentially to zero after the capacitor is connected to a resistor

$$RI = -Q/C$$

$$\frac{\mathrm{d}Q}{\mathrm{d}t} = -\frac{1}{RC}Q \,.$$

If the charge on the capacitor was Q_0 at time $t = 0$, the solution of this differential equation is:

$$Q = Q_0 e^{-t/RC} \,.$$

The charge on the capacitor thus decreases exponentially and goes to zero (The reader is supposed to be familiar with this.).

In the context of "oscillations and waves", we will concentrate on oscillating electrical circuits. An oscillating electrical circuit usually consists of at least one capacitor and an inductor. If the two elements are connected in series so as to form a closed loop, Kirchhoff's law gives:

$$\frac{Q}{C} = -L\frac{\mathrm{d}I}{\mathrm{d}t} = -L\frac{\mathrm{d}^2 Q}{\mathrm{d}t^2}$$

$$\frac{\mathrm{d}^2 Q}{\mathrm{d}t^2} = -\frac{1}{LC}Q \,.$$

We can write this in the same way as was done for the mechanical system:

$$\ddot{Q}(t) = -\frac{1}{LC}Q(t) \,. \tag{2.26}$$

If we compare Eq. (2.26) with Eq. (2.9), we see that they are completely analogous. The coefficient on the right-hand side is k/m for the mechanical system, and $1/LC$ in the electrical analogue, but they are both positive constants.

This is oscillation once more, and we know that the overall solution is:

$$Q = Q_0 \cos(\omega t + \phi)$$

where $\omega = 1/\sqrt{LC}$. Q_0 and ϕ are two constants whose values are fixed on the basis of the initial state ($t = 0$) of the system.

It may be worth reflecting on why there must be two initial conditions to obtain a specific solution for the LC circuit as compared to the RC circuit. In the RC circuit, the current is uniquely given if the charge is given. We can then decide, by means of a snapshot, either the charge or the voltage, will vary with time (assuming that R and C are known). For the LC circuit, this is not the case. There we must know, for example, both charge and current at a particular instant, or the charge at two adjacent times, to determine the further development. The reason is that we can not deduce power from one charge (or voltage) alone. The difference in physical descriptions for the RC and (R)CL circuit is reflected mathematically by the difference between a first-order and a second-order differential equations.

An electrical circuit in practice contains some kind of loss/resistance. Let us take the simplest example, namely that the loss is due to a constant series resistance R in the closed loop. If Kirchhoff's law is used again, we get the following differential equation:

$$\frac{Q}{C} = -RI - L\frac{dI}{dt} = -R\frac{dQ}{dt} - L\frac{d^2Q}{dt^2}$$

or

$$\frac{d^2Q}{dt^2} + \frac{R}{L}\frac{dQ}{dt} + \frac{1}{LC}Q = 0 . \tag{2.27}$$

This is a homogeneous second-order differential equation that can be solved using the characteristic polynomial:

$$a^2 + \frac{R}{L}a + \frac{1}{LC} = 0$$

whose solution is:

$$a_\pm = -\frac{R}{2L} \pm \sqrt{\left(\frac{R}{2L}\right)^2 - \frac{1}{LC}} .$$

The general solution to the differential equation is:

$$Q = Q_{0,1}e^{-\frac{R}{2L}t + \left(\sqrt{(\frac{R}{2L})^2 - \frac{1}{LC}}\right)t} + Q_{0,2}e^{-\frac{R}{2L}t - \left(\sqrt{(\frac{R}{2L})^2 - \frac{1}{LC}}\right)t} . \tag{2.28}$$

We note that for $R = 0$, we recover Eq. (2.26), whose solution is

$$\begin{aligned} Q &= Q_{0,1}e^{(\sqrt{-1/LC})t} + Q_{0,2}e^{-(\sqrt{-1/LC})t} \\ &= Q_{0,1}e^{i(\sqrt{1/LC})t} + Q_{0,2}e^{-i(\sqrt{1/LC})t} \\ &= Q_0 \cos(\omega t + \phi) . \end{aligned}$$

where $\omega = 1/\sqrt{LC}$. We see again that there are two constants to be determined by means of the initial conditions.

When $R \neq 0$, we get an exponentially decreasing term $e^{-(R/2L)t}$ multiplied by either an oscillating term or a second exponentially decreasing term, depending on whether $(R/2L)^2$ is less or greater than $1/LC$. When $(R/2L)^2 = 1/LC$, the term under the radical in Eq. (2.28) becomes zero, which corresponds to what we have seen previously with two coincident roots. In such a case, the overall solution turns out to of the same form as Eq. (2.17). Again, it is natural to talk about sub-critical, critical and supercritical damping, similar to a mechanical pendulum.

We have seen that electrical circuits are described by equations completely analogous to those for a mechanical pendulum. Other physical phenomena show similar oscillating behaviour.

Common to all the systems examined above is the *equation for oscillatory motion*, which can be stated, in its simplest form, as

$$\frac{d^2 f}{dt^2} + c_1 \frac{df}{dt} + c_2 f = 0$$

where c_1 and c_2 are positive constants.

2.8 Energy Considerations

Let us calculate the energy and its time development in electrical circuits. We limit ourselves to a loss-less oscillating system, that is, we take $R = 0$. The solution of the differential equation is then:

$$Q = Q_0 \cos(\omega t + \phi)$$

where $\omega = \frac{1}{\sqrt{LC}}$. Q_0 and ϕ are two constants whose values are determined by using the initial conditions ($t = 0$) of the system.

The energy stored in the capacitor at any particular time is given by:

$$E_C = \frac{1}{2} QV = \frac{1}{2} \frac{Q^2}{C} .$$

The instantaneous energy is thus:

$$E_C(t) = \frac{1}{2} \frac{[Q_0 \cos(\omega t + \phi)]^2}{C}$$

$$= \frac{1}{2} \frac{Q_0^2}{C} \cos^2(\omega t + \phi) \,.$$

From electromagnetism we know that the energy stored in an inductor is given by the expression:

$$E_L = \frac{1}{2} L I^2 = \frac{1}{2} L \left(\frac{dQ}{dt} \right)^2 .$$

Substituting the expression for Q from the general solution, the instantaneous energy in the inductance is found to be

$$E_L(t) = \frac{1}{2} L \left[\frac{d[Q_0 \cos(\omega t + \phi)]}{dt} \right]^2$$

$$= \frac{1}{2} L Q_0^2 \omega^2 \sin^2(\omega t + \phi) \,.$$

Since $\omega = \dfrac{1}{\sqrt{LC}}$, the expression can also be written as:

$$E_L(t) = \frac{1}{2} \frac{Q_0^2}{C} \sin^2(\omega t + \phi) \,.$$

The total energy, found by summing the two contributions, is thus:

$$E_{tot}(t) = E_C(t) + E_L(t)$$

$$= \frac{1}{2} \frac{Q_0^2}{C} \left[\cos^2(\omega t + \phi) + \sin^2(\omega t + \phi) \right]$$

$$E_{tot}(t) = \frac{1}{2} \frac{Q_0^2}{C} \,.$$

We notice that the total energy remains constant, i.e. time-independent. Although the energy of the capacitor and inductor varies from zero to a maximum value and back in an oscillatory fashion, these variations are time shifted by a quarter period, making the sum independent of time. The energy "flows" back and forth between the capacitor and inductor. A time shift between two energy forms seems to be a characteristic feature of all oscillations. Simple oscillations are often solutions of second-order differential equation, but oscillations may also originate from phenomena that have to be expressed mathematically in different way.

For the mechanical system, potential energy (from the conservative spring force) and kinetic energy are the two energy forms. You are recommended to perform a similar calculation as we have done in this section for the mechanical system to see that the result is indeed analogous to what we found for the electrical system (This is the theme for a calculation task in the end of this chapter.).

The energy calculations we have just completed apply only if there is no loss in the system. If loss due to resistance (the equivalent of friction) is preset, the energy will of course decrease over time. The energy loss per unit time pattern will depend on the extent of damping (supercritical, critical or sub-critical), but in general, the energy loss will follow an exponential decline.

2.9 Learning Objectives

The title of the book is "Physics of Oscillation and Waves", but just about all basic theory of *oscillations* is presented already in this chapter and Chap. 3. Nevertheless, the basic ideas from these two chapters will resurface many times when we refer to waves. We therefore think that a thorough study of this chapter and Chap. 3 will pay handsome dividends when the reader moves to later chapters.

After working through this chapter you should be able to
- Know that a harmonic oscillatory motion can be expressed mathematically in a variety of ways, both with sines and/or cosine functions, or in complex form (using Euler's formula). One goal is to recognize the different forms and to be able to go mathematically from any of these representations to another.
- Know that oscillations may occur in systems affected by a force that tries to bring the system back to equilibrium. Mathematically, this can be described easily in simple cases:

$$\ddot{z} = -kz$$

where x is the displacement from the equilibrium position and k is a real, positive number.

- Know that any oscillation must contain the two terms given in the equation in the previous paragraph, but that other terms may also be included.
- Know how physical laws/relationships are combined by deriving the second-order differential equation for both a mechanical and an electrical system.
- Know that in order to find a unique solution to the above-mentioned equation, two independent initial conditions must be imposed and suggest at least a few different choices of initial conditions.
- Be able to derive and solve the equation of oscillatory motion both for free and damped oscillation with linear damping. This means that you must be able to distinguish between supercritical, critical and sub-critical damping, and to outline graphically typical features for different initial conditions.
- Be able to deduce the equation for oscillatory motion also for a nonlinearly damped system and find the solution numerically (after studying Chap. 4).
- Be able to explain why the superposition principle does not apply when nonlinear terms are included in the equation of motion.

2.10   Exercises

Remark:

For each of the remaining chapters, we suggest concepts to be used for student active learning activities. Working in groups of two to four students, improved learning may be achieved if the students discuss these concepts vocally together.

The purpose of the comprehension/discussion tasks is to challenge the student's understanding of phenomena or formalism. Even for these tasks, it may be beneficial for learning that students discuss the tasks vocally in small groups.

The "problems" are more traditional physics problems. However, our apperception is that the correct answer alone is not considered a satisfactory solution. Full marks are awarded only if the correct answer is supplemented with sound arguments, underlying assumptions, and approaches used for arriving at the answer.

Suggested concepts for student active learning activities: Kinematics, dynamics, amplitude, phase, frequency, harmonic, second-order differential equation, general solution, particular solution, initial conditions, phasor, damping, characteristic polynomial, supercritical/critical/sub-critical damping, superposition, linear equation.

Comprehension/discussion questions

1. Make a sketch similar to Fig. 1.1, which shows a time plot for one oscillation, but also draw the time course for another oscillation with the same amplitude and initial phase term, but a different frequency compared to the first one. Repeat the same for the case where the amplitudes are different, while the phase and

frequency are the same. Finally, present the third variant of such sketches (Find out what is meant by this.).

2. What demands must we make for a force to be able to form the basis for oscillations?

3. If a spring is cut in the middle, what will be the spring constant for each part compared to that for the original spring? How large is the time period for a mass at the end of the half-spring compared with the period of the mass in the original spring?

4. Suppose we have a mass in a spring that oscillates up and down with a certain time period here on earth, and that the spring and the mass are brought to the moon. Will the time period change?

5. Suppose we do as in the previous task, but take a pendulum instead of a mass and spring. Will the time period change?

6. A good bouncing ball can bounce up and down many times against a hard horizontal surface. Is this a harmonic motion (as we have used the word)?

7. In the text, a rather vague statement is made about a judicious choice of mass and maximum extension of the spring to achieve an approximately harmonic oscillatory motion. Can you give examples of what conditions will be unfavourable for a harmonic motion?

Problems

8. Show mathematically that the total energy of an oscillating mass–spring system (executing up and down movement only) is constant in time if there is no friction present (Remember that changes in potential energy in the gravitational field disappear if you take the equilibrium position of the plot as the starting point for the calculations.).

9. It is sometimes advantageous to describe dynamics by plotting velocity versus position, instead of position versus time, as we have done so far. Create such a plot for a mass that swings up and down at the end of a spring (plot in *phase plane*). What is the shape of the plot?

10. Make a plot in the phase plane (see previous task) for the movement of a bouncing ball that bounces vertically up and down on a hard surface (practically without loss). What is the shape of the plot? Comment on similarities/differences between the plots in this and the previous task.

11. A spring hangs vertically in a stand. Without any mass, the spring is 30 cm long. We attach a 100 g ball at the lower end, stretch the spring by pulling the mass (and then releasing it) and find, after the ball has come to rest, that the spring has become 48 cm long. We then pull the ball 8.0 cm vertically downwards, keep the ball steady, and then let go. Find the oscillation period of the ball. Write a mathematical expression that can describe the oscillatory movement. Find the maximum and minimum force between the ball and the spring.

12. An oscillating mass in a spring moves at a frequency of 0.40 Hz. At time $t = 2.0$ s, its position is $+2.4$ cm above the equilibrium position and the velocity of the mass is -16 cm/s. Find the acceleration of the mass at time $t = 2.0$ s. Find a mathematical description appropriate to the movement.

13. A mass m hangs in a massless spring with spring constant k. The amplitude is A. How big is the displacement relative to the equilibrium point when the kinetic energy is equal to half of the potential energy?

14. An oscillatory motion can be described by the equation $z(t) = A \cos(\omega t + \phi)$ where $A = 1.2\,\text{m}$, the frequency $f = \omega/(2\pi) = 3.0\,\text{Hz}$, and $\phi = 30°$. Find out how this oscillatory motion can be formally specified when we (a) do not use the phase term, but only a combination of sine and cosine terms, and (b) when using a complex description based on Euler's formula.

15. Another oscillatory motion is given at $y(t) = \Re\{(-5.8 + 2.2\text{i})e^{\text{i}\omega t}\}$. Convert the equation to the same form as Eq. (2.1) and convert further until it has the same form as Eq. (2.1).

16. Show that the period of a mathematical pendulum with small amplitude is given by $T = 2\pi \sqrt{L/g}$ where L is the length of the pendulum and g is the acceleration due to gravity. Hint: Use the relation $\tau = I\alpha$ where τ is the torque, I the moment of inertia (mL^2) and α is the angular acceleration, to show that the equation of motion is $\ddot{\theta}(t) = (g/L)\sin\theta$ and then use the usual approach for sines at small angles.

17. A mass weighing $1.00\,\text{N}$ is hung at the end of a light spring with spring constant $1.50\,\text{N/m}$. If we let the mass swing up and down, the period is T. If instead we let the mass settle down and pull it to the side and release it, the resulting movement will have a period of $2T$ (the amplitude in the second case is very small). What is the length of the spring without the mass? (You may need the expression in the previous assignment.)

    Note: We recommend strongly that you make a real mass/spring system with a length so that the period of the sidewise pendulum oscillation is twice the period for the vertical mass–spring pendulum. Start the movement of the system by a pure vertical displacement of the mass, and release it from rest at this position. Watch the movement. You may be surprised! What you witness is an example of a so-called parametric oscillator.

18. Show that the energy loss for a damped pendulum where the frictional force is $F_f = -bv$ is given by $dE/dt = -bv^2$. Here, b is a positive number (friction coefficient) and v is the velocity (Start from the mechanical energy of the system, $E = E_{\text{potential}} + E_{\text{kinetic}}$.).

19. An object of $m = 2.0\,\text{kg}$ hangs at the end of a spring with the spring constant $k = 50\,\text{N/m}$. We ignore the mass of the spring. The system is set in oscillations and is damped. When the velocity of the mass is $0.50\,\text{m/s}$, the damping force is $8.0\,\text{N}$.

    (a) what is the system's natural oscillation frequency f (i.e. if the damping was not present)?

    (b) Determine the frequency of the damped oscillations.

    (c) How long does it take before the amplitude is reduced to 1% of the original value?

Chapter 3
Forced Oscillations and Resonance

Abstract In this chapter, we study a mechanical system forced to oscillate by the application of an external force varying harmonically with time. The amplitude of the oscillations, which is shown to depend on the frequency of the external force, reaches its peak value when the frequency of the applied force is close to the natural frequency of the system, a phenomena called resonance. However, details depend on the energy loss in the system, a property described by a quality factor Q, and the phase difference is described by so-called phasors. Emphasis is placed on how the system behaves when the external force starts and vanishes. Numerical calculations facilitate the analysis. At the end, some relevant details concerning the physiology of the human ear are briefly mentioned.

3.1 Introductory Remarks

The words "resonance" and "resound" are derived from the Latin root *resonare* (to sound again). If we sing with the correct pitch, we can make a cavity to sing along and, to somehow, augment the sound we emitted. Nowadays, the word is used in diverse contexts, but it always has the connotation of an impulse causing reverberation in some medium. When we tune the radio to receive weak signals from a transmitter, we see to it that other, unwanted signals, also captured by the radio antenna at the same time, are suppressed. It may seem like pure magic. The physics behind such phenomena is straightforward when we limit ourselves to the simplest cases. If we dig a little deeper, we uncover details that make our inquiry much more demanding and exciting.

3.2 Forced Vibrations

The Foucault pendulum in the foyer of the Physics building at the University of Oslo oscillates with the same amplitude year after year, although it encounters air resistance, which, in principle, should have dampened its motion. This is because

© Springer Nature Switzerland AG 2018
A. I. Vistnes, *Physics of Oscillations and Waves*, Undergraduate Texts in Physics,
https://doi.org/10.1007/978-3-319-72314-3_3

the bob at the end of the pendulum receives a small electromagnetic push each time it passes the lowest point. When that happens, a small red LED lights up. The push comes exactly at the time the bob is moving away from the equilibrium point. In this way, the time period is almost completely determined by the natural oscillation period of the pendulum itself (determined by the length of the pendulum and acceleration due to gravity).

In other contexts, "the pushes" come at a rate different from the natural rate of oscillation of the system. Electrons in an antenna, the diaphragm of the loudspeaker, the wobble of a boat when waves pass by, are all examples of systems being forced by a vibratory motion energized by an external force that varies in time independently of the system in motion. Under such circumstances, the system is said to be executing *forced oscillations*.

In principle, an external time-dependent force can vary in infinitely many ways. The simplest description is given by a harmonic time-varying force, i.e. as a sinusoid or cosinusoid. In the first part of the chapter, we assume that the harmonic force lasts for a "long time" (the meaning of the phrase will be explained later).

If we return to the mechanical pendulum examined earlier and confine ourselves to a simple friction term and a harmonic external force, the movement can be described analytically.

For a mechanical system, the starting point is Newton's second law (see Chap. 2): The sum of the forces equals the product of mass with acceleration:

$$F \cos(\omega_F t) - kz(t) - b\dot{z}(t) = m\ddot{z}(t)$$

where $F \cos(\omega_F t)$ is the external force that oscillates with its own angular frequency ω_F. If we put

$$\omega_0^2 = k/m ,$$

(angular frequency of the freely oscillating system), the equation can also be written as follows:

$$\ddot{z}(t) + (b/m)\dot{z}(t) + \omega_0^2 z(t) = (F/m) \cos(\omega_F t) . \qquad (3.1)$$

This is an inhomogeneous second-order differential equation, and its general solution may be written as:

$$z(t) = z_h(t) + z_p(t)$$

where z_h is the general solution of the corresponding homogeneous equation (with F replaced by zero) and z_p is a particular solution to the inhomogeneous equation itself.

We have already found in Chap. 2 the general solution of the corresponding homogeneous equation, so the challenge is to find a particular solution.

We know that the solution of the homogeneous equation decreases with time to zero. Therefore, after a long time from start, the movement will be dominated by the external periodic force.

It becomes natural then to investigate if a particular solution may have the form:

$$z_p(t) = A \cos(\omega_F t - \phi) \qquad (3.2)$$

where A is real.

Here, we have to discuss the choice of the sign of the phase term ϕ. Assume ϕ to be positive. In that case, we have: If F is maximum at time $t = t_1$ (for example, $\omega_F t_1 = 2\pi$), the displacement $z_p(t)$ will reach its maximum value at a time $T = t_2$ (with $t_2 > t_1$), i.e. at a time later than when F was at its maximum.

We then say that the output $z_p(t)$ is *delayed* with respect to the applied force.

When the expressions for $z_p(t)$ and $F(t)$ are inserted into Eq. (3.1) and the terms are rearranged, the following result is obtained:

$$(\omega_0^2 - \omega_F^2) \cos(\omega_F t - \phi) - (b/m)\omega_F \sin(\omega_F t - \phi) = F/(Am) \cos(\omega_F t) .$$

If we use the trigonometric identities for the sines and cosines of difference of angles (see Rottmann), we find:

$$(\omega_0^2 - \omega_F^2)\{\cos(\omega_F t) \cos \phi + \sin(\omega_F t) \sin \phi\} - (b/m)\omega_F \{\sin(\omega_F t) \cos \phi - \cos(\omega_F t) \sin \phi\}$$

$$= F/(Am) \cos(\omega_F t) .$$

Upon collecting the terms with $\sin(\omega_F t)$ and $\cos(\omega_F t)$, we get:

$$\left[(\omega_0^2 - \omega_F^2) \cos \phi - F/(Am) + (\omega_F b/m) \sin \phi\right] \cos(\omega_F t)$$

$$+ \left[(\omega_0^2 - \omega_F^2) \sin \phi - (\omega_F b/m) \cos \phi\right] \sin(\omega_F t) = 0 .$$

Since $\sin(\omega_F t)$ and $\cos(\omega_F t)$ are linearly independent functions of t, the above equation can be satisfied only if each term within the square brackets vanishes separately. This conclusion gives us two equations which can be used for the determination of the two unknowns, namely A and ϕ.

Equating to zero the terms within the square brackets multiplying $\sin(\omega_F t)$, we find:

$$(\omega_0^2 - \omega_F^2) \sin \phi = (\omega_F b/m) \cos \phi .$$

The phase difference between the output and the applied force can be expressed as:

$$\cot \phi = \frac{\cos \phi}{\sin \phi} = \frac{\omega_0^2 - \omega_F^2}{\omega_F b/m} . \tag{3.3}$$

We see that when $\omega_F = \omega_0$, $\cot \phi = 0$, which means that $\phi = \pi/2$ or $3\pi/2$. Since $\cot \phi$ changes from a positive to negative value when ω_F passes ω_0 from below, only the choice $\phi = \pi/2$ is acceptable.

When we set the expression with the square brackets multiplying $\cos(\omega_F t)$ to zero, we get:

$$(\omega_0^2 - \omega_F^2) \cos \phi - F/(Am) - (b\omega_F/m) \sin \phi = 0 .$$

We use the expression $\sin x = \pm 1/\sqrt{1 + \cot^2 x}$ from Rottmann (and a corresponding expression of cos) together with Eq. (3.3).

After a few intermediate steps, we get the following expressions for the amplitude of the required oscillations:

$$A = \frac{F/m}{\sqrt{(\omega_0^2 - \omega_F^2)^2 + (b\omega_F/m)^2}} . \tag{3.4}$$

It is time now to sum up what we have done:

When a system obeying an inhomogeneous linear second-order ordinary differential equation is subjected to a harmonic force that lasts indefinitely, a particular solution (which applies "long after" the force is applied) is itself a harmonic oscillation *of the same frequency* that is phase shifted with respect to the original force, as given in Eq. (3.2). "Long after" refers to a time many time constants $1/\gamma$ long, where γ is proportional to the damping of the system. We refer to the exponential decaying term $e^{-\gamma t}$ in the solution of the homogeneous differential equation discussed in the previous chapter.

The amplitude of the oscillations is then given by Eq. (3.4), and the phase difference between the output and the applied force (or the input) is given by Eq. (3.3). Figure 3.1 shows schematically how the amplitude and phase vary with the frequency of the applied force. The frequency of the force is given relative to the frequency of the oscillations in the same system if there was no applied force or no friction/damping.

We see that the amplitude is greatest when the frequency of the applied force is nearly the same as the natural frequency of oscillation in the same system when the applied force and damping are *both absent*. We call this phenomenon *resonance*, and it will be discussed in more detail in the next section.

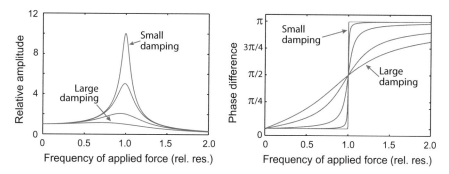

Fig. 3.1 The amplitude of a forced oscillation (*left*) and the phase difference between the output and the applied force (*right*) as a function of the frequency of the applied force

The phase ϕ appearing in Eq. (3.2) is approximately equal to $\pi/2$ at resonance; that is, the output is lagging behind (in phase) by about $\pi/2$ with respect to the applied force. For the spring oscillation, it means that the force is greatest in the upward direction when the pendulum has its highest speed and passes the equilibrium point on the way upwards.

Away from resonance, the phase difference is less than (greater than) $\pi/2$ when the applied frequency is lower than (higher than) the "natural" frequency. These relationships can be summarized so that the pendulum "is impatient" and tries to move faster when the applied force changes too slowly relative to resonance frequency ("natural frequency"). The movement of the pendulum depends more and more on the force when the force changes too quickly in relation to resonant frequency. *The phase difference is an important characteristic of forced fluctuations.*

3.3 Resonance

One sees from Eq. (3.4) that the amplitude of the forced oscillations varies with the frequency of the applied force. When the frequency is such that the amplitude is greatest, the system is said to be at *resonance*.

It may be useful to reflect a little about what is needed to get the largest possible output, which corresponds to the highest possible energy for the system.

Let us start with the mechanical mass–spring oscillator again. We then have a mechanical force that works on a moving system. We remember from mechanics that the work done by the force is equal to the magnitude of the force multiplied by how far the system moves under the action of the force. For a constant force, the power delivered by the force equals the power multiplied by the velocity of the system experiencing the force. Force and velocity are vectorial forces, and it is their dot product that counts (Remember $P = \vec{F} \cdot \vec{v}$ from the mechanics course.).

Fig. 3.2 A close-up view of
the relative amplitude in a
forced oscillation as a
function of the frequency of
the applied force. Note the
numbers along the axes

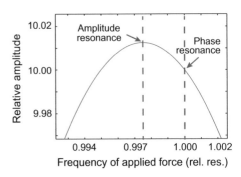

In our case, the force will deliver the greatest possible power to the system
if the power has the highest value while the pendulum bob has the highest
possible velocity. Force and velocity must work in the same direction. This
will happen if the force, for example, is the greatest, while the bob passes the
equilibrium position on the way up. This corresponds to the position is phase
shifted $\pi/2$ by force. To achieve such a state, the external force must swing
with the *resonance frequency*.

So far, we have been somewhat imprecise when we have discussed resonance.
Strictly speaking, we must differentiate between two nuances of the term resonance,
namely *phase resonance* and *amplitude resonance*. The difference between the two
is often in practice so small that we do not have to worry about it.

Phase resonance is said to occur when the phase difference between the applied
force and the output equals $\pi/2$. This happens when the frequency of the applied force
(input frequency) coincides with the natural frequency of the (undamped) system.

A close-up view of Fig. 3.1 shown in Fig. 3.2 shows that the amplitude is greatest
at a slightly lower frequency than the natural frequency. The small but significant
difference is due to a detail we mentioned when we discussed damped harmonic
motion in the previous chapter. In the presence of damping, the oscillation frequency
is slightly lower than the natural frequency. The frequency at which amplitude is
greatest indicates *amplitude resonance* for the system. The two resonance frequencies
are often quite close to each other, as already mentioned.

Let us find mathematical expressions for the two resonance frequencies.

The amplitude resonance frequency can be found by differentiating the expression
for the amplitude given by Eq. (3.4) (a common procedure for finding extreme values).
We calculate the ω_F angular frequency at which:

$$\frac{dA}{d\omega_F} = 0 .$$

We find that

$$\omega_F = \sqrt{\omega_0^2 - \frac{b^2}{2m^2}} \ .$$

If we want to state the frequency rather than the angular frequency, we use the expression:

The amplitude resonance frequency is:

$$f_{\text{amp.res.}} = \frac{1}{2\pi} \sqrt{\omega_0^2 - \frac{b^2}{2m^2}} \tag{3.5}$$

where $\omega_0 = \sqrt{k/m}$.

The phase resonance frequency is:

$$f_{\text{ph.res.}} = \frac{1}{2\pi} \omega_0 \ . \tag{3.6}$$

We observe that the two resonance frequencies coincide only when $b = 0$ (no damping).

3.3.1 Phasor Description

We will now consider forced oscillations in an electrical circuit. First, we will proceed in much the same manner as adopted in dealing with the mechanical system examined above, but eventually we will go over to an alternative description based on phasors. The system is a series RCL circuit with a harmonically varying voltage source $V_0 \cos(\omega_F t)$, as shown in Fig. 3.3. The differential equation for the system then becomes [compare with Eq. (2.27)]:

$$L \frac{d^2 Q}{dt^2} + R \frac{dQ}{dt} + \frac{1}{C} Q = V_0 \cos(\omega_F t) \ . \tag{3.7}$$

Fig. 3.3 A series RCL circuit driven by a harmonically varying applied voltage. The labels $+$, I, and Q indicate the signs chosen for our symbols

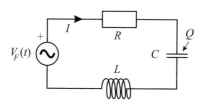

This is an inhomogeneous equation, whose solution is found in the same way as for its mechanical counterpart considered above. The solution consists of a sum of a particular solution and the general solution of the homogeneous equation (with $V_0 = 0$). The solution of the homogeneous equation is already known, and it only remains for us to find a particular solution. We try a similar solution as for the mechanical system, but adopt a complex representation:

$$Q_p(t) = Ae^{i\omega_F t} \tag{3.8}$$

where A can be a complex number.

At the same time, a complex exponential form is chosen for the externally applied voltage:

$$V(t) = V_0 \cos(\omega_F t) \rightarrow V_0 e^{i\omega_F t} . \tag{3.9}$$

It goes without saying that the real part of the expressions are to be used for representing physical quantities.

Inserting the expressions for $Q_p(t)$ and $V(t)$ into Eq. (3.7), and cancelling the common factor $e^{(i\omega_F t)}$, we get:

$$-L\omega_F^2 A + iR\omega_F A + \frac{1}{C}A = V_0 .$$

Solving the equation for A, we get:

$$A\left(-L\omega_F^2 + iR\omega_F + \frac{1}{C}\right) = V_0$$

$$A = \frac{V_0}{\frac{1}{C} - L\omega_F^2 + iR\omega_F} .$$

A again becomes a complex number (except when $R = 0$).

The instantaneous current in the RCL circuit is found by applying Ohm's law to the resistor:

$$I = \frac{V_R}{R} = \frac{dQ}{dt} .$$

If we wait long enough for the solution of the homogeneous equation to die out, only the particular solution remains, and the current is then given by the expression:

$$I = \frac{dQ_p}{dt} = Ai\omega_F e^{i\omega_F t}$$

Simple manipulations lead one to the following expression:

$$I(t) = \frac{V_0}{R + i(L\omega_F - \frac{1}{C\omega_F})}e^{i\omega_F t} . \tag{3.10}$$

This expression should be compared with V_F, the voltage applied to the circuit, which in complex form is given by:

$$V_F(t) = V_0 e^{i\omega_F t} \ .$$

It follows from Eq. (3.10) that if $R = 0$ the current will be phase shifted $90°$ relative to applied voltage. If in addition $L = 0$, the current will *lead* the voltage by $90°$. However, if $\omega_F L$ is much larger than $1/(\omega_F C)$ (C "shorted"), the current will be offset $90°$ *after* the voltage (In a calculation exercise at the end of the chapter you are asked to show this.).

If $R \neq 0$, but $L\omega_F - \frac{1}{C\omega_F} = 0$, the current and voltage will be in phase, and $I = V_0/R$. This corresponds to $\omega_F = \frac{1}{\sqrt{LC}}$, which was named phase resonance above.

The connection between R, C, L, current and phase can be elegantly illustrated by means of phasors. We have already mentioned phasors, but now we extend the scope by drawing in multiple rotating vectors at the same time. Figure 3.4 shows an example.

Both currents and voltages are displayed in the same plot. We start with a vector that represents the current generated by the applied voltage. Then we draw vectors representing voltages across the resistor, capacitor and inductor resulting from the current flow. The vector which shows the voltage across the capacitor will then be $90°$ after the vector showing the current, the voltage across the resistor will have the same direction as the current and the voltage across the inductance will be $90°$ ahead of the current. The total voltage across the serial link of R, C and L should then be the vector sum of the three voltage phasors and correspond to the applied voltage. We see that the phase difference between current and voltage will be between $+90°$ and $-90°$.

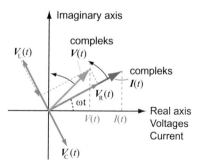

Fig. 3.4 Example of phasor description of an RCL circuit subjected to a harmonically varying voltage. The current at any time (anywhere in the circuit) is the x components of the vector $I(t)$, while the voltage across the various circuit components is given by the x component of the vectors $V_R(t)$, $V_C(t)$ and $V_L(t)$, and their sum is $V(t)$. See the text for details

Fig. 3.5 A time plot in which the current slightly leads the applied voltage

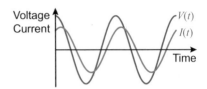

Phasor diagrams can also be based on quantities other than those we have chosen here. One variant is to use complex impedances that are added vectorially. The strength of phasor diagrams is that we can easily understand, for example, how the phase differences change with frequency. The depiction in Fig. 3.4 applies only to a particular applied angular frequency ω_F. If the angular frequency increases, the voltage across the capacitor decreases while the voltage across the inductance will increase. Phase resonance occurs when the two voltage vectors are exactly the same size (but oppositely directed) so that their sum is zero.

Figure 3.5 shows the time development of voltage and current in a time plot. The current in the circuit is slightly leading the applied voltage. For a series RCL circuit with an applied voltage, this means that the applied frequency is lower than the resonant frequency of the circuit.

Note that phasors can be used only after the initial rather complicated oscillatory pattern is over, and we have a steady sinusoidal output corresponding to the particular solution of differential equation.

3.4 The Quality Factor Q

In the context of forced oscillations, it is customary to characterize oscillating systems with a *Q-factor* or *Q-value*, where the symbol Q, not to be confused this with the charge Q in an electrical circuit, stands for "quality", which is why the Q-factor is also called the quality factor. The factor tells us something about how easy it is to make the system oscillate, or how long the system will continue to oscillate after the driving force is removed. This is more or less equivalent to how small loss/friction is in the system.

The quality factor for a spring oscillator is given by:

$$Q = \frac{m\omega_0}{b} = \sqrt{\frac{mk}{b^2}}. \tag{3.11}$$

We see from this formula that the smaller the value of b, the larger is the quality factor Q.

Figure 3.6 shows how the oscillation amplitude varies with the frequency of the applied force for four different quality factors. A Q-value of 0.5, in this case, cor-

Fig. 3.6 When the
frequency of the applied
force changes relative to the
system's own natural
frequency, the amplitude will
be greatest when the two
frequencies are nearly equal.
The higher the quality factor
Q (i.e. smaller loss), the
higher the resonance
amplitude

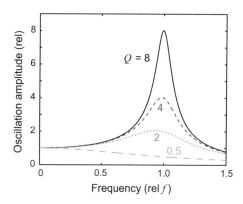

responds to critical damping and we see no hint of any resonance for such a large
damping.

There are two traditional ways of defining Q. The first is:

$$Q \equiv 2\pi \frac{\text{stored energy}}{\text{energy loss per period}} = 2\pi \frac{E}{E_{\text{loss-per-period}}} . \qquad (3.12)$$

This definition implies a particular detail which few people are familiar with, but
which is extremely important for forced oscillations in many contexts. Once we have
achieved a steady state (when the applied force has been working for long and is still
present), the loss of energy will be compensated by the work done on the system by
the applied force. We see from Eq. (3.12) that a system with a high Q-value loses
only a tiny part of the total energy per period.

Suppose now we turn off the applied force. Then the system will oscillate at the
amplitude resonance frequency $\omega' = \sqrt{\omega_0^2 - (b/2m)^2} \approx \omega_0$", and the energy will
eventually disappear. It will take the order of $Q/(2\pi)$ periods before the energy is
used up and the oscillations end. Let us look a little more closely at this.

Loss of energy per period is a slightly unfamiliar quantity. Let us consider first
P_{loss}, which is "energy loss per second" with the unit watt. We know that after the
force has been removed, the loss will be given by:

$$P_{\text{loss}} = -\frac{dE}{dt} . \qquad (3.13)$$

Then we can approximate the loss of energy over a period of time T with:

$$E_{\text{loss-per-period}} = P_{\text{loss}} T .$$

Using the definition given in Eq. (3.12), we get:

$$P_{\text{loss}} = \frac{2\pi}{TQ} E .$$ (3.14)

Combining Eqs. (3.13) and (3.14) and the relation $\omega = 2\pi/T$, we get a differential equation governing the time development of the stored energy after the removal of the driving force. The equation is:

$$P_{\text{loss}} \equiv -\frac{dE}{dt} = \frac{\omega_0}{Q} E .$$

The solution is:
$$E(t) = E_0 e^{-\omega_0 t/Q} .$$

The energy falls to $1/e$ of the initial energy after a time

$$\Delta t = \frac{Q}{\omega_0} = \frac{QT}{2\pi} .$$ (3.15)

We see that the amplitude of oscillation decreases in a neat exponential manner after the removal of an applied oscillatory force, with the time constant given in Eq. (3.15).

It can be shown that nearly the same time constant describes the *growth* of the output after the application of oscillating force. Obviously, the time course is not as simple because it depends, apart from other factors, on whether or not the frequency of the applied force equals the resonant frequency of the circuit (see Fig. 3.7). Nevertheless, if it takes an interval of the order of 10 ms for an oscillation to die out after an applied force is removed, it will also take nearly the same interval to build a steady amplitude after we switch on the applied force.

One might think that the time constant (and thus the Q-value) of the system could be found by referring to the thin red line in Fig. 3.7 and noting how long it takes from the moment the force is removed till the output falls to $1/e$ of the value just before the power was turned off. It turns out, however, that the number so inferred is twice the expected value! The difference can be traced to the fact that the time constant deduced in Eq. (3.15) applies to how *energy* changes over time, whereas Fig. 3.7 shows amplitude and not energy. *The energy is proportional to the square of the amplitude.* Note that the stationary amplitude after the force has worked for a while is greatest at the resonance frequency!

The curves in Fig. 3.7 show that after an applied force is turned on, the amplitude of the oscillations increases, without becoming infinite. Sooner or later, the loss in energy is as large as the power applied through the oscillating force. After equilibrium

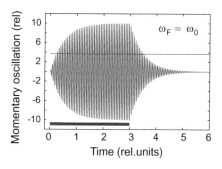 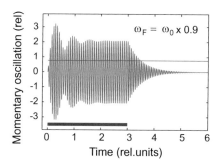

Fig. 3.7 Two examples of the build-up of oscillations in an oscillating system after an external sinusoidal force is coupled and subsequently removed (the force acts only during the interval indicated by a thick red line at the bottom). The frequency of the applied voltage is equal to the resonant frequency on the left and slightly lower on the right. While the force is present, the system oscillates with the frequency of the force. After the force has ceased, the circuit oscillates with its own resonance frequency. The thin red line marks the value $1/e$ times the maximum amplitude just before the applied force was removed. The Q-factor of the circuit is 25

a steady state is achieved, the amplitude of the oscillations will remain constant as long as the applied force has constant amplitude.

The mathematical solution of an inhomogeneous differential equation for an oscillating system subjected to an oscillatory force with given initial conditions is rather tedious. However, it is possible to find such a solution exactly using, for example, Maple or Mathematica. However, we have used numerical solutions in the preparation of Fig. 3.7; it is a rational approach since complex differential equations can often be solved numerically about as easily as simple differential equations. More about this in the next chapter.

In experimental context, a different and important definition of the Q-value is often used instead of that in Eq. (3.12). If we create a plot that shows *energy* (NOTE: *not* amplitude) in the oscillating system as a function of frequency (as in Fig. 3.8), the Q-value is defined as:

$$Q = \frac{f_0}{\Delta f} \tag{3.16}$$

where the half-width Δf, shown in the figure, compared to the resonance frequency f_0.

This relationship can be shown to be in accordance with the relationship given in Eq. (3.12), at least for high Q-values.

The definitions given in Eqs. (3.12) and (3.16) apply to all physical oscillating systems, not just the mechanical ones.

Fig. 3.8 The Q-value can also be defined from a graphical representation of *energy* stored in the oscillating system as a function of frequency. The Q-value is then given as the resonance rate f_0 divided by the half value Δf

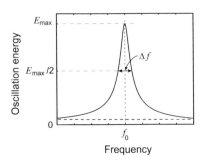

For the most interested: It is now possible to make a remarkable observation: A resonant circuit responds significantly to frequencies within a frequency band of width

$$\Delta f = \frac{f_0}{Q} .$$

However, the circuit needs a certain amount of time

$$\Delta t = \frac{Q}{\omega}$$

to build-up the response if we start from zero. It takes about the same time also for a response that is already built to die out.

The product of Δf and Δt comes out to be:

$$\Delta t \, \Delta f = \frac{Q}{\omega} \frac{f_0}{Q}$$

$$\Delta t \, \Delta f = \frac{1}{2\pi} . \tag{3.17}$$

Multiplying this expression with Planck's constant h, and using the quantum postulate that the energy of a photon is equal to $E = hf$, we get:

$$\Delta t \, \Delta E = \frac{h}{2\pi} . \tag{3.18}$$

This expression is almost identical to what is known as Heisenberg's uncertainty relationship for energy and time. There is a factor 1/2 in front of the term after the equality sign, but such a factor will depend on how we choose to define widths in frequency and time.

There are certain parallels between a macroscopically oscillating system and the relationships we know from quantum physics. In quantum physics, Heisenberg's uncertainty relationship is interpreted as an "uncertainty" in time and energy: we cannot "measure" the time of an event more accurately than what is implicit in the relationship

$$\Delta t = \frac{h}{2\pi \, \Delta E}$$

provided that we do not change the energy of a system by more than ΔE.

Our macroscopic variant applies irrespective of whether we do measurements or not, but measurements will of course reflect the relationship that exists. We will return to this relationship later in the book, but in the form of Eq. (3.17) instead of (3.18).

"Inertia" in a circuit is important for what we can do with measurements. For a high Q oscillation cavity in the microwave region (called a "cavity"), we can easily achieve Q-values of 10,000 or more. If such a cavity is used in pulsed microwave spectroscopy, it will take of the order of 60,000 periods to significantly change the energy in the cavity. If the microwave frequency is 10 GHz (10^{10} Hz), the time constant for energy changes will be of the order of 6 μs. If we study relatively slow atomic processes, this may be acceptable, and the sensitivity of the system is usually proportional to the quality factor. However, if we want to investigate time intervals lasting only a few periods of the observed oscillations, we must use cavities with much lower Q-value. More will be said about this in the next chapter.

3.5 Oscillations Driven by a Limited-Duration Force

So far, we have considered a system that is influenced by an oscillating force lasting "infinitely long", or a force that has lasted for a long time and ends abruptly. In such a situation, we can determine a quality factor Q experimentally in terms of the frequency response of the system as shown in Fig. 3.8 and Eq. (3.16). Relative oscillation energy (relative amplitude squared) must be determined after the system has reached the stationary state, i.e. when the amplitude no longer changes with time.

How will such a system behave if the oscillatory force lasts only for a short time? We will now investigate this matter.

When we introduce a limited-duration force (a "temporary force"), we must choose how the force should be started, maintained and terminated. For a variety of reasons, we want to avoid sudden changes, and have chosen a force whose overall amplitude follows a Gaussian shape, but follows, on a finer scale, a cosinuosidal variation. Mathematically, we shall describe such a force by the function:

$$F(t) = F_0 \cos[\omega(t - t_0)]e^{-[(t-t_0)/\sigma]^2} . \tag{3.19}$$

where σ indicates the duration of the force (the time during which the amplitude falls to $1/e$ of its maximum value). ω is the angular frequency of the underlying cosine function, and t_0 is the time at which the force has the maximum amplitude (peak of the pulse occurs at time t_0). The oscillating system is assumed to be at rest before the force is applied.

Figure 3.9 shows two examples of temporary forces with different durations. Here, the force has a frequency equal to 100 Hz (period $T = 10$ ms). In the figure on the left, σ is equal to 25 ms, i.e. $2.5 \times T$, and successive peaks have been marked (from maximum onwards until the amplitude has decreased to $1/e$) to highlight the role played by the size of σ. In the figure to the right, $\sigma = 100$ ms, i.e. $10 \times T$; again, the markings give an indication of the relationship between ω (or rather the frequency or period) and σ.

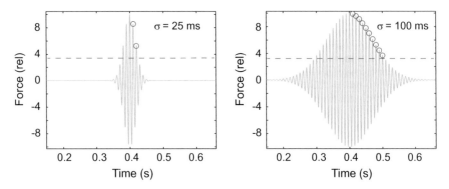

Fig. 3.9 The force $F(t)$ for centre frequency $100\,\text{Hz}$ and pulse width σ equal to 0.025 and 0.10 s. See the text for further explanations

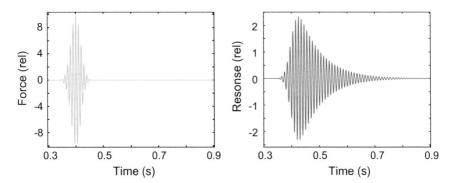

Fig. 3.10 The temporal response of the system (*right*) due to the applied force shown in the left part of the figure

We would now like to study how an oscillating system will behave when it is subjected to a temporary force. Based on Fig. 3.7, we expect the response to be quite complicated. Since it is not easy to make headway analytically, we have opted for numerical calculations instead.

Figure 3.10 shows the time course for one temporary force along with the response of the system. For simplicity, the frequency of the force has been made equal to the resonant frequency of the system, and according to the initial conditions chosen, the system is at rest before the force is applied.

Figure 3.10 shows some interesting features. The system attempts, but fails to keep pace with the force as it grows. We see that the peak of the response (amplitude) occurs a little later than the time at which the force reached its maximum value.

The force adds some energy to the system. When the force decreases as quickly as it does in this case, the system cannot get rid of the supplied energy at the same rate as that at which the force decreases. Left with surplus energy after the vanishing of the force, the system executes damped harmonic oscillations at its own rate. It may

Fig. 3.11 Dependence of the maximum amplitude on the duration of the applied force (σ). Note the logarithmic scale on both axes

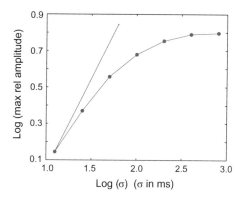

be mentioned that σ here is 25 ms and that the Q-factor of the oscillating system is chosen to be 25, which corresponds to a decay time for the energy for the oscillations of 40 ms.

It may be useful to point out some relationships between various parameters:

- How much energy can be delivered to the system within a given time depends on the force (proportionality?).
- The amount of energy that can be delivered, for a given input of force, will depend on how long the force works.
- The loss of energy is independent of the strength of the force after it has disappeared.
- The loss of energy is proportional to the amplitude of the oscillations.

As mentioned, we expect the amplitude to increase when the force lasts longer and longer, but the precise relationship is not self-evident. In Fig. 3.11 are shown calculated results for the maximum amplitude attained by the system for different σ values. ω always corresponds to the resonance frequency of the system. The figure has logarithmic axes to get a large enough range of σ. The straight line represents the case that the amplitude increases linearly with σ (duration of the force).

We see that for too small σ (the power lasting only a few oscillation periods), the maximum amplitude increases approximately proportionally with the duration of the force. When the force lasts longer, this does not apply anymore, and beyond a certain limit, the amplitude of the oscillation does not increase, however long the duration of the pulse may be. This is due to the fact that at the given amplitude, the loss is as large as the energy supplied by the power.

If the amplitude of the force is increased, the amplitude of the oscillations will also increase, but so will the loss. It is therefore found that the duration of the force required to obtain the maximum amplitude is approximately independent of the amplitude of the force.

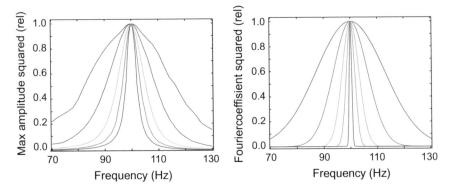

Fig. 3.12 The frequency response (actually only maximum amplitude) of the oscillating system for different durations (σ) of the force pulse (*left part*). The σ values used are respectively 25, 50, 100, 200, 400 and 800 ms (*from blue/widest to red/narrowest curves*). In the right part of the figure, corresponding frequency analyses of the force pulses themselves are shown. See the text for further explanations

3.6   Frequency Response of Systems Driven by Temporary Forces *

There is an unexpected consequence of using short-term "force pulses". We will address this topic already now,[1] but will return to it more than once in other parts of the book. Full understanding of the phenomenon under discussion is possible only after a review of Fourier analysis (see Chap. 5).

In Fig. 3.8, we showed how large an oscillation energy (proportional to amplitude squared) a system gets if it is exposed to a harmonic force with an "infinitely long" duration. The oscillation energy achieved was plotted as a function of the frequency of the applied force. A plot like this is usually called "*frequency response*" of the system, and the curve can be used to determine the Q-factor of the oscillating system from Eq. (3.16). The narrower the frequency response, the higher the Q-factor.

It is natural to determine the frequency response also for the case when the force lasts only a short time. The maximum energy system achieves as a result of the power is plotted as a function of the centre frequency of the power in a similar manner as in Fig. 3.8, and the result is given in the left part of Fig. 3.12. Relative energy is proportional to the square of the amplitude of the oscillations.

It turns out (left part of Fig. 3.12) that the frequency response of the system becomes different with temporary "force pulses" than with a harmonic force of infinitely long duration (as shown in Fig. 3.8). The frequency response becomes

---

[1]This sub-chapter is for the most interested readers only.

wider and wider (spreading over ever greater frequency range on both sides of the resonant frequency) as duration of the force pulse becomes shorter and shorter.

If, on the other hand, we apply longer and longer "force pulses", the frequency response of the system will reach a limiting value. There is a lower limit for the width of the curve, and thus a maximum limit for the calculated Q-factor. In general, the term Q-factor is used only for this limiting value. For shorter power pulses, the frequency response is specified rather than the Q-value.

However, it is possible to make a frequency analysis of the *temporary force pulse itself.* We will find out how this is done in Chap. 5 when we come to review Fourier analysis. To provide already now a rough idea of what a frequency analysis entails, it will be enough to say that such an analysis yields information about the frequency content of a signal, and tell us "which frequencies will be needed to reproduce the signal at hand".

The right part of Fig. 3.12 shows the frequency analysis of the "*force* as a function of time" for the same σ values as in the left part of the figure. The figure actually shows a classical analogy to Heisenberg's uncertainty relationship also known as the time-bandwidth product . We already found this in Eq. (3.17), and we will return to this in Chap. 5.

The two halves of Fig. 3.12 can be condensed into a single plot, and the result will then be as shown in Fig. 3.13.

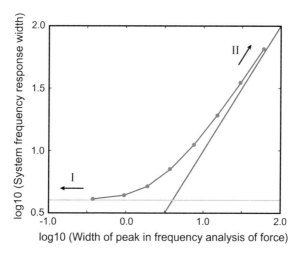

Fig. 3.13 The correlation between the frequency response of a system and the frequency of the driving force when the duration of the force changes. There are two border cases. In one case (I) the force lasts so long that the response depends only on the system itself (how much loss it is, and thus which Q-value it has). In the other case (II), the system's loss is so low in relation to the working time of the influence that the response to the force depends only on the force itself (how short time it lasts). The system's features have the least to say for the response

Based on these observations, we can say that:

- The quality factor is a parameter/quantity which *characterizes the oscillating system*. The smaller the loss in the system, the higher the Q-factor and the narrower frequency response, well and mark for harmonic forces that last long.
- When the force lasts for a short time (few oscillations) the frequency of the force is poorly defined. When an oscillating system is subjected to such a force, the frequency response is dominated by the *frequency characteristic of the power itself* and, to a lesser extent, the system itself.

Figure 3.13 is of some interest in the debate about whether Heisenberg's uncertainty relationship is primarily due to the perturbing influence of measurement on a system, or to the system itself. We do not delve into this issue here, but the result suggests that each point of view has some merit.

3.7   Example: Hearing

Finally in this chapter, we will say a little about our hearing and the mechanisms behind the process. Forced oscillations occupy the centre state in the present section, while other aspects associated with hearing will be treated in Chap. 7.

In our ears (see Figs. 3.14, 3.15 and 3.16), sound waves in the air cause oscillations at different frequencies in the auditory canal, tympanic membrane (eardrum), auditory ossicles (three tiny bones in the middle ear that conduct sound from the tympanic membrane to the inner ear), and the cochlea ("snailhouse")—a system of fluid-filled ducts which makes up the inner ear.

It is the inner ear that is of particular interest for us here, since it exemplifies resonance phenomena and demonstrates how ingenious our hearing sense is. Figure 3.15

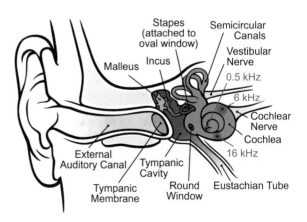

Fig. 3.14 Anatomical structures of the human ear. Inductiveload, CC BY 2.5, [1]

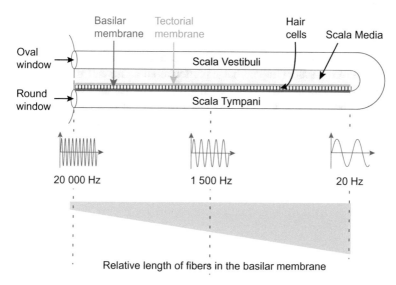

Fig. 3.15 The inner ear has a three-channel structure that stretches almost three rounds from bottom to top. This figure indicates how this would look like if we stretched out the insides of the cochlea. See the text for details

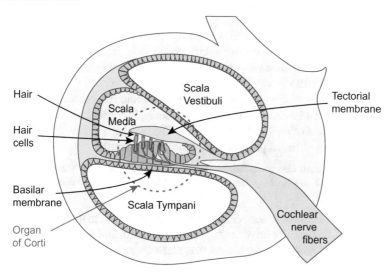

Fig. 3.16 Details on the anatomical structure of the basilar and tectorial membrane and their close connection through the organ of Corti. Note the hair cells that translate mechanical strain to electric signals. The organ of Corti structures are found along the full length of the basilar membrane with the result that it is an impressive number of nerve cells going from each ear to the brain

illustrate a "stretched out" cochlea with the fluid-filled ducts scala vestibuli from the oval window to the top of the cochlea and scala tympani from the top back to the round window (which is facing the air filled space of the middle ear).

One wall of the scala tympani has a particular structure called the basilar membrane, and weakly connected to the wall along the scala vestibuli we find the tectorial membrane. These membranes will oscillate when the ear picks up a sound signal.

Between the basilar and tectorial membranes, we find "hair cells" that respond to pressure. The amplitude of the oscillations is picked up by these hair cells, and the information is transmitted through the nerves to the brain (via different signal processing centres along the way).

It is a fascinating structure of cells named Organ of Corti (see Fig. 3.16) that translate pressure changes into electrical signals in nerves. Figure 3.16 also indicates how the third duct inside the cochlea, the air filled scala media, is a part of the total structure.

From our perspective, the important part is the basilar membrane. Earlier in this chapter, forced oscillations have been analysed. By way of a trial, that analysis will be applied to oscillations in the basilar membrane, which extends diametrically across the conical cavity of the cochlea in the inner ear (see Figs. 3.15 and 3.16).

The membrane can vibrate, just like the belly (top plate) of a violin, in unison with the pressure variations generated by the sound. The membrane, however, changes character from the outer to the inner parts of cochlea. The relative length of some fibres in the basilar membrane varies from the outer to the inner part as indicated in Fig. 3.15. As a result, if we hear a dark sound (low frequency), only the inner part of the basilar membrane will vibrate. If we hear a light sound (high frequency), only the outer part will vibrate. This is a fabulous design that allows us to hear many different frequencies at the same time as separate audio impressions. We can hear both a bass sound and a disk rhythm simultaneously, because the two sound stimuli excite different parts of the basilar membrane. The hair cells and nerve endings pick up vibrations from different parts of the membrane in parallel.

It was the biophysicist Georg von Békésy from Budapest who found out how the basilar membrane works as a "position-frequency map". He received the Nobel Prize in Physiology and Medicine for this work in 1961.

The basilar membrane is a mechanical oscillation system that behaves in a manner similar to the externally driven mass–spring oscillator and RCL circuit. Different parts of the membrane have properties that make them responsive to different frequency ranges. We can assign different Q-values to different parts of the basilar membrane.

Based on what we have learned in this chapter, we should expect that even if we hear a sound that delivers a harmonic force with a well-defined frequency on the eardrum, the basilar membrane will vibrate not at one position only along the basilar membrane, but over a somewhat wider area. Since we have "parallel processing" of the signals from the hair cells, the brain can still "calculate" a fairly well-defined centre frequency.

If, however, we listen to shorter and shorter sound pulses, we expect that wider and wider parts of the basilar membrane will be excited. This would make it harder for the brain to determine which centre frequency the sound pulse had. This means that it is harder to determine the pitch of a sound when the sound lasts very shortly.

When musicians play fast passages on, for example, a violin they can falter *a little* with the pitch without the error coming to the notice of a listener. If they stumbled as much with longer lasting tones, their slips will not escape the attention of the audience.

When the sound pulse lasts only one period (and this period, for example, corresponds to 1000 Hz), we only hear a "click". It is impossible to tell which frequency was used to create the sound image itself.

On the other hand, it is easier to perceive the direction of the audio source of a click than the source of a sustained sound. The ability to determine the time fairly precisely when a sound occurs, along with the fact that we have two ears, is very important in order to determine the direction the incoming sound (Nevertheless, it should be mentioned that there are other mechanisms to determine where a sound comes from.).

According to Darwin, our ears are the result of millions of years of natural selection that was beneficial for the survival of our species. The ear has become a system where there is an optimal relationship between the ability to distinguish between different frequencies and the ability to follow fairly quick changes over time. Resonance, time response and frequency response are very important details to understand our hearing.

An interesting detail with regard to hearing relies on phase sensitivity. Nerve impulses (they are digital!) cannot be transmitted over nerve fibres with a repetition rate much higher than about 1000 Hz. It is therefore impossible for the ear to send signals to the brain with a better time resolution than about 1 ms. This means that the ear cannot, in principle, provide information about the phase of a sound vibration for frequencies higher than a few hundred hertz (Some disagree and claim that we can follow phases up to 2000 Hz.). The prevalent view is that sound impression become indifferent to the phase of the various frequency components of a sound signal.

3.8 Learning Objectives

After working through this chapter you should be able to:

- Set up the differential equation for a system subject to forced harmonic oscillations and find an analytical solution for this when the friction term is linear.
- Find a numerical solution of the aforementioned differential equation also for nonlinear friction terms and for nonharmonic forces (after having been through Chap. 4).

- Derive mathematical expressions for resonance frequency, phase shift and quality factor for a single mechanical oscillating system or an electrical oscillating circuit.
- Set up a phasor diagram to explain typical features of an RCL circuit for different frequencies of an applied voltage.
- Know the time course of the oscillations in a circuit, as an externally applied force begins and when it ends and how the time course is affected by the Q-factor.
- Know how the response to an oscillating system changes when the force lasts for a limited period of time.
- Know the coarse features of the anatomy of the ear well enough to explain how we can hear many pitches all at the same time.
- Know that in a mechanical system we cannot get both high frequency-selectivity and high time resolution simultaneously.

Both for the mechanical and electrical oscillating system examined so far, we end up with an equation where the second derivative of a quantity along with the quantity itself is included. It may lead to the opinion that all oscillations must be described by a second-degree differential equation.

However, there are also oscillations that are normally described by two or more coupled first-order differential equation and a significant time delay between the "force" and "the response" in the differential equations. In biology, such relationships are not uncommon.

3.9 Exercises

Suggested concepts for student active learning activities: Forced oscillation, resonance, phasor, phase difference, quality factor, initial and terminal transient behaviour, frequency response, simultaneous multiple frequency detection, basilar membrane, cochlea, inner ear.

Comprehension/discussion questions

1. For a mass–spring oscillator, the phase difference between the applied force and the amplitude of the bob change with the frequency of the applied force. How is the phase difference at the resonance frequency and at frequencies well below and well above it?
2. How does the phase difference between the applied force and the *velocity* vary for a mass–spring oscillator exposed to a harmonic force?
3. It is often easier to achieve a high Q-value in a oscillating system with a high resonance frequency than with a low one. Can you explain why?
4. If our hearing (through natural selection could distinguish much better between sound at nearby frequencies than we are able to achieve, what would the disadvantage have been?

5. We operate with two almost equal resonant frequencies. What are their characteristics? Is it possible for these frequencies to coincide?

6. What would happen to an oscillating system *without* damping if it was exposed
   to a harmonic applied force at the resonant frequency? What would happen if the
   applied force had a frequency slightly different from the resonance frequency?

7. In several laboratories attempting to detect gravity waves, oscillating systems
   with suitable resonance frequencies and Q-values are used as detectors. For
   example, a resonance frequency of about 2–4 kHz is chosen when one wants
   to detect gravity waves due to instability in rotating neutron stars. What is the
   motivation behind using an oscillating system as a detector for this purpose?

8. For a mechanical system, the phase shift $\pi/2$ between the amplitude and the
   applied force was explained by the fact that such a phase shift corresponds to
   the force supplying the maximum power to the system (maximum force applied
   over the longest possible way). Explain in a similar manner the phase shift also
   for the electrical RCL circuit with a harmonically varying applied voltage.

9. Attempt to explain the phase shift for the RCL series circuit with applied voltage
   in case the frequency is far less and far greater than the resonant frequency of
   the circuit alone. Based on how the impedance of a capacitor and the impedance
   of an inductance change with frequency.

10. How can the oscillations that led to the collapse of the Tacoma Narrows Bridge
    in Washington, USA, in 1940 be explained as a forced oscillation? Do you think
    the Q-value was big or small? (May be relevant to watch one of the movies
    featured on YouTube.)

11. An AC voltage $V(t) = V_1 \cos(\omega_F t)$ is applied to an electrical oscillating circuit,
    ω_F is equal to the resonance (angular) frequency of the circuit. After a long
    time, the oscillations in the circuit stabilize and the amplitude of the current
    fluctuations is I_1. An interval of duration t_1 elapses between the connection of
    the AC voltage to the circuit and the current reaching the value $0.9 \times I_1$. We
    then remove the voltage and let the circuit come to rest. We then reconnect to
    the AC voltage, but now with twice the amplitude: $V(t) = 2V_1 \cos(\omega_F t)$.
    (a) How large is the current in the circuit (relative to I_1) a long time after the AC
    voltage was reconnected?
    (b) How long does it take for the amplitude of the current in the circuit to reach
    90% of the limiting, long-time value?
    (c) What do we mean by the expression "long-time value" in this context?

Problems

12. In the case of old-fashioned radio reception in the medium wave range, we
    used circuitry consisting of an inductance (coil) and capacitance (capacitor) to
    discriminate between two radio stations. The radio stations occupied 9 kHz on
    the frequency band, and two radio stations could be as close as 9 kHz. In order for
    us to choose one radio station from another, the receiver had to have a variable
    resonant circuit that suited one radio station, but not another. The frequency of
    the Stavanger transmitter was 1313 kHz. Which Q-factor did the radio receiver's

Fig. 3.17 Sensitivity curve
of a "single-photon detector"

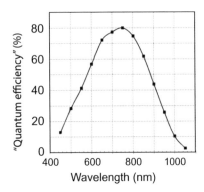

resonant circuit need? [These considerations are still applicable in our modern
times, although digital technology makes certain changes.]

13. Figure 3.17 shows "sensitivity curve" for a "single-photon detector". Let us con-
    sider this curve as a sort of resonance curve, and try to estimate how long a contin-
    uous electromagnetic wave (light) will have to illuminate the detector to achieve
    maximum/stationary response in the detector? (Imagine a similar response as
    in Fig. 3.7.) The frequency of the light can be calculated from the relationship
    $\lambda f = c$ where λ is the wavelength, f the frequency and c the velocity of light.

14. Search the web and find at least ten different forms of resonance in physics.
    Enter a web address, where we can read a little about each of these forms of
    resonance.

15. Derive the expressions given in Eq. (3.11) from Eq. (3.12) and other expressions
    for an oscillating mass–spring oscillator.

16. The Q-value for an oscillating circuit is an important physical parameter.
    (a) Give at least three examples of how the Q-value influences the func-
    tion/behaviour of a circuit.
    (b) Describe at least two procedures as to how the Q-value can be determined
    experimentally.
    (c) If we use a temporary force, it is more difficult to determine the Q-value
    experimentally. Explain why.

17. A series RCL circuit consists of a resistance R of $1.0\,\Omega$, a capacitor C of $100\,nF$,
    and an inductance L of $25\,\mu H$.
    (a) Comparing Eq. (3.7) (slightly modified) with Eq. (3.1), we realize that these
    equations are completely analogous. Just by replacing a few variables related
    to the mechanical mass–spring oscillator, we get the equation for an electrical
    series RCL circuit. Using this analogy, we can easily reshape the expressions
    for phase shift [Eq. (3.3)], amplitude [Eq. (3.4)], Q-value [Eq. (3.11)] and the
    expressions for phase resonance and amplitude resonance for the mass–spring
    oscillator, to corresponding formulas for a series RCL circuit. Determine all
    these terms for a series RCL circuit.
    (b) Calculate the resonant frequencies (both for phase and amplitude resonance)

of the circuit (based on amplitudes of charge oscillations, not current oscillations).

(c) Calculate the Q-value of the circuit.

(d) What is the difference in phase between the applied voltage and current in the circuit at phase resonance and at a frequency corresponding to $\omega_0 + \Delta\omega/2$ in Eq. (3.16)?

(e) How wide is the frequency response of the circuit for a "long-lasting" applied voltage?

(f) How "long" must the applied voltage actually last for the circuit to reach an almost stationary state (that amplitude no longer changes appreciably with time)?

(g) Assume that the circuit is subjected to a force pulse with centre frequency equal to the resonance frequency and that the force pulse has a Gaussian amplitude envelope function [Eq. (3.19)] where σ has a value equal to twice the time period corresponding to the centre frequency of the circuit. Estimate the width of the frequency response to the circuit with this force pulse.

Reference

1. Inductiveload, https://commons.wikimedia.org/wiki/File:Anatomy_of_Human_Ear_with_Cochlear_Frequency_Mapping.svg. Accessed April 2018

Chapter 4
Numerical Methods

Abstract The purpose of this chapter is to provide a brief introduction as to how a first- or second-order differential equation may be solved to the desired precision by using numerical methods like Euler's method and fourth-order Runge–Kutta method. Emphasis is placed on the difference between an analytical and a numerical solution. Movement of a pendulum for an arbitrary amplitude is calculated numerically to exemplify how easily some problems can be solved by numerical methods. Methods for solving partial differential equations are also described, but are not used until a later chapter. The importance of testing, reproducibility and documentation of successive program versions are discussed. Specimen programs are given at the end of the chapter.

4.1 Introductory Remarks

During my student days (1969–1974), Norway's largest computer had a memory capacity (RAM) of 250 kB and it filled a whole room. We made programs by punching holes in a card, one card for each line (see Fig. 4.1). The pack of cards was carried carefully to a separate building; Abel's House (it was a disaster to drop the pack). A waiting period of a few hours up to a whole day passed before we could collect the result in the form of a printout on perforated pages. A punching error meant that a card had been punched again so that the wrong card in the stack could be exchanged with the new card. This was followed by a new submission and another waiting period. Guess if debugging a program took an eternity! Today, the situation is totally different. Everyone owns a computer. Program development is incomparably easier and far less time-consuming than in earlier times. And numerical methods have become a tool as natural as analytical mathematics.

But all tools have one thing in common: training is needed in how they are to be used. In this chapter, our primary concern will be to see how the equation of motion for an oscillating system and the wave equation can be solved in a satisfactory manner. It is not enough to read how things can be done. Practice is needed for acquiring the requisite skills and mastering the routine.

© Springer Nature Switzerland AG 2018   59
A. I. Vistnes, *Physics of Oscillations and Waves*, Undergraduate Texts in Physics,
https://doi.org/10.1007/978-3-319-72314-3_4

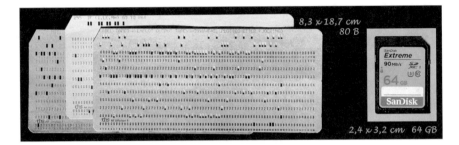

Fig. 4.1 Examples of punch cards, along with a modern memory device (sizes indicated) with storage capacity equivalent to 800 million punch cards (which would have weighed 1900 tons!). The memory device weighs about 0.5 g

Parts of the chapter were written by David Skålid Amundsen as a summer job for CSE 2008. Amundsen's text has since been revised and expanded several times by Arnt Inge Vistnes.

4.2 Introduction

When in the "old days" (i.e. more than 30 years ago), we investigated the motion of a mathematical or physical pendulum in a lower-level physics course, we had to be content with "small amplitudes". At that time, with only the rudiments of analytical mathematics in our toolkit, we could only proceed by imposing the approximation of small displacements, which implied that the movement is a simple harmonic motion. Larger amplitudes are much more difficult to handle analytically, and if we consider complicated friction as well, there is simply no analytical solution to the problem.

Once we have learned to use numerical methods of solution, it is often almost as easy to use a realistic, nonsimplified description of a moving system as an idealized simplified description.

This book is based on the premise that the reader already knows something about solving, for example, differential equations with the aid of numerical methods. Nevertheless, we make a quick survey of some of the simplest solution methods so that those who have no previous experience with numerical methods would nonetheless be able to keep pace with the rest. After the quick review of some simple methods, we spend a little more time on a more robust alternative. Additionally, we will say a little about how these methods can be generalized to solve partial differential equations.

It should be mentioned here that the simplest numerical methods are often good enough for calculating, for example, the motion of a projectile, even in the presence of air resistance. However, the simplest methods often accumulate errors and give quite a bad result for oscillatory motion. In other words, it is often necessary to use some advanced numerical methods in dealing with oscillations and waves.

This chapter is structured along the following lines:

First, a quick review of the simplest numerical methods used for solving differential equations is given. Secondly, the fourth-order Runge–Kutta's method is presented. This first part of the chapter is rather mathematical. Then comes a practical example, and finally, we will include examples of program codes that can be used for solving the problems given in later chapters.

4.3   Basic Idea Behind Numerical Methods

In many parts of physics, we come across the second-order ordinary differential equations:

$$\ddot{x} \equiv \frac{d^2 x}{dt^2} = f\big(x(t), \dot{x}(t), t\big) . \tag{4.1}$$

with the initial conditions $x(t_0) = x_0$ and $\dot{x}(t_0) = \dot{x}_0$. The symbol $f\big(x(t), \dot{x}(t), t\big)$ means that f (for the case when x is the position variable and t the time) is a function of time, position and velocity.

In mechanical systems, differential equation often arises when Newton's second law is invoked. In electrical circuitry containing resistors, inductors and capacitors, it is often Kirchhoff's law together with the generalized Ohm's law and complex impedances that are the source of differential equations.

When we solve second-order differential equations numerically, we often consider the equation as a combination of two coupled first-order differential equations. We rename then the first derivative and let this be a new variable:

$$v \equiv \frac{dx}{dt} .$$

The two coupled first-order differential equations then becomes:

$$\frac{dx}{dt} = v\big(x(t), t\big) ,$$

$$\frac{dv}{dt} = f\big(x(t), v(t), t\big) .$$

We will shortly see some simple examples of this in practice.

4.4 Euler's Method and Its Variants

We can solve a first-order differential equation numerically by specifying a starting value for the solution we are interested in, using our knowledge of the derivative of the function to calculate the solution for a short time Δt afterwards. We then let the new value act as a new initial value to calculate the value that follows Δt after this (that is, at $t = 2\Delta t$). We repeat the process until we have described the solution in as many points n as we are interested in.

The challenge is to find out how we can determine the next value from what we already know. It can be done in a crude or refined method. The easiest method is perhaps Euler's method. It is based on the well-known definition of the derivative:

$$\dot{x}(t) = \lim_{\Delta t \to 0} \frac{x(t + \Delta t) - x(t)}{\Delta t} .$$

If Δt is sufficiently small, we can manipulate this expression and write:

$$x(t + \Delta t) \approx x(t) + \Delta t \dot{x}(t) .$$

Assume the initial values are given by $(x_n, \dot{x}_n, t_n)$. Then follows the discrete version of our differential equation (named "difference equations"):

$$x_{n+1} = x_n + \dot{x}_n \Delta t .$$

By using such an update equation for both $x(t)$ and $\dot{x}(t)$, we get the familiar Euler method (in our context for the solution of second-order differential equation):

$$\dot{x}_{n+1} = \dot{x}_n + \ddot{x}_n \Delta t$$
$$x_{n+1} = x_n + \dot{x}_n \Delta t .$$

Thus, we have two coupled difference equations.

Figure 4.2 outlines how the method works. This is the most common way to make such an illustration, but in my view it only gives a superficial understanding. What happens when the discrepancy between the correct solution and the numerical solution becomes bigger and bigger? Here are some details we should know.

Figure 4.3 looks similar to Fig. 4.2, but is illustrating a different message. The mid-blue blue curve (bottom) shows how a projectile thrown obliquely will proceed with an initial velocity of 1.0 m/s in the horizontal direction and 3.0 m/s in the vertical direction. The calculation is based on an analytical solution to this simple problem.

Fig. 4.2 Euler's simple method of calculating a function numerically. The top (blue) curve is the exact analytic solution. The lower (red) curve is calculated using Euler's simple method, while the middle curve is calculated using the midpoint method. The time step is the same in both cases and is chosen very large to accentuate the differences

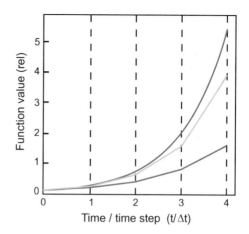

Fig. 4.3 Calculation of an oblique throw using Euler's method with a large time step. Each new point calculated is the starting point for a new solution of the original differential equation. See the text for details

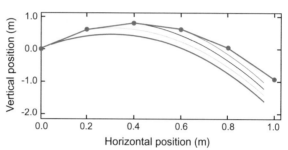

The figure also shows a plot of the solution found by using Euler's method (red curve) with very large time steps (0.2 s). Even after the first step, the calculated new position is quite far from what it should be.

After the first step, new values have been calculated for position and speed in both horizontal and vertical directions. These values are now plugged into the differential equation. If we had calculated the path for exactly these values, we would have got the solution given by a green curve (next to bottom). *This is a different solution of the differential equation than we started with!*

Not even now, we manage to follow this new solution closely since the time step is so big and when we use Euler's method once more, we get a position (and velocity) quite far from the second solution of the differential equation we started with.

We keep going along this route. For each new time step, we get a new solution of differential equation, and in our case, the error, being systematic, becomes bigger and bigger after each time step.

It can be shown that if we reduce the time step significantly (!) compared to that used in Fig. 4.3, the solution will be far better than in the figure. Nevertheless, it is not always enough to only reduce the size of the time step.

First of all, we cannot make the step size so small that we run into trouble with inputting numbers accurately on a computer (without having to use extremely time-consuming techniques). When we calculate $x_{n+1} = x_n + \dot{x}_n \Delta t$, the contribution

$\dot{x}_n \Delta t$ must not always be so small that it can only affect the least significant digit of x_{n+1}.

Another limitation lies in the numerical method itself. If we make systematic errors which accumulate at each time step, no matter how small the time steps are, we also get problems. Then we must use other numerical methods instead of this simplest variant of Euler's method.

An improved version of Euler's method is called the *Euler-Cromer method*. Assume that the starting values are $(x_n, \dot{x}_n, t_n)$. The first step is identical to Euler's simple method:

$$\dot{x}_{n+1} = \dot{x}_n + \ddot{x}_n \Delta t .$$

However, the second step in the Euler-Cromer method differs from that in the simpler Euler version: To find x_{n+1}, we use $\dot{x}_{n+1}$ and not $\dot{x}_n$ as we do in Euler's method. It provides the following update equation for x:

$$x_{n+1} = x_n + \dot{x}_{n+1} \Delta t .$$

The reason that the Euler-Cromer method works and that it often (but not always) works better than Euler's method is not trivial, and we will not go into this. Euler's method often causes the energy of the modelled system to become an unconserved quantity that slowly but steadily increases. This problem becomes dramatically reduced with the Euler-Cromer method, which in most cases works better.

Another improvement over Euler's method, which is even better than the Euler-Cromer method, is *Euler midpoint method*. Instead of using the gradient *at the beginning* of the step, and using this for the entire interval, we use the gradient *in the middle* of the interval. By using the slope at the midpoint of the interval, we will usually get a more accurate result than using the slope at the beginning of the interval when we are looking for the average growth rate.

In Euler's midpoint method, we first use the gradient at the beginning of the interval, but instead of using this value for the entire interval, we use it for half the interval. Then we calculate the gradient at the middle of the interval and use this for the entire interval. Mathematically, this is done by using the same notation as before:

$$\dot{x}_{n+\frac{1}{2}} = \dot{x}_n + f\left(x_n, \dot{x}_n, t_n\right)\tfrac{1}{2}\Delta t ,$$
$$x_{n+\frac{1}{2}} = x_n + \dot{x}_n \tfrac{1}{2}\Delta t .$$

Here, $\dot{x}_{n+\frac{1}{2}}$ and $x_{n+\frac{1}{2}}$ are the values of the unknown function and its derivative at the midpoint of the interval. The update equation for the entire range will be as follows:

$$\dot{x}_{n+1} = \dot{x}_n + f\left(x_{n+\frac{1}{2}}, \dot{x}_{n+\frac{1}{2}}, t_{n+\frac{1}{2}}\right)\frac{1}{2}\Delta t ,$$

$$x_{n+1} = x_n + \dot{x}_{n+\frac{1}{2}} \Delta t .$$

4.5 Runge–Kutta Method

In Euler's method, we found the next value by using the slope at the beginning of the chosen step. In Euler's midpoint method, we used the slope in the middle of the chosen step. In either case, it is quite easy to imagine that for some functions we will be able to get a systematic error that will add up to a significant total error after many subsequent calculations have been carried out. It can be shown that the error we make becomes significantly less if we switch to using more refined methods for finding the next value. One of the most popular methods is called the fourth-order Runge–Kutta method. A total of four different estimates of the increase, one at the beginning, two in the middle and one at the end are then used to calculate the average increase in the interval. This makes the Runge–Kutta method much better than Euler's midpoint method, and since it is not much harder to program, this is often used in practice.

Let us see how the fourth-order Runge–Kutta method works and how it can be used to solve a second-order differential equation (At the end of the chapter one will find a pseudocode and the full code for a program that uses the fourth-order Runge–Kutta method.).

4.5.1 Description of the Method

The Runge–Kutta method is not really difficult to understand, but you probably have to read the details that are included twice to see it. We will first provide a mathematical review and then try to summarize the method using a figure (Fig. 4.4). Let us begin with a few words about the mathematical notation. Consider the differential equation given below:

$$\ddot{x}(t) = f\left(x(t), \dot{x}(t), t\right) . \tag{4.2}$$

For the damped mass–spring oscillator considered in Chap. 2 (where t does not appear explicitly), this equation will take the following form:

$$\ddot{z}(t) = -\frac{b}{m}\dot{z}(t) - \frac{k}{m}z(t) . \tag{4.3}$$

Suppose we are at the point $(x_n, \dot{x}_n, t_n)$ and that the duration of the time step is Δt. In what follows we will find estimates for x_n, $\dot{x}_n$ and $\ddot{x}_n$, and it will be convenient to replace $\dot{x}_n$, and $\ddot{x}_n$ by v_n and a_n, respectively. An additional numerical index will be used to indicate the ordinal position of an estimate (first, second, etc.). With this notation, the kth estimate of a quantity χ_n ($\chi = x, v = \dot{x}, a = \ddot{x}$) will be represented by the symbol $\chi_{k,n}$.

We can find the first estimate of $\ddot{x}_n$ by using Eq. (4.1):

$$a_{1,n} = f(x_n, \dot{x}_n, t_n) = f(x_n, v_n, t_n) .$$

At the same time, the first derivative is known at the beginning of the time step:

$$v_{1,n} = \dot{x}_n = v_n .$$

The next step on the route is to use Euler's method to find $\dot{x}(t)$ and $x(t)$ in the middle of the step:

$$x_{2,n} = x_{1,n} + v_{1,n}\frac{\Delta t}{2} ,$$

$$v_{2,n} = v_{1,n} + a_{1,n}\frac{\Delta t}{2} .$$

Furthermore, we can find an estimate of the second derivative at the midpoint of the step by using $v_{2,n}$, $x_{2,n}$ and Eq. (4.2):

$$a_{2,n} = f(x_{2,n}, v_{2,n}, t_n + \Delta t/2) .$$

The next step now is to use the new value for the second derivative at the midpoint in order to find a new estimate of $x(t)$ and $\dot{x}(t)$ at the midpoint of the step using Euler's method:

$$x_{3,n} = x_{1,n} + v_{2,n}\frac{\Delta t}{2} ,$$

$$v_{3,n} = v_{1,n} + a_{2,n}\frac{\Delta t}{2} .$$

With the new estimate of $x(t)$ and $\dot{x}(t)$ at the midpoint of the step, we can find a new estimate for the second derivative at the midpoint:

$$a_{3,n} = f(x_{3,n}, v_{3,n}, t_n + \Delta t/2) .$$

Using the new estimate of the second derivative in addition to the estimate of the first served in the middle range, we can now use Euler's method to estimate $x(t)$ and

$\dot{x}(t)$ at the end of step. This is done as follows:

$$x_{4,n} = x_{1,n} + v_{3,n}\Delta t \, ,$$

$$v_{4,n} = v_{1,n} + a_{3,n}\Delta t \, .$$

Finally, in the same way as before, we can estimate $\ddot{x}(t)$ at the end of the step using these new values:

$$a_{4,n} = f(x_{4,n}, v_{4,n}, t_n + \Delta t) \, .$$

We can now calculate a weighted average of the estimates, and then we get reasonable estimates of the average values of the first and second derivatives in the step:

$$\overline{a_n} = \tfrac{1}{6}\left(a_{1,n} + 2a_{2,n} + 2a_{3,n} + a_{4,n}\right) \, , \tag{4.4}$$

$$\overline{v_n} = \tfrac{1}{6}\left(v_{1,n} + 2v_{2,n} + 2v_{3,n} + v_{4,n}\right) \, . \tag{4.5}$$

Using these averages, which are quite good approximations to the mean values of the slopes over the entire step, we can use Euler's method of finding a good estimate of $x(t)$ and $\dot{x}(t)$ at the end of the step:

$$x_{n+1} = x_n + \overline{v_n}\Delta t \tag{4.6}$$
$$v_{n+1} = v_n + \overline{a_n}\Delta t \tag{4.7}$$
$$t_{n+1} = t_n + \Delta t \tag{4.8}$$

These are equivalent to the initial values for the next step.

In the Runge–Kutta method (see Fig. 4.4), we extract much more information from the differential equation than in Euler's method. This makes the Runge–Kutta method significantly more stable than Euler's method, Euler-Cromer method and Euler's midpoint method. The Runge–Kutta method does not make an excessive demand on the resources of a computer, but it is relatively simple to program. The Runge–Kutta method, in one or other variant, is therefore often the method we first turn to when we want to solve ordinary differential equations numerically.

Programming of the basic part of the Runge–Kutta method is done almost once and for all. It is usually only a small file that changes from one problem to another. The file specifies exactly the differential equations that will be used in exactly the calculations that will be performed. See example code later in the chapter.

Fig. 4.4 Summary of the
fourth-order Runge–Kutta
method. See text

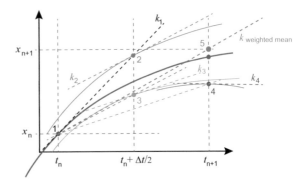

Some concentration is required to fully understand Fig. 4.4: In point 1 (x_n, t_n), the slope is k_1. We follow the tangent line at point 1 for half a step to point 2 (pink). This point is based on another solution of differential equation (thin pink line) than the one we seek. We calculate the slope k_2 at point 2 for this solution (pink dotted line). We then draw a line from point 1 again, but now with the gradient we found at point 2. Again we only go half the step length and find point 3 (green). There is yet another solution of the differential equation that goes through this point (thin green line). We calculate the slope k_3 at point 3 for this solution (dotted green line). We then draw a line through point 1 again, but now with the slope we just found. Now we go all the way up to point 4 (brown). Again there is a new solution of the differential equation that goes through this point. We calculate the slope k_4 of this solution at point 4.

The final step is to calculate the weighted mean of four different slopes and use this from the starting point 1 in the figure a full time span Δt to get the estimate (point 5) for the change of our function in the current time interval. The result is relatively close to the correct value (compare point 5 by a red dot in the figure).

4.6   Partial Differential Equations

Many physical problems are described by partial differential equations, perhaps the most well known are Maxwell's equations, Schrödinger equation and wave equation. The term "partial differential equation" means that the unknown function depends on two or more variables, and that derivatives with respect to these occur in the differential equation.

There are several methods for solving partial differential equations, but a key concept is finite differences. It is about replacing the differentials in the differential equation with final differences. Consider the simple differential equation

$$\frac{\partial y}{\partial x} = K \frac{\partial y}{\partial t} . \tag{4.9}$$

The simplest way to convert the derivatives in this equation into difference quotients is to use the definition of the derivative, as we have done before. The above equation will then become

$$\frac{y(x + \Delta x, t) - y(x, t)}{\Delta x} = K \frac{y(x, t + \Delta t) - y(x, t)}{\Delta t}.$$

This equation can be solved for $y(x, t + \Delta t)$, which gives

$$y(x, t + \Delta t) = y(x, t) + \frac{\Delta t}{K \Delta x} [y(x + \Delta x, t) - y(x, t)].$$

Suppose that $y(x, t)$ is known at a time $t = t_0$ for the interesting interval in x. The right-hand side of the above equation gives $y(x, t_0 + \Delta t)$, the value of the function at a later time $t + \Delta t$. However, note that we also need the value of the function at a different x from that appearing on the left-hand side. This means that we will encounter a problem when we come to calculating the value of the function at a point x near the outer limit of the region over which the calculation is to be performed. From the equation above, we see that we need to know what the function was at the next x coordinate at the last instant, and at the extreme x point, this is not feasible.

This means that, in order to find a unique solution to our problem, we must know the *boundary conditions*, that is, the state of the system at the boundary of the region of interest. These must be specified before the calculations can even begin.

Note: Initial and boundary conditions are two different things and must not be mixed together. Initial conditions specify the state of the system at the very beginning of the calculations and must also be used here. Boundary conditions specify the state of the system at the endpoints of the calculations at *all* time.

The finite differences introduced above are, however, rarely used, since they can be replaced by something that is better and not much more difficult to understand. Instead of using Euler's method in the above differentials, Euler's midpoint method, which significantly reduces the error in the calculations, is used. If we do this, the discretization of Eq. (4.9) leads to the following result:

$$\frac{y(x + \Delta x, t) - y(x - \Delta x, t)}{2\Delta x} = K \frac{y(x, t + \Delta t) - y(x, t - \Delta t)}{2\Delta t}.$$

It is not hard to understand that the result will now be better, for instead of calculating the average growth through the current point and the next point, the average growth is used through the previous and next point. In the same way as before, this equation can be solved with regard to $y(x, t + \Delta t)$, and the result will be:

$$y(x, t + \Delta t) = y(x, t - \Delta t) + \frac{\Delta t}{K \Delta x} [y(x + \Delta x, t) - y(x - \Delta x, t)] \ .$$

We see that we get the same problem with boundary conditions as above; in fact, an extra boundary condition is needed, even at the beginning of the x grid. Since this is a problem that concerns a spatial dimension, we need to set two boundary conditions to make the solution unique (there are two boundaries). To use the first one the update equation must therefore take into account the other boundary as well.

In the same way as we replaced first derivative with a finite difference quotient, the nth derivative can be approximated in the same way. An example is the second derivative that can be approximated with the following difference quotient:

$$f''(x) \approx \frac{f(x + \Delta x) - 2f(x) + f(x - \Delta x)}{\Delta x^2} \ . \tag{4.10}$$

Proof

$$f''(x) \approx \frac{f(x + \Delta x) - 2f(x) + f(x - \Delta x)}{(\Delta x)^2} \tag{start}$$

$$= \frac{[f(x + \Delta x) - f(x)] - [f(x) - f(x - \Delta x)]}{(\Delta x)^2} \tag{4.11}$$

$$= \frac{1}{\Delta x} \left[\underbrace{\frac{f(x + \Delta x) - f(x)}{\Delta x}}_{\approx f'(x)} - \underbrace{\frac{f(x) - f(x - \Delta x)}{\Delta x}}_{\approx f'(x - \Delta x)} \right] \tag{4.12}$$

$$= \frac{f'(x) - f'(x - \Delta x)}{\Delta x} \ . \tag{end}$$

This expression is nothing more than the definition of the derivative; thus, it is a proof of the validity of Eq. (4.10). The expressions make it clear why we must know the value of the function at three points (at least) in order to be able to calculate a second derivative.

As with the ordinary differential equations, we can move on and use methods that provide an even better result.

There are a number of methods available for different parts of physics. Interested refer to special courses/books in numerical calculations.

4.7 Example of Numerical Solution: Simple Pendulum

Let us take a concrete example, namely a pendulum that can swing with arbitrary large amplitudes (up to $\pm\pi$) without collapsing (i.e. the suspending rod is "rigid"). We expect all mass to be in a tiny ball (or bob) at the end of the rod.

Mechanics tell us that the force that pulls the pendulum along the path towards the equilibrium point is

$$F_\theta = -mg \sin \theta$$

where θ denotes the angular amplitude. If the length of the rod is L, the moment of this force around the pivot (suspension point) is:

$$\tau = -mgL \sin \theta \ .$$

The torque applied around the pivot can also be written as:

$$\tau = I\alpha = I\ddot{\theta} \ .$$

Here $\alpha = \ddot{\theta}$ is the angular acceleration and I the moment of inertia about the axis of rotation (which passes through the pivot and is perpendicular to the plane in which motion takes place). By using our simplifying assumptions for the pendulum, we have:

$$I = mL^2$$

which leads to the differential equation for the motion of the bob:

$$mL^2\ddot{\theta} = -mgL \sin \theta \ ,$$

$$\ddot{\theta} = -\frac{g}{L} \sin \theta \ .$$

In an elementary mechanics course, this equation is usually solved by assuming that the angle θ is so small that $\sin \theta \approx \theta$. The solution then turns out to be a simple harmonic motion with swing frequency (angular frequency) given by:

$$\omega = \sqrt{\frac{g}{L}} \ .$$

The approximation $\sin \theta \approx \theta$ was made to use analytical methods. This approach was not absolutely necessary in just this particular case, because we *can* solve the original differential equation analytically also for large angles by utilizing the series expansion of the sinus function. However, it is by far easier to use numerical methods.

The result of numerical calculations where we use fourth-order Runge–Kutta method is shown in Fig. 4.5. We see that the motion is near harmonic for small angular amplitudes, but very different from a sinusoid for a large swing amplitude.

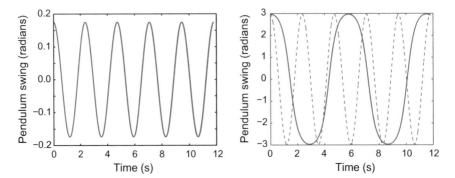

Fig. 4.5 A pendulum swings harmonically when the amplitude is small, but the swinging motion changes considerably when the swing angle increases. The swing period changes as well. See also the text

Moreover, the period has changed a lot. Note that in the right-hand part of the figure, we have chosen a motion where the pendulum *almost* reaches the "right-up" direction both "forward" and "return" (swing angle near $+\pi$ and $-\pi$).

If we wanted to include friction in the description of the pendulum motion, it would represent a more complex expression of the effective force than we had in our case. For nonlinear description of friction, there is no analytical solution.

Since the main structure of a numerical solution would be the same, irrespective of our description of the effective force acting on the system, the more complicated physical conditions can often be handled surprisingly easily with numerical solution methods (see Fig. 4.7 in one of the tasks in the problem section below).

This is an added bonus of numerical solutions: the force that works—and thereby the actual physics of the problem—becomes more central in our search for the solution! What force produces which result? Numbers are numbers, and there is no need to figure out different—occasionally intricate—analytical methods and tricks especially adapted for each functional representation of the force. The focus is where it should be: basically, the effective force, the governing differential equation, the pertinent initial condition(s), and the results that emerge from the analysis.

4.8 Test of Implementation

It is so easy to make a mistake, either in analytical calculations or in writing a computer program for obtaining numerical solutions. We have examples of many disasters in such contexts.

It is therefore very important to test the results of numerical solutions to detect as many errors as we can. It is often easier said than done! We often use numerical methods because we do not have any analytical methods to fall back on.

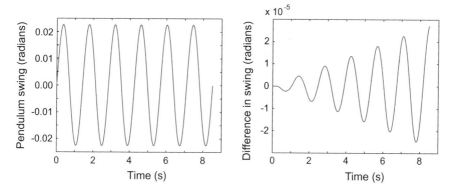

Fig. 4.6 Comparison between analytical and numerical solution of a shuttle movement. For explanations: See the text

In the case of the simple pendulum, there happens to be a trick up our sleeve. There is an analytical solution that is approximately correct for *small* amplitude. For *this* special case, we can test if the numerical solution becomes nearly the same as the analytical. If there is a serious disagreement between these two solutions, there must be an error somewhere.

That the numerical solution is close to its analytical counterpart in this special case, is unfortunately not a proof that the program is flawless! The implementation of the program beyond the special case may give incorrect results. Here it is necessary to consider the physical predictions: Do they seem reasonable or otherwise? It is often impossible to be absolutely sure that a computer program is completely correct. Within numerical analysis, there are special techniques that can be used in some cases. We cannot go into these. The main point is that we must be humble and alert to the possibility of errors and try to test the implementation of numerical methods every time we develop a computer program.

As an example, we will now try to check the program we used in the calculations that led to Fig. 4.5. We will use only the small amplitude case in our test.

In Fig. 4.6, the results of the numerical calculations (red curve) are shown on the left with an analytical solution (dashed blue curve) for the special case when the pendulum swing is small (maximum ± 0.023 rad). There is no perceptible difference between the two curves.

Plotting analytical and numerical solutions in the same figure are a common way to check that two solutions are in agreement with each other. However, this is a very rough test, because there is limited resolution in a graphical representation. In the right part of the figure, we have chosen a better test. Here, the *difference* between analytical and numeric results is plotted, and we see that there were certainly some differences, although we did not see this in the left part.

We can now see that the difference is increasing systematically. After six periods, the difference has increased to 2.7×10^{-5} rad. Is this an indication that our computer program is incorrect?

We know, however, that the analytical solution is *itself* only an approximation, and the smaller the swing angle, the smaller will be the error in the approximation. We can then reduce the amplitude and see what happens. Calculations show that if the amplitude is reduced to 1/10 of what we have in the figure, the maximum difference is reduced after six periods to 1/1000 of the earlier value. If we reduce the amplitude to 1/100 of the original, the maximum difference is reduced to 10^{-6} of the original difference. We see that numerical and analytical solutions are becoming more and more similar and in a way that we would expect. If we take a look at the series development for the sine function, it gives us a further clue that our results are what we would expect.

We can then feel reasonably sure that the program behaves as it should for small angular displacements, and that it seems to handle larger angles as it should, at least as long as they remain small.

There is also another test we often have to do in connection with numerical calculations. We chose to use 1000 steps within each period in the calculations whose results are plotted in Figs. 4.5 and 4.6. For calculations that span very many periods, we cannot use such small time steps. If we go down to, e.g., 100 calculations per period, the result will still be acceptable usually (depending on what requirements we impose), but if we go down to, say 10 steps per period, the result will almost certainly depend markedly on the choice of the step size. We often have to do a set of calculations to make sure that the "resolution" in the calculations is appropriate and manageable (neither too high nor too low).

4.9 Reproducibility Requirements

Today it is easy to change a program from one run to another. Ironically, this presents extra challenges that need to be taken seriously. When we make calculations to be used in a scientific article, a master's thesis, a project assignment, and almost in any context where our program is used, we must know the exact program and parameters that are used if the results are to have full value. In experimental physics, we know that it is important to enter in the laboratory journal all details of how the experiments have been performed. The purpose is that it should be possible to test the results we get. This is essential for reproducibility and for achieving so-called intersubjectivity (that the result should be independent of which person actually executes the experiment), which is extremely important in science and development.

In experimental work, one occasionally succumbs to the temptation of not jotting down all relevant details while the experiment is underway. Being interested primarily in the result, we think that when we have come a little further and got even better results, *then* we would write down all the details. Such practice often causes some frustration at a later date, because suddenly we discover that an important piece of information was never actually noted. At worst, the consequence of this lapse may be that we have to repeat the experiment, and hunt for the conditions under which the previous experiment, the results of which proved to be particularly interesting, was performed.

Modern use of numerical methods can in many ways be compared to experimental work in the laboratory. We test how different parameters in the calculation affect the results, and we use different numerical methods in a similar manner as we use different measuring instruments and protocols in experiments. This means that there are stringent requirements for documentation for those who use numerical methods as for the experimentalist.

In order to comply with this requirement, we should incorporate good habits in the programming. One way we can comply with reproducibility requirements is to do the following:

- In the program code, insert a "version number" for your application.
- In the result file you generate, the version number must be entered automatically.
- Every time you change the program in advance of a calculation that you would like to make, the version number must be updated.
- Each version of the program (actually used in practice) must be saved to disk so that it is always possible to rerun an application with a given version number.
- Parameters which are used and which vary from run to run within the same version of the program must be printed to a file along with the result of the run.

If we keep to these rules, we will always be able to return and reproduce the results obtained in the past. It is assumed here that the results are independent of the computer used for the calculations. If we suspect that a compiler or an underlying program or an operating system might malfunction, it may be appropriate to provide additional information about this along with the results (in a result file).

In the specimen programs given in this book, the lines needed for documentation of parameters and version number are, for the most part, *not* included in the code. The reason is that the program pieces provided here are intended primarily for showing how the calculations can be performed.

4.10   Some Hints on the Use of Numerical Methods

In our context, it is often necessary to create relatively small computer programs to get a specific type of calculation. There is usually no need to have the fancy interface to select parameters and fancy presentations of the results as it is for commercial programs. We need to do a specific task, and the program is usually not used by many, or very often. This is the starting point for the tips that follow.

Many of the issues we encounter in this book are related to the integration of differential equations that describe the processes we are interested in. The following hints are partly influenced by this preoccupation.

Planning

Before we get to the computer, we should have a clear notion of what we want to achieve. We must have already established the differential equation that describes the process of our interest and have pondered over the parameters that are to be included in the calculations. Current parameter values and initial values need to be looked up or chosen by ourselves.

It may be useful to outline how we can partition the program into main components, each of which has its separate function. We also have to decide the order in which we will work through the various parts of the program and have thoughts of how we can test the different parts individually and together.

It is also natural to ask: Do we want to provide parameters while the program is running or is it sufficient to insert them into the program code before the program starts? How will we take care of the results? Should it be in the form of plots or animations or numbers are printed on screen, or should the final results be written to file(s) for later processing?

Writing of Code

There should be a one-to-one correspondence between the mathematical description of a problem (algorithm) and the code. It applies to variables, formulas, etc.

It is recommended to adhere to the programming language guidelines, such as "PEP 8—Style Guide for Python Code" or "MATLAB Style Guidelines 2.0".

Try to collect the code lines where parameters are given special values already as part of the code. This makes it easier to change parameters for later runs. Reset arrays or give arrays values.

Put together all expressions of fixed constants which will be used in that part of the program that is most frequently run, in order to avoid more calculation operations than necessary in a loop. For example, it is a good idea to create a parameter

```
coeff = 4.0*3.141926*epsilon0*epsilonR*mu0*muR
```

and use this coefficient in a loop that is recalled many times, instead of having to repeat all these multiplications each time the loop is run (the parameters in this example have been selected randomly).

A code should be broken up into logical functions. In Python, multiple functions can be added to one and the same file. In Matlab, various functions are often allocated to separate files (although it is actually possible to use a similar layout in Matlab as in Python).

Generalize when you are writing a program, unless it seems inadvisable. For example, when integrating an expression, a general integral of $f(x)$ is programmed and then a special f is chosen as its argument. This requires frequent use of functions. Do not overdo it though, because it obstructs a survey and the readability of the program.

Testing and Debugging

Make an effort to construct test problems for checking that the implementation is correct. Functions should be tested as they are written. Do not postpone testing until code writing is finished!

There are several types of errors that may occur. Some errors are detected by the compiler. Read the error message carefully to see how such errors can be corrected.

Other errors appear when running the program. For example, we can end up in an infinite loop and must terminate the program manually. It is not always easy to find out where in the program code such a fault is located. It is then useful to add dummy print-to-screen here and there in the code so we can locate that line in the code where the problem occurs.

While we are going through program development and testing, it is important to save the program several times along the way, and preferably change names sometimes, in order to avoid a potential catastrophe. Then we will not have to start all over again if you lose everything in a file.

Check that the program provides the correct result for a simplified version of the problem, where there is also an analytical solution. This is crucial!

Repeat the calculations using different resolutions (often given by Δt) to see how many points are needed to get a good match with the analytical answer or to verify that the result depends only to a small extent on moderate changes in resolution.

Forms of Presentation

Plot the results or present them in some other form. Save data to file if desired.

Simple plots are often sufficient, but we can rarely read precise details from a plot, at least not without having chosen a very special plot that displays just what we want to show. Sometimes, the choice of linear or logarithmic axes in a plot is crucial for whether we discover interesting relationships or not.

Make sure that the axes in the plot are labelled properly that symbol sizes and line thicknesses and other details in the presentation meet the expected requirements.

In reports, articles and theses, one is a requirement that numbers and text along the axes of the plots must be readable without the use of magnifying glass (!) *in the final size the characters have in a document.* This means that numbers and letters should have a size between 9 and 12 pt in *final size*, and indexes may be even a bit smaller).

When using Matlab, it is a good idea to save figures which do *not* fill the entire screen (use default display of figures on screen). Then the font size will be sufficiently large even if the figure is reduced to approximately the same format as used in this book. However, if the image size is reduced too much, the font size in the final document will become too small. You can choose, for example, line thickness and font size in plots generated by Matlab and Python. The following code piece indicates some of the possibilities that exist (the example is in Matlab, but there are similar solutions in Python):

```
...
axes('LineWidth',1,'FontSize',14,'FontName','Arial');
plot(t,z,'-r','LineWidth',1);
xlabel('Time (s)','FontSize',16,'FontName','Arial');
...
```

Learn good habits as early as possible—it will pay off in the long run!

Reproducibility

When we believe that the program as whole works as it should, we can finally embark upon the calculations for the particular project we are occupied with. Reproducibility requirements must be adhered to when the program now receives a solemn version number, and the program code must be saved and not changed without a new version number.

Files that document later runs must be preserved in a manner similar to a laboratory record.

4.11   Summary and Program Codes

Summary of the Chapter

Let us try to summarize the key points in our chapter:

- A second-order differential equation can be considered equivalent to two coupled first-order differential equations.
- In a single differential equation, we replace the derivative df/dt with the differential quotient $\Delta f/\Delta t$. Starting from this approximate equation and initial conditions, we can successively calculate all subsequent values of $f(t)$. This method is called Euler's method. The method often gives large errors, especially when we are dealing with oscillations!
- There are better methods for estimating the average slope of the function during the step Δt than just using, as we in Euler's method, the derivative at the beginning of the interval. One of the most practical and robust methods is called fourth-order Runge–Kutta method. In this method, a weighted average of four different calculated increments in the interval Δt is used as the starting point for the calculations. The method often provides good consistency with analytical solutions where these exist, also for oscillatory phenomena. However, we must be aware that this method is not exempt from error, and for some systems it will not work properly.
- For second-order ordinary differential equations, such as the equation for oscillation, we can find the solution if we know the differential equation and the initial conditions. For the second-order partial differential equations, for example, a wave equation, we must *in addition* know the so-called boundary conditions not only at the start but also throughout the calculations. This makes it often far more difficult to solve partial differential equations than ordinary other order diffusions.
- It is valuable to compare numerical calculations and analytical calculations (where these exist) to detect errors in our programming. However, even if the conformity is good in such special cases, there is no guarantee that the numerical solutions will be correct also for other parameter values (where analytical solutions are not available).

- The program code is divided into an appropriate number of separate functions that have their own task. In this way, the logical structure of the program will clarify. Some features can be made so general that they can be reused in many different contexts. For example, we can create one general Runge–Kutta function that calls for a more specialized function that contains the appropriate differential equation (where only the last small function will vary from problem to problem).
- Since we can easily change programs and parameters, it is a big challenge to keep track of how the computer program looked and what parameters we used when we made calculations and arrived at results we would use. Some systematic form of documentation is imperative, where program, input parameters and results can be linked to each other in a clear way.

Pseudocode for Runge–Kutta Method *

The input to this function is x[n-1], v[n-1] and t[n-1] and returns x[n] and v[n].

1. Use the input parameters in order to find the
   acceleration, a1, in the start of the interval.
   The speed in the start of the interval, v1, is given as
   an input parameter.
   x1 = x[n-1]
   v1 = v[n-1]
   a1 = ...

2. Use this acceleration and speed to find an estimate for
   the speed (v2) and position in the middle of the interval.
   x2 = ...
   v2 = ...

3. Use the new position and speed to find an estimate for
   the acceleration, a2, in the middle of the interval.
   a2 = ...

4. Use this new acceleration and speed (a2 and v2) to find
   a new estimate for position and speed (v3) in the middle
   of the interval.
   x3 = ...
   v3 = ...

5. Use the new position, speed and time in the middle of
   the interval to find a new estimate for the acceleration,
   a3, in the middle of the interval.
   a3 = ...

6. Use the last estimate for the acceleration and speed in
   the middle of the interval to find a new estimate for the
   position and speed (v4) in the END of the interval.
   ```
   x4 = ...
   v4 = ...
   ```

7. Use the last estimate for position and speed to find an
   estimate for the acceleration in the END of the interval, a4.
   ```
   a4 = ...
   ```

8. A mean value for speed and acceleration in the interval
   is calculated by a weighted, normalized sum:
   ```
   vMiddle = 1.0/6.0 * (v1 + 2*v2 + 2*v3 + v4)
   aMmiddle = 1.0/6.0 * (a1 + 2*a2 + 2*a3 + a4)
   ```

9. Finally, use these weighted mean values for speed and
   acceleration in the interval to calculate the position
   and speed in the end of the interval.
   The function return this position and speed.
   ```
   x[n]  = ...
   v[n] = ...
   return x[n], v[n]
   ```

Matlab Code for Runge–Kutta Method

Important

The code of most of the example programs in this book is available (both for Matlab and Python) at a "Supplementary material" web page. At the same web page,
files required for solving some of the problems are available as well as a list of
reported errors, etc. The address for the "Supplementary material" web page is
http://www.physics.uio.no/pow.

```
function [xp,vp,tp] = rk4x(xn,vn,tn,delta_t,param)

% Runge-Kutta integrator (4th order)
%**************************************************************
% This version of a 4th order Runge-Kutta function for Matlab
% is written by AIV. Versjon 09282017.
% This function can be used for the case where we have two
% coupled difference equations
%     dv/dt = ffa(x,v,t,param)
%     dx/dt = v  NOTE: This part is taken care of automatically
%                in this fuction.
% Input parameters: x,v,t can be position, speed and time,
% respectively. delta_t is the step length in time.
% param is a structure in Matlab (in Python it is called a
```

```
% class). It contains various parameters that is used to
% describe the actual second order differential equation.
% It MUST contain the name of the function that contains
% the differential equation. The class "param" the user has
% to define.

% Input argumentents (n: "now")
%    [xn,vn,tn,delta_t,param] = values for x, v and t "now".
% Output argumentets (p : "n plus 1")
%    [xp,vp,tp] = new values for x, v and t after one step in
%    delta_t.
%************************************************************

ffa = eval(['@' param.fn]);    % Picks up the name of the
% Matlab-code for the second derivative. Given as a text
% string in a structure param.

half_delta_t = 0.5*delta_t;
t_p_half = tn + half_delta_t;

x1 = xn;
v1 = vn;
a1 = ffa(x1,v1,tn,param);

x2 = x1 + v1*half_delta_t;
v2 = v1 + a1*half_delta_t;
a2 = ffa(x2,v2,t_p_half,param);

x3 = x1 + v2*half_delta_t;
v3 = v1 + a2*half_delta_t;
a3 = ffa(x3,v3,t_p_half,param);

tp = tn + delta_t;
x4 = x1 + v3*delta_t;
v4 = v1 + a3*delta_t;
a4 = ffa(x4,v4,tp,param);

% Returns (estimated) (x,v,t) in the end of the interval.
delta_t6 = delta_t/6.0;
xp = xn + delta_t6*(v1 + 2.0*(v2+v3) + v4);
vp = vn + delta_t6*(a1 + 2.0*(a2+a3) + a4);
tp = tn + delta_t;
return;
```

The Function that Contains the Differential Equation

```
function dvdt = forced(y,v,t,param)

%***********************************************************
% This function is calculating the accelleration of a
% mass-spring oscillator that is influenced by an external
% periodic force that last only for a limited time interval.
% The trivial first order diff.eq. dx/dt = v is taken care
% of automatically in rk4x. The function "forced" is used
% by a RK4 function, but the necessary parameters are
% defined by the main program (given separately).
% Written by AIV. Versjon 09282017.

% Input parameters:
%   y = position
%   v = speed
%   t = time
% Output parameters:
%   dvdt = Left side of an equation in a difference equation
%   for v.

%***********************************************************

% The external periodic force last from the start of
% calculation until the time is param.end. See the main
% program for explanations of the other param items.

if (t < param.end)
    dvdt = - param.A*v - param.B*y + param.C*cos(param.D*t);
else
    dvdt = - param.A*v - param.B*y;
end;
return;
```

Example:
Matlab Program that Uses the Runge–Kutta Method

A program for calculating forced mechanical oscillations (spring pendulum) is given below. It shows how Runge–Kutta method is used in practice if we program the Runge–Kutta routine itself.

```
function forcedOscillations17

% An example program to study how forced oscillations which
% start with a mass-spring oscillator with no motions. The
% external force is removed after a while. The program calls
% the functions rk4r.m which is also using the function
% forced.m.
```

```
global param;

% Constants etc (see theory in previous chapters) in SI units
omega = 100;
Q = 25;
m = 1.0e-2;
k = m*omega*omega;
b = m*omega/Q;
F = 40;
time = 6.0; % Force only present halv of this time, see later
% Parameters used in the calculations (rk4.m, tvungen.m)
param.A = b/m;
param.B = omega*omega;
param.C = F/m;
param.D = omega*1.0;  % If this value is 1.0, the angular
                      % frequency of the force equals the
                      % angular frequency for the system.
param.end = time/2.0;
param.fn = 'forced'; % Name of Matlab file for 2. derivative

% Choose number steps and step size in the calculations
N = 2e4;                     % Number calculation points
delta_t = time/N;            % Time step in the calculations

% Allocate arrays, set initial conditions
y = zeros(1,N);
v = zeros(1,N);
t = zeros(1,N);
y(1) = 0.0;
v(1) = 0.0;
t(1) = 0.0;

% The loop where the calculations actually are done
for j = 1:N-1
    [y(j+1), v(j+1), t(j+1)]=rk4x(y(j),v(j),t(j),delta_t,param);
end;

% Plot the results
plot(t,y,'-b');
maxy = max(y);
xlabel('Time (rel units)');
ylabel('Position of the mass (rel. units)');
axis([-0.2   time   -maxy*1.2   maxy*1.2]); % want some
     % open space arround the calculated results

% We should also have compared our results with the analytical
% solution of the differential equation in order to verify
% that our program works fine. Not implementet in this
% version of the program.
```

Using Matlab's Built-in Runge–Kutta Function *

Finally, here is a specimen program for calculating damped oscillations, if we use Matlab's built-in solver of ordinary equations (ode) using the fourth-order Runge–Kutta method. First, we enter the main program we called *dampedOscill.m* (the name is insignificant here) and then follows a small application snap *ourDiffEq.m* that the main application calls. Matlab's equation solver requires a small additional function that specifies the current differential equation as such and that is the one given in vaarDiffLign.m.

```
function dampedOscill
% Program for simulation of damped oscillations.
% Written by FN. Version 09282017
% Solves two copuled differential equations
% dz/dt = v
% dv/dt = - coef1 v - coef2 z

clear all;

% Defines the physical properties for the oscillator
% (in SI units).
b = 3.0;  % Friction coefficient
m = 7.0;  % Mass
k = 73.0; % Spring constant
% Reminder:
%      Overcritical damping : b > 2 sqrt(k m)
%      Critical damping :     b = 2 sqrt(k m)
%      Undercritical damping: b < 2 sqrt(k m)

coef1 = b/m;
coef2 = k/m;

% Initialconditions (in SI-units)
z0  = 0.40; % Position rel. equilibrium point
v0  = 2.50; % Velocity

% Time we want to follow the system [start, end]
TIME = [0,20];

% Initial values
INITIAL=[z0,v0];

% We let Matlab perform a full 4th order Runge-Kutta
% integration of the differential equation. Our chosen
% differential equation is specified by the function
% ourDiffEq.

% T is time, F is the solutions [z v], corresponding to the
```

```
% running variable t (time) and f is the running variable
% [z(t) v(t)] that Matlab use through the calculations.
% Matlab chooses itself the step lengths in order to give
% proper accuracy. Thus, the calculated points are not
% equidistant in time!

[T F] = ode45(@(t,f) ourDiffEq(t,f,coef1,coef2),TIME, INITIAL);

% Plot the results, we choose to only  plot position vs time.
plot(T,F(:,1));

% length(T) % Option: Write to sceen how many points Matlab
%                actually used in the calculation. Can be useful
%                when we compare with our calculations with our
%                own Runge-Kutta function.

% We should also compare our results with the analytical
% solution of the differential equation in order to verify
% that our program works fine. Not implementet so far...
```

Our Own Differential Equation

Here comes the small function that gives the actual differential equation (in the form of two coupled difference equations):

```
function df = ourDiffEq(~,f,coef1,coef2)

% This function evaluate the functions f, where f(1) = z and
% f(2) = v. As the first variable in our input parameters we
% have written ~ since time does not enter explicitly in our
% expressions.

df = zeros(2,1);

%The important part: The first differential equation: dz/dt = v
df(1) = f(2);

% The second differential equation: dv/dt = -coef1 v - coef2 z
df(2) = -coef1*f(2)-coef2*f(1);
```

4.11.1   Suggestions for Further Reading

The following sources may be useful for those who want to go a little deeper into this material:

- Hans Petter Langtangen: A Primer on Scientific Programming with Python. 5th Ed. Springer, 2016.
- http://en.wikipedia.org/wiki/Semi-implicit_Euler_method (accessed 01.10.2017)
  - `http://en.wikipedia.org/wiki/`
    `Numerical_partial_differential_equations`

4.12   Learning Objectives

After working through this chapter, you should be able to:
- Know that a second-order differential equation can be considered equivalent to two coupled first-order differential equations.
- Solve a second-order differential equation numerically using the fourth-order Runge–Kutta method.
- Explain why numerical methods can handle, more frequently than analytical methods, complex physical situations, such as nonlinear friction.
- Point to some factors that could cause numerical calculations to fail.
- Explain in detail why the fourth-order Runge–Kutta method usually works better than Euler's method.
- Make a reasonably good test that a computer program that uses numerical solution methods works as it should.
- Put into practice your practical experience in using numerical methods to integrate an ordinary differential equation or a partial differential equation.
- Know and have some practical experience working out a computer program with several functions that interact with each other and could explain the purpose of such a partitioning of code.
- Know and have some experience with troubleshooting and know some principles that should be used to avoid postponing comprehensive troubleshooting until most of the code is written.
- Know how we can proceed to consolidate documentation of programs and parameters associated with the calculated values.
- Know why it is a good idea to save a computer program under a new name just as it is, while one is going through modifications to the program.

4.13   Exercises

Suggested concepts for student active learning activities: Discretizing, algorithm, numerical method, Euler's method, Runge–Kutta's method, accuracy, coupled differential equations, partial differential equation, documentation for programming activities.

Comprehension/discussion questions

1. Why does the fourth-order Runge–Kutta method usually work better than Euler's method?
2. Figure 4.7 shows the result of calculations of a pendulum motion for the case that there is some friction present. The figure shows position (angle) as a function of time (left part) and angular velocity as a function of position (angle) in the right part (also called a phase plane plot). The two upper figures result from an initial condition where the pendulum at time $t = 0$ hangs straight down, but at the same time has a small angular velocity. The lower figures result from an initial condition which is the same as for the upper part, but that the initial angular velocity is a good deal greater than in the first case.
   Explain what the figures say about the motion (try to bring as many interesting details as possible). How would the figure look if we increased the initial angular velocity even more than the one we have in the lower part of the figure?
3. Try to outline the working steps involved in analytical calculations of an oblique projectile throw with or without friction (or planetary motion around the sun). What do we spend most of the time on, and what do we concentrate on when we inspect the calculation afterwards? Attempt to outline the work plan for a numerical calculation and how we examine the result of such a calculation. What are the pros and cons of each method? Also try to incorporate physical understanding of the mechanisms of motion.

Problems

Remember: A "Supplementary material" web page for this book is available at http://www.physics.uio.no/pow.

4 The purpose of this composite task is to create your own program to solve different order differential equations using the fourth-order Runge–Kutta method (RK4) and to modify the program to cope with new challenges. Feel free to get extra help to get started! Specific assignments are as follows:
   (a) Write a computer program in Matlab or Python that uses RK4 to calculate the damped harmonic motion of a spring pendulum. The program should consist of at least three different parts/functions following a similar scheme outlined in Sect. 4.7. You should not use Matlab's built-in Runge–Kutta function. The program should be tested for the case: $m = 100$ g, $k = 10$ N/m, and the friction

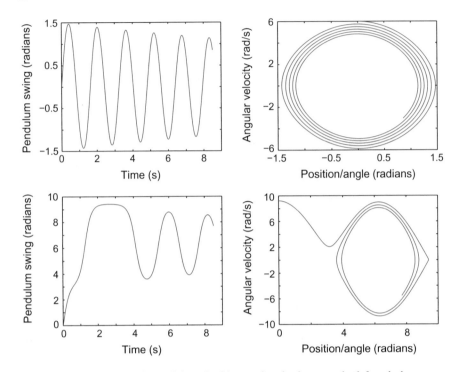

Fig. 4.7 Motion of a simple pendulum. Position vs time is shown to the left and phase space presentation of the motion to the right. See the text for a detailed description

is assumed to be linear with the coefficient of friction $b = 0.10$ kg/s. Initial terms are $z(0) = 10$ cm and $[dz/dt]_{t=0} = 0$ m/s. Conduct a test of which time steps are acceptable and check if there is agreement between numerical calculations and analytical solution. Put correct numbers, text and units along the axes of the plots. Add a copy of your code.

(b) Modify the program a little and change some parameters so that you can create a figure similar to Fig. 2.5 that shows the time course of the oscillation when we have subcritical, critical and supercritical damping. Explain how you chose the parameters. [We assume that the tests you did in (a) with respect to time resolution and comparison with analytical solutions do not need to be repeated here.]

(c) Modify the program so that it can also handle forced vibration (may last for the entire calculation period). Use $m = 100$ g, $k = 10$ N/m, $b = 0.040$ kg/s and $F = 0.10$ N in Eq. (3.1). Try to get a plot that corresponds to the initial part of each of the time courses we find in Fig. 3.7.

(d) Use this last version of the program to check that the "frequency response" of the system (à la Fig. 3.8) comes out to be correct, and that you can actually read the approximate Q value of the system from a plot made by you.

5. Write your own program to calculate the time development of a damped oscillator using the fourth-order Runge–Kutta method. Test that it works by comparing the results for analytical solution and numerical solution for a case in which they should be identical. How large is the error in the numerical solution for the position (relative to maximum amplitude)? If you choose the time step Δt, we ask you to test at least two to three different options for Δt to see how much this choice means for accuracy.

6. Carry out calculations of forced oscillations for a variety of different applied frequencies and check that the quality factor expression in Chap. 2 corresponds to the frequency curve and the alternative calculation of Q based on the half-value and centre frequency.

7. Study how fast the amplitude grows by forced oscillations when the applied frequency is slightly different from the resonant frequency. Compare with the time course at the resonance frequency. Initial conditions: the system starts at rest from the equilibrium point.

8. Find out how the calculations in the previous tasks have to be modified if, for example, wanted to incorporate an additional term $-cv^2 \times (\vec{v}/v)$ for the friction. Feel free to comment on why numerical methods have a certain advantage over analytical mathematical methods alone.

9. This task is to check if the superposition principles apply to a swinging spring pendulum with damping, first in the case that the friction can be described only with a $-bv$, that the friction must be described by $-bv - sv^2$, or rather: $-bv - s|v|v$ to take account of the direction (see Chap. 2 where this detail is mentioned). In practice, the task involves making calculations for one swing mode, then for another, and then checking if the sum of solutions is equal to the solution of the sum of states.

   The physical properties of the spring pendulum are characterized by $b = 2.0$, $s = 4.0$, $m = 8.0$ and $k = 73.0$, all in SI units. Make calculations first with the initial conditions $z_0 = 0.40$ and $v_0 = 2.50$, and then the initial conditions $z_0 = 0.40$ and $v_0 = -2.50$. Add the two solutions. Compare this sum with the solution of differential equation when the initial conditions are equal to the sum of the initial conditions we used in the first two runs. Remember to check the superposition principle both for runs where $-s|v|v$ is present and where it is absent. Can you draw a preliminary conclusion and put forward a hypothesis about the validity of the superposition principle based on the results you have achieved?

   Note: In case you use Matlab's built-in solver, the times will not match the two runs. You must then take into account the time series corresponding to one run and use interpolation when the addition of the result for the second run is to be performed. Below is an example of how such an addition can be made. Ask for help if you do not understand the code well enough to use it or something similar in your own program.

```
% Addition of two functions Z1(t) and Z2(t'), where t is
% elements in T1 and t' in T2. The two series have the same
% start value (and end value), but is different elsewhere.
% n1 = length(T1) and n2 = length(T2). The function only
```

```
% works for n2>=n1. Modify the code if that is not the case.

% Use T1 as basis for for the summation
Z12(1)=Z1(1)+Z2(1);
for i = 2:n1
    % Find index to the last point in T2 less than T1(i)
    j = 1;
    kL = -1;
    while kL<0
        if (T2(j)<T1(i)) j=j+1;
        else;
            kL=j-1;
        end;
    end;
    % The first point in T2 is then larger or equal the
    % T1(i) index:
    kH = kL+1;
    % Summation of the two solutions (linear interpolation)
    Z12(i) = Z1(i)+Z2(kL) + (Z2(kH)-Z2(kL))...
    *(T1(i)-T2(kL))/(T2(kH)-T2(kL));
end;
```

4.13.1 An Exciting Motion (Chaotic)

11. Let us look at a nonharmonic "swing" that is beyond analytical mathematics. We consider a ball that is bouncing vertically up and down influenced by gravity, and we assume, for the sake of simplicity, that there is no loss. The special aspect here is that the floor oscillates vertically and has much greater mass than the bouncing ball so that the motion of the floor is not affected by the ball.

The velocity of the floor is described as $u(t) = A\cos(\omega t) = A\cos(\phi(t))$. The ball has a speed of v_i down just before it hits the floor, but according to mechanics, the speed $v_{i+1} = v_i + 2u(t)$ will rise soon after the ball has hit the floor. We assume that the ball bounces so high in relation to the amplitude of the floor that we can make the approximation that the time the ball uses from leaving the floor until it hits the floor again is independent of the position of the floor and depends only on the speed the ball had when it last left the floor. This time is $\Delta t_i = 2v_i/g$ where g is the acceleration due to gravity. Note that Δt varies from bounce to bounce.

With these approximations, the phase difference between the floor oscillation and the oscillations of the ball until their next encounter is:

$$\Delta\phi_i = \Delta t_i\,\omega = \frac{2\omega}{g}v_i \equiv \gamma_i \qquad (4.13)$$

where γ is a "normalized velocity" that depends on the constants g and ω and varies as v_i. The term "velocity" is a little misleading, but since g and ω are both constant in our context, γ varies linearly with the velocity of the floor at the instant the ball hits it. When $\gamma_i = 2\pi$, the bounce will equal exactly one period in the oscillation of the floor.

We can then set up the following algorithm to calculate a new bounce based on the knowledge of the previous bounce in the following way:

$$\phi(n+1) = [\text{modulo } 2\pi](\phi(n) + \gamma(n)) \tag{4.14}$$

where $[\text{modulo } 2\pi]$ means that we take the modulo of what we calculate (to ensure that ϕ is in the range of $[0, 2\pi >)$. And further:

$$\gamma(n+1) = \gamma(n) + \alpha \cos(\phi(n+1)) \tag{4.15}$$

where $\alpha \propto A$.

In this description, we operate with "normalized velocity" $\gamma(n)$, which is proportional to the initial velocity of each bounce, and with $\phi(n)$, which is the phase of the floor motion just as nth bounce begins. The quantity α is proportional to the amplitude of the floor, and for simplicity we will choose an amplitude corresponding to $\alpha = 1.0$.

We will plot the results in a form of phase plot, but not quite. We let the phase of the oscillation $\phi(n)$ lie along the x-axis and "normalized velocity" $\gamma(n)$ along the y-axis.

Create a plot showing *points* $(\phi(n), \gamma(n))$ for N number of bounces. During the test you can, for example, take $N = 2 \times 10^3$, but when the program works without errors, you may want to expand this to e.g. $N = 2 \times 10^6$ if the calculation time is still acceptable.

Remember to allocate space to the "phi" and "gamma" array before you enter the loop using the algorithm in Eqs. (4.14) and (4.15).

Note: Do not connect the points with lines! Plotting of the points can be done in Matlab, for example, as follows:

```
plot(phi,gamma,'r','MarkerSize',2);
```

Try the following initial conditions for (phi, gamma): $(0.0, 1.0)$, $(\pi/2, 0.0)$, $(1.4, 1.71)$, $(1.4, 1.75)$. Also try other initial values to create a picture of various movements that may occur. Try to describe in words different forms of motion.

Chapter 5
Fourier Analysis

Abstract In this chapter, the first major challenge is to understand the difference between two descriptions of a signal: one in the time domain and another in the frequency domain. We initially use a gradual increase in complexity to help the reader grasp the difference. We then use phasors in order to introduce positive and negative frequencies, a detail that is encountered later. The formal mathematical Fourier transform and inverse transform are then introduced as well as Fourier series. The remainder of the chapter is devoted to discrete Fourier transform in the form of fast Fourier transform (FFT). All exact details on intervals in time and frequency are stated with great care. Important details like aliasing/folding and sampling theorem are given. We also analyse a time-limited oscillating signal and get our first encounter with the bandwidth theorem, and a theme we will recur to in several later chapters of this book.

5.1 Introductory Examples

5.1.1 A Historical Remark

Fourier transformation and Fourier analysis bear close resemblance to the medieval use of epicycles for calculating how planets and the sun moved relative to each other. That gives us an inkling of how powerful Fourier analysis is, but at the same time it reminds us that Fourier analysis can sometimes hinder a deeper understanding of the phenomena around us. Several later chapters in this book are based on a good understanding of Fourier transformation, including the awareness of the danger to think and argue almost in the same manner as in the Middle Ages.

5.1.2 A Harmonic Function

Before delving into the details about Fourier transformation, it will be useful to take a look at Chap. 2. We saw that a harmonic function can be written in several different

© Springer Nature Switzerland AG 2018

A. I. Vistnes, *Physics of Oscillations and Waves*, Undergraduate Texts in Physics,

https://doi.org/10.1007/978-3-319-72314-3_5

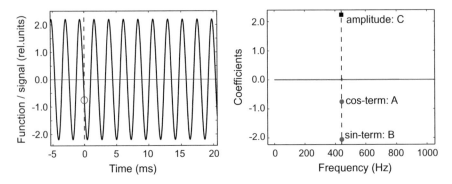

Fig. 5.1 Section of a harmonic function plotted, in the left part, as a function of time ("time domain") and, in the right part, as a function of frequency ("frequency domain"). See text for other details

ways:

$$z(t) = C \cos(\omega t + \phi) = A \cos(\omega t) + B \sin(\omega t) = \Re \left\{ \mathscr{D} e^{i \omega t} \right\}. \qquad (5.1)$$

$\Re \{\}$ means that we take the real part of the complex expression within the braces, and $\mathscr{D}$ is a complex number.

In the left part of Fig. 5.1, we have plotted a section of an arbitrary harmonic function of time. Amplitude C is 2.2 in some unspecified units and the frequency $f = 440$ Hz, which corresponds to the period $T \approx 2.27$ ms $\approx 1/440$ s. We chose the phase shift $\Phi = 110°$. This means that the value of the function is neither zero nor at the maximum at time $t = 0$.

The three parameters C, $\omega = 2\pi f$ and ϕ specify the function $z(t) = C \cos(\omega t + \phi)$ unambiguously. Using the identities in Chap. 2, this function can also be expressed as $A \cos(\omega t) + B \sin(\omega t)$. In that case, $A = C \cos \phi \approx -0.76$ and $B = -C \sin \phi \approx 2.06$. The three parameters that specify the function completely are A, B and ω.

Usually we plot a function of time as has been done in the left part of Fig. 5.1. However, we can also display the function graphically in an altogether different way, which is done in the right part of the figure. Here we have *frequency* along the x-axis and the coefficients A and B along the y-axis, and colour coding has been used to distinguish A from B. Since we have *time* along the x-axis in the left part of Fig. 5.1, we call this a "time-domain" representation of the function. For the right part, the frequency is along the x-axis, and we therefore call this a "frequency-domain" representation. Both representations contain (under certain assumptions) the same information.

In the frequency-domain picture, we have also displayed C. Occasionally we are interested only in amplitudes and not phases. Then $C = \sqrt{A^2 + B^2}$ is useful, and C is always positive (or zero). However, C and ω alone are not sufficient to determine the function unambiguously—phase information is missing.

If we use the last expression in Eq. (5.1), we can also specify the function as follows:

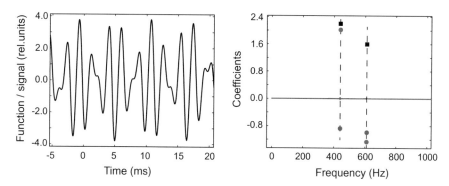

Fig. 5.2 A segment of a function that is a sum of two harmonic functions with frequencies 440 and 610 Hz plotted, one the left, as a function of time ("time-domain picture") and on the right as a function of frequency ("frequency-domain picture"). The colour coding is the same as in the previous figure. See text for other details

$$z(t) = \Re\left\{\mathscr{D}e^{i\omega t}\right\}. \tag{5.2}$$

It is important to remember that $\mathscr{D}$ is a complex number, and that $\mathscr{D} = A - iB$ so that $\mathscr{D}$ is the detail in Eq. (5.2) that contains the information about the phase of the harmonic function. The amplitude C is the absolute value of the complex number $\mathscr{D}$.

If you do not remember all the details in Chap. 2 which are used in transforming one version to another in Eq. (5.1), it is recommended that you revise that section now. In the rest of this chapter, we will use the rendering given in Eq. (5.2), and it is very important to fully understand this expression.

At present we need to refer only to the mathematics in Chap. 2. We will show that, by using a so-called Fourier transform, we can generate the plot in the right part of Fig. 5.1 completely automatically. The prime purpose of this introductory part is to find out what are meant by the terms "time-domain picture" and "frequency-domain picture".

5.1.3 Two Harmonic Functions

Let us see now what happens when we have a sum of two harmonic functions. The time-domain picture is given in the left part of Fig. 5.2. Since we have generated this function ourselves, we know that it is described by

$$z(t) = C_1 \cos(\omega_1 t + \phi_1) + C_2 \cos(\omega_2 t + \phi_2) \tag{5.3}$$

where all the six parameters appearing above are known.

We can also use the alternative form:

$$z(t) = A_1 \cos(\omega_1 t) + B_1 \sin(\omega_1 t) + A_2 \cos(\omega_2 t) + B_2 \sin(\omega_2 t) \qquad (5.4)$$

where A_1, A_2, B_1 are B_2 are to be found by using C_1, ϕ_1, C_2 and ϕ_2, and, since the frequencies ω_1 and ω_1 are known, we can make a frequency plot corresponding to this function. Such a plot is shown in the right part of the figure.

Someone who did not know how the function was generated, and obliged to evaluate it only from the time plot in the left part of Fig. 5.2, would find it difficult to say with certainty that this a sum of only two harmonic signals. It would be quite a challenge to determine the amplitudes and phases.

However, with the help of Fourier transformation, which is the subject of this chapter, we can use the time plot to calculate, automatically, A_1, A_2, B_1, B_2, ω_1, and ω_2 and we can confirm that there are no other contributions to the signal. You may now appreciate how useful Fourier analysis can be!

We recall the rendering based on Euler's formula and complex coefficients. For two harmonic functions, this takes the form:

$$z(t) = \Re \left\{ \mathscr{D}_1 e^{i\omega_1 t} + \mathscr{D}_2 e^{i\omega_2 t} \right\}. \qquad (5.5)$$

It is important to realize that all three form of writing in Eqs. (5.3), (5.4) and (5.5) are equivalent.

Since the coefficients $\mathscr{D}_1$ and $\mathscr{D}_2$ can be determined by Fourier transformation, they are commonly called *Fourier coefficients* of the $z(t)$ function.

5.1.4   Periodic, Nonharmonic Functions

In the last example, the signal was nonperiodic. In many parts of physics, we deal with periodic functions. An example is shown in Fig. 5.3. Looking at this feature in the time-domain picture, it is hard to understand that such a signal can be described in a relatively simple way.

Since we have generated the signal ourselves, we know how it was constructed. The signal is made as a sum of six harmonic functions, each of which is described by a set of $[A_i, B_i, \omega_i]$-values. In order to get a periodic signal, each ω_i was taken as $n\omega_0$, an integral multiple of the lowest value ω_0, called "the fundamental frequency". In our case, $\omega_0 = 610\,\text{Hz}$ and $n = 1, 2, \ldots 6$. The right part of Fig. 5.3 shows how the frequency-domain picture in this case looks like.

It is pleasing to note that even in this case we succeeded, thanks to a Fourier transformation, in analysing the $z(t)$ signal directly, and in finding how the signal was composed. It would be almost impossible to extract these details without Fourier transformation, as there are 18 different parameters to be determined. We will come back to the details later.

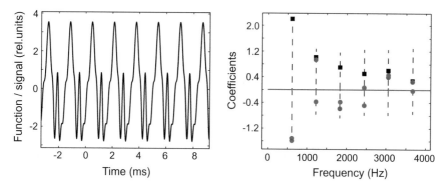

Fig. 5.3 Time-domain picture on the left shows a section of a periodic, nonharmonic function and on the right is shown the corresponding frequency-domain picture. See text for other details

It turns out that the more a periodic signal differs from a pure sinusoid, the more harmonic functions (higher n values) are needed for describing it.

We remind the reader that if we choose Euler's formula and complex coefficients, a *periodic* function would look like this:

$$z(t) = \Re \left\{ \sum_{n=1}^{N} \mathscr{D}_n e^{in\omega_0 t} \right\}.$$

In our case $N = 6$.

5.1.5 Nonharmonic, Nonperiodic Functions

In the end, we look at something rather odd. We have seen in the three previous examples that it is possible to make many different signals by combining harmonic functions with different amplitudes and phases. As we shall see immediately, an *arbitrary* function, including nonharmonic and nonperiodic functions, can be written as a sum of harmonic functions as follows:

$$z(t) = \sum_{n=1}^{N} C_n \cos(\omega_n t + \phi_n) = \Re \left\{ \sum_{n=1}^{N} \mathscr{D}_n e^{i\omega_n t} \right\} \tag{5.6}$$

for some large N. Occasionally, we have to use a very large number of frequencies in the description of a function. We can then replace the summation by an integral with a continuous function $\mathscr{D}(\omega)$ that specifies the coefficients:

$$z(t) = \Re \left\{ \int_{\omega=0}^{+\infty} \mathscr{D}(\omega) e^{i\omega t} \right\} \tag{5.7}$$

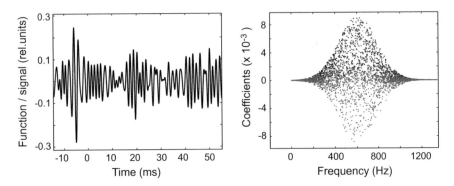

Fig. 5.4 Left part is a "time-domain picture" of a nonperiodic, nonharmonic function, and on the right is the "frequency-domain picture" of the same function. See text for other details

In Fig. 5.4, we have created a signal that is built by adding more than 3000 harmonic functions with frequencies lying in a wide band centred around 610 Hz. The amplitude varies randomly, but the largest amplitudes occur only for frequencies in the broad region near 610 Hz. The phases are random. The sum signal is then both nonharmonic and nonperiodic, as indicated in the time plot on the left. An analysis similar to that we have done in the previous examples gives the coefficients (and amplitudes) indicated in the right part of the figure.

5.2 Real Values, Negative Frequencies

It is a little tiresome that when we use the functional form given in Eq. (5.2), we always have to find the real value $\Re$ of the complex expression inside the braces on the right. There is a useful trick to get around this problem.

The basic element is this equation is the exponential term $e^{i\omega t}$ and Euler's formula $e^{i\omega t} = \cos(\omega t) + i \sin(\omega t)$. This relation is often illustrated through phasors.

The function $z(t) = C \cos(\omega t + \phi)$ can be described by a phasor which at time t has an orientation as shown in Fig. 5.5. The phasor rotates in a positive direction (anticlockwise) with the angular frequency ω, and it is always the component along the x-axis (the real axis) that indicates the value of $z(t)$.

If we now create a vector of the same length C, but always reflected about the x-axis relative to the previous one, rotating in the *negative* direction (clockwise), the sum of this phasor and the previous will always be along the x-axis. There will be no imaginary contribution!

The maximum value of the sum of the two vectors will be equal to $2C$, so we need to enter a factor of 1/2 to correct for this. The maximum of the sum vector occurs every time $\omega t + \phi$ is an integer multiple of 2π.

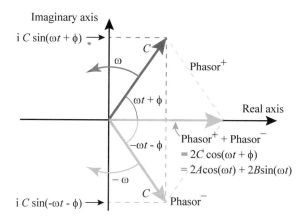

Fig. 5.5 Common phasor description (*in red*) of a harmonic function $C\cos(\omega t + \phi)$ at time t. A second phasor is also drawn (*in green*), which is the reflection of the original phasor about the x-axis, and rotates therefore the opposite way. Adding the two vectors, we get a resultant (*blue*) that always lies along the real axis, but has twice the length we are interested in

We have now put sufficient pictorial flesh on algebraic bones to make the following formula palatable:

$$C\cos(\omega t + \phi) = \frac{1}{2}\left\{\mathscr{D}e^{i\omega t} + \mathscr{D}^*e^{-i\omega t}\right\} = \frac{1}{2}\left\{\mathscr{D}e^{i\omega t} + c.c.\right\} \qquad (5.8)$$

where the asterisk in $\mathscr{D}^*$ and "c.c." stands for "complex conjugate".

We see that by introducing "negative frequencies", we can avoid having to take the real value of the complex function $\mathscr{D}e^{i\omega t}$.

Fourier analysis uses the connection given in Eq. (5.8), which means that what was said in the introductory examples was not the whole truth. If we actually do a Fourier analysis of the first harmonic function we examined, the frequency-domain picture will have the appearance shown in the right part of Fig. 5.6. We receive contributions from -440 to $+440$ Hz. The coefficients in front of the cosine term have the same value for positive and negative frequency, but only half of the coefficient A in Eq. (5.1). However, the coefficients in the sine term, which correspond to the imaginary axis of the phasor diagrams, have changed sign when we go from positive to negative frequency. Here too the factor 1/2 comes in. The same also applies to the C's since $C = \sqrt{A^2 + B^2}$.

All Fourier analysers of real signals have in principle this positive and negative division, where the coefficients are complex conjugate of each other. A little later, under the heading "folding", we will see that the negative frequencies appear in a rather odd way in the so-called fast Fourier transform (FFT).

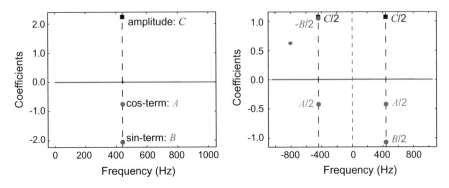

Fig. 5.6 Frequency-domain picture obtained when we work with only the positive frequencies on the left. In that case, we must ourselves extract the real part of the expression in Eq. (5.5) if we use this representation. With normal Fourier transform of real signals, half of the coefficients $\mathscr{D}(\omega)$ are apportioned to the frequency ω and the other half to the frequency $-\omega$; furthermore, the coefficient at a negative frequency is the complex conjugate of the corresponding coefficient at positive frequency

5.3 Fourier Transformation in Mathematics

So far in this chapter, we have seen several examples of how a continuous signal or function of time can be written as a sum (or integral) of harmonic functions. This actually applies in general, as was shown by the French mathematician and physicist Joseph Fourier (1768–1830).[1]

We would like to write Fourier's relation in the following manner:

Let $f(t)$ be an integrable function of t (usually time) as a continuous parameter. In physics, $f(t)$ is often a real function, but mathematically it may be complex. The function $f(t)$ can then be described as an integral of harmonic functions as the limiting value of a sum:

$$f(t) = \int_{-\infty}^{\infty} F(\omega)\mathrm{e}^{\mathrm{i}\omega t}\,d\omega. \qquad (5.9)$$

Here $F(\omega)$ corresponds to Fourier coefficients and is called the "Fourier transform of f". $F(\omega)$ forms the so-called frequency-domain picture of the function, while $f(t)$ represents the time-domain picture.

On comparing with Eqs. (5.6), (5.7) and (5.8), we see that we have now changed the notation to $z(t) \rightarrow f(T)$ and $\mathscr{D}(\omega) \rightarrow F(\omega)$ and we have availed ourselves of negative frequencies by allowing the integration to go from minus infinity to plus infinity. If $f(t)$ is a real function, $F(\omega) = F^*(-\omega)$.

---

[1] Fourier is also known to have demonstrated/explained the global warming effect in 1824.

The challenge now is to find $F(\omega)$, and this is where Fourier lends us a helping hand of giant proportions. He introduced Fourier transformation in analytical mathematics:

Given $f(t)$, a new function $F(\omega)$ (the Fourier transform of f) can be calculated as follows:

$$F(\omega) = \frac{1}{2\pi} \int_{-\infty}^{\infty} f(t) e^{-i\omega t} dt. \tag{5.10}$$

The parameter ω is the angular frequency if t represents time. Both t and ω are continuously variables.

You may have come across Fourier transformation in an earlier course in mathematics. In mathematics, the transformation is often linked to the inner product between two functions, and one defines a basis of sine and cosine functions and uses Gram–Schmidt process on a function to find its Fourier transform. Here, we choose a more practical approach in our context.

It may seem difficult to understand that Eq. (5.10) will work as we would like it to, but let us look at some basic properties in analytical mathematics.

The harmonic functions $\sin(\omega t)$ and $\cos(\omega t)$ together form a complete set of integrable functions that can describe any other integrable function. The functions $\sin(\omega_1 t)$ are orthogonal to $\sin(\omega t)$ when $\omega \neq \omega_1$, all $\sin(\omega t)$ are orthogonal to all $\cos(\omega t)$. This is embodied in the familiar expression of the delta function:

$$\delta(\omega_1 - \omega) = \frac{1}{2\pi} \int_{-\infty}^{\infty} e^{-i(\omega_1 - \omega)t} dt. \tag{5.11}$$

As an example, we now allow $f(t)$ to be the simple harmonic function in Eq. (5.1), but for the sake of simplicity, skip the details of finding the real value. We then write:

$$f(t) = \mathscr{D} e^{i\omega_1 t}.$$

Substitution in Eq. (5.10) gives:

$$F(\omega) = \frac{1}{2\pi} \int_{-\infty}^{\infty} \mathscr{D} e^{i\omega_1 t} e^{-i\omega t} dt,$$

$$F(\omega) = \mathscr{D} \times \frac{1}{2\pi} \int_{-\infty}^{\infty} e^{i(\omega_1 - \omega)t} dt.$$

We recognize the last part as the delta function, and the result is that $F(\omega)$ is zero everywhere except when $\omega_1 = \omega$ where $F(\omega_1) = \mathscr{D}$. We therefore see that, in this case, Eq. (5.10) does indeed work as desired.

Equation (5.10) gives what we call the Fourier transform of the function $f(t)$. In our context, it amounts to exchanging the time-domain description of a function with one in the frequency domain.

Equation (5.9) gives what we call an *inverse* Fourier transformation. It takes us from the frequency-domain representation of a function to a picture in the time domain.

Note that in a Fourier transform we integrate over time and the exponent has a minus sign in front. In the inverse transformation, we integrate over frequency and the exponent has a plus sign in front. Also note that the factor $1/(2\pi)$ is only used in one transformation, as we have chosen to express the two equations that, in part, belong together. Another choice is to use a $1/\sqrt{2\pi}$ in both Eqs. (5.10) and (5.11).

Remarks: Several reasons account for why Fourier transformation became popular in mathematics and physics. There are many simple mathematical relationships for harmonic functions. This means that if we have to deal with a troublesome function $f(t)$ and do not know how to handle it directly, we can use Fourier transformation as an intermediate step in the calculation. By Fourier transforming the awkward function, we obtain a linear sum (or integral) of harmonic functions. We can then perform mathematical operations on this alternative expression and use inverse Fourier transformation on the result to retrieve the result we actually wanted. Fourier transformation is therefore used extensively in analytical mathematics for, among other purposes, solving differential equations.

We know from mathematics that there are several complete sets of functions (e.g. polynomials), and in different parts of physics, we prefer to choose a basis set that is best adapted for the particular system under consideration. Fourier transformation utilizes probably the most widely used basis set of functions; unfortunately, it is also applied in situations where it is not particularly beneficial.

5.3.1 Fourier Series

A special case in Fourier transformation is of particular interest, especially when we study Chap. 7 to analyse sound from musical instruments. If $f(t)$ is *a periodic function* with period T, Fourier transformation can be made more efficient than through the general transformation in Eq. (5.10). The transformation can be specified by an infinite but *discrete* set of numbers, called Fourier coefficients, $\{c_k\}$, the index k being a natural number between minus and plus infinity(!).

The Fourier coefficients are calculated by integrating over a single period T:

$$c_k = \frac{1}{T} \int_{t_0}^{t_0+T} f(t) e^{-ik\omega_1 t} dt \qquad (5.12)$$

where $\omega_1 = (2\pi/T)$, that is to say, the angular frequency corresponding to a function that has *exactly one period in the time interval T*, and k is an integer.

Since in this case $f(t)$ is periodic, the lower limit for integration (t_0) can be chosen freely in principle. It is supposed that $f(t)$ is piecewise smooth and continuous, and that $\int |f(t)|^2 dt < +\infty$ when the integration is over an interval of length T.

The inverse transformation is then given by the relation:

$$f(t) = \sum_{k=-\infty}^{+\infty} c_k e^{ik\omega_1 t} \qquad (5.13)$$

where, once again, $\omega_1 \equiv 2\pi/T$ corresponds to a frequency that has precisely one sine period within the interval T.

Should $f(t)$ be real, it is easy to see that the symmetry properties of the sine and cosine functions lead to the relation

$$f(t) = a_0 + \sum_{k=1}^{\infty} \{a_k \cos(k\omega_1 t) + b_k \sin(k\omega_1 t)\} \qquad (5.14)$$

where

$$a_k = c_k + c_{-k} = \frac{2}{T} \int_{t_0}^{t_0+T} f(t) \cos(k\omega_1 t) dt, \qquad (5.15)$$

$$b_k = i(c_k - c_{-k}) = \frac{2}{T} \int_{t_0}^{t_0+T} f(t) \sin(k\omega_1 t) dt. \qquad (5.16)$$

Take note of the factor 2 in the last two expressions! The reason for this factor is the simple recognition that the mean of both $\sin^2$ and $\cos^2$ is 1/2 and another factor of 2 that was explained above when we mentioned the inclusion of negative frequencies.

Equation (5.14) along with the expressions (5.15) and (5.16) are as precious as gold! They show that any *periodic signal* with period T can be written as a sum of harmonic signals *having exactly integral number of cycles within the period T*.

5.4   Frequency Analysis

Hitherto there has been a lot of mathematics and little physics in this chapter. It is therefore high time to give a few examples of the practical use of Fourier transformation.

Fourier transformation is widely used for so-called frequency analysis where we determine which frequency components are present in a signal. We often call the frequency-domain picture a "frequency spectrum". The frequency spectrum is useful because it often gives a "fingerprint" of the physical processes that lie behind the signal under consideration.

The number of sunspots increases and decreases over time regularly with an approximately 11-year cycle, we are often told. What is the basis for such an assertion? We can plot the number of sunspots per year over a number of years. We then get a curve like the left part of Fig. 5.7 where the curve corresponds to the $f(t)$ function in the theory above. This is the so-called time picture.

In the right part of Fig. 5.7, an extract of the results is shown after a Fourier transformation of the data in the left part. Actually, the results after a Fourier transformation are complex numbers. However, if we are not interested in getting $A \cos(\omega t)$ and $B \sin(\omega t)$ separately for the different frequencies, but are rather interested in the amplitude $C = \sqrt{A^2 + B^2}$, we choose to plot the absolute value of the complex numbers. It is the absolute values that are plotted in the right part of Fig. 5.7.

The peaks near the middle of the figure correspond to a harmonic function with a frequency of 0.09 or 0.10 per year. Since a frequency of 0.09–0.10 per year corresponds to a period of approximately 10–11 years, we get a satisfactory confirmation that the sunspots in the 300 years analysed have a considerable periodicity at 10–11 years. At the same time, the noise in the plot shows that the indicated time period is more poorly defined than what we find for example in the movement of a shuttle!

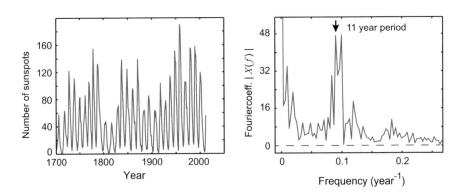

Fig. 5.7 Left part shows the number of sunspots that appeared annually over the past three hundred years. The right part shows an excerpt from the corresponding Fourier transformed functions (absolute values of $\{c_k\}$-s in Eq. (5.12)). The sunspots data were accessed on 30.1.2012 from http://sidc.be/silso/datafiles

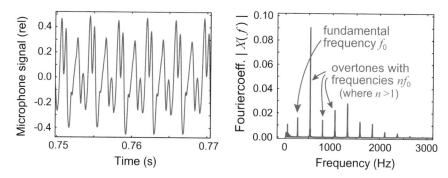

Fig. 5.8 An example of sound from a flute displayed both in the time domain and frequency domain. Amplitudes in the frequency domain are given as absolute values of $\{c_k\}$-s in Eq. (5.12))

In this book, we often use Fourier transformation to analyse sound. For example, Fig. 5.8 shows a time-domain picture and a frequency-domain picture for a audio signal from a transverse flute. The figure also shows relative amplitudes in the frequency spectrum. We then lose the phase information, but the "strength" of the different frequency components shows up well.

The spectrum consists mainly of a number of peaks with different heights. The peak positions have a certain regularity. There is a frequency f_0 (might have been called f_1), the so-called *fundamental tone*, such that the other members of a group of lines have approximately the frequencies kf_0, where k is an integer. We say that the frequencies kf_0 for $k > 1$ are *harmonics of the fundamental tone* and we refer to them as *"overtones"*.

The frequency spectrum shows that when we play a flute, the air will not vibrate in a harmonic manner (like a pure sine). The signal is periodic, but has a different time course (shape) than a pure sinusoid. A periodic signal that is not sinusoidal (harmonic) will automatically lead to overtones in the frequency range. It is a result of pure mathematics.

The reason that it does not become a pure sinusoid is that the physical process involved in the production of the sound is complicated and turbulence is involved. There is no reason why this process should end up in a mathematically perfect harmonic audio signal. For periodic fluctuations with a time course very different from a pure sinusoid, there are many overtones. The ear will perceive the vibrations as sound different from that which has fewer harmonics.

Different instruments can be characterized by the frequency spectrum of the sound they generate. Some instruments provide fewer overtones/harmonics, while others (e.g. oboe) provide many!

The frequency spectrum can be used as a starting point also for synthesis of sound: Since we know the intensity distribution in the frequency spectrum, we can start with this distribution and make an inverse Fourier transform to generate vibrations that sound like a flute.

It must be noted, however, that our sound impression is determined not only by the frequency spectrum of a sustained audio signal, but also by how the sound starts and fades. In this context, Fourier transformation is of little help. Wavelet transformation of this type of sound discussed later in the book is much more suitable for such an analysis.

A tiny detail at the end: In Fig. 5.8, we also see a peak at a frequency near zero. It is located at 50 Hz, which is the frequency of the mains supply. This signal has somehow sneaked in with the sound of the flute, perhaps because the electronics have picked up electrical or magnetic fields somewhere in the signal path.

It is important to be able to identify peaks in a frequency spectrum that corresponds to the fundamental frequency and its harmonics, and features which do not fit into such a line-up.

5.5 Discrete Fourier Transformation

A general Fourier transformation within analytical mathematics given by Eq. (5.10) is based on a continuous function $f(t)$ and a continuous Fourier coefficient function $F(\omega)$.

In our modern age, experimental and computer-generated data are only quasi-continuous. We sample a continuous function and end up with a function described only through a finite number of data points. Both the sunspot data and the audio data we just processed were based on a finite number of data points. Assume that N data points are registered ("sampled") sequentially with a fixed time difference Δt. The total time for data sampling is T, and the sampling rate is $f_s = 1/\Delta t$. Data points have values x_n where $n = 0, \ldots, N - 1$. The times corresponding to these data points are then given as:

$$t_n = \frac{T}{N}n \quad \text{for} \quad n = 0, 1, \ldots (N - 1).$$

Based on the N numbers we started with, we cannot generate more than N independent numbers through a Fourier transformation. The integral of Eqs. (5.10) and (5.9) must then be replaced by summation sign and the sum extends over a finite number of data points in both the time domain and the frequency domain.

A side effect of discrete Fourier transformation is that when we Fourier transform N data points x_n taken at times $t_0, t_1, \ldots, t_{N-1}$, the result in practice is the same as if we had one periodic signal which was defined from minus to plus infinity, with period T.

We have seen in the theory of Fourier series that for periodic signals only discrete frequencies are included in the description. These are:

$$\omega_k = \frac{2\pi}{T}k \quad \text{for} \quad k = \dots, -2, -1, 0, 1, 2, \dots.$$

When we record the function at only N instants, as mentioned above, the data cannot encompass a frequency range with infinitely many discrete frequencies. It is only possible to operate with N frequencies, namely

$$\omega_k = \frac{2\pi}{T}k \quad \text{for} \quad k = -\frac{N-1}{2}, -\frac{N-1}{2}+1, \dots, -2, -1, 0, 1, 2, \dots, \frac{N-1}{2}-1, \frac{N-1}{2}.$$

Note that the highest frequency included is

$$f_{max} = \frac{\omega_{max}}{2\pi} = \frac{1}{2}\frac{N-1}{T} = \frac{1}{2}\frac{N-1}{N}f_s \approx \frac{f_s}{2}$$

for a sufficiently large N. Here f_s is the sampling frequency.

In the original Fourier transformation, $e^{-i\omega t}$ entered as a factor in the integrand. For N discrete data points, this is replaced by the following expressions:

$$-i\omega t \rightarrow -i\omega_k t_n = -i\frac{2\pi}{T}k \times \frac{n}{N}T = -i\frac{2\pi k n}{N}. \tag{5.17}$$

The discrete Fourier transformation is thus given by the formula:

$$X_k = \frac{1}{N}\sum_{n=0}^{N-1} x_n e^{-i\frac{2\pi}{N}kn} \tag{5.18}$$

for $k = 0, \dots, N-1$. If the set x_n consists of values given in the time domain, X_k will be the corresponding set of values in the frequency domain.

Note that here we indicate that k runs from 0 to $N-1$, which corresponds to frequencies from 0 to $\frac{N-1}{N}f_s \approx f_s$, while earlier we let k be between $-(N-1)/2$ and $+(N-1)/2$, corresponding to frequencies from $\approx -f_s/2$ to $\approx +f_s/2$. Since we only operate with sine and cosine functions with an integral number of wavelengths, it does not matter whether we use one set or the other. We come back to this page when we mention folding or **aliasing**.

Further, take note of the factor $1/N$ in this expression. This factor is advantageous for the variant of Fourier transformation we will use, because then we get a simple correlation between Fourier coefficients and amplitudes, as in the introductory sections of the chapter.

Through the expression in Eq. (5.17), we have shown that the expression for the discrete Fourier transform in Eq. (5.18) is based squarely on the same expression as we had in the original Fourier transformation. The difference is that in the discrete case we operate with a function described at N points and that only N frequencies are included in the description.

The inverse discrete Fourier transformation naturally looks like this:

$$x_n = \sum_{k=0}^{N-1} X_k e^{i\frac{2\pi}{N}kn} \qquad (5.19)$$

for $n = 0, \ldots, N - 1$.

5.5.1 Fast Fourier Transform (FFT)

Discrete Fourier transformation will be our choice when we use Fourier transformation in this book. We could have written a program ourselves to complete the procedure given in Eqs. (5.18) and (5.19), but we will not do that. It would not be a particularly effective program if we used the expressions directly. There exists nowadays a highly effective algorithm for discrete Fourier transformation that utilizes the symmetry of the sine and cosine functions in a highly effective way to reduce the number of computational operations. Efficiency has contributed greatly to the fact that Fourier transformation is widely used in many subjects, not least physics.

The algorithm was apparently discovered already in 1805 by Carl Friedrich Gauss, but fell into oblivion (it was of little interest as long as we did not have computers). The algorithm was launched in 1965 by J. W. Cooley and J. Tukey, who worked at Princeton University. Their four-page article "An algorithm for the machine calculation of complex Fourier series" in *Math. Comput.* 19 (1965) 297–301, belongs to the "classic" articles that changed physics.

In Matlab and Python, we make use of Cooley and Tukey's algorithm when we apply *FFT* ("fast Fourier transform") or *IFFT* ("inverse fast Fourier transform"). With this method, it is advantageous that the number of points N is exactly one of the numbers 2^n where n is an integer. Then we will fully utilize the symmetry of the sine and cosine functions.

5.5.2 Aliasing/Folding

When using FFT, we need to take care of a particular detail. We previously saw that it was beneficial to introduce negative frequencies in Fourier transformation.

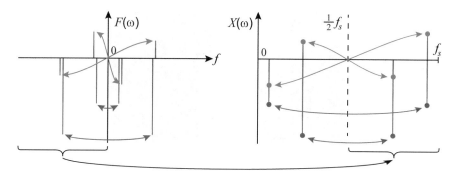

Fig. 5.9 Left part: A spectrum obtained by a continuous Fourier transformation of an infinite signal contains all frequencies between $-\infty$ and $+\infty$, but it is, in fact, a reflection and complex conjugation about the zero frequency (provided that the original signal was real). The real part of the Fourier transformed function is marked in red, the imaginary in blue (We have shifted the real ones relative to the imaginary points in the left part so that the sticks became distinct.). **Right part**: By discrete Fourier transformation of a signal, the information for negative frequencies (left part of the figure) is moved to the range *above* half the sampling frequency. Due to symmetries in sine and cosine functions, this also actually corresponds to signals with the frequencies $f_s - |f_{\text{negative}}|$. For this reason, FFT also receives a reflection/folding and complex conjugation in the analysis of real signals, but this time around half the sampling rate $f_s/2$. The part of the plots that have a light background colour contains all the information in the Fourier transformed signal of a real function since the other half is just the complex conjugate of the first

For a continuous Fourier transform of a real function $f(t)$, we saw that $F(\omega_0) = F^*(-\omega_0)$, that is, the Fourier transform at an angular frequency is the complex conjugate of the Fourier transform at the negative angular frequency. The same also applies to FFT. The data points after a Fourier transform with FFT are nevertheless arranged differently. The lower half of the frequency axis, which represents negative frequencies, is simply moved so that it is above (to the right of) the positive points along the frequency axis (see Fig. 5.9).

When we perform inverse Fourier transformation with IFFT, it is expected that the negative frequencies are positioned in the same way as they are after a simple FFT.

5.6 Important Concrete Details

5.6.1 Each Single Point

In Eq. (5.18), mathematically speaking, only a set of $\{x_n\}$ with N *numbers* can be transformed into a new set X_k with N *numbers* and back again. All the numbers are unlabelled.

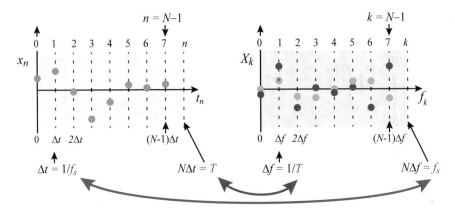

Fig. 5.10 A function sampled $N = 8$ times (*left*) along with the Fourier transform of the function (*right*) consisting of $N = 8$ complex numbers. The real values are given by red circles and the imaginary values by blue. Each point corresponds to a small time and frequency range (*left and right, respectively*). Note the relationship between the sampling rate f_s and Δt and in particular the relationship between T and Δf. In order to get a high resolution in the frequency range in the frequency range, we have to sample a signal for a sufficiently long time T

We, the users, must connect physics with the numbers. Let us explore what the indexes k, n and the number N represent.

We imagine that we make N observations of a physical quantity x_n over a limited time interval T (a single example is given in the left part of Fig. 5.10). If the observations are made at instants separated by an interval Δt, we say that *sampling rate* (or sampling frequency) is $f_s = 1/\Delta t$. The relationship between the quantities is as follows:

$$N = Tf_s = T/\Delta t.$$

This is an important relationship that we should know by heart!

Note that each sampling corresponds to a very small time interval Δt. In our figure, the signal in the *beginning* of each time interval is recorded.

Fourier transformation in Eq. (5.18) gives us the frequency-domain picture (right part of Fig. 5.10). The frequency-domain picture consists of N complex numbers, and we *must* know what they represent in order to properly utilize Fourier transformation! *Here are the important details*:

- The first frequency component specifies the *mean* of all measurements (corresponding to frequency 0). The imaginary value is always zero (if f is real).

- The second frequency component indicates how much we have of a harmonic wave with a period of time T equal to the entire sampling time. The component is complex, which allows us to find amplitude and phase for this frequency component.
- Amplitudes calculated by using only the lower half of the frequency spectrum must be multiplied by 2 (due to the folding) to get the correct result. This does not apply to the first component (mean value, frequency zero).
- The next frequency components indicate contributions from harmonic waves with exactly 2, 3, 4,...periods within the total sampling time T.
- The previous points tell us that the difference in frequency from one point in a frequency spectrum to the neighbouring point is $\Delta f = 1/T$.
- Assuming that the number of samples N is even, the first component after the centre of all the components will be purely real. This is the component that corresponds to a harmonic oscillation of $N/2$ complete periods during the total sampling time T. This corresponds to a frequency equal to half of the sampling rate f_s mentioned above.
- All the remaining frequency components are complex conjugates of the lower frequency components (assuming that $f(t)$ is real). There is a "mirroring" around the point just above the middle of the numbers (mirroring about half the sampling rate). We do not get any new information from these numbers, and therefore we often drop them from the frequency spectrum.
- Since the mirroring occurs around the first point *after* the middle, the first point will not be mirrored (the point corresponding to the average value, the frequency 0).
- The last frequency in a frequency spectrum is $f_s(N-1)/N$ since the frequency ranges are half open.

Why, one may wonder, do we calculate the top $N/2 - 1$ frequency components when these correspond to "negative frequencies" in the original formalism (Eq. (5.10)). As long as f is real, these components are of little/no worth to us.

However, if f happens to be complex, as some users of Fourier transformation take it to be, these last, almost half of the components, are as significant as the others.

This is related to Euler's formula and phases. As long as we look at the real value of a phasor, it corresponds to the $\cos(\omega t + \phi)$ term, and it is identical regardless of whether ω is positive or negative. We can distinguish between positive and negative rotational speed of a phasor only if we take into account both the real and imaginary part of a complex number.

5.6.2 Sampling Theorem

As mentioned above, the top half of Fourier coefficients correspond to negative frequencies in the original formalism. However, we suggested that because of the

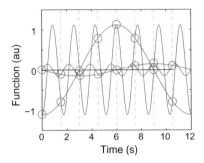

Fig. 5.11 Harmonic functions with frequencies f_k and f_{N-k} (here $k = 1$) have exactly the same value at the times for which the original function was defined (assuming that $k = 0, 1, \ldots, N - 1$). Therefore, we cannot distinguish between the two for the sampling rate used. In order to distinguish functions with different frequencies, the rate must be at least twice as high as the highest frequency component. If you carefully consider the curves with the highest frequency in the figure, you will see that there are fewer than two samples per period for these

symmetry of the sine and cosine functions, it is also possible to consider these upper coefficients as coefficients of frequencies above half the sampling frequency (except that we get problems with the factor 1/2 mentioned earlier).

We can illustrate this by picking out two sets of Fourier coefficients from a Fourier transform of an arbitrary signal. We have chosen to include the relative coefficients for $k = 1$ (red curves) along with $k = N - 1$ and the imaginary coefficients for $k = 1$ and $k = N - 1$ (blue curves). The result is shown in Fig. 5.11.

The functions are drawn at "all" instants, but *the times where the original function is actually defined* is marked with vertical dotted lines. We then see that the functions of very different frequencies still have the exact same value at these times, although the values beyond these times are widely different. This is in accordance with equation

$$e^{i\frac{2\pi}{N}kn} = e^{-i\frac{2\pi}{N}k(N-n)} \tag{5.20}$$

for k and $n = 1, \ldots, N - 1$ in the event that these indices generally range from 0 to $N - 1$.

The two functions $\cos(\omega_1 t)$ and $\cos[(N - 1)\omega_1 t]$ are thus identical *at the discrete times $t \in \{t_n\}$ our description is valid* (ω_1 corresponds to one period during the time we have sampled the signal.). Similarly, for $\cos(2\omega_1 t)$ and $\cos[(N - 2)\omega_1 t]$ and beyond for $\cos(3\omega_1 t)$ and $\cos[(N - 3)\omega_1 t]$, etc. Then there is really no point in including the upper part of a Fourier spectrum, since all the information is actually in the lower half (Remember, this only applies when we transform a real function.).

Looking at the argument we see that at the given sampling rate, we would get exactly the same result when sampling continuous signal $\cos[(N - m)\omega_1 t]$ as if the continuous signal was $\cos(m\omega_1 t)$ (m is an integer). After the sampling, we cannot determine if the original signal was one or the other of these two possibilities—unless we have some additional information.

The additional information we need, we must supply ourselves through experimental design! We must simply ensure that there are no contributions with frequencies above half the sampling frequency of the signal we sampled. If so, we can be sure that the signal we sampled was $\cos(m\omega_1 t)$ and not $\cos[(N - m)\omega_1 t]$. This means that we must sample at least twice per period for the highest frequency that is present in the signal (see Fig. 5.11).

This is an example of a general principle:

If we want to represent a harmonic function in an unambiguous manner by a limited number of measurements, the target density (measurement frequency, sampling frequency) must be so large that we get at least two measurements within each period of the harmonic signal. The "Nyquist–Shannon Sampling Theorem" says this more succinctly:

The sampling frequency must be at least twice as high as the highest frequency component in a signal for the sampled signal to provide an unambiguous picture of the signal.

If the original signal happens to contain higher frequencies, these must be filtered by a low-pass filter before sampling to make the result unambiguous.

It is strongly recommended that you complete the second *problem* at the back of the chapter. Then you can explore how folding arises in practice, and how we can be utterly deceived if we are not sufficiently wary.

5.7 Fourier Transformation of Time-Limited Signals

It follows from Eq. (5.10) that a Fourier transform can be viewed as a sum (integral) of the product of the signal to be transformed with a pure sine or cosine:

$$F(\omega) = \frac{1}{2\pi} \int_{-\infty}^{\infty} f(t) e^{-i\omega t} dt.$$

$$F(\omega) = \frac{1}{2\pi} \int_{-\infty}^{\infty} f(t) \cos(\omega t) dt - i \times \frac{1}{2\pi} \int_{-\infty}^{\infty} f(t) \sin(\omega t) dt.$$

We assumed, without stating explicitly, that the signal we analysed lasted forever. Such signals do not exist in physics. It is therefore necessary to explore characteristic features of Fourier transformation when a signal lasts for a limited time.

We choose a signal that gradually becomes stronger, reaches a maximum value and then dies out again. Specifically, we choose that the amplitude change follows a so-called *Gaussian envelope*. Figure 5.12 shows two different signals (red curves), one lasting a very short time, and another that lasts considerably longer. Mathematically,

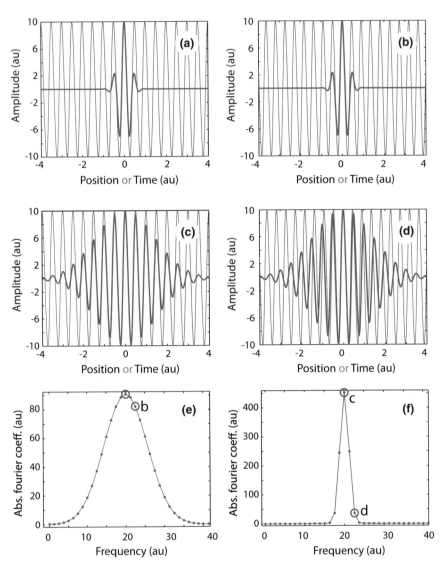

Fig. 5.12 Fourier transformation of a cosine signal multiplied with a Gaussian function. Only a small part of the total frequency range is shown. See the text for details

the signal is given as:

$$f(t) = C \cos[\omega(t - t_0)] e^{-[(t-t_0)/\sigma]^2}$$

where σ gives the duration of the signal (the time after which the amplitude has decreased to $1/e$ of its maximum). ω is the angular frequency of the underlying cosine function, and t_0 is the time at which the signal has maximum amplitude (the peak of the signal occurs at time t_0).

In panels **a** and **b**, in Fig. 5.12 the signal is of short duration (small σ), but in panels **c** and **d** it lasts a little longer (σ five times as large as in **a** and **b**).

Panels **a** and **c** show, in addition to the signal pulse (in red), the cosine signal with a frequency equal to $\omega/2\pi$ (thinner blue line). In panels **b** and **d**, the cosine signal has 10% higher frequency, which explains why we will calculate X_k at two adjacent frequencies.

We see that the integral (sum) of the product between the red and blue curves in **a** and **b** will be about the same. On the other hand, we see that the corresponding integral of **d** must be significantly smaller than the integral of **c** since the signal we analyse and the cosine signal get out of phase a little bit away from the centre of the pulse in **d**. When the phases are opposite, the product becomes negative and the calculated integral (the Fourier coefficient) becomes smaller.

If we make a Fourier transform ("all frequencies") of the red curve itself in **a** (the short-duration signal) and take the absolute value of the Fourier coefficients, we get the result shown in **e**. The Fourier transform of the signal in **c** (the longer lasting signal) is displayed in the lower right corner of **f**. We can see that Fourier transformation captures the predictions we could make from visual examinations of **a–b**.

Note that the short-duration signal yielded a broad frequency spectrum, while the signal with several periods in the underlying cosine function gave a narrower frequency range. This is again a manifestation of the principle we have observed in the past, which has a clear resemblance to Heisenberg's uncertainty relationship. In classical physics, this is called *time-bandwidth theorem* or *time-bandwidth product*: The product of the width (duration) of a signal in the time domain and the width of the same signal in the frequency domain is a constant, whose precise value depends on the shape of the envelope of the signal.

$$\Delta t \; \Delta f \geq 1.$$

The actual magnitude of the number on the right-hand side depends on how we define the widths Δt and Δf. We will later find in the chapter the same relationship with the number 1 replaced by 1/2, but then we use a different definition for the Δ's.

Figure 5.12 illustrates important features of Fourier analysis of a signal. More precisely, the following applies:

In a frequency analysis, we can distinguish between two signal contributions with frequencies f_1 and f_2 only if the *signals* last longer than the time $T = 1/(|f_1 - f_2|)$.

Even for signals that last a very long time, in experimental situations, we will have to *limit the observation of signal* for a time T. If we undertake an analysis of this signal, we will only be able to distinguish between frequency components that have a difference in frequency of at least $1/T$.

The difference we talk about means in both cases that there must be a difference of at least one period within the time we analyse (or the time the signal itself lasts) so that we can capture two different signal contributions in a Fourier transform. Suppose we have N_1 periods of one signal in time T and N_2 periods of the second signal. In order to be able to distinguish between the frequencies of the two signals, we must have $|N_1 - N_2| \geq 1$. [Easily derived from the relationship $T = 1/(|f_1 - f_2|)$.]

5.8 Food for Thought

The relationships in the time and frequency domains we see in Fig. 5.12 can easily lead to serious misinterpretations. In **a**, we see that the oscillation lasts only a very short time (a few periods). The rest of the time the amplitude is simply zero (or we could set it exactly to zero with no notable difference in the frequency spectrum).

What does Fourier transformation show? From the panel **e**, we can see that there are about 30 frequency components that are clearly different from zero. This means that we must have of the order of 30 different sine and cosine functions *which last all the time* (even when the signal is zero) to describe the original signal. We see this by writing the inverse Fourier transform in a way that should be familiar to us by now:

$$x_n = \sum_{k=0}^{N-1} \left[\Re(X_k) \cos(\omega_k t_n) - \Im(X_k) \sin(\omega_k t_n) \right] \tag{5.21}$$

for $n = 0, \ldots, N - 1$. $\Re$ and $\Im$ stand, as before, for the real and imaginary parts, respectively.

There are some who conclude that the oscillation, when it appears to be zero, is not *really* zero but simply the sum of about 30 different sine and 30 different cosine functions throughout. This is nonsense!

It is true that we can describe the time-limited oscillation in panel **a** using all of these sine and cosine functions, but this is a pure mathematical view that has little to do with physics. Notwithstanding that, there is a good deal of physics and physical reality that goes hand in hand with the width of the frequency spectrum. However, there are other ways to make this point without invoking the presence of something physical when the amplitude is actually equal to zero. In Chap. 14, we will acquaint ourselves with the so-called wavelet transformation, and then this will become much clearer.

In my own field of research, quantum optics, we see how unfortunate this type of short circuit is. Some say that we must "use many different photons" to create a light pulse and that each photon must have the energy $E = hf$ where h is Planck's constant and f frequency. Then a layer of physical reality is added to each Fourier coefficient, but one should focus more on what is physics and what is mathematics.

An important point here is that all time information about a signal disappears as soon as we take the absolute value of Fourier coefficients. As long as we retain complex Fourier coefficients, the time information remains intact, but is often very well hidden. The time information is scattered throughout the Fourier spectrum. Only a full inverse transformation (with complex Fourier coefficients!) from the frequency domain to the time domain can retrieve the temporal information. Fourier transformation, and in particular a frequency spectrum, has therefore limited value for signals that are zero during certain periods or completely change character otherwise during the sampling time.

Also in another context, a Fourier analysis can lead to unfortunate conclusions. Figure 5.13 shows the Fourier transform of a periodic motion. In essence, this figure resembles Fig. 5.8, which shows the frequency spectrum of sound from a transverse flute, with fundamental tone and harmonics. On that occasion, we said that the reason we get overtones is that the signal, though periodic, is not a pure sinusoid.

Some persons speak of the higher harmonics in another way. For example, they say "when we play a flute, air vibrates not only at one particular frequency, but at multiple frequencies simultaneously". Though common, such phraseology is problematic.

If we say that "several frequencies are present simultaneously" in the motion that lies at the back of the Fourier spectrum in Fig. 5.13, the statement accords poorly with the underlying physics! The figure was made this way: we first calculated a planet's path around the sun. The path was described by a set of coordinate as a function of

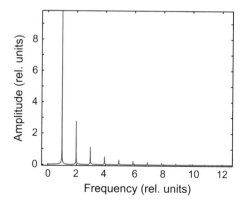

Fig. 5.13 Fourier transformation of a periodic motion. See the text for explanation

time $[x_i(t), y_i(t)]$. Figure 5.13 is simply the Fourier transform of $\{x_i(t)\}$ for a time that is much longer than the solar orbital period of the planet under consideration.

The reason we get a series of "harmonics" in this case is that planetary motion is periodic but not a pure sinusoid. We know that Fourier transformation is based on harmonic basis functions, and these correspond to circular motion. But if we think in terms of "several frequencies existing at the same time", it is tantamount to saying that the movement of the planet must be described with multiple circular movements occurring at the same time! In that case, we are back to the Middle Ages!

Some vicious tongues say that if computers, equipped with an arsenal of Fourier transform tools, had been around in Kepler's time, we would still have been working with the medieval *epicycles* (see Fig. 5.16). From our Fourier analysis in Fig. 5.13, we see that we can replace the ellipse with a series of circles with appropriate amplitudes (and phases). However, most people would agree that it makes more sense to use a description of planetary motion based on ellipses and not circles. I wish we were equally open to dropping mathematical formalism based on Fourier analysis also in some other contexts.

Fourier analysis can be performed for virtually all physical time variables, since the sine and cosine functions included in the analysis form a complete set of basis functions. Make sure you that you do *not* draw the conclusion that "when something is feasible, it is also beneficial". In the chapter on wavelet transformation, we will come back to this issue, since in wavelet analysis we can choose a set of basis functions totally different from everlasting sines and cosines. We can sum up in the following words:

> *Fourier transformation is a very good tool, but it has more or less the same basis as the medieval description of planetary movements. It is perfectly possible to describe planetary paths in terms of epicycles, but such an approach is not particularly fruitful. Similarly, a number of physical phenomena are described today by Fourier analysis where this formalism is not very suitable. It can lead to physical pictures that mislead more than they help us. Examples may be found in fields which include quantum optics.*

5.9   Programming Hints

5.9.1   Indices; Differences Between Matlab and Python

Strings such as $\{x_n\}$ and $\{X_k\}$ are described as arrays in the parlance of numerical analysis. It is important to remember that in Python, the indexes start with 0, while in Matlab they start with 1. In $\{X_k\}$, $k = 0$ and 1 correspond, respectively, to the frequency 0 (constant) and the frequency $1/T$. In Matlab, their counterparts are indices 1 and 2.

The expression for a discreet Fourier transform in Python will then be as follows:

$$X_k = \frac{1}{N} \sum_{n=0}^{N-1} x_n e^{-i\frac{2\pi}{N}kn} \tag{5.22}$$

for $k = 0, \ldots, N-1$.

On the other hand, the expression for a discreet Fourier transform in Matlab takes the following form:

$$X_k = \frac{1}{N} \sum_{n=1}^{N} x_n e^{-i\frac{2\pi}{N}(k-1)(n-1)} \tag{5.23}$$

for $k = 1, \ldots, N$.

For the inverse discrete Fourier transformation, similar remarks apply.

5.9.2 Fourier Transformation; Example of a Computer Program

```
% A simple example program which aim is to show how Fourier
% transform may be implemented in practice i Matlab. The
% example is a modification of an example program at a
% tutorial page at Matlab.

Fs = 1000;                % Sampling frequency
delta_t = 1/Fs;           % Time between each sampling
N = 1024;                 % Number of samples
t = (0:N-1)*delta_t;      % Time description

% Create an artificial signal as a sum of a 50 Hz sine and a
% 120 Hz cosine signal, plus a random signal:
x = 0.7*sin(2*pi*50*t) + cos(2*pi*120*t);
x = x + 1.2*randn(size(t));

plot(Fs*t,x)              % Plot the time domain representation
title('The signal in time domain')
xlabel('time (millisec)')

X = fft(x,N)/N;           % Fast Fourier Transformation

freqv = (Fs/2)*linspace(0,1,N/2);   % The frequency range

% Plot the absolute value of the frequency components in the
```

```
% frequency domain representation. Plot only frequencies up to
% half the sampling frequency (drop the folded part).
figure;                  % Avoids overwriting the previous plot
plot(freqv,2*abs(X(1:N/2))) % Plots half the frequency spectrum
title('Absolute value of the frequency domain representation')
xlabel('Frequency (Hz)')
ylabel('|X(freq)|')
```

5.10 Appendix: A Useful Point of View

There are big differences between how we physicists use and read the contents of mathematical expressions. In this appendix, I would like to give an example of a way of thinking that has been useful to me whenever I have wondered why some Fourier spectra look as they do.

We start with the mathematical expression shown below:

$$F(\omega) = \frac{1}{2\pi} \int_{-\infty}^{\infty} f(t) e^{-i\omega t} dt \tag{5.24}$$

or the discrete variant of the same expression:

$$X_k = \frac{1}{N} \sum_{n=0}^{N-1} x_n e^{-i\frac{2\pi}{N}kn} = \frac{1}{N} \sum_{n=0}^{N-1} x_n \cos(\omega_k t_n) - i \times \frac{1}{N} \sum_{n=0}^{N-1} x_n \sin(\omega_k t_n) \tag{5.25}$$

where $\omega_k = \frac{2\pi}{T}k$, $t_n = \frac{T}{N}n$, and T is the total sampling time. Then we simply have a sum of single products $x_n \cos(\omega_k t_n)$ (or sines) with many n. The integral or the sum we get by adding a lot of such numbers (with a scaling that we need not discuss here).

If now $\{x_n\}$ is simply a cosine function with the same frequency and phase as $\cos(\omega_k t_n)$, the products of these two terms will always be equal to or greater than zero, being a $\cos^2$ function. Then the sum will be big and positive.

If $\{x_n\}$ is a cosine function with a frequency different from that of $\cos(\omega_k t_n)$, the two cosine functions will sometimes be in phase, yielding a positive product, but at other times with an opposite phase, resulting in a negative product.

Due to the factor $\frac{1}{N}$, the sum of all product terms will be close to zero if we get many periods of positive and negative contributions in all.

Based on this argument, we find that the Fourier transform of a single harmonic function when we have integration with limits plus minus infinity is considered a δ-function. But what will happen when the test function is simply zero everywhere except for a limited length of time T where the function is a simple harmonic function?

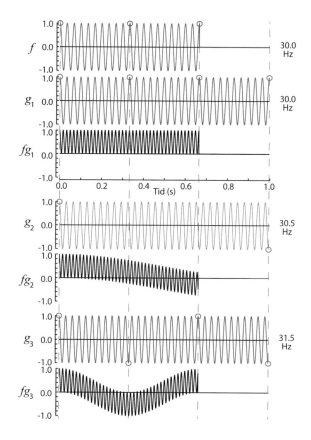

Fig. 5.14 Two functions included in the integral in Fourier integral Eq. (5.24). The f function to be analysed is shown in blue. It is different from zero only for a limited period of time. $g_k(t)$, which corresponds to $\Re\{e^{-i\omega_k t}\}$, is shown in red, and the product of the two functions in black. Three different frequencies ω_k are selected. See text for comments

Figure 5.14 shows a section of the function:

$$f(t) = \cos(\omega_a t) \qquad \text{for } t \in [0, T] \quad \text{and } 0 \text{ otherwise.} \qquad (5.26)$$

In the figure $T = 2/3$ s.

The time interval T in the figure is just sufficient to cover the entire window where $f(t)$ differs from zero. Also shown are plots $g(t) = \Re\{e^{-i\omega_k t}\} = \cos(\omega_k t)$ for three different choices of the analysing frequency ω_k, and the corresponding plots of the product functions $f(t)g(t)$.

The integral of the product function now receives contributions only in the time interval where f is different from zero. We get full contribution from the entire range when $\omega_k = \omega_a$. We see that the integral (sum of all values of product function) also becomes positive in the middle case where the difference between ω_k and ω_a is so small *in relation to the length of time interval* that the phase of f and the phase of $\cos(\omega_k)$ is always less than π.

In the bottom case, we have chosen an analysing frequency of ω_k which is such that

$$(\omega_k - \omega_a)T = 2\pi.$$

Because of the symmetry, we see that the integral here vanishes, but we realize that we would get a certain positive or negative value if we had chosen the frequency difference (in relation to T) as we did in this case.

What has this example shown us? In the first part of the chapter, we explained that when $f(t) = \cos(\omega_a t)$ for all t, the Fourier integral will be null in absolutely all cases where $\omega_k \neq \pm\omega_a$. In Fig. 5.14, we see that when the function we analyse lasts for a limited time T, the two frequencies may be slightly different and yet we may receive contributions to the Fourier integral. The contribution will be greatest when $(\omega_k - \omega_a)T < \pi$.

It should be noted that we can rename the quantities as follows: $(\omega_k - \omega_a) \equiv 2\pi\Delta f$ and $T \equiv \Delta t$. In that case, we get that the Fourier integral will have an appreciable value so long as

$$\Delta f \, \Delta t < 1/2.$$

This is again a relation analogous to the Heisenberg uncertainty relation.

We can repeat the same type of calculations of the f g–product function for many different ω_k relative to ω_a and add up positive and negative contributions over T, the interval we wish to integrate over. Examples of such calculations are shown in Fig. 5.15. When the two frequencies are identical, the area below the middle curve becomes the maximum, which corresponds to the peak value in the (real part) of the Fourier spectrum. The area may be positive or negative depending on whether the mean value of the f g–product function is above or below zero. Time intervals where the f g–product is positive is marked with blue background colour and intervals with negative product with red. The integral is just the sum of positive and negative areas in these plots. In case of **3** and **5**, the total area is equal to zero (as much positive as negative), while in case of **4** the total area is negative.

Deeper red or blue colour is used to mark the areas that are not balanced by corresponding area with the opposite sign. We see then that the deepest red-marked area in case **4** is greater in absolute value than the deepest blue area in case **6**, reflecting that the peak in the area near the **6** mark in the lower part of the figure is less than the (absolute) value of the peak in the area near the **4** mark.

Figure 5.15 indicates that the frequency spectrum of a portion of a harmonic function has a broad and sharp peak in the middle, and characteristic oscillations with smaller and smaller amplitude the farther away from the peak one moves.

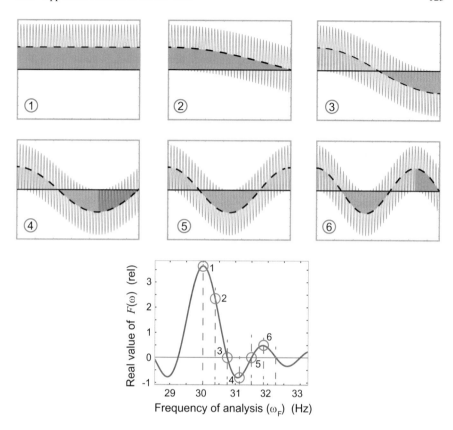

Fig. 5.15 Integrand (the $f\,g$–product) in the real part of Fourier calculations for different choices of the analysing frequency. The real part of a section of the Fourier spectrum of the function appearing in Eq. (5.26) is given at the bottom. See also the text for details

Remarks: In Chap. 13, we will see that the frequency spectrum in Fig. 5.15 appears again when we consider the diffraction image that emerges when we transmit a laser beam with visible light through a narrow gap. Within optics, there is a separate field called Fourier optics.

5.10.1 Program for Visualizing the Average of Sin–Cos Products

```
function sincosdemo
% Program to visualize the average of sine/cosine product

N = 2000;        % Number points in the description
T = 1.0;         % The time we describe the signal (1 sec)
t = linspace(0,T*(N-1)/N,N);   % Make the timeline t
```

```
freq1 = 100.0;   % One frequency is kept constant
freq2 = 100.0;   % 1) Try to vary this from 102 to e.g. 270
                 % 2) Try also values where |freq2-freq1|< 2.0
omega1 = 2*pi*freq1; % Calculate angular frequencies
omega2 = 2*pi*freq2;
f = cos(omega1*t);   % Try also sin( )
g = cos(omega2*t);

plot(t,f.*g,'-b');   % Plot the product of f and g
xlabel('Time (s)');
ylabel('Signal (rel.units)');
null = zeros(N,1);
hold on;
plot(t,null,'-r');   % Draw also the zero line

integral = sum(f.*g);   % Drop the normalization 1/N
integral        % Write the "integral" (sum) to screen.
```

5.10.2 Program Snippets for Use in the Problems

Snippet 1: Here is a piece of code that shows how to read data from an audio file in Matlab:

```
s = 'piccoloHigh.wav';   % File name (the file must be in
                         % the same folder as your program)
N = 2^16;
nstart = 1;   % First element number you want to use in
              % your audio file
nend = N;     % last element number you want to use
[f,Fs] = audioread(s, [nstart nend]);
%   sound(f,Fs);  % Play back the sound if you want
                  % (then remove %) for control purposes
g = f(:,1);   % Pick a mono signal out of the stereo
              % signal f
X = (1.0/N)*fft(g);   % FastFourierTransform of the
                      % audio signal
Xa = abs(X);   % Calculate the absolute value out of the
               % freqency domain representation
```

Snippet 2: Recording a sound from the PC's microphone.

```
T = 2.0;      % Duration of sound track in seconds
Fs = 11025;   % Chosen sampling frequency (must be
              % supported by the system)
N = Fs*T;
t = linspace(0,T*(N-1)/N,N);   % For x-axix in plot
recObj = audiorecorder(Fs, 24, 1);
deadtime = 0.13;  % Delay. Trick due to Windows-problems
recordblocking(recObj, T+3*deadtime);
myRecording = getaudiodata(recObj);
stop(recObj);
Nstart = floor(Fs*deadtime);
Nend = Nstart + N -1;
y = myRecording(Nstart:Nend,1);
s = sum(y)/N;   % Removes the mean value
y = y-s;
plot(t,y,'-k');
title('Time domain representation');
xlabel('Time (sec)');
ylabel('Microphone signal (rel units)');
```

N.B. The code for sampling the sound does not work perfectly and sometimes leads to irreproducible results. This is because the sound card is also under control of Windows (or other operating system), and the result depends on other processes in the computer. Those who are particularly interested are referred to specially developed solutions via "PortAudio" (www.portaudio.com).

Snippet 3: One possible method to make an animation.

```
function waveanimation1
clear all;
k = 3;
omega = 8;
N = 1000;
x = linspace(0,20,N);
y = linspace(0,20,N);
p = plot(x,y,'-','EraseMode','xor');
axis([0 20 -2.5 2.5])
for i=1:200
    t = i*0.01;
    y = 2.0*sin(k*x-omega*t);
    set(p,'XData',x,'YData',y)
    drawnow
    pause(0.02);  % This is to slow down the animation
end
```

Pieces of code will be transferred from the *Problems* in several chapters to the "Snippet subsection" at the "Supplementary material" web page for this book available at http://www.physics.uio.no/pow.

5.11 Learning Objectives

After working through this chapter, you should know that:

- An integrable time-dependent continuous function can be transformed by continuous Fourier transformation into a "frequency-domain picture", which can then be uniquely transformed with an inverse Fourier transformation back to the starting point.

- A discrete function can be transformed by a discrete Fourier transform into a "frequency-domain picture", which can then be uniquely transformed with a discrete inverse Fourier transform back to the starting point.

- Only integers are included in a mathematical/numerical implementation of a Fourier transformation. We must manually keep track of the sampling times and the frequencies of the elements in the Fourier spectrum. We must also take account of normalization of the numerical values (e.g. whether or not we should divide/multiply the numbers after transformation by N), as different systems handle this differently.

- The frequency-domain picture in a discrete Fourier transformation consists of complex numbers, where the real part represents cosine contributions at the different frequencies, while the imaginary part represents the sine contributions. The absolute value of the complex numbers gives the amplitude of the contribution at the relevant frequency. The arctan of the ratio between imaginary and real parts indicates the phase of the frequency component (relative to a $\cos(\omega t + \phi)$ description).

- For a real signal, the last half of the Fourier coefficients are complex conjugate of the first half, and "mirroring" occurs. Therefore, we usually use only the first half of the frequency spectrum.

- In a discrete Fourier transform, the first element in the data string X_k corresponds to a constant (zero frequency), second element to the frequency $1/T$, third to frequency $2/T$, etc. Here T is the total time function/signal we start with is described above (total sampling time). It is necessary to sample for a long time if we are to get a high resolution in the frequency picture.

- If a signal is "sampled" with a sampling frequency f_s, we will only be able to process signals with frequencies below half the sampling frequency in an unambiguous manner.

- In order to avoid "folding" problems, a low-pass filter must be used to remove signal components that may have a frequency higher than half the sampling frequency. For numerical calculations, we have to make sure that the "sampling rate" is high enough for the signal we are processing.

- Fourier transformation is a great aid in studying stationary time-varying phenomena in much of physics. For example, Fourier transformation is extensively used in analysis and synthesis of sound.
- It is possible to implement Fourier transformation of (almost) any signal, but it does not mean that Fourier transformation is useful in every situation!
- Fourier transformation is (almost) suitable only for analysing signals that have more or less the same character throughout the sampling time. For transient signals that change character greatly during sampling time, a Fourier spectrum sometimes may be more misleading than useful.
- Normally when Fourier transformation is performed numerically, we use ready-made functions within the programming package we use. If we create the code ourselves, the calculations take an unduly long time (unless we code the actual "fast Fourier transform" algorithm). Calculations are most effective if the number of points in the description is 2^n.

5.12 Exercises

Suggested concepts for student active learning activities: Periodic/nonperiodic function, Fourier transformation, time domain, frequency domain, frequency analysis, fundamental frequency, harmonic frequencies, sampling, sampling frequency, folding, aliasing, sampling theorem, time-bandwidth product, classical analogue to Heisenberg's uncertainty relationship, high-pass/low-pass filters, stationary signal.

Comprehension/discussion questions

1. In a historical remark first in this chapter, we claimed the Fourier transformation and Fourier analysis bear close resemblance to the medieval use of epicycles for calculating how planets and the sun moved relative to each other (see Fig. 5.16). Discuss this claim and how Fourier analysis may lead to unwanted conclusions if it is used in an uncritical manner.
2. How can we make a synthetic sound by starting from a frequency spectrum? Would such sound simulate in a good way the output of a proper instrument?
3. For CD sound, the sampling rate is 44.1 kHz. In the case of sound recording, we must have a low-pass filter between the microphone amplifiers and sampling circuits that remove all frequencies above 22 kHz. What could happen to the sound during playback if we did not take this rule seriously?
4. After a fast Fourier transform (FFT), we often plot only a part of all the data produced. Mention examples of what may influence our choices.
5. Suppose that you Fourier analyse sound from a CD recording of an instrument and find that the fundamental tone has a frequency of 440 Hz. Where do you find the folded frequency?

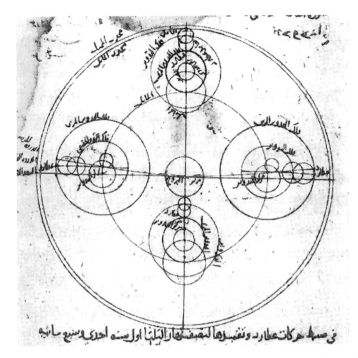

Fig. 5.16 A drawing of epicycles in an old Arabic document written by Ibn_al-Shatir [1], Public Domain

6. What are the resemblances between Fourier series and a discrete Fourier transform? Discuss the difference between periodic and nonperiodic signals.
7. Describe in your own words *why* the Fourier transform of a cosine function that lasts for a limited time T is different than if the cosine function had lasted from minus to plus infinity.
8. Consider Fig. 5.17 and tell us what it means to you.

Problems

Remember: A "Supplementary material" web page for this book is available at http://www.physics.uio.no/pow.

9. Show both mathematically and in a separate programming example that the first point in a digital Fourier transform of a signal is equal to the average value of the signal we started with.
10. Use the computer program provided in "computer software example" on page xxx to explore how folding works in practice. Let the signal be:

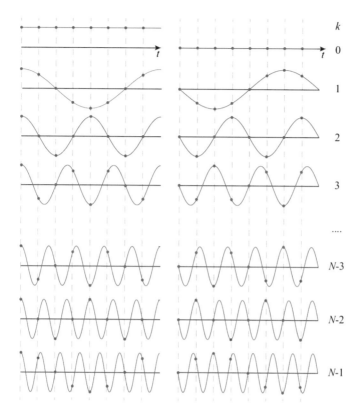

Fig. 5.17 Use this figure in Problem 8

```
freq = 100.0;  % Frequency in hertz
x = 0.8 * cos(2*pi*freq*t);  % Signal is a simple cosine
```

and run the program. Be sure to zoom in so much that you can check that the frequency in the frequency spectrum is correct.

Then run the program, setting the frequency (one by one) equal to 200, 400, 700, 950, 1300 (Hz). Do you find a pattern in where the peaks come out in the frequency spectrum?

11. Some people claim that the moon phases influence everything from the weather to the mood of us humans. Check if you can find indications that the temperature (maximum and/or minimum daily temperature) varies slightly with the moon phases (in addition to all other factors)?

The data for the place (and period) you are interested in can be downloaded from api.met.no. Alternatively, you can use an already downloaded and slightly simplified file *tempblindern10aar.txt* from the web pages providing supplementary material for our book. The file gives the temperature of Blindern, Oslo, Norway in the period 1 January 2003 through 31 December 2012. The fourth column in the file provides minimum temperatures, while the fifth column provides the maximum values.

Explain carefully how you can draw a conclusion as to whether or not the moon phase affects the temperature.

Below you will find some lines of Matlab code that shows how data may be read from our file into a Matlab program (the data file has five columns, and we use the last two of them):

```
filename = 'tempBlindern10years.txt';
fileID = fopen(filename, 'r');
A = fscanf(fileID, '%d %d %f %f %f',[5,inf]);
minT = A(4,:);
maxT = A(5,:);
plot(minT,'-r');
hold on;
plot(maxT,'-b');
```

12. Collect sunspot data from the Web and create an updated figure similar to our Fig. 5.7. Pay particular attention to getting correct values along the axes of the frequency-domain representation. Is there a correspondence between the peak heights in the time-domain picture and the amplitude of the frequency spectrum? Below are some lines of Matlab code showing how data can be read into a Matlab program (two columns):

```
filename='sundata.txt';
fileID = fopen(filename, 'r');
A = fscanf(fileID,'%f %f', [2,inf]);
plot(A(1,:),A(2,:),'-b');
```

13. Pick up a short audio signal from a CD, a wav file or record sound from a microphone (use, for example, one of the program snippets a few pages before this one). The sampling rate is 44.1 kHz. Save $2^{14} = 16384$ data points (pairs of points if it is a stereo signal, but use only one of the channels). Perform a "fast Fourier transformation" and end up with 16384 new data points representing the frequency spectrum. How do you change your program from point number to frequency in Hz along the x-axis when the frequency spectrum is to be plotted?

14. What is the resolution along the x-axis of the plot in the previous task? In other words, how much change in frequency do we get by moving from one point in the frequency range to the next? Would the resolution be the same even if we had used only 1024 points as the starting point for Fourier transformation?

15. Write a program in Python or Matlab (or any other programming language) that creates a harmonic signal with exactly 13 periods within 512 points. Use the built-in FFT function to calculate a frequency range. Will this be what you expected? Feel free to let the signal be a pure sinusoidal signal or a sum of sine and cosine signals.

16. Modify the program so that the signal now has 13.2 periods within the 512 points. How does the frequency spectrum look now? Describe as well as you can!

17. Modify the program to get 16 full periods of *FIRKANTsignal* within $2^{14} = 16384$ points. How does the frequency spectrum look now? Find on the Internet how the amplitude of different frequency components should be for a square signal and verify that you get nearly the same output from your numerical calculations.

18. Modify the program so that you get 16 full *sagtenner* (triangular signal) within the 1024 points. Also describe this frequency spectrum!

19. In an example in Chap. 4, we calculated the angular amplitude of a physical pendulum executing large displacements. Perform these calculations for 3–4 different angular amplitudes and carry out a Fourier analysis of the motion in each case. Comment on the results.

20. AM radio (AM: Amplitude Modulated). Calculate how the signal sent from an AM transmitter looks like and find the frequency spectrum of the signal. It is easiest to do this for a radio signal on the long wave band (153–297 kHz). Let the carrier have the frequency $f_b = 200$ kHz and choose the speech signal to be a simple sine with frequency (in turn) $f_t = 440$ Hz and 4400 Hz. The signal should be sampled at a sampling rate of $f_s = 3.2$ MHz, and it may be appropriate to use $N = 2^{16} = 65536$ points. The AM signal is given by:

$$f(t) = (1 + A\sin(2\pi f_s t)) \times \sin(2\pi f_b t)$$

where A is the normalized amplitude of the audio signal (the loudest sound that can be sent without distortion is $A = 1.0$). Use a slightly smaller value, but please test how the signal is affected by A).

Plot the AM signal in both the time domain and the frequency domain. Select appropriate segments from the full data set to focus on what you want to display. Remember to set correct timing along the x-axis of the time-domain plot and correct frequency scale along the x-axis of the frequency spectrum.

Each radio station on the medium and long wave may extend over only a 9 kHz frequency band. What are the consequences for the quality of the sound being transmitted?

21. FM radio (FM: Frequency Modulated). Calculate how the signal sent from an FM transmitter looks like and find the frequency spectrum of the signal. Use the same parameters as in the previous task (although in practice, no long wave FM is used). The FM signal can be given as follows:

```
f(t) = sin(phase(t)); % Principally (!)
```

where the phase is integrated by means of the following loop:

```
phase(1) = 0.0;
for i=1:(N-1)
    phase(i+1)=phase(i) + \cdots
    omega_b*delta_t*(1.0 + A*sin(omega_t*t(i)));
end;
```

where "omega_b" and "omega_t" are the angular frequencies of the carrier and the speech signal, respectively. The time string "t(i)" is assumed to be calculated in advance (distance between the points is "delta_t", which is determined by the sampling frequency).

A is again a standard amplitude for the audio signal, which also includes the so-called degree of modulation. You can choose, for example, $A = 0.2$ and 0.7 (in turn), and see how this affects both the time-domain picture and the frequency-domain picture.

Plot the FM signal in both the time domain and the frequency domain according to the same guidelines as in the previous task (Hint: It may be easiest to plot the case where the voice frequency is 4400 Hz and that $A = 0.7$.).

Are there any clear differences in how the frequency-domain picture appears for FM signals compared to AM signals?

22. Use inverse Fourier transformation to generate a simple sinusoid and play the sound on your computer. Use the inbuilt *sound* or *wavplay* function (program snippet 1 a few pages ahead of this on indicates how). Specifically, the following is recommended: Use the CD sampling rate $f_s = 44100$ Hz and $2^{16} = 65536$ points. The values of the signal f must not exceed the interval $[-1, +1]$. Attempt to make sound with frequencies 100, 440, 1000 and 3000 Hz. You may want to make a signal consisting of several simultaneous sinusoids too? Remember to scale the total signal before using *wavplay* or *sound*.

23. Read the audio file "transient.wav" and perform Fourier transformation to obtain the frequency spectrum. The audio file is available from the web pages providing supplementary material for our book, the sampling rate is $f_s = 44100$ Hz. Use 2^{17} points in the analysis. You may use program snippet 1 a few pages ahead of this one for reading the file.

If you listen to the sound and then consider the frequency-domain picture, I hope that you would pause and reflect on what you have done. Fourier analysis is sometimes misused. What is the problem with the analysis performed on the current audio signal?

24. (a) Perform a frequency analysis of sound from a tuba and from a piccolo flute (audio files available from the course's web pages). The sampling frequency is 44100 Hz. Use, e.g., 2^{16} points in the analysis. Plot the absolute value of the frequency spectrum (see program below). Determine the pitch of the tone on a tempered scale using the Fig. 5.18. Remember to get correct values along the frequency axis when plotting the frequency spectrum and zoom in to get a fairly accurate reading of the fundamental tone frequency.

(b) The frequency spectrum shows varying degrees of harmonics as described in this chapter (we will return to this in later chapters). Zoom into the time signal so much that you get a few periods. Does the signal look like a harmonic signal, or is it far more irregular than a sinus? (Comparison must be done only when considering 3–8 periods in the audio signal.) Does there appear to be some kind of connection between how close the time signal is to a pure sinusoid and the number of harmonics in the frequency spectrum?

Fig. 5.18 Tone scale for a tempered scale as we find it on a piano. Frequencies for the tones are given. The figure is inspired from [2], but completely redrawn

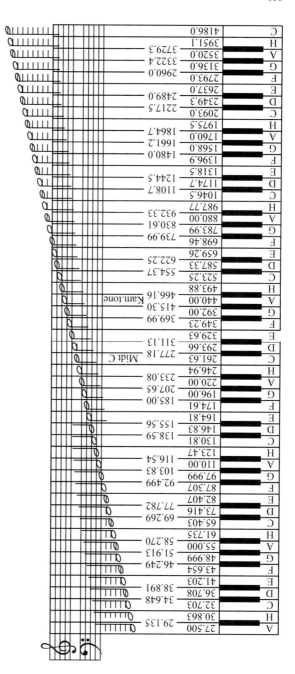

(c) Attempt to include data only for such a small time interval that there is only room for one period in the signal. Carry out the Fourier transform of this small portion of the signal (need not have 2^n data points). Do you find a connection between the Fourier spectrum here compared to the Fourier spectrum when you used a long time string containing many periods in the audio signal?

(d) For one of the audio files, you are asked to test that an inverse Fourier transform of the Fourier transform brings us back to the original signal. Remember that we must keep the Fourier transform as complex numbers when the inverse transform is carried out. Plot the results.

(e) Perform an inverse Fourier transform on the *absolute value* of the Fourier transform signal. Describe the difference between the inverse of the complex Fourier transform and the one you found now. Try to give the reason for the difference.

25. "Open task" (i.e. very few guidelines and hints are given): Fourier transformation can be used in digital filtering. Explain the principle and how this can be done in practice. Create a small program that performs self-selected digital filtering of a real audio file, where it is possible to listen to the sound both before and after filtering (Be scrupulous in describing the details of what you do!).

References

1. Ibn_al-Shatir. https://en.wikipedia.org/wiki/Ibn_al-Shatir, Accessed April 2018
2. Unknown. http://amath.colorado.edu/outreach/demos/music/MathMusicSlides.pdf. Accessed Feb 18 2012

Chapter 6
Waves

Abstract Waves, viewed as phenomena extended in both time and space, are introduced in this chapter. The mathematical wave equation is presented together with the concepts of wavelength, period and wave velocity. Also, the mathematical expressions of a wave, in both real and complex notation, are presented, as well as the concepts of transverse and longitudinal waves. The transverse equation of motion of a string, as well as the longitudinal movement of air (or water) molecules when a sound wave passes through the compressible medium, is shown to follow a wave equation.

6.1 Introduction

Everyone has seen circular waves propagating along a water surface (see Fig. 6.1). We are so used to the phenomenon that we barely notice it.

But have you really understood the magic of waves? How come that the wave migrates along the surface of the water without any matter moving at the wave speed? If we throw a ball from point A to point B, the ball moves spatially with all its mass from A to B. But when a wave moves from A to B, there is no corresponding mass that is transported from A to B. What in heaven's name is causing the wave to move along?

Waves are generated when a vibration at one place in space somehow affects the neighbouring area so that it too starts to vibrate, causing in turn another neighbouring area to begin to vibrate, and so on. When we describe this interaction and focus on the explanation of the physics that lies behind wave motion, we study the dynamics of the system. Nevertheless, we start in the same way as in Chap. 2 with "kinematics", that is, with the mathematical description.

A wave can be visualized in three ways:

- We can take a snapshot ("flash image", a high-speed flash photograph) of how the wave looks at a selected time in different parts of space (as a function of position).
- We can record the amplitude as a function of time at *one* place in space as the wave passes this location and plot the result.

© Springer Nature Switzerland AG 2018

A. I. Vistnes, *Physics of Oscillations and Waves*, Undergraduate Texts in Physics,
https://doi.org/10.1007/978-3-319-72314-3_6

Fig. 6.1 Waves that form on water

- We can use a "movie" (animation) that shows how the wave spreads in space as time goes by.

Figure 6.2 shows examples of the first two viewing modes. Imagine standing on a pier and watching waves rolling gently in front of you. You can take a picture of the waves and get something that corresponds to the left part of Fig. 6.2. Take another picture a moment later, and you will see that the wave has moved a little (as indicated in the figure).

Imagine that there is a vertical pole in the water. The water surface then moves up and down the post, and you can record the height as a function of time. This corresponds to the right part of Fig. 6.2. If there are two pegs that stand a little apart, the water surface will not be on top simultaneously on both pins, in general.

For a *harmonic* wave (with a form like that of a sine or cosine function), the first two modes of view will both look like harmonic oscillation: the first as harmonic oscillation as a function of position, the other as harmonic oscillation as a function of time. We know from before that a harmonic oscillation is a solution of a second-order differential equation. If we consider how the wave looks like a function of position (at one point), the result f must be a solution of the differential equation:

$$\frac{\mathrm{d}^2 f}{\mathrm{d}x^2} = -C_x f \ .$$

If we regard the wave as a function of time as it passes one place in the room, the result must be a solution of the differential equation:

$$\frac{\mathrm{d}^2 f}{\mathrm{d}t^2} = -C_t f \ .$$

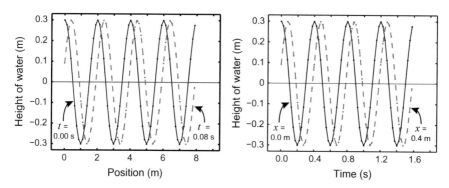

Fig. 6.2 A wave can be considered as a function of position at a certain time, or as a function of time for a particular position. See the text for details

In these equations, x indicates the position and t the time, and C_x and C_t are positive real constants that differ in the two cases. In contrast, the amplitude of the harmonic wave f is the same quantity in each viewing mode. We therefore use the same symbol f in both equations. The amplitude may be, for example, the air pressure of sound waves, or the electric field strength of electromagnetic waves or the height of surface waves at sea.

When we realize that the wave f is the same, regardless of whether we consider the wave as a function of position in space or as a function of time, we can combine the two equations and get:

$$\frac{d^2 f(x, t)}{dt^2} = \frac{C_t}{C_x} \frac{d^2 f(x, t)}{dx^2} .$$

In the above notation, the dependence of the amplitude on space and time has been explicitly indicated. And when a function depends on more than one independent variable, we use *partial differentiation* and write:

$$\frac{\partial^2 f(x, t)}{\partial t^2} = \frac{C_t}{C_x} \frac{\partial^2 f(x, t)}{\partial x^2} . \tag{6.1}$$

Upon renaming the quotient C_t/C_x as v^2, the above equation takes the form:

$$\frac{\partial^2 f(x, t)}{\partial t^2} = v^2 \frac{\partial^2 f(x, t)}{\partial x^2} . \tag{6.2}$$

This equation is called the "wave equation".

Since the C's were positive real constants, v must be real (and positive).

Remark: We take a short detour to recall what we mean by partial differentiation.

Suppose we have a function $h = h(kx - \omega t)$ and that we wish to find the partial derivative of this function with respect to x. We define a new variable $u = kx - \omega t$ and, using the chain rule, we find:

$$\frac{\partial h}{\partial x} = \frac{dh(u)}{du} \times \frac{\partial u}{\partial x} .$$

It is the second factor on the right-hand side where the implications of partial differentiation strike us first. We have:

$$\frac{\partial u}{\partial x} = \frac{\partial (kx - \omega t)}{\partial x} .$$

Both x and t are variables, but when we calculate the partial derivative with respect to x, we will treat t as a constant! Consequently, we get:

$$\frac{\partial (kx - \omega t)}{\partial x} = k .$$

Similarly, we can go on to deduce the partial derivative with respect to t. In this case, we treat x as constant.

Partial derivatives thus represent the derivative of the function, assuming that all variables are kept constant, except for the one with respect to which the derivative is to be calculated.

We will come across the "wave equation" Eq. (6.2) quite a few times in the book. It may therefore be useful to try to understand it forthwith.

When we discussed oscillations in Chap. 1, we saw that if we know the starting position and the starting speed, e.g. for a shuttle, we can unambiguously calculate how the oscillation will be in the future (so long as the differential equation governing the motion is known).

For waves, it is totally different. Even when we have the exact same wave equation and the very same initial conditions, there are infinitely many different solutions. The reason is that the waves spread out into space, and the shape of the volume the wave is confined in will affect the wave even though the basic differential equation is the same. This is easy to understand if we think of swells rushing towards land. The wave will show enormous local variations, all depending on the landscape, with its rocks and protrusions and recesses. Solving the wave equation therefore requires that we know the initial as well as the boundary conditions. And since there are infinitely many boundary conditions we can imagine, there will also be infinitely many solutions. But once we have specified both initial conditions and complete set of boundary conditions, there is a unique solution.

Since there is such an incredibly wide variety of waves, we often have to resort to simplified solutions to extract at least some typical features. Some such solutions are actually serviceable approximations to real waves in special cases. The most common simplified solution is called *plane wave*, and we will take a closer look at it now.

6.2 Plane Waves

A wave is said to be plane when its amplitude is constant throughout a plane that is normal to the direction of propagation in space. If a wave in three-dimensional space travels in a direction parallel to the x-axis, a planar wave will have an identical amplitude, at any selected instant, throughout an infinite plane perpendicular to the x-axis.

For a plane sound wave that moves in the x-direction, this will in practice mean that, at any time whatsoever, the local air pressure has a maximum everywhere along a plane perpendicular to the x-axis. We call such a plane a "wavefront". For plane waves, the wavefront is plane.

Mathematically, a plane harmonic (monochromatic) wave can be described as:

$$f(x, t) = A \cos(kx - \omega t) . \tag{6.3}$$

In this context, k is called *wavenumber* and ω, the angular frequency. If we keep the time constant, for example, at $t = 0$, and start at $x = 0$, we move a wavelength when $kx = 2\pi$. The wavelength λ is therefore precisely this value of x, so that:

$$\lambda = \frac{2\pi}{k} .$$

In a similar manner, we can keep the position constant, for example, by setting $x = 0$, and starting at $t = 0$. We find then that, if we want to change the time function by a period, the time must increase by $\omega t = 2\pi$. This time difference is called the *time period T*, and we get:

$$T = \frac{2\pi}{\omega} .$$

It may be added that the word "wavenumber" comes from k indicating the number of wavelengths within the chosen unit of length ("how many wave peaks are there in a metre?"), but multiplied by 2π.

We can also apply a similar idea to the angular frequency ω. In that case, we can say that the ω is a "(time) period" which indicates how many periods we have within the chosen unit of time ("how many periods of vibration are there in one second?"), but multiplied by 2π.

The unit for wavenumber measurement is inverse metre, i.e. m^{-1}. Angular frequency unit is actually inverse second, that is, s^{-1}, but in order to reduce the likelihood of confusion with frequency, we often give angular frequencies in *radians per second*.

6.2.1   Speed of Waves

Let us find out how fast the wave travels in the x-direction. Imagine following a peak that corresponds, let us say, to the value 6π for the argument in the cosine function, in which case we will have

$$kx - \omega t = 6\pi \ ,$$

$$x = \frac{\omega}{k} t + \frac{6\pi}{k} \ .$$

We differentiate the expression for position with respect to time, so that we may see how quickly this point moves, and we obtain

$$\frac{dx}{dt} \equiv v = \frac{\omega}{k} \ .$$

The velocity with which the wave travels is thus equal to the ratio between angular frequency and wavenumber. We can rephrase this relation in terms of the wavelength and time period as:

$$v = \frac{2\pi/T}{2\pi/\lambda} = \frac{\lambda}{T} \ .$$

But we know that the frequency is given as the inverse of the period, i.e. $v = 1/T$. If we insert this in the last equation, we get a well-known relationship:

$$v = \lambda v \ . \tag{6.4}$$

The velocity of a plane wave given in Eq. (6.3) is thus the wavelength multiplied by the frequency (Eq. 6.4). *This is a very important relationship!*

6.2.2   Solution of the Wave Equation?

So far, we have only *asserted* that Eq. (6.3) satisfies the wave equation. We will now verify this, and we get by double differentiation of Eq. (6.3):

$$\frac{\partial^2 f(x, t)}{\partial t^2} = -\omega^2 f(x, t)$$

and

$$\frac{\partial^2 f(x,t)}{\partial x^2} = -k^2 f(x,t) .$$

We observe that:

$$\frac{\partial^2 f(x,t)}{\partial t^2} = \frac{\omega^2}{k^2} \frac{\partial^2 f(x,t)}{\partial x^2}$$

or:

$$\frac{\partial^2 f(x,t)}{\partial t^2} = v^2 \frac{\partial^2 f(x,t)}{\partial x^2} . \tag{6.5}$$

We see that the plane wave given in Eq. (6.3) satisfies the wave equation, but what about the initial and boundary conditions? Well, here some difficulties arise. If a planar wave should be able to form and remain so, we must initiate a wave that actually has infinite extent and the same amplitude and initial variation in time throughout this infinite plane. There must also be no boundary conditions that affect the wave at any point. If all of these requirements were met, the plane wave would remain plane, but we realize that this is physically unattainable.

However, if we start by considering a wave many, many wavelengths away from the location where it was generated—for example, sunlight as it reaches earth—the so-called wavefront will be quite flat as long as we only consider the light over, for example, a one square metre flat surface normal to the direction of light. If we then follow the light a few metres further, the wave will behave approximately like a plane wave in this limited volume. But if reflected light reaches this volume, we will not have a plane wave anymore!

Remark: The wavefront of light from the sun will in fact not be plane, as indicated above. Due to the angular size of the sun in relation to the wavelengths of visible light, the spatial coherence length is short and the wavefronts irregular. This will be discussed in detail in Chap. 15.

Plane waves are therefore just an idealization that we can never achieve in practice. The plane-wave description can nevertheless provide a relatively good account over a limited volume when we are far away from bits and bobs that can affect the wave in one way or another.

By "far away" one means that the distance is large relative to the wavelength, from the source of the waves, and from boundaries that distort the wave.

6.2.3 Which Way?

We found above that a plane wave described by the equation

$$f(x, t) = A \cos(kx - \omega t)$$

has a velocity $v = +\omega/k$. That is, the wave propagates in positive x-direction as time passes. With a little practice, we can infer this directly from the argument of the cosine function: if we stay at the same place on a wave (e.g. a peak), the argument must remain unchanged as time increases. And increasing the time t, we can achieve the constancy of the argument only if we compensate by letting x also increase. In other words, the peak of the wave moves towards larger x-values as time increases.

By using similar reasoning, we can easily show that a wave described by:

$$f(x, t) = A \cos(kx + \omega t)$$

propagates towards lower x-values as time increases. Pictorially, for those of us who are accustomed to the x-axis increasing to the right, we can say that the waves described in the first of these ways (with the minus sign) move to the right, and waves described in the other way (with the plus sign) move leftward.

Note that the speed of the wave does not describe speed in the same way as the speed of a ball after it is thrown. The speed of the ball, a physical body, is defined as the time derivative of the position of the ball. For the wave, speed is defined as a more *abstract quantity*, for example, the time derivative of the position in space where the wave has its maximum value. For a sound wave in air, the velocity of the wave is equal to the velocity of, say, a point in space where the local air pressure has a maximum. This can be described as the speed of a "wavefront". We will come back to more complicated relationships later.

Fig. 6.3 Snapshot of "amplitude" (y *in red*), the time derivative of the amplitude ($\dot{y}$ *in blue*) and the double derivative of the amplitude ($\ddot{y}$ *in green*) in different positions along a wave. The wave as such moves to the right (*top*) and to the left (*bottom*). The dashed red curve shows where the wave is a short time after its current location (*solid curve*)

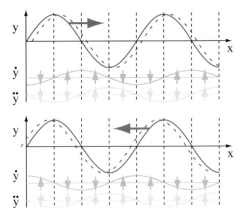

Figure 6.3 shows a snapshot of "amplitude", the single and double time derivatives of the amplitude at all positions along a wave. The wave as such goes to the right or to the left as the arrows show (top). Note that for a right-going wave, the time derivative of the amplitude will lie a quarter period "in front" of the "amplitude" and the second time derivative a quarter of period "in front" of the first time derivative. For a leftward wave, exactly the same applies, but "in front" of the wave must now mean to the left of the wave.

Let's try to concretize these considerations, but choose a wave on a string (e.g. the one we get just after we swing one end of a long horizontal string up and down a few times). The "amplitude" in this case is very concrete because it simply indicates the position of the string at the spot where the amplitude is measured. The time derivative of the amplitude will then say the vertical velocity to the point along the string we consider, and the double time derivative for this point will then be the vertical acceleration of this point. The wave itself moves in the horizontal direction.

Note that for any arbitrary point along the string, the sign of acceleration at all times is the opposite of the position relative to the equilibrium point. Thus, *the effective force on every element of the wave is always pointing towards the equilibrium state/position.* We hope you realize that this is just as it should be (based on what we learned in Chap. 2).

6.2.4 Other Waveforms

So far, we have considered harmonic waves, i.e. waves with sinusoidal shape. Can waves of another shape satisfy the wave equation?

Let us investigate a wave described by:

$$g(x, t) = G(kx - \omega t) .$$

where G can have any shape (but G must be a differentiable function). We introduce a new variable $u = kx - \omega t$, partially differentiate, use the chain rule and get for the left-hand side of Eq. (6.5):

$$\frac{\partial^2 g(x, t)}{\partial t^2} = \frac{d^2 G(x, t)}{du^2} \left(\frac{\partial u}{\partial t} \right)^2$$

$$= \omega^2 \frac{d^2 G(x, t)}{du^2} .$$

For the right-hand side, a similar differentiation gives:

$$\frac{\partial^2 g(x, t)}{\partial x^2} = k^2 \frac{d^2 G(x, t)}{du^2} .$$

We see that $g(x, t)$ indeed satisfies the wave equation, assuming that k and ω are real constants.

That is, any wave that can be described by a differentiable function and a single argument $(kx - \omega t)$, where k and ω are constant is a solution of the wave equation.

6.2.5 Sum of Waves

What if we have a sum of two different functions, one of which has a slightly different combination of k and ω than the other. The sum function is then given by:

$$g(x, t) = G_1(k_1x - \omega_1t) + G_2(k_2x - \omega_2t)$$

$$= G_1(u_1) + G_2(u_2) .$$

Partial differentiation with respect to time gives:

$$\frac{\partial^2 g}{\partial t^2} = \omega_1^2 \frac{d^2G_1(u_1)}{du_1^2} + \omega_2^2 \frac{d^2G_2(u_2)}{du_2^2} ,$$

and partial differentiation with respect to position gives:

$$\frac{\partial^2 g}{\partial x^2} = k_1^2 \frac{d^2G_1(u_1)}{du_1^2} + k_2^2 \frac{d^2G_2(u_2)}{du_2^2} .$$

If these functions are to satisfy the wave equation

$$\frac{\partial^2 g}{\partial t^2} = v^2 \frac{\partial^2 g}{\partial x^2} .$$

We must require that the time derivative should equal v^2 times the second spatial derivative. We assume that this demand can be met, and then get the insertion and arrangement of the terms:

$$(\omega_1^2 - v^2k_1^2) \frac{d^2G_1(u_1)}{du_1^2}$$

$$= -(\omega_2^2 - v^2k_2^2) \frac{d^2G_2(u_2)}{du_2^2} .$$

Since G_1 and G_2 can be chosen freely, this equation cannot be satisfied in general unless

$$(\omega_1^2 - v^2 k_1^2) = (\omega_2^2 - v^2 k_2^2) = 0$$

and this implies that

$$v = \frac{\omega_1}{k_1} = \frac{\omega_2}{k_2}$$

and one is led to conclude that the two waves must travel with the same velocity!

We have now established that *the sum of two (or more) waves travelling at the same speed will satisfy the wave equation if each of the sub-waves does.*

We have also shown that *if a wave consists of several components that move with different speeds, we will not be able to describe the time development of the wave by using a single wave equation.* Then the waveform will change as the wave moves (an effect we call *dispersion* in Chap. 8).

6.2.6 Complex Form of a Wave

We can use complex description for a wave in the same way we did it for oscillations.

A plane harmonic wave in the x-direction can be descried in terms of complex quantities as:

$$f(x,t) = A e^{i(kx - \omega t + \phi)} . \tag{6.6}$$

Similarly, we can describe a plane harmonic wave travelling along an arbitrary direction $\mathbf{k}/|\mathbf{k}|$, where $\mathbf{k}$ is a so-called wave vector, in the following manner:

$$f(\mathbf{r},t) = A e^{i(\mathbf{k}\cdot\mathbf{r} - \omega t + \phi)} . \tag{6.7}$$

where $\mathbf{k} \cdot \mathbf{r}$ is the dot product between the position vector and the wave vector.

Since f should normally be real, we must either take the real part of the above expressions or seek some other safeguard. An elegant and common way to avoid this problem is to add the complex conjugate ("c.c.") of the expression and divide by 2:

$$f(\mathbf{r},t) = \frac{1}{2} A e^{i(\mathbf{k}\cdot\mathbf{r} - \omega t + \phi)} + \text{c.c.} . \tag{6.8}$$

This form of representation can be used for both real and complex A.

6.3 Transverse and Longitudinal

There are several types of waves. One classification is based on which direction "the amplitude" has in relation to the direction of propagation of the wave. But since the "amplitude" can be almost anything, and not necessarily something that moves in space, such considerations often turn out to be misleading. It is safer to base the classification on the symmetry properties of the wave, and we shall attempt to do this in what follows.

For sound waves, "the amplitude" is a pressure change. For sound waves in air, this is a pressure change in air, and likewise for sound in other materials. Pressure changes occur locally because air molecules move in the same (or opposite) direction as the direction of propagation of the wave.

It is the local rotational symmetry axis of the air pressure that determines the direction of propagation of the wave. By saying that one means that the local air pressure varies in the same manner irrespective of which direction we take to be the normal to the direction in which the wave travels (thus, we have cylindrical symmetry). Such a wave is called *longitudinal* (lengthwise).

However, air molecules do not move from, say a speaker to my ear, when I listen to music. It is tempting to say that each air molecule fluctuates (statistically) back and forth relative to an equilibrium point. The problem, however, is that there is no equilibrium point because Brownian displacements of the air molecules are usually greater than the displacements caused by the passage of sound. However, the movement due to the sound is "collective" for many air molecules, while individual movements are more chaotic. That way, the sound wave can survive after all. The amplitude of the oscillations due to the sound wave alone is usually much smaller than one millimetre (even smaller for sound in metals).

Transverse waves are the other main type of waves. The best-known example is that of electromagnetic waves. When the physicists at the beginning of the nineteenth century realized that light had to be described by waves (and not as particles as Newton had convinced physicists to believe for over a hundred years), they had trouble explaining polarization. The reason is that they assumed that light waves were longitudinal, as they thought all waves to be. Only when Fresnel suggested that the light waves were transverse, were they able to fathom polarization.

A transverse wave has an "amplitude" perpendicular to the wave propagation direction (transverse: "turned across"). By that we mean that *the physical parameter we call "the amplitude" does not have local rotational symmetry about the axis indicating the direction of wave motion.* There is no cylindrical symmetry.

For electromagnetic waves, the electric and magnetic field is "the amplitude". Electrical and magnetic fields are vectors and have a direction in space. That an electromagnetic wave is transverse means that the electric and magnetic field are in a direction perpendicular to the direction along which the wave propagates. Then the rotation symmetry is automatically broken. (It is sufficient with symmetry breaking within a limited volume in space, of the order one half of the wavelength in all directions.)

Note that there is no relocation of anything material across an electromagnetic wave! Many imagine that there is something that moves across an electromagnetic wave, similar to the water level in a surface wave of water. That is wrong. If we depict electric fields as vector arrows at points along the propagation direction, then the arrows will extend and retract. But these arrows are mere aids for thought and have no existence of their own. They only indicate the size and direction of the abstract quantities electric and magnetic fields at the different positions in space. We will discuss common misconceptions when we treat electromagnetic waves in Chap. 9.

Some waves (proclaim to) have a portmanteau character, lying between longitudinal and transverse. Surface waves on water are an example. Here, water molecules move back and forth in the direction of propagation, as well as up and down in a perpendicular direction.

6.4 Derivation of Wave Equation

We have previously given a mathematical expression for a wave and arrived (through quasi-reverse reasoning) at a differential equation with solutions displaying wave behaviour. We will now start with a physical system and derive the pertinent wave equation. We will do this for oscillations on a string and for sound waves in air/liquid. It is considerably more difficult to derive an equation for surface waves in water, and we will just settle for an approximate solution without a derivation. Subsequently, we will also deduce the equation for an electromagnetic wave. Surface waves on water and electromagnetic waves will be discussed in later chapters.

6.4.1 Waves on a String

The starting point is a wave along a string. We consider a small segment of the string, more specifically a segment that is small in relation to the effective wavelength. Figure 6.4 shows the segment along with forces that work on it. The wave is assumed to propagate in the horizontal direction (x-direction), and the equilibrium position of the string when there are no waves on it is also horizontal. The wave is assumed to be purely transverse so that the result is solely in the vertical direction of the figure (y-direction). It should be noted that the amplitude in the vertical direction is *exceedingly*

Fig. 6.4 Forces that act on a small segment of a string suffering transverse motion. See the text for details

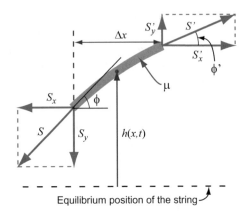

small in relation to the length of the piece under consideration. We expand the vertical scale in the figure to get some visual help when important relationships are to be entered.

It is assumed that the stiffness of the string is so small that the forces S and S' that work at each end of the string are *tangential* aligned along the string.[1] The mass centre for the segment will still change position $h(x, t)$ relative to a mean position (the equilibrium position of the string when there is no wave). The movement of the segment must be described by Newton's second law.

Newton's second law will be applied separately to the horizontal and vertical directions, and we take the horizontal first. Since the string is assumed to have a purely transverse movement, the centre of mass of the string segment does not move (notably) in the x-direction. Consequently, the sum of forces in the horizontal direction must be equal to zero, in other words:

$$S_x = S \cos \phi = S' \cos \phi' = S'_x .$$

This is accomplished automatically (to second order in ϕ) if $S = S'$, since ϕ is a very small angle (remember, according to Taylor's theorem, $\cos \phi \approx 1 - \phi^2 + \cdots$).

Newton's second law, when applied in the y-direction, gives:

$$\sum F_y = m a_y . \tag{6.9}$$

The string has a linear mass density (mass per length) equal to μ, and the length of the segment is Δx. The mass of the segment is therefore $m = \mu \Delta x$.

Let $h(x, t)$ denote the position of the midpoint of the segment relative to the equilibrium position when there is no wave on the string. Also, since $S \approx S'$, it follows from Eq. (6.9):

---

[1] For sufficient small wavelengths, this approximation cannot be used. The limiting wavelength depends on the stiffness of the material of the string.

$$S \sin \phi' - S \sin \phi = \mu \Delta x \left(\frac{\partial^2 h}{\partial t^2} \right)_{\text{midpoint}}. \tag{6.10}$$

The subscript of the last parenthesis indicates that the double derivative of the centre of mass is calculated in the middle of the Δx range, i.e. in the middle of the segment. Since the ϕ and ϕ' angles are very small, a Taylor expansion provides:

$$\sin \phi \approx \phi \approx \tan \phi$$

and likewise for ϕ'; further, $\sin \phi$ can be replaced by $\tan \phi$ in the above expression. But the tangent indicates the slope, which can also be written as $\partial h / \partial x$. Since there is an increase both at the beginning and the end of the segment, we get:

$$\sin \phi' - \sin \phi \approx \left(\frac{\partial h}{\partial x} \right)_{(x+\Delta x)} - \left(\frac{\partial h}{\partial x} \right)_x.$$

This can be rephrased as:

$$\frac{\left(\frac{\partial h}{\partial x} \right)_{(x+\Delta x)} - \left(\frac{\partial h}{\partial x} \right)_x}{\Delta x} \Delta x \approx \left(\frac{\partial^2 h}{\partial x^2} \right)_{\text{midpoint}} \Delta x.$$

Make sure that you recognize the second derivative in the above expression! If this expression is inserted in Eq. (6.10), one obtains:

$$S \left(\frac{\partial^2 h}{\partial x^2} \right)_{\text{midpoint}} \Delta x \approx \mu \Delta x \left(\frac{\partial^2 h}{\partial t^2} \right)_{\text{midpoint}}.$$

Since both derivatives refer to the same point (midpoint), this index can now be dropped. Upon cancelling Δx and carrying out some straightforward manipulations, one is led to the result:

$$\frac{\partial^2 h}{\partial t^2} \approx \frac{S}{\mu} \frac{\partial^2 h}{\partial x^2}.$$

The desired equation follows as soon as we replace the sign for approximate equality with an equality sign:

$$\frac{\partial^2 h}{\partial t^2} = \frac{S}{\mu} \frac{\partial^2 h}{\partial x^2}. \tag{6.11}$$

We have shown that the transverse motion of a strand can be governed by the wave equation. The speed of the wave is easily deduced:

$$v = \sqrt{\frac{S}{\mu}} \, . \tag{6.12}$$

One solution of this equation is:

$$h(x, t) = A \cos(kx - \omega t + \phi)$$

where A is the amplitude, k the wavenumber, ω the angular frequency and ϕ an arbitrary phase angle. In the first instance, all four quantities can be chosen freely, apart from the fact that k and ω must conform to the relation:

$$v = \sqrt{\frac{S}{\mu}} = \frac{\omega}{k} \, .$$

In other words, there are three degrees of freedom in the wave motion, and it is perhaps most common to choose these as amplitude, frequency and phase (phase indicates in practice the choice of zero point for time). The initial conditions determine these, but the boundary conditions too play an enormous role, and they can cause the solution in practice to become a standing wave even if the initial conditions alone indicate something completely different.

Before we leave the wave equation that describes the movement of a string, it may be useful to recall the starting point for our derivation:

- Newton's second law holds.
- The wave is purely transverse.
- The force acting at each end of a segment of the string is tangentially directed (i.e. a purely geometric assumption).
- The angle between the tangent line to the string at any point and the equilibrium line is very small all along the string.
- Only when the angle between the tangent line to the string and the equilibrium line is different at each end of a segment of the string, do we get a net force that performs work on this segment. This corresponds to the fact that there must be a *curvature* on the segment under consideration for it to experience a net force.

Based on these simple assumptions, one is able to infer that a delicate interplay between forces, position and time is responsible for propagating the wave along the string. You are advised to think about what this interaction is in fact. What is actually propelling the wave? What makes the amplitude increase, and what causes it to diminish? It is not only Mona Lisa who conceals something intriguing!

In Chap. 8, we return to the basic requirements for a wave to move. We then base ourselves on numerical methods because these provide extra insight precisely into this context.

6.4.2 Waves in Air/Liquids

Derivation of the wave equation for movement in air/liquids is more complicated than the case considered in the last section. One reason for this is that we now work with a three-dimensional medium. To make the derivation manageable, we limit ourselves to a plane, longitudinal wave, which in effect allows positional changes to be described in terms of only one spatial dimension (plus time). Even so our presentation will be only approximate, but will hopefully reveal the two main mechanisms behind the waves in air and liquids: (1) the mechanical properties of a compressible medium and (2) Newton's second law.

Mechanical properties

In our context, the most important property of air and liquids is that they are relatively *compressible*; that is, it is possible to compress a certain amount of gas or liquid to a smaller volume than it originally had. Air can be compressed relatively more easily than liquids and liquids relatively more easily than solids. (This was why we did not discuss compressibility of the vibrating string in Sect. 6.4.1.] Figure 6.5 illustrates the nomenclature used in the following derivation.

Suppose that a small amount of gas/liquid with volume V expands or compresses to a new volume of $V + dV$ as a response of a change in the pressure from P to $P + dP$. It is assumed that dV and dP may be positive or negative, but their magnitudes are always small relative to V and P, respectively.

Fig. 6.5 A gaseous volume element can be compressed slightly if the external pressure increases. If dP is positive, dV will be negative

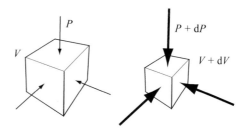

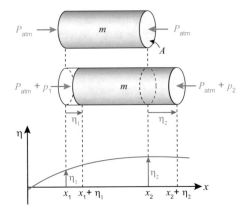

Fig. 6.6 With longitudinal movement of a gas or liquid volume, position, pressure and volume will change, but only in one spatial dimension (here in the x-direction). In the upper part of the figure, the gas volume is at equilibrium, but in the lower part a snapshot of a dynamic situation is given where the same amount of material has moved and changed volume compared to the upper part. Note that p, x and η are functions of both time and space, while the cross section A and the mass of gas or liquid are fixed

The ability of a material to withstand volume changes when the pressure is increased is called the "bulk compressibility module" for the material. It is defined in the following manner:

$$K = -\frac{dP}{dV/V} .$$
(6.13)

The unit of both the bulk compressibility module and pressure is pascal (abbreviated Pa) where 1 pascal = 1 Pa = 1 N/m^2

Let us now apply this point of view to sound waves. Figure 6.6 is based on a situation where pressure changes and movements of a gas volume occur only in one direction, namely the x-direction. For a given value of x, there are no changes in pressure when we move in the y- or z-direction.

We choose to follow the movement of an arbitrary cylinder-shaped portion of the continuous medium and assume that a negligible number of molecules will be exchanged between this cylinder and the surroundings while we consider the system. The cross section of the cylinder will not change in time, but the cylinder will move in the $\pm x$-direction and will change in length (volume) with time.

Figure 6.6 shows our limited volume at equilibrium (no sound) and at an arbitrary time of a dynamic situation. The cross section A in the yz-plane is fixed, and the positions of the bounding surfaces of the cylinder are x_1 and x_2 in the equilibrium and $x_1 + \eta_1$ and $x_2 + \eta_2$ in the dynamic situation.

In our attempt to arrive at a wave equation, we must find relationships between positions, pressures and volumes and use Eq. (6.13) for the model in Fig. 6.6. We choose to write the expression in a slightly different form:

$$dP = -K\frac{dV}{V} .$$ (6.14)

where K is the "bulk compressibility modulus".

We apply the symbols in Fig. 6.6 in Eq. (6.14) and get:

$$\frac{p_1 + p_2}{2} = -K\,\frac{(x_2 + \eta_2 - x_1 - \eta_1)\,A - (x_2 - x_1)\,A}{(x_2 - x_1)\,A} .$$

where the mean value of the pressure changes at the two ends of the cylinder is chosen for the dP term.

We would like to go to the limit where $\Delta x = x_2 - x_1$ goes to zero. This is strictly not permitted given the assumptions about the size of the chosen cylinder of gas or liquid.

We need to make an "acoustic approximation" characterized by the following:

- $\Delta x = x_2 - x_1$ is large relative to the average length of the free paths of air molecules between collisions with other air molecules in their chaotic movement.
- Δx is small relative to the wavelength of the sound waves.
- The displacements η are small compared to Δx, which means that the sound is weak.

The first point ensures that the gas or liquid volume under consideration is reasonably well separated from neighbouring areas. It also ensures that there are a large number of molecules within the selected volume, so that we can disregard individual molecules and treat the contents of the volume as quasi-continuous.

The next point ensures that pressure differences between the end faces are small compared to the pressure variation when the wave passes. The last point just ensures that the displacement of the gas volume is small compared with the length of the gas volume.

Under these approximations, the statement

$$dP = \frac{p_1 + p_2}{2} = p(x, t) .$$

is justified and yields:

$$p = -K\frac{\eta_2 - \eta_1}{x_2 - x_1} .$$

They also justify that the right side of this expression can be approximated to

$$p = -K\frac{\partial \eta}{\partial x} \;.$$

(6.15)

This equation provides a relation between pressure and displacement.

Newton's second law

We will now apply Newton's second law on the system. The mass of the volume element is equal to the mass density ρ_0 multiplied by the equilibrium volume. If the positive x-axis is also defined as the positive direction for F and the acceleration a, one can write:

$$\sum F = ma$$

where we sum over the forces acting on the volume element in the x-direction.

Applied to our cylinder of air or liquid:

$$(P_{Atm} + p_1)A - (P_{Atm} + p_2)A = A(x_2 - x_1)\rho_0 \frac{\partial^2 \eta}{\partial t^2}$$

where the acceleration is the double time derivative of the displacement of the gas volume. Rearranging the terms give:

$$p_1 - p_2 = (x_2 - x_1)\rho_0 \frac{\partial^2 \eta}{\partial t^2}$$

$$\frac{p_2 - p_1}{x_2 - x_1} = -\rho_0 \frac{\partial^2 \eta}{\partial t^2} \;.$$

Again, we would happily have gone to the $\Delta x \to 0$ limit, but with the limitations we have imposed, that is not permissible. However, as long as we adhere to each of the last two points of acoustic approximation, the left-hand side would not change significantly if we made Δx smaller. We get roughly

$$\frac{\partial p}{\partial x} = -\rho_0 \frac{\partial^2 \eta}{\partial t^2} \;.$$

(6.16)

Then, substitution of Eq. (6.15) into Eq. (6.16) and rearranging of terms give:

$$\frac{\partial^2 \eta}{\partial t^2} = \frac{K}{\rho_0}\frac{\partial^2 \eta}{\partial x^2} \;.$$

(6.17)

We have thus arrived at the wave equation for this system as well. We then realize that if small volumes of air (or liquid) molecules are displaced in an oscillatory manner, as indicated in the derivation, the result of the displacement will spread as a wave. The speed of the wave is given by:

$$v = \sqrt{\frac{K}{\rho_0}} \, . \tag{6.18}$$

In other words, the speed of sound increases if the gas/liquid is hard to compress, but decreases with the mass density of the medium through which sound travels.

The expression for the wave velocity bears close resemblances to the comparable expression for a wave on a string (Eq. 6.12). The wave velocity was then $v = \sqrt{S/\mu}$ where S was the force trying to bring the string back to equilibrium. In our case, K is a measure of the force (pressure) trying to bring the volume back to equilibrium. For the string, μ is the mass per length, while in our case ρ_0 is the mass per volume.

It should be borne in mind that the foregoing derivation is based on a number of approaches. If one uses a more rigorous approach, one arrives not at the simple swing equation but at a nonlinear equation that can only be solved numerically. However, for weak sounds and for normal air pressure, the solution of the latter equation would be quite close to that found by using the simpler wave equation.

We have chosen to ignore another aspect of sound waves in air. When a gas expands or contracts, there is also a change in temperature. We implicitly assumed that there has been no exchange of thermal energy from different volume elements in the gas or the liquid through which the sound wave is transmitted—an adiabatic approach. It can be justified for weak sounds, and that is precisely what we have treated above.

It is quite common to use gas laws instead of the definition of compressibility modulus in the derivation of wave equation for sound waves through a gas. Our choice was dictated by the consideration that this chapter should be comprehensible to those without significant knowledge of statistical physics and thermodynamics. In addition, we think that the concept of compressibility is useful for understanding the underlying mechanism of wave propagation in gases and liquids.

Remark: It is interesting to note that the speed of sound in air is lower (but still not very much lower) than the median of the speed of air molecules between intermolecular collisions on account of their chaotic thermal movement. For nitrogen at room temperature and atmospheric pressure, the maximum in the probability distribution of the molecular speed is about 450 m/s. Those interested in the topic can read more about this under "Maxwell–Boltzmann distribution", e.g. in Wikipedia.

6.4.3 Concrete Examples

The calculation we made to deduce the wave equation for movements in air and liquids is quite rough. We started out with Newton's second law, used the validity of what that lies in the definition of the compressibility modulus, plus some other less significant details, and came to the wave equation. Can such an easy description provide a useful estimate of the speed of sound?

Let's try to calculate the sound speed in water. The compressibility modulus for water (at about atmospheric pressure) is given as $K = 2.0 \times 10^9$ Pa. The density of water is $\rho \approx 1.0 \times 10^3$ kg/m^3. If these values are entered into the expression of the sound speed in Eq. (6.18), the result is:

$$v_{water} \approx 1.43 \times 10^3 \text{ m/s}$$

The literature value for sound velocity in water is 1402 m/s at 0 °C and 1482 m/s at 20 °C. In other words, the conformity is actually good!

Let us then try to calculate the sound speed in air. Then a problem arises because the compressibility modulus is usually not given as a general table value, since the value depends on what pressure we consider. Instead, we start with the gas law:

$$PV^\gamma = \text{constant} \qquad (\gamma = C_p/C_v),$$

where C_p is the specific heat capacity at constant pressure, and C_v is the specific heat capacity at constant volume. It is assumed that the changes in volume and pressure take place so that we do not supply energy to the gas (adiabatic conditions). For sound with normal intensity, this requirement is reasonably well satisfied, but not for very loud sound.

A general differentiation of the gas law gives:

$$dP \, V^\gamma + P \, d(V^\gamma) = 0$$

$$V^\gamma dP + \gamma V^{\gamma-1} dV P = 0 \,.$$

Upon combining this with Eq. (6.13), one gets

$$K = -\frac{dP}{\frac{dV}{V}} = \gamma P \,.$$

The ratio of the two specific heats for air is known to be

$$\gamma = \frac{C_p}{C_v} = 1.402 \,.$$

Since a pressure of one atmosphere equals 101,325 Pa, it follows that the value of the bulk modulus for air under atmospheric pressure (under adiabatic conditions) is:

$$K = 1.402 \times 101{,}325 \approx 1.42 \times 10^5 \text{ Pa.}$$

Standard tables show that the mass density of air at atmospheric pressure and about 20 °C is $\rho = 1.293\,\text{kg/m}^3$. With all relevant data at hand, we are able to deduce the speed of sound in air:

$$v_{\text{air}} = 331\,\text{m/s}.$$

The value turns out to be 344 m/s.

Not all the data used above refer to 20 °C and one atmosphere pressure. No wonder, then, that the calculated and experimental values are not in complete agreement. Nevertheless, the calculated value is "only" about 4% too low. It indicates that our calculations and the formula found for the speed of sound in gases/liquids are reasonably good.

Remarks: The tables also provide the data for the bulk modulus for metals, and by using the same formula (derived for gases and liquids), we get values that are close to the tabulated values but the discrepancy is larger than that for air and water. For example, we calculate the speed of sound in steel to be 4510 m/s, whereas the actual value is 5941 m/s. For aluminium, the calculation leads to 5260 m/s, but the experimental value is 6420 m/s.

We should also bear in mind that in metals sound is able to propagate as a transverse wave instead of or in addition to a longitudinal wave, for example when the metal piece is shaped as a rod. The speed of a transverse sound wave in a metal depends on the rigidity of the metal, with the result that transverse waves often have lower speeds than longitudinal waves. If we strike a metal rod, we usually get transverse and longitudinal waves at the same time, and the latter usually have a higher frequency (after the standing waves have developed).

6.4.4 Pressure Waves

In the above derivation, we saw that the effective motion of small volumes of gas or liquid can follow a wave equation. It is interesting to see how much displacement is undergone by the small volumes of fluids when a wave passes, but usually it is more interesting to describe the wave in the form of *pressure changes*. Sound waves are usually detected with a microphone, and the microphone is sensitive to small variations in the pressure. The transition can be carried out as follows.

A possible solution of the wave equation Eq. 6.17 is as follows:

$$\eta(x,t) = \eta_0 \cos(kx - \omega t) \,. \tag{6.19}$$

When switching to pressure waves, we use the definition of the compressibility modulus again, more specifically Eq. (6.15) that was derived earlier:

$$p(x,t) = -K \frac{\partial \eta(x,t)}{\partial x} \,.$$

By combining Eqs. (6.19) and (6.15), one gets:

$$p(x, t) = kK\eta_0 \sin(kx - \omega t) \equiv p_0 \sin(kx - \omega t) . \qquad (6.20)$$

The result shows that wave motion in a compressible medium can be described both as displacements of tiny volumes of the medium or as pressure variations. There is a phase difference between these waves, and a fixed relationship between the amplitudes. If the amplitude of displacement of the tiny volumes (with thicknesses significantly less than the wavelength) is η_0, the amplitude of the pressure wave is $k K \eta_0$.

6.5 Learning Objectives

After working through this chapter, you should be able to:
- Write down the standard wave equation (for a plane wave).
- Explain amplitude, wavenumber, wavelength, period, frequency, phase, wave velocity and the formula $f\lambda = v$.
- Give a mathematical expression for a harmonic plane wave as well as any arbitrarily shaped wave, which moves in a specified direction. For a harmonic plane wave, you should also be able to provide a mathematical description based on Euler's formula.
- Explain how a wave can be visualized either as a function of time or as a function of position.
- Explain the difference between a longitudinal and a transverse wave, and give at least one example of each.
- Derive the wave equation for a transverse vibration on a string.
- Know the main steps in the derivation of the wave equation for a pressure wave through, for example, air (sound wave).
- Calculate approximately the speed of sound in water using material/mechanical properties for water.

6.6 Exercises

Suggested concepts for student active learning activities: Wave velocity, amplitude, wavelength, plane wave, wave equation, transverse, longitudinal, Taylor expansion, compressible medium, compressibility modulus.

Comprehension/discussion questions

1. Present an example of the equation for oscillatory motion and an example of the wave equation. What types of information should we have in order to find a concrete solution of each of these two types of differential equations?
2. Does the velocity of waves as described in Eq. (6.6) depend on the amplitude? Explain the answer.
3. During thunderstorms, we usually see the lightning before we hear the thunder. Explain this. Some believe that we can determine the distance between us and the lightning by counting the number of seconds between our seeing the lightning and hearing the thunder. Can you find the connection?
4. Suppose that a long string hangs from a high ceiling almost down to the floor. Suppose that the string is given a transverse wave motion at the lower end and that the wave then rises to the ceiling. Will the wave speed be constant on the way up to the ceiling? Explain the answer.
5. If you stretch a rubber band and pluck it, you hear a kind of tone with some pitch. Suppose you stretch more and pluck again (have a go at it yourself!). How is the pitch now compared to the previous one? Explain the result. (hint: the length of a vibrating string is equal to half the wavelength of the fundamental tone.)
6. When we discussed sound waves, we said (with a modifying comment) that each air molecule swings back and forth relative to an equilibrium point. This is in a way totally wrong, but still the picture has a certain justification. Explain.
7. The difference between a longitudinal and a transverse wave is linked in the chapter to symmetry. How?
8. Finally, in Sect. 6.4.1, an overview was given of the essential assumptions made in the derivation of the wave equation for motion along a string. Attempt to set up a corresponding list for the derivation of the wave equation in air/water.
9. Our derivation of the wave equation for a pressure wave in a fluid is rather lengthy and full of details. In spite of this, can you actually point out the physical mechanisms that determines the speed of sound in air or water?
10. Discuss sound waves with regard to energy.
11. For surface waves on water: can you determine, if you know the height of the water surface at *one point* on the surface as a function of time, (a) where the wave comes from, (b) wavelength and (c) whether the height (amplitude) is the result of waves from one or more sources? Use your own experience and the photograph in Fig. 6.1.

Problems

12. Check whether the function $y(x, t) = A \sin(x + vt)$ satisfies the wave equation.
13. What characterizes a plane wave? Mention two examples of waves that are not plane and give an example of an (approximate) plane wave.
14. State a mathematical expression for a plane moving in the negative z-direction.
15. Is this a plane wave: $S = A \sin(\mathbf{k} \cdot \mathbf{r} - \omega t)$? Here $\mathbf{k}$ is the wave vector which points in the direction of wave propagation at the point $\mathbf{r}$, and $\mathbf{r}$ is an arbitrarily

chosen position vector, ω is the angular frequency and t the time. A is a real scalar. Justify your answer.

16. Explain in your own words how we can see from the mathematical expressions that a wave $A\cos(kx - \omega t)$ moves towards larger x-values as time passes, while the wave $B\cos(kx + \omega t)$ moves opposite the way.

17. A standing wave can be expressed as $g(x, t) = A\sin(kx)\sin(\omega t)$. Show by direct substitution that a standing wave is also a solution of the wave equation for $v = \omega/k$ (we will return to standing waves in Chap. 7).

18. What is the wavelength of a 100 Hz sound wave in air and in water?

19. When we take ultrasound images of foetuses, hearts, etc., the image quality depends on the wavelength not being more than about 1 mm. Sound waves in water/tissues have a speed of about 1500 m/s. What frequency must the ultrasound have? Is the word "ultrasound" an apt term?

20. How long is the wavelength of FM broadcast at 88.7 MHz? And what wavelength does your mobile phone have if it operates on 900, 1800 or 2100 MHz?

21. A young human ear can hear frequencies in the range of 20–20,000 Hz. What is the wavelength in air at each of these limits? (The speed of sound in air is about 340 m/s.)

22. A 2 m metal string weighing 3×10^{-3} kg is held under tension roughly like a guitar string. Clamped at one end, it is stretched slightly above a table surface and bent over a smooth round peg at the edge of the table (see Fig. 6.7); the other end of the string is attached to a freely hanging object weighing 3 kg, which provides the tension.
(a) Calculate the speed of a transverse wave along the horizontal part of the string.
(b) Would the velocity of the wave change if we change the length of the horizontal part of the string (i.e. how much of the 2-m-long string is located between the clamped point and the round edge)?
(c) How long should the horizontal part of the string be in order that it may vibrate at 280 Hz if you pluck at it? (Hint: Assume that the string is then half a wavelength long.)
(d) How heavy should the bob be in order to make the frequency twice than that in the previous task (assuming that the length does not change)?

23. Write a program in Matlab or Python that samples the sound signal reaching the microphone input of a PC when a microphone is connected and plot the signal with the correct timing along the x-axis. You may use program snippet 2 in the end of Chap. 5 for part of the program.
Sample the sound as you sing a deep "*aaaaaa*". Is the sound wave harmonic? What is its frequency?

24. In Fig. 6.8, there is a wave along a string (a small section) at three neighbouring times. Base your answer on the figure and explain:

- What is the direction of the net force acting on each of the eight segments of the string at time t (ignore gravity).

Fig. 6.7 Experimental setup
in the following problem

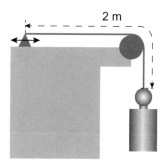

Fig. 6.8 A wave along a
piece of a string at three
neighbouring instants

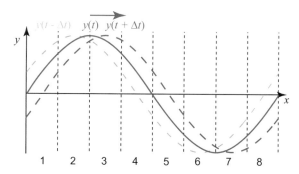

- Explain in detail your arguments for finding the force, especially for segments 2, 4, 5, 6 and 7.
- What is the direction of the velocity of each of these segments at time t?
- At first, it may seem that there is a conflict between force and velocity. Explain the apparent conflict.
- The last point is related to the difference between Aristotle's physics and Newton's physics. Do you know the difference?
- How does the energy vary for an element along the string when the wave passes by?
- Elaborate on the expression "the wave brings with it the energy".

25. Make an animation of the wave $A \sin(kx - \omega t)$ in Matlab or Python. Choose yourself values for A, k, ω, and the ranges for x and t. Once you have got this animation working, try to animate the wave $A \sin(kx - \omega t) + A \sin(kx + \omega t)$. Describe the result.

   You may use program snippet 3 in the end of Chap. 5 for part of the program (also available at the "Supplementary material" web page for this book is available at http://www.physics.uio.no/pow).

26. Read the comment article "What is a wave?" By John A. Scales and Roel Snieder in Nature vol. 401, 21 October 1999 page 739–740. How do these authors define a wave?

Chapter 7
Sound

Abstract The prime theme of this chapter is reflection of (sound) waves at interfaces between two media with different wave velocities (or different acoustical impedances). Such reflections are used in, apart from other situations, ultrasound imaging in medicine, e.g. of foetuses during pregnancy. If a wave moves in an extended medium with reflective boundaries at both ends, a wave of arbitrary shape will go back and forth repeatedly with a fixed time period determined by the wave velocity and the distance between the reflecting ends. We argue that this lies at the core of musical instruments, and not pure standing waves, used as the paradigm in most physics textbooks. We then present the tone scale and go on to define sound intensity, both physically and in relation to human hearing. The chapter ends with a discussion of beats, Doppler shifts and shock waves.

7.1 Reflection of Waves

Sound waves are reflected by a concrete wall, and light waves by a mirror, whereas the waves on a guitar string are reflected at the ends where the string is clamped. Reflection of waves under different circumstances is a topic that we would encounter time and again in this book. Mathematically, reflections are treated by the introduction of so-called boundary conditions. As mentioned earlier, the same differential equation for wave motion can arise in various contexts, yet the solutions differ markedly because the boundary conditions are not identical. The first and perhaps the simplest illustration of this is wave motion along a string of finite length, where physical conditions at the "boundaries" (the ends of the string) play a decisive role in the wave motion.

Suppose that we have a taut string, one end of which is attached to a large mass, and that we produce a transverse "pulse" by providing a sharp blow to the other end (see Fig. 7.1). The pulse will move along the string at the rate $\sqrt{S/\mu}$, where S is the tension and μ the mass per unit length. The shape of the pulse is preserved.

When the pulse reaches the clamped end of the string, the amplitude at this end must necessarily vanish. This means that the pulse close to this end will be compressed and the force across the string will increase significantly. Since the endpoint

© Springer Nature Switzerland AG 2018

A. I. Vistnes, *Physics of Oscillations and Waves*, Undergraduate Texts in Physics,

https://doi.org/10.1007/978-3-319-72314-3_7

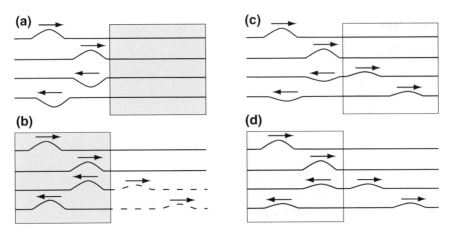

Fig. 7.1 A transverse wave reaches an interface between two media. The wave is drawn for four successive instants. In **a**, the wave goes from a region of low impedance (white) to one of much larger impedance (yellow). The wave is fully reflected and the result gets the opposite sign of the incoming. In **b**, the impedance at the interface decreases (the wave comes from a high impedance region and meets a low impedance region). Here, reflection of energy is near total, but if we consider only amplitudes, the effect is more modest (indicated by the dotted line). Panels **c** and **d** illustrate a case where the amplitudes of the reflected and transmitted wave are equal

cannot move, the compressed string experiences a force in the opposite direction, which creates an imbalance between the amplitude and the travel speed, compelling the pulse to turn back along the string. However, the wave that travels backwards will have an amplitude opposite to that of the original (incoming) pulse (case **a** in Fig. 7.1). No energy is lost (to a first approximation) since loss on account of friction requires the frictional force to work over a certain distance, while we have assumed that the end point is completely fixed.

Another extreme is that where the end is free to move. This can be achieved, for example by holding the string at one end and allowing it to fall freely downwards and let the end move freely in air (disregarding air resistance). However, this is not a good model, since the tension in the string is not defined. A much better model is a string of large linear mass density (mass per unit length) connected at the free end to a string of significantly smaller linear mass density, and subject the entire structure to a fairly well-defined tensile force. It will be convenient to call the former a thick and the latter a thin string.

A pulse transmitted along the thick string will move normally until it reaches the boundary between the two strings. The disturbance that reaches the thin string will give it a significantly greater impact than if the string were of a uniform density. There is again a mismatch between amplitude and velocity, resulting in reflection, but the result in this case is a reflected pulse with same amplitude as the original pulse. In this case, however, some of the wave (and energy) will also propagate along the thin string. If the thin part has a significantly smaller density, almost all energy will be reflected (case **b** in Fig. 7.1).

The terms "a massive structure" and "a thinner or thicker string" (signifying linear mass density) are not sufficiently precise word, and it is better, when one is discussing production and transmission of sound, to use the term "acoustic impedance", defined below:

Acoustic impedance is defined as acoustic pressure (sound pressure) divided by acoustic volume flow rate (details in the next subsection).

Meanwhile, we will content ourselves with qualitative descriptions, but will continue to employ the term "impedance", even though our understanding of acoustic impedance is still vague. Based on this understanding, the rules for reflection and transmission of waves at an interface can be enunciated as follows:

It can be shown both experimentally and theoretically that:
- Waves that strike an interface beyond which the impedance of the medium *increases*, split so that the reflected part is of the opposite sign to that of the incident wave. The transmitted wave has the same sign as that of the incoming wave.
- Waves that strike an interface beyond which the impedance of the medium *decreases*, split so that the reflected part is of the same sign as that of the incident wave. The amplitude of the transmitted wave also has the same sign as that of the incident wave.
- The fraction that is reflected or transmitted depends on the relative impedance change in relation to the impedance of the medium the wave originates from. If there is no impedance change, nothing is reflected; if the relative impedance change is infinitely large, all energy is reflected.

In Fig. 7.1, the waveform at the instant the wave strikes the interface is not shown on purpose, but a detailed profile can be constructed using the method outlined in Fig. 7.2. In the figure, total reflection is illustrated. Correct waveform before the interface is drawn, and we let an upward pulse approach the interface. A downward virtual pulse is also made to travel towards the interface with the same velocity. The virtual pulse has an equal and opposite amplitude to that of the incoming pulse if total reflection occurs against a medium with very high impedance (e.g. by attaching a string to a massive structure). The shape of the actual wave during and after reflection is found by adding the waveforms of the original and virtual pulses (the resultant is indicated by the thick dotted line in the figure). Eventually, only the virtual wave survives in the region to the left of the interface, and further wave evolution follows the motion of the virtual pulse alone.

This model can be easily be modified to deal with the case in which the wave approaches a medium with much lower impedance, which leads to almost total

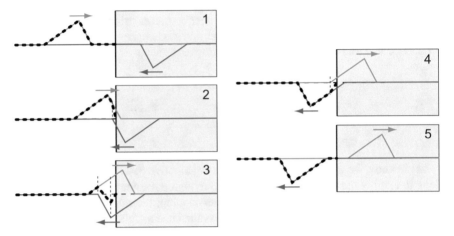

Fig. 7.2 A model for portraying the time development of the waveform of the reflected part of a transverse wave reflected at an interface between two media. See the text for details

reflection with no change in the sign; it can also be modified to handle cases where a part of the wave is reflected and a part transmitted.

We will return to a more detailed description of reflection and transmission of electromagnetic waves when they meet an interface between two media.

7.1.1 Acoustic Impedance *

We will in this chapter speak about "acoustic impedance" rather loosely. However, for the sake of those who wish to acquire greater familiarity with acoustic impedance, this subsection provides a slightly more detailed description. Go ahead to the next sub-chapter if you are not interested in spending more time on this topic at this point.

The notion of acoustic impedance arose when we discussed reflection of waves at the interface of two media. Let us delve a little deeper into this issue. There are several variants of acoustic impedance.

"Characteristic acoustic impedance" Z_0 is defined as:

$$Z_0 = \rho c \tag{7.1}$$

where ρ is the mass density of the medium (kg/m^3), and c is the speed (m/s) of *sound* in this medium. Z_0 depends on the material and its units are Ns/m^3 or Pa s/m.

The characteristic impedance of air at room temperature is about 413 Pa s/m. For water, it is about 1.45×10^6 Pa s/m, i.e. about 3500 times larger than the characteristic impedance of air.

Differences in characteristic acoustic impedance determine what fraction of a wave is transmitted and what fraction is reflected when a "plane wave" reaches a plane interface between two media.

The big difference in characteristic acoustic impedance between air and water means that sound in the air will be transmitted into water only to a small extent, and sound in water will penetrate into air only to small extent. Most of the sound will be reflected at the interface between air and water.

In Chap. 6, we found that the sound speed in air or water was given (using c instead of v) as:

$$c = \sqrt{K/\rho}$$

where K is the modulus of compressibility and ρ is the mass density. Upon eliminating ρ by using the definition of characteristic impedance in Eq. (7.1), we get:

$$Z_0 = K/c . \tag{7.2}$$

This expression gives us another idea of what influences the *characteristic* acoustic impedance. For a particular system, e.g. a musical instrument, an different measure is often used:

"Acoustic impedance" Z is defined as:

$$Z = \frac{p}{vS} \tag{7.3}$$

where p is the sound pressure, v is the particle speed (over and above the contribution of thermal movements) and S is the pertinent cross-sectional area (e.g. the mouthpiece of a trumpet).

There is a close analogy between acoustic impedance and impedance in electromagnetism. For this reason, the definition of acoustic impedance is often compared with Ohm's law, and Z is sometimes called "sound resistance" or "audio impedance".

If you wish to learn more about acoustic impedance, the following article might be of interest: "What is acoustic impedance and why is it important?" available on: http://www.phys.unsw.edu.au/jw/z.html (accessed May 2018).

7.1.2 Ultrasonic Images

Characteristic acoustic impedance will change with mass density and the modulus of compressibility [see Eqs. (7.1) and (7.2)]. Precise correspondence is not so easy to obtain from these equations since the speed of sound also depends on the same quantities.

Nevertheless, there are differences in the characteristic acoustic impedance of, e.g. blood and heart muscle. The characteristic acoustic impedance of a foetus and foetal fluid are different. Therefore, if we send sound waves to the body, some of

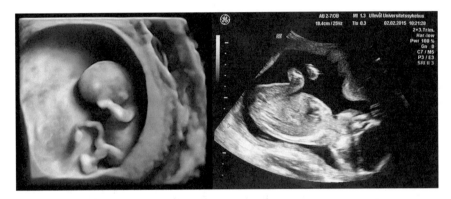

Fig. 7.3 Ultrasound images of two foetuses. On the left is a 3D image of a foetus about 11 weeks old. On the right is a sectional image (2D) of a foetus about 18 weeks old. Reproduced with permission from the owners (private ownership)

the sounds will be reflected from the interfaces between blood and heart muscle, and between the placenta and the foetus.

However, there is a huge difference between the characteristic acoustic impedance of air and body. In order to get sound efficiently in and out of the body during an ultrasound examination, a gel is applied on the skin, which reduces friction and acts as a conductor of the ultrasonic waves from the ultrasound probe. This material should have approximately the same characteristic acoustic impedance as the tissue the sound is going to enter.

After reflection at interfaces between different impedances, the sound will be captured as an echo, provided that the original sound pulse has already ceased before the echo returns. By analyzing the echo as a function of time delay, we will be able to determine distances. And if we can send sound in well-defined directions, we will also be able to form images of what is inside the body. Figure 7.3 shows a pair of ultrasonic images of a foetus.

Much interesting physics goes into the design of the sound probe in ultrasound surveys. We can control the beam in two directions by causing interference between many independent transmitters on the surface of the sound probe. Control of the sound beam is achieved by systematically changing the phase of the sound for each single transducer on the ultrasound probe. Focusing for the sake of reducing diffraction can also be done by similar tricks. We will return to this in later chapters.

It should be added that there are major similarities between ultrasound surveys, for example, of foetuses and mapping of the seabed for oil exploration. In the latter case, a number of sounders (and microphones) are used along a long cable towed along the seabed. Echo from different geological layers in the ground with different acoustic impedances is the starting point for finding out where to expect oil and where there is no oil and how deep the oil lies.

Many physicists in this country, educated at NTNU, UofO or other institutions, have helped develop ultrasound and seismic equipment. The Norwegian company

Vingmed has been a world leader in developing ultrasound diagnostics equipment. Vingmed has now been purchased by General Electric, but Norwegian scientists trained in physics and/or informatics still play an important role in the development. Similarly, we have taken an active part in seismic surveys as well. A great deal of interesting physics lies behind these methods, and these principles are sure to find other applications in the years to come. Perhaps *you* will become one of the future inventors by exploiting these ideas?

7.2 Standing Waves, Musical Instruments, Tones

7.2.1 Standing Waves

When a persistent wave travels along a taut string that is firmly attached to a massive object at one end, the wave will be reflected from the endpoint and travel backwards along the string with an amplitude opposite to that of the incoming wave. If there is negligible loss, the incident and (an equally strong) reflected wave will add to each other (superposition principle). Let the incoming wave be a harmonic wave described in the following form:

$$y(x, t) = A \cos(\omega t + kx)$$

for $x \geq 0$. That is to say, the wave comes from "the right" (large x) and is *moving toward the origin*. The string is tied to a massive object at the origin, which gives rise to a reflected wave that can be described by the equation:

$$y_r(x, t) = -A \cos(\omega t - kx) .$$

We have chosen to describe the waves in a somewhat unusual way to ensure that the amplitude at the origin is exactly the same for incoming as reflected wave, but with the opposite sign. The two contributions will then cancel each other exactly at the origin.

The superposition principle allows us to express the resultant of the incoming and reflected waves as follows:

$$y_{sum} = A \cos(\omega t + kx) + \left[- A \cos(\omega t - kx) \right]$$

for $x \geq 0$.

We have the trigonometric identity

$$\cos a - \cos b = -2 \sin \left(\frac{a + b}{2} \right) \sin \left(\frac{a - b}{2} \right) .$$

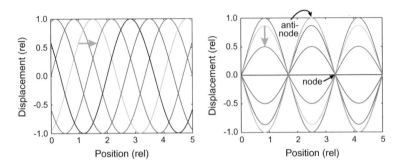

Fig. 7.4 A travelling (*left*) and a standing wave (*right*) as a function of position at various times. The green arrow shows how the wave changes from one point of time (*green curve*) to a subsequent time (*blue curve*). Pay particular attention to how we use the words "anti-node" and "node" in the standing wave

By using the above identity for our sum of an incoming and a totally reflected wave on a string, we find:

$$y_{\text{sum}} = -2A \sin(kx) \sin(\omega t) . \tag{7.4}$$

In this expression, we have taken account of the fact that the reflection occurs against a medium with greater impedance, so that the wave is reflected with the opposite sign.

The important point about Eq. (7.4) is that *the coupling between position and time is broken*. Maximum amplitude in a given position is achieved at *times* for which $\sin(\omega t) = \pm 1$, and these times have nothing to do with position. Similarly, the *positions* where the maximum amplitude occurs is determined solely by the term $\sin(kx)$, which does not change with time. These characteristic features are displayed in Fig. 7.4.

Remark: In the foregoing account, we have assumed that the incoming wave is harmonic, but beyond this we have *not* imposed any requirements on the three main parameters needed to describe a wave: amplitude, phase and frequency. Regardless of the values chosen for the three parameters, standing waves will result after a total reflection as described above, but *this holds only for a pure harmonic wave*!

Standing waves are an important phenomenon when one is dealing with harmonic waves; they can arise with sound waves, water waves, radio waves, microwaves and light—indeed, for all approximately harmonic waves!

The addition of several harmonic waves which do not all have the same phase cannot give rise to standing waves with fixed nodes and anti-nodes like those shown in the right part of Fig. 7.4.

7.2.2 Quantized Waves

Identical endpoints

Suppose that a string is clamped at *both ends* and that we manage to create a wave pulse similar to the one in Fig. 7.1. The wave pulse will be reflected each time it reaches an end, and the wave will travel to and fro indefinitely, provided that there is no energy loss. The same sequence repeats over and over again with a time period T, which is the time taken by the wave to go back and forth once.

The time period equals the total distance back and forth divided by the velocity of the wave; that is:

$$T = \frac{2L}{v}$$

where L is the distance between the two identical ends of the string. The frequency of the periodic movement comes out to be $f = 1/T$, or

$$f = \frac{v}{2L} . \tag{7.5}$$

If we use the general relationship $\lambda f = v$ for a wave, we can assign a kind of wavelength λ to the wave along the string:

$$\lambda = 2L .$$

This will in general not be a harmonic wave.

It is somewhat artificial to speak of the wavelength *inside* the instrument, but it becomes meaningful as soon as we consider the sound created by the instrument in the surrounding air.

Note that these relationships are generally applicable and are *not* limited only to harmonic waves!

A wave moving along a string clamped at *both* ends will have a "quantized" time course with a period given by the time the wave needs to travel back and forth along the string once.

The same regularity will hold also, to take another example, for an air column inside a flute (where there is low impedance at both ends).

Figure 7.5 attempts to highlight reflection of waves for two equal endpoints. The figure shows the wave pulse as a function of position at different instants of time (*left*). After an interval equal to the time period, we are back to the same situation as

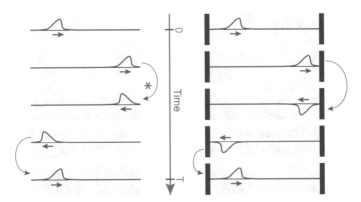

Fig. 7.5 A wave pulse with an arbitrary amplitude travels without loss between two reflecting ends. The wave pulse may be a pressure pulse in air (to the *left*) or a mechanical transverse wave pulse on a string (to the *right*). There are no change in sign of the wave pulse in the first case, but the sign is changed for the latter kind of reflection. See explanations in the text

we had at the beginning of the interval. In the case a flute, a (weak) sound wave will emerge from the flute every time the internal wave reaches the open end (marked with an asterisk in the figure). The player then has to add to the wave pulse at correct time once every period to compensate for the loss to the surroundings. The time period of emitted sound will be the same as the time taken by the wave peak to make one round trip inside the flute.

One should note that it is completely possible to admit more wave peaks within the fundamental period of time we have considered so far. Figure 7.6 attempts to specify a hypothetical case where there are three identical wave peaks evenly distributed over the fundamental time period. The frequency of the sound emerging from a flute sustaining such a wave will be three times the fundamental frequency. It is important to remember that the to and fro movement of a wave does not affect the movement of other waves, even though the total amplitude is a sum of the several independent contributions (assuming that a linear wave equation describes the movement).

Nonidentical endpoints

For a wave moving in a musical instrument where one end of an air column has a high impedance and the other a low impedance, the conditions are different than when the impedance is the same at both ends. An example is an organ pipe sealed at one end, the other end being open to ambient air. In such a case, a wave reflected from the low impedance end will continue in the opposite direction with unchanged amplitude, while the wave amplitude will change sign when reflection occurs at the high impedance end.

In such a situation, the wave as a whole will experience a sign reversal by travelling up and down the pipe once. If the wave makes a second round trip, its sign will change again. This means that a wave must make two journeys back and forth twice (cover a distance $4L$) for it to repeat itself. Figure 7.7 provides an illustration of this principle.

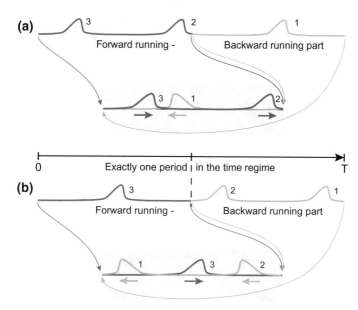

Fig. 7.6 A wave that travels back and forth between two identical ends a distance L apart will have a fundamental frequency $v/2L$. However, it is possible to add more than one wave peak to the fundamental wave. In this figure, three equivalent peaks equally spaced within the $2L$ distance are depicted at one instant of time (**a**) and at a slightly later time (**b**). Wave peaks travelling to the right are indicated by red, and peaks travelling to the left by blue. The resulting wave is always the sum of these two contributions. In this case, the sound will have a frequency three times the fundamental

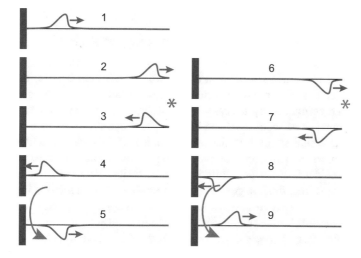

Fig. 7.7 A wave peak must travel twice back and forth in order to regain its initial amplitude when one end of an air string has a high acoustic impedance and the other end low. The massive blocks drawn to the left indicate high acoustic impedance. At this end, the wave undergoes a sign reversal upon reflection

The period for a wave that is reflected at two unlike ends is

$$T = \frac{4L}{v} .$$

The corresponding frequency is

$$f = \frac{v}{4L} . \tag{7.6}$$

The wave can also have any shape whatever; it is only the duration of the period that counts in this context.

As in the case when both ends have the same impedance, we can have an integral multiple of the fundamental frequency, but with one notable exception. We cannot have an even number of identical sequences during the fundamental period of time (you can verify this yourself by making a drawing similar to Fig. 7.7). We can therefore have only an odd multiple of the fundamental frequency given in Eq. (7.6). We usually write this in the form:

$$f = \frac{(2n - 1)v}{4L} \tag{7.7}$$

where $n = 1, 2, 3, \ldots$

7.2.3   Musical Instruments and Frequency Spectra

Some musical instruments, such as a drum, provide transient sounds, while other instruments emit more or less persistent "tones". A tone can be characterized as deep/dark or high/light. The pitch height depends on the frequency of the fundamental tone. The sound of an instrument can be "sampled" and displayed as a time series (a plot of the signal strength in the time domain). The frequency content can be determined experimentally, for example, by Fourier transformation of the time series.

Pure sinusoidal form occurs rarely in the time series of sounds from real instruments. Why is it so difficult to generate harmonic waves from a musical instrument?

It becomes easy to understand that the waveform is not harmonious when we look into the mechanism for the production of sound in a musical instrument. When we pluck a guitar string, it becomes obvious that we are unable to produce a perfect sinusoidal wave. The deviation from a sinusoidal shape will depend on where the string was plucked. This can be easily seen from a Fourier analysis of the sound, since the intensity distribution among the different harmonics depends on where the string is plucked.

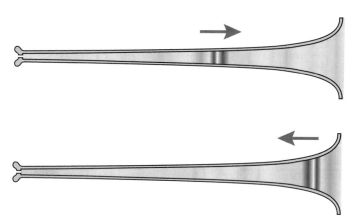

Fig. 7.8 In a musical instrument, a wave goes back and forth with the speed of sound in air and is reflected at each end of the instrument. If we analyse the sound signal with Fourier transformation, we can get many harmonics in addition to the fundamental tone. The harmonics are *not* independent of each other, and their existence only means that the pressure wave is not harmonious

We know that when someone plays the trumpet, the air passes through the tightened lips of the player in small puffs, and it is obvious that these puffs will not lead to sinusoidal variations for the resultant pressure waves (illustrated in Fig. 7.8). In a clarinet or oboe or a transverse flute, we create air currents and vibrations where turbulence plays an important role. The air eddies are nonlinear phenomena and will not lead to sinusoidal timescales for the pressure waves. It is therefore quite natural that the pressure waves in the instrument do not become harmonic. Nonharmonic waves inevitably lead to more harmonics in the frequency spectrum, something already pointed out in the chapter on Fourier transformation. There is no mystery about it.

Nonlinear effects are present in virtually all musical instruments. For string instruments, the vibration and rotation of the string affect in return the contact between the (violin) bow and the string. This results in continuous small changes in the vibration pattern, even though the salient features last long. It is the nonlinearity that gives life to the sound of the instrument and makes it difficult to generate synthetic sound that is as lively as that which emanates from musical instruments.

When the sound waves in the instrument are almost periodic but do not have sinusoidal time periods, the frequency spectrum will consist of several discrete peaks separated by the fundamental tone frequency. How should we determine the tone of the sound? It is the fundamental tone that determines the pitch we perceive with our hearing.

Curiously enough, it is possible that a frequency spectrum may lack the peak corresponding to the fundamental tone and still our ear will perceive the pitch of the fundamental tone. Figure 7.9 shows a small segment of a time signal from a tuba, an instrument that plays low-frequency tones. The time display shows a periodic signal, but a waveform that is far from a pure sine. The frequency spectrum shows a number of peaks, and it is clear that the peaks have frequencies that are integer multiples of

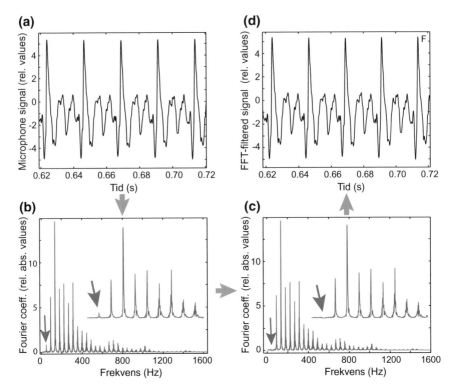

Fig. 7.9 It may happen that the intensity of the fundamental tone is much less than any of the harmonics. In such cases, we can remove the fundamental tone completely without the time signal changing noticeably. Also when we listen to the sound, the pitch will be determined by the periodicity of the time signal rather than the frequency of the harmonics

the fundamental frequency. However, the intensity of the fundamental frequency is quite small.

For the sake of a little amusement, the fundamental tone was completely removed from the frequency spectrum[1] and an inverse Fourier transform was calculated over the entire frequency domain. The result was a time profile visually indistinguishable from the original time signal of the tuba sound (see Fig. 7.9). If we listen to the filtered signal, we do not hear any difference either (at least not easily).

Let us use Fig. 7.9 to point out an important message regarding our perception of the pitch of a tone. *The fundamental tone is found by requiring all peaks in the frequency spectrum to have frequencies equal to an integral multiple of the fundamental frequency. The fundamental frequency does not have to be present.*

---

[1] Both from the positive and the negative half of the frequency domain, due to folding in Fourier transform.

There can also be "holes" in the series of harmonics (e.g. only third, fifth and seventh harmonics exist). Nevertheless, only one frequency satisfies the requirement that the frequency spectrum consists of components that are harmonics of the fundamental.

7.2.4 Wind Instruments

When we pluck on a guitar string, the resulting wave will travel back and forth along the string, and the total movement (at each time and each instant) will be the sum of the forward and reverse wave.[2] However, the energy imparted to the string by the act of plucking eventually changes to sound that disappears in the surroundings and it also heats the string, since it bends a bit here and there and is not fully elastic. The oscillations of the string will die out in a matter of seconds, which is several hundred times longer than the time the wave needs to make one round trip along the string.

A wind instrument (such as a flute, trumpet, clarinet, oboe) is a little different from a guitar string. With such an instrument, a musician can keep a steady volume of the sound for a long time—until he/she has to pause for breath. For wind instruments, therefore, we have a (quasi)-steady-state excitation of the instrument as long as we keep blowing air into it.

In a trumpet, 40–50% of the energy in a wave disappears when the wave reaches the funnel-like opening of the instrument. This means that only 50–60% of the sound energy of the wave is reflected, and the musician must supplement the reflected wave to uphold a steady state situation.

The pace at which the musician blows air must have proper timing relative to the reflected waves, in order to get good sound intensity. This may seem like a difficult task, but sound waves reflected from the end of the instrument back to the mouthpiece of a brass wind instrument impress on the musician's lips, making it easy to provide new puffs at the right time. Finesse is achieved by tightening and shaping the lips and how forcefully the musician squeezes air through the lips.

For a flute, the reflected wave will affect the formation of new air eddies, which ensures proper timing also for such instruments.

There is some leeway with respect to timing (a slightly higher or lower frequency of air blows than that corresponding to the wave speed and the length of the instrument), but too great a departure will not lead to a sonorous success, because new air blows will not work in unison with the reflected waves.

---

[2]There are details to the movement of a wave on a guitar string not mentioned here. These are easier to understand when we use numerical methods to calculate wave movements in Chap. 8.

7.2.5 Breach with Tradition

In this subsection, we will discuss a traditional way of presenting the physics behind different instruments. The purpose is to show that such an approach can easily lead to misunderstandings.

Figure 7.10 shows a traditional representation of what characterizes the physics behind an organ pipe. The organ pipe is chosen as a concrete example. The same conditions apply to all instruments with corresponding impedance termination at the end of an air column. The illustration focuses on the notion of standing waves, as described in Eq. (7.4). The wavelength is determined by the requirement that there be either an anti-node or a node in the standing waves at the end of the air column inside the instrument ("anti-node" corresponds to maximum and "node" to zero amplitude). We must distinguish between the pressure and the displacement, since we know that in a harmonic sound wave there is a phase difference of 90° between the two.

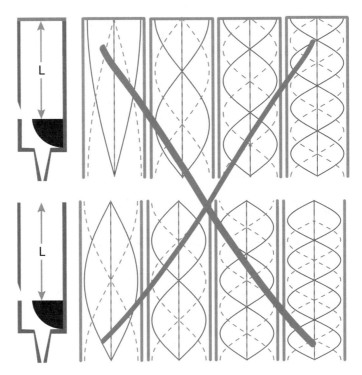

Fig. 7.10 Ostensible displacement amplitude (dashed blue curve) and pressure amplitude (solid red curve) for sound waves in an open and closed organ pipe (or instrument with corresponding acoustic endpoints). This is a standard illustration found in most textbooks in this field. However, the figure is liable to misunderstanding, which is why it has been marked with a large cross. See the text for details

For an instrument that is closed at one end and opens in the other, there will be an anti-node for pressure and a node for displacement at the closed end. The converse holds for open ends, that is, a node for the pressure and anti-node the displacement.

The conventional account is correct, provided that there is a pure harmonic wave in the instrument (only one frequency component), which provides a perfect standing wave. The problem is that the proviso is seldom met in practice!

This is compensated by drawing standing waves also for the higher harmonics, and the figure shows how these waves appear in addition to the wave of the fundamental frequency. One is given the impression that one need only add the separate contributions to get the correct result.

However, the recipe cannot work. There are phase differences between the harmonic frequency components in a Fourier analysis of the sound. These phase differences are vital for reproducing the original time profile of the sound. The phase differences are conspicuously *absent* in Fig. 7.10.

The phase difference means that there will be no standing wave inside the instrument! It becomes meaningless to talk about anti-nodes and nodes inside the instrument. Application of these terms at the endpoints does have a certain justification. However, in our explanatory model, it is more natural to associate this with the rules for reflection of waves.

An the open ends, the air molecules move more easily than inside the pipe. Impedance outside the pipe being lower than that inside, we demand that waves reflected at such an interface do not change sign upon reflection. This means that there is maximum movement of the air molecules at ends that are open.

Similarly, air molecules will find it difficult to move against a massive wall, for example, at the close ends of a closed organ pipe. Accordingly, waves reflected at the closed end will have a sign opposite to that of the incoming wave, with the result that the displacement of the molecules at the boundary becomes zero.

For the pressure wave, the argument is reversed.

We are led to the same conclusion, but for the end faces only, whether we base our argument on reflection of waves or on standing waves for the fundamental frequency, but there is disagreement everywhere else.

Animation

It may be instructive to see how a wave evolves inside a wind instrument. We can make an animation in a simple way and the procedure is shown in Fig. 7.11. We have chosen an animation based on a wave travelling back and forth inside the instrument with negligible loss (but still sufficient to permit detection of the sound emitted by the instrument). We have also chosen a situation where both ends have the same impedance, and the impedance is lower at the ends than inside the instrument, so that a wave is reflected without a sign reversal.

We have chosen a real audio signal from an instrument (French Horn) and picked exactly one period of the signal. The starting point and endpoint are arbitrary, and the signal has been divided so that one half indicates the wave amplitude at different positions for the part of the wave moving toward the opening (where some of the

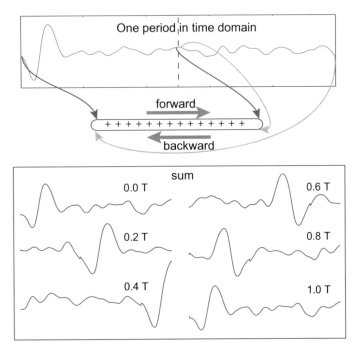

Fig. 7.11 Total sound pressure at different positions within an instrument can be found by summing the sound pressure of the forward and backward waves. One whole period of the sound output from the instrument must be divided equally between the forward and backward waves, as shown at the bottom. By moving forward the waveform (cyclically), keeping pace with the wave, we create an animation of total sound pressure vs position with the passage of time. We see that it is meaningless to talk of standing waves in a case where the frequency spectrum has many harmonics (the time profile is very different from a that of pure sinusoid)

sounds are released). The other half represents the reflected wave. The part that has just been reflected is very close to the opening of the instrument. The part that was reflected half a period earlier has travelled all the way back to the mouthpiece of the instrument.

The wave inside the instrument can be found by adding the forward wave and the backward wave at each point.

Animation is achieved by cyclic stepwise movement of the waveform, each step or frame representing a later instant. The last point of the forward wave becomes, as we move to the next instant, the first point in the backward, while the last point in the backward wave becomes the first point in the forward wave.

Figure 7.11 shows some examples of how the wave looks at six instants (separated by a fifth of the period). We can follow the dominant wave peak and see that it first travels towards the open end of the instrument, but is reflected and moves away from this end during the next half of a period, at the end of which a new reflection occurs.

The animation is meant to show that there is really nothing here that would evoke standing waves, as described by Eq. (7.4) and in the left part of Fig. 7.4. The model based on a wave that travels back and forth provides by far the more faithful description of the state of affairs.

The upshot of the foregoing discussion is that it is more appropriate, when one is describing the process underlying the production of sound in a musical instrument, to speak of a "trapped moving wave" than of a "standing wave". Joe Wolfe at The University of New South Wales, Australia, focuses on trapped moving waves in his outstanding Web pages about audio and musical instruments (see references at the end of the chapter).

Concrete examples of quantization

It may be useful to look at some concrete examples of quantization (or lack of quantization) of frequencies from different "musical instruments".

For a 94-cm-long brass tube (internal diameter about 15 mm) two series of measurements were made. In the first, one end of the tube was placed just next to a speaker where a pure tone was played with a tunable frequency. At the other end of the tube, a small microphone was placed for monitoring the signal strength. When the frequency was varied from about 150 to about 1400 Hz, resonances (sound intensity at the location of the microphone) were observed at frequencies of approximately 181, 361, 538, 722, 903, 1085 and 1270 Hz. This corresponds to nf_1 ($n = 1, 2, \ldots, 7$), with f_1 calculated from a tube open at both ends [Eq. (7.5)].

When we used the tube as a makeshift trumpet, we could turn (by tightening our lips more from one variant to the next) generating sound with frequencies (ground tone) of about 269, 452, 622, 830 and 932 Hz, that is, to say completely different frequencies than the resonant frequencies at both ends open! The frequencies here correspond approximately to $\frac{1}{2}nf_1$ ($n = 3, 5, 7$), with f_1 pertaining to a tube open at both ends [Eq. (7.5)]. This is in perfect accord with the frequencies predicted by Eq. (7.7) for an instrument closed at one end and opened at the other.

For a trumpet, the situation is a little different. The trunk of the tube in a trumpet results in nonlinear effects because the effective length of the tube is slightly different for different frequency sounds. The tract also causes the sound to come into the surroundings in a more efficient manner than with instruments made hundreds of years ago. The mouthpiece also has complicated acoustic features, but we will not go into these details here.

In Fig. 7.12 is shown an example of a continuous sound from a trumpet, both considered in the time domain and the frequency domain. In this case, the fundamental tone and higher harmonics are present at the same time, and the amplitude ratio between them appears in the frequency domain (often called "frequency spectrum" or "Fourier spectrum").

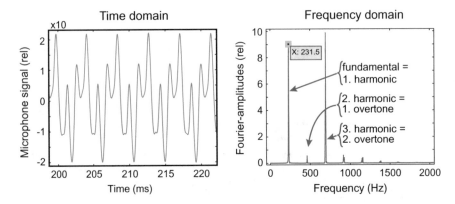

Fig. 7.12 Example of time frame and frequency picture of the sound of a B trumpet playing a "C" note (which is really a B, see next sub-chapter). It is obvious that the time signal is not a pure sinusoid, but a mixture of several frequencies, as revealed by the frequency spectrum. Note that the fundamental frequency is a part of the harmonic range, while the fundamental frequency is not counted in the numbering of so-called overharmonics

Note the asymmetry in the time frame of the sound from the trumpet. The maximum peak is found once again to be as large as the negative peak half a period later. Similar to the second largest peak. This corresponds well with the picture that a wave peak undergoes a sign change after one round trip, but the wave peak returns to the original after two round trips.

It is this asymmetry of the signal itself that causes integer harmonics to almost disappear in the Fourier analysis of the sound, as we see in the right part of Fig. 7.12.

Examples of nonquantization

The importance of reflection of waves and wave velocity within the instrument for obtaining a given (quantized) frequency can be grasped by referring to Fig. 7.13. Here, we have sampled the sound of a mouthpiece from a trumpet (removed from the trumpet itself) while the musician has changed the tightening of the lips slightly up and down. The time signal is analysed by a form of time-resolved Fourier transformation (wavelet analysis, which we will return to later in the book). In such a diagram, peaks in the frequency spectrum are shown as a function of time. We can see that the pitch of the fundamental tone here can be varied continuously. There is no quantization, because there is no reflection of the sound waves of some importance. Frequency is determined exclusively by the rate of air blows through the lips, and here there is no physical process that could impose quantization on frequency.

The harmonics also appear in the wavelet diagram, but since we use a logarithmic frequency scale (*y*-axis), it does not become the same distance between the different harmonics.

It is interesting to note that we get many harmonics even when the mouthpiece is used alone. This means that it is the slightly chaotic opening/closing of the lips that

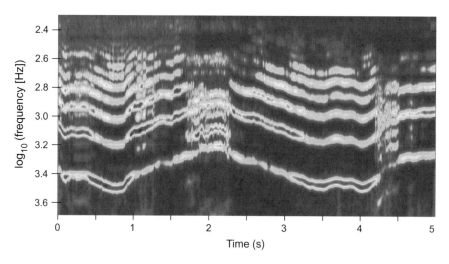

Fig. 7.13 An advanced form of time-resolved Fourier transformation (wavelet transformation) of the sound from the mouthpiece of a trumpet. Time is measured along the x-axis, the logarithm of the frequency along the y-axis. The intensity of the frequency spectrum is highlighted with colours. See also the text for comments

Fig. 7.14 A slice of the time picture of the sound from a mouthpiece shows that the sound pressure does not vary harmoniously with time. Note that the asymmetry seen in the trumpet signal in the left part of Fig. 7.12 is no longer present, even though the same mouthpiece was used in both cases

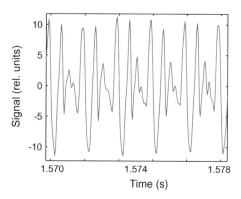

make sure the airflow does not acquire a sinusoidal time course. This is confirmed by Fig. 7.14, which shows a small section of the time signal from the mouthpiece sound. In other words, it is not the trumpet itself that creates the harmonics. What matters more is the action of the tight lips whereby small air puffs are ejected in a rather erratic manner. On the other hand, back and forth passage of the wave in the trumpet leads to the quantization of whatever tones are emitted by the instrument.

Later in the book, we will use wavelet transformation for further analysis of sound. It will then be seen that Fourier transformation often furnishes a picture that lacks life and nuance. In reality, the harmonics do not exist at the same intensity all the

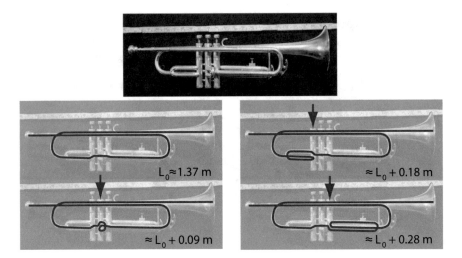

Fig. 7.15 The air column in a trumpet is slightly funnel shaped from the mouthpiece to the outer opening. Valves permit changing the length of the air column. For a B-trumpet (the fundamental tone is a B when no valves are pressed) the length of the air column is about as long as specified

time. The intensity distribution of the harmonics varies as shown in Fig. 7.13. This is one reason why sound from real musical instruments has often more life in it than synthetically produced sound.

7.2.6 How to Vary the Pitch

It is interesting to see how we may change the pitch in different instruments. For a guitar, it is obvious that we should change the length of the vibrating part of the string. Since the *tension* is largely unchanged when we press a string against a fret in the neck of the guitar, the velocity of the waves remains unchanged. When we choose to reduce the *length* of the string, the time taken by a wave to go back and forth decreases proportionally, and the frequency rises according to the relationship $f = v/2L$.

In a brass wind instrument, such as a trumpet, the length of the air column in the instrument changes when the valves are pressed. For a trumpet, when the middle valve is pressed, the air is diverted to a small extra loop. If only the first valve is pressed, the extra loop is about twice as long if only the middle valve is pressed, and if only the third valve is pressed, the extra loop is about three times as long. In Fig. 7.15, the data for the effective air column length for different single valves are given in bold letters. Several valves can be pressed simultaneously, and then the total air extension will equal the sum of all the additional loops that are inserted.

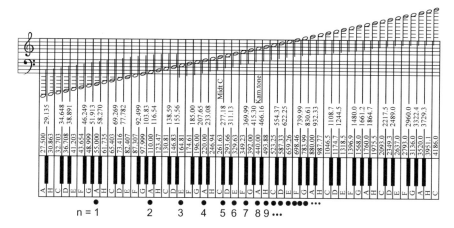

Fig. 7.16 Tones on a piano along with calculated frequency on a tempered scale. The figure is inspired from [1] but completely redrawn

7.2.7 Musical Intervals

In the Western musical tradition, tones are graded on a scale of 12 semitones, which together span a frequency range over which the frequency increases by a factor of 2.0. This means that for a tone C there is a new tone C with a base tone of a frequency twice as large as that of the first one. The tone range is called *an octave*.

The semitones (or half-tones) in-between are chosen so that there is a constant frequency ratio between a tone and the lower semitone. Since there are 12 such steps to achieve an octave, it follows that the ratio of the frequency of one tone and the lower semitone must be

$$2^{1/12} \approx 1.0595$$

provided that all steps are equal. A scale defined in this way is called *tempered*. Figure 7.16 shows the frequencies on a tempered scale if we assume that one-stroke A should have a frequency of 440.00 Hz.

Two tones from, for example, a violin can together sound particularly pleasant if their frequency ratio equals an integer fraction (where only numbers up to 5 are included). The ratio between the frequency of an E relative to the C below on a tempered scale is about 1.260. This is close to 5:4, and the leap is called a (major) *third*. Similarly, the frequency of an F relative to the C is equal to 1.335, which is close to 4:3, and the jump is called *fourth*. Finally, we may note that the relationship between a G and the C below is 1.4987 which is very close to 3:2, a leap called *fifth*.

It is feasible to create a scale where the tones are *exactly* equal to the integer fractions mentioned above for selected tones. Such a scale is called "just". Certain combinations of tones sound more melodious than on a tempered scale, but the drawback is that we cannot transpose a melody (displace all the tones by a certain number of semitones) and maintain the melodious character.

In Fig. 7.16, some interesting inscriptions can be seen at the bottom. If we start with a low A with frequency 55 Hz ($n = 1$), the frequency of the first overtone ($n = 2$) will be twice as large (110 Hz). The difference between the fundamental tone and the first overtone is a whole octave.

The frequency of the second overtone ($n = 3$) will have $3 \times 55\,\text{Hz} = 165\,\text{Hz}$, which almost corresponds to an E, and the third overtone ($n = 4$) will have the frequency $4 \times 55\,\text{Hz} = 220\,\text{Hz}$, which is the next A. This amounts to two overtones within one and the same octave.

Continuing in the same vein, one sees that there are four overtones within the next octave and eight within the following octave. In other words, the higher harmonics will eventually stay closer than the semitones. That is why we can almost play a full scale without the use of valves in a lur, by forcing the instrument to emit sound mostly at the higher harmonics.

On a trumpet, the fundamental tone (which corresponds to $n = 1$) is achieved if the lips are pressed together only moderately. The frequency of the fundamental tone can be increased in leaps (n increases) by tightening/pressing the lips more and more. The air that escapes through the lips will then come in a tighter bunches than when the lips are more relaxed.

In Fig. 7.12, we saw that the frequency of the fundamental tone for a B trumpet was about 231.5 Hz. This should be a B, and those familiar with the tone scale will know that a B is the semitone that lies between A and H. From Fig. 7.16, we see that this is as it should be. By slightly varying lip tension, the tone from the trumpet can be changed quite a bit (even I can vary the frequency between about 225 and 237 Hz for the current B). Good musicians take advantage of this fine-tuning of the pitch when they play.

7.3   Sound Intensity

Sound may be so weak that we do not hear it, or so powerful as to become painful. The difference is in the *intensity* of the sound, and the sound intensity is defined as:

Sound intensity is the time-averaged energy transported per unit time and area in the direction of the sound.

Alternatively, the sound intensity can be defined as the time-averaged energy per unit area and time flowing normally across a surface in the direction of propagation of the wave.

Sound intensity is measured in units of watt per square metre: W/m^2.

It is also possible to work with an "instantaneous" intensity (as opposed to the time average), but this will depend on both position and time. For sound waves, the local instantaneous intensity I_{ins} will be given by:

$$I_{ins}(\vec{r}, t) = \vec{p}(\vec{r}, t) \cdot \vec{v}(\vec{r}, t) \tag{7.8}$$

where $\vec{p}$ is the local pressure (strictly, the pressure deviation relative to the mean) and $\vec{v}$ is here *the local velocity of air molecules* at the same place and time (*not* the speed of sound!).

Remark: A useful rule of thumb will now be derived. Instead of looking at the amount of energy, we may consider what work the sound wave is able to perform. Work is force times distance, and the force that can work on a cross-sectional area A is the local pressure in the sound wave multiplied by the area (actually the excess or deficit pressure in the sound wave multiplied by the area).

Work is "force times distance", and if the wave moves a distance Δx in a time Δt, then it follows that:

$$\text{Instantaneous intensity} = \frac{\text{work that can be performed}}{\text{area and time}}$$

$$I_{ins} = \frac{p A \Delta x}{A \Delta t}$$

$$I_{ins} = p \frac{\Delta x}{\Delta t} \approx pv$$

which is the desired rule of thumb.

In the previous chapter, a harmonic sound wave was described in terms of η, the local displacement of the molecules, by the following equation:

$$\eta(x, t) = \eta_0 \cos(kx - \omega t)$$

where η_0 is the maximum displacement relative the equilibrium position (in addition to the thermal movements!).

The speed of the molecules executing in the motion is the time derivative of the displacement η:

$$\frac{\partial \eta}{\partial t} = \omega \eta_0 \sin(kx - \omega t) .$$

It was also shown that the same wave can also be described as a pressure wave by using the equation:

$$p(x, t) = k K \eta_0 \sin(kx - \omega t) .$$

where K is the compressibility module for the medium in which the sound wave is moving.

The instantaneous intensity will now be the product of the local velocity of the molecules and the local pressure as described in Eq. (7.8). The wave is assumed to be longitudinal and moving in the x-direction, so that velocity and pressure have the same direction. Accordingly:

$$I_m = p \frac{\partial \eta}{\partial t} = k\omega K \eta_0^2 \sin^2(kx - \omega t) . \tag{7.9}$$

The wavenumber k and the angular velocity ω must satisfy the relation

$$v = \frac{\omega}{k} = \sqrt{\frac{K}{\rho}}$$

where v now stands for the speed of sound, K is the modulus for bulk elasticity and ρ the mass density.

Whence follows the expression for the time-averaged intensity:

$$I = \frac{1}{2}k\omega K \eta_0^2 = k\omega K \eta_{rms}^2 = 4\pi^2 \frac{K}{v}(f\eta_{rms})^2$$

since the time-averaged value of $\sin^2$ equals $1/2$. Here, η_{rms} is the root mean square displacement of the air molecules, or $\eta_{rms} = \eta/\sqrt{2}$. [The reader is reminded that we are speaking here of the collective displacement of the molecules *over and above* the thermal motion of "individual" molecules.]

It will be useful to eliminate K, the bulk modulus for compressibility, and use the amplitudes of displacement and pressure, together with mass density, sound speed, wavelength and frequency. After some trivial manipulation of the above expression, one can show that:

$$I = \frac{(p_{rms})^2}{\rho v} \tag{7.10}$$

where p_{rms} is the root mean square deviation of the pressure fluctuation, ρ is the mass density of air and v is now the speed of sound in air.

Further, it can be shown that:

$$I = 4\pi^2 \rho v (f\eta_{rms})^2 \tag{7.11}$$

where λ is the wavelength of the sound in air, that is, $\lambda = v/f$ where f is the frequency of the sound.

Equation (7.10) shows that sound with different frequencies will have the same intensity if the pressure amplitude is the same.

Equation (7.11) shows that sounds of the same intensity, but different frequencies, have displacement amplitudes η_{rms} which are inversely proportional to the frequency, hence proportional to the wavelength.

It is much easier to measure pressure fluctuations than displacements of molecules. Therefore, Eq. (7.10) is the version that finds practical applications when sound intensities are to be measured and reported.

Before looking at some examples of intensity values, let us return briefly to Eq. (7.9). The equation shows the instantaneous value of energy transport as a function of position and time. The expression is always positive (since $\sin^2 > 0$). It is an important characteristic of waves! The molecules that propagate the wave swing back and forth, but their mean position remains fixed, and does not move with the wave (apart from thermal movement). Yet, energy is transported onward from the source of the wave, and it normally never returns to the source.

It is of some interest therefore to integrate over time all energy transmitted from the source to the wave. We can do that by looking, for example, at total energy per time going through a spherical shell around the source of the waves. The unit for such integrated intensity is watts.

A human voice during normal conversation produces a total power of about 10^{-5} W. If one shouts, the power may amount to about 3×10^{-2} W. In other words, the production of a usable sound wave does not require an unusual expenditure of power.

The figures for the human voice may seem strange when we know that a stereo system can produce powers at 6–100 W. Of course, a stereo system used at 100 W produces a sound far more powerful than human voice can provide. Nevertheless, the difference in intensities of sound from a human voice and a stereo system is striking.

The reason for the big difference is that only a small part—a few per cent for ordinary speakers—of the power supplied to the speakers is converted into sound energy. For special horn speakers, the efficiency can reach up to about 25%. The rest of the energy is converted to heat.

7.3.1 Multiple Simultaneous Frequencies

In the derivation of Eq. (7.10), we assumed a single harmonic wave. We will now consider waves with many different frequencies occurring simultaneously?

We must distinguish between correlated and uncorrelated waves. If we send one and the same harmonic signal simultaneously to two stereo speakers, the sound waves from the two sources will be correlated. At some places in the room, the waves will be added constructively. The amplitude can be twice as large as that from a single speaker, in which case the intensity would increase by a factor of four. Elsewhere in the room, the waves will be added destructively and, in the extreme case, will nullify each other. The intensity at such a place would be zero.

For uncorrelated waves (no temporal coherence, see Chap. 15), there will be no fixed pattern of intencifying an nullifying waves at various positions in the room. It will change all the time. For those cases, the following applies:

When we measure sound intensities, the contributions are usually uncorrelated. The sound intensity is then equal to the sum of the intensities of the separate contributions.

7.3.2 Audio Measurement: The Decibel Scale dB(SPL)

Sound intensity can be specified in watts per square metre, as described above. However, it is not a convenient scale. One reason for this is that human hearing has a more logarithmic than linear response. This means that the ear perceives changes in volume based on percentage change compared to the existing sound level. Increases the sound intensity from 10^{-5} to 10^{-4} W/m^2, the change is perceived to be approximately as large as when the sound intensity increases from 10^{-3} to 10^{-2} W/m^2.

Therefore, a logarithmic scale for sound intensity, the so-called decibel scale, has been introduced. The sound intensity I relative to a reference intensity I_0 is given in the number of decibels as follows:

$$\beta = L_I = (10\text{ dB}) \log \frac{I}{I_0} . \tag{7.12}$$

The unit "bel" is named after Alexander Graham Bell, the inventor of the telephone. The prefix "deci" comes from the factor of 10 that is introduced to get simple working values. The decibel scale is used in many parts of physics, not just when we deal with sound intensity.

In principle, we can choose any reference value and can say, for example, that the sound intensity 10 m away from the speakers in the example above is 26 dB higher than the sound intensity 200 m away (check that you understand how the number 26 arises).

In some contexts, it becomes necessary to specify sound intensity on an absolute scale. This can be achieved by using a well-defined reference value specified on an absolute scale. For sound, the following absolute scale is often used:

$$L_{I abs} = 10\text{ dB(SPL)} \log \frac{I}{I_{abs.ref}} = 10\text{ dB(SPL)} \log \frac{p^2}{p^2_{abs.ref}} . \tag{7.13}$$

SPL stands for *sound pressure level* and the reference value is 1000 Hz audio with sound pressure $p_{rms} = 20\,\mu$Pa (rms). This sound pressure corresponds approximately to an intensity of 10^{-12} W/m^2 and represents about the lowest intensity a 1000 Hz sound may have for a human being to perceive it. This corresponds to approximately the sound intensity 3 m away from a flying mosquito.

It is amazing that the displacements $\eta(x, t)$ of tiny volumes of air molecules for such a weak sound wave is only of the order an atomic diameter. Our ear is a very sensitive instrument!

Although dB(SPL) has been chosen with reference to human hearing, this is nevertheless a purely physical measure of intensity, based solely on W/m^2. dB(SPL) can be used for virtually all frequencies, regardless of whether a human being can hear the sound or not.

The conversion from intensity to the square of sound pressure is given by Eq. (7.10).

In practice, the term SPL is frequently omitted when sound intensity is specified. This is unfortunate, for when one says that the sound intensity is 55 dB, the statement is, in principle, incomplete because the reference has not been specified. If it had been stated instead that the sound intensity is 55 dB(SPL), it would have implied that the reference level is as indicated above, and that the sound level has been specified on an absolute scale.

7.3.3 Sound Intensity Perceived by the Human Ear, dB(A)

Several factors must be taken into account when sound intensities are specified. The definition in Eq. (7.13) is based on a reference sound with a frequency of 1000 Hz. However, we hear sound in a wide range of frequencies, and the ear does not perceive sound with different frequencies as equally intense, even if the number of watts per square metre remains unchanged. We find it harder to hear sounds of frequencies which are lower and higher than sound with average frequencies. The dB(SPL) decibel scale refers to intensity values of sound, irrespective frequencies. In order to get a measure of perceived loudness of a sound, we need to take into the consideration the properties of the human ear.

Figure 7.17 shows equal-loudness contours for different frequencies, that is, the physical intensity in dB(SPL) required to give the same perceived loudness as the frequency varies. Several curves are recorded, since the relative change in frequency varies somewhat with how loud the sound is initially.

The unit phon device indicates the intensity of pure tones. 1 phon corresponds to 1 dB(SPL) at the frequency 1000 Hz. The sound intensity corresponding to a given number of phon varies greatly with the frequency of the pure tones. For example, we see from Fig. 7.17 that a pure 20 Hz sound of 100 dB(SPL) volume is perceived to be equally intense as a pure 1000 Hz sound of 40 dB(SPL). We further see that the sound intensity at 100 Hz must be about 25 dB(SPL) to be audible. Furthermore, an audio intensity of 40 dB(SPL) at 1000 Hz corresponds to the intensity of 55 dB(SPL) for sound of 10,000 Hz.

The curves, issued by the International Organization for Standardization (ISO), were updated in 2003. The year indicates that it is not easy to determine such curves as long as there are significant individual variations. People with obvious hearing deficiencies are probably not used when the data for such curves are collected!

Fig. 7.17 Sound levels at different frequencies giving approximately the same sensation of loudness (see text). Lindosland, Public Domain, Modified from original [2]

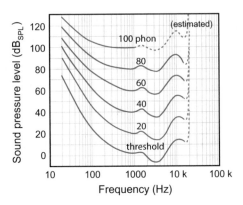

Fig. 7.18 Weighting curves used to indicate the perceived strength of a signal that has many different frequencies at the same time. The graphs give rise to dB(A)-scale, dB(B)-scale, etc. Lindosland, Public Domain, Modified from original [3]

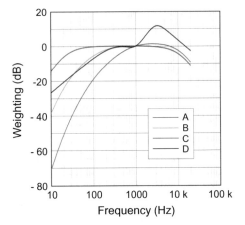

It goes without saying that the decibels scale as presented in Eq. (7.13) cannot be used to indicate *perceived* sound intensity in humans, which becomes particularly demanding when the sound is composed of multiple frequencies. For this reason, an intensity measure is introduced so that different frequencies are weighted according to how intense the sound appears to the *ear*. There are various weight functions, giving rise to dB(A)-scale, dB(B)-scale, etc. Figure 7.18 shows examples of the most common weighting curves.

The curves show that low frequencies count much less than average frequencies when dB(A)-scales are to be determined, as compared to a pure dB-scale as defined in Eqs. (7.12) or (7.13).

The reason for employing different weight functions is based on the phon curves in Fig. 7.18. If the intensity is high, the ear weights various frequencies a little differently than if the intensity is low. dB(A) is best suited for mean and low-intensity levels, whereas, for example, dB(C) or dB(D) is best suited for measurements at high intensities.

Concrete example of calculation

Let us go through an example of the slightly more appropriate procedure that needs to be used when sound contains multiple frequencies.

Suppose that a sound consists of a pure 100 Hz signal and a pure 1000 Hz signal and that the signals are uncorrelated. Assume that, taken individually, the two components are of equal strength on the dB(SPL) scale, for example, 80 dB(SPL) each. The sound intensity of the composite signal on a dB(SPL) scale would then be:

$$L = 10 \text{ dB(SPL)} \ \log \frac{p_{tot}^2}{p_{abs.ref}^2} = 10 \text{ dB(SPL)} \log \frac{p_{100\,Hz}^2 + p_{1000\,Hz}^2}{p_{abs.ref}^2}$$

$$= 10 \text{ dB(SPL)} \ \log 2 \frac{p_{1000\,Hz}^2}{p_{abs.ref}^2} = 3 + 80 \text{ dB(SPL)} = 83 \text{ dB(SPL)} \ .$$

However, in a dB(A) scale, the calculation would go like this: The contribution from the 1000 Hz signal should be weighted with a weight factor 1.0, that is, effectively as 80 dB(SPL). However, the contribution from the 100 Hz signal is to be weighted by a factor of -20 dB, that is, we must subtract 20 dB from the 80 dB the sound would have on a dB(SPL) scale, because it is placed on a dB(A)-scale. 80 dB(SPL) corresponds to

$$\frac{p^2}{p_{abs.ref}^2} = 10^8$$

and 60 dB(weighted) corresponds to

$$\frac{p^2}{p_{abs.ref}^2} = 10^6 \ .$$

The sum comes out to be:

$$L = 10 \text{ dB(A)} \log \frac{p_{tot,\,weighted}^2}{p_{abs.ref}^2} = 10 \text{ dB(A)} \log \left(\frac{p_{100\,Hz,\,weighted}^2}{p_{abs.ref}^2} + \frac{p_{1000\,Hz,\,weighted}^2}{p_{abs.ref}^2} \right)$$

$$= 10 \text{ dB(A)} \log(10^6 + 10^8) = 80.04 \text{ dB(A)} \ .$$

In other words, sound at 100 Hz contributes hardly anything to the perceived intensity as compared with sound at 1000 Hz.

We often see tables with sound intensities in different circumstances, and a typical example is shown below:

Audibility threshold at 1000 Hz . . .0 dB(A)
Whispering   20 dB(A)
Quiet radio at home   40 dB(A)
Conversation   60 dB(A)
General city traffic   70 dB(A)
Loud music   100 dB(A)

It is most common in such overviews to use the dB(A) scale, but presented just as "dB". In principle, we should state the intensities in dB(A), dB(B), etc., instead of just dB, to point out, first, that the values refer to an absolute scale, and second, that the contributions from different frequencies have been weighted, in order to show the *perceived sound intensity* and not a measurement of sheer physical intensity.

For our ear to experience that the sound level has doubled, the sound intensity must increased by 8–10 dB(A).

As for large sound intensities, we know that:

85 dB(A)   prolonged exposure can lead to hearing loss
120 dB(A)   acute exposure can cause hearing loss
130 dB(A)   causes pain ("Pain threshold")
185 dB(A)   causes tissue damage.

Data like these vary from source to source and must be taken with a pinch of salt. It is clear, however, that loud noise can destroy the hairs in contact with the basilar membrane in the inner ear (see Chap. 3). Too many persons regret that they were tempted to listen to such powerful music that hearing impairment became permanent. Also note that with very powerful sound, ordinary tissue is torn apart and shredded, so that the body as such degenerates completely. Powerful sound is not something to play with!

7.3.4 Audiogram

We can test our hearing by visiting an audiologist, or by using available computer programs and the computer's sound card (but the accuracy is often dubious). In fact, there even are smartphone apps for this type of test. The result of a hearing test is often displayed as a so-called audiogram, and an example is given in Fig. 7.19. An audiogram is constructed such that if a person has normal hearing, her/his audiogram should be a horizontal straight line at the 0 dB level (or at least within the blue band the figure between -20 and $+10$ dB). If the person has impaired hearing for some frequencies, the curve will be below the 0 level. The distance from the null line indicates the difference in the sensitivity of the test person at a particular frequency compared with the norm.

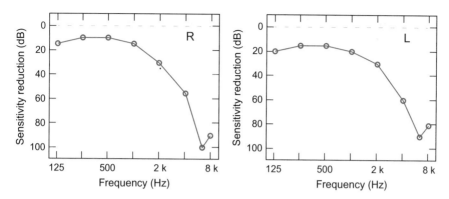

Fig. 7.19 Example of an audiogram recorded by an audiologist. The curves show age-related hearing loss in a 68-year-old man. **R** and **L** stand for the right and left ear, respectively. Normal hearing is within the blue area between -10 and $+20$ dB

Figure 7.19 shows that the person tested has normal hearing for 500 and 1000 Hz in the left ear but has impaired hearing loss for all other frequencies. The hearing loss is 80–90 dB in both ears at 8 kHz. This means that the person is practically deaf at high frequencies. This is an example of age-related hearing impairment. It is small wonder that older people have trouble understanding conversations between people because the most important frequency range in this context is between 500 and 4000 Hz.

Remarks: You have previously worked with Fourier transformation of sound. If the Fourier transform with appropriate calibration provides a measure of the sound intensity at different frequencies, you should be able to calculate dB(A) values, dB(B) values, etc. using the curves in Fig. 7.18. As you can see, you can create your own sound-measuring instrument! (But calibration must be done!)

dBm

Finally, another dB-scale will be defined that is widely used in physics, namely the dBm scale. This is an absolute scale where I_0 is selected equal to 1 mW. The dBm scale is used in many parts of physics, often associated with electronics, but rarely when the sound level is reported. The scale is generally used to specify radiated power from, for example, an antenna. If a source yields 6 dBm, it means that the radiated power is

$$10^{6/10} \text{ mW} = 4 \text{ mW} .$$

7.4 Other Sound Phenomena You Should Know

7.4.1 Beats

When we listen to two simultaneous sounds with approximately the same frequency, it may sometimes appear that the *strength* of the sound varies in a regular manner. Such a phenomenon is called "beating" or "producing beats". The word "beat" is used because the resulting sound appears to the listener as a regular beat.

Mathematically, this can be displayed in approximately the same way as in the expression of a standing wave. However, for our new phenomenon, it is not interesting to follow the wave's propagation in space. The interesting thing is to consider how the sound is heard at one spot in the room.

The first step is to add two sinusoidal oscillations:

$$y_{\text{sum}} = A\cos(\omega_1 t) + A\cos(\omega_2 t) .$$

This sum is mathematically equivalent to a formula similar to that found earlier:

$$y_{\text{sum}} = 2A\cos\left[\tfrac{1}{2}(\omega_1 + \omega_2)t\right]\cos\left[\tfrac{1}{2}(\omega_1 - \omega_2)t\right] .$$

If the two (angular) frequencies are nearly equal, a mean and differential value can be inserted as $\overline{\omega}$ and $\Delta\omega$ in the formula, which yields the following result:

$$y_{\text{sum}} = 2A\cos(\overline{\omega}t)\cos\left(\frac{\Delta\omega}{2}\right)t . \tag{7.14}$$

This expression is mathematically speaking a product of a "mean frequency oscillation" factor and a "difference frequency oscillation" factor, which is nearly independent on each other.

If the frequency differences are too small to be distinguished by the ear, the mean frequency oscillation factor $\cos\overline{\omega}t$ in Eq. (7.14) will correspond to approximately the same auditory experience as if only one of the two sounds was present. The difference frequency oscillation factor $\cos(\tfrac{1}{2}\Delta\omega t)$, however, oscillates with a much lower frequency than the original. For example, if we listen to two equally sounds simultaneously, with frequencies of 400 and 401 Hz, the difference frequency oscillation factor will be a $\cos(\pi t)$. Once a second, this factor will vanish, and the total sound will disappear. The listener will experience a sound of almost the same frequency as that of a single original sound, but with a volume fluctuating at a frequency of 1 Hz. This pulsation in the volume is known as "beating".

Fig. 7.20   When two sound
signals with nearly the same
frequency are added, the
intensity of the resulting
signal will vary in time in a
characteristic manner

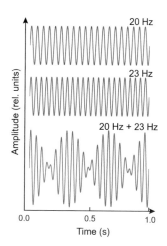

Figure 7.20 shows an example of beating. There are two signals with 20 and 23 Hz respectively, and we follow each of the signals and their sum over a period of one second. We see that in the sum signal there are three "periods" with strong and weak sound within the interval we consider. Note the factor $\cos(\frac{1}{2}\Delta\omega t)$ in Eq. (7.14), and that *half* of the difference of the two frequencies (that are added) corresponds, in our case, to 1.5 Hz. Why does one see three "periods" in the intensity of the beat plot in Fig. 7.20? This is a detail you should notice and understand, because it creeps into several different contexts (Hint: How many times is a sine curve equal to zero during one period?).

There are more puzzles to the beat sound phenomenon. A Fourier analysis of the signal described by Eq. (7.14) gives two peaks corresponding to ω_1 and ω_2 only. There are *no* peak corresponding to the difference frequency. Why do we then experience beating and not two simultaneous sounds with slightly different pitch?

If the difference in the two frequencies is increased, we will eventually hear two separate tones and no beat. Thus, the beat phenomenon is a result of our ear and further processing in the brain. Detailed explanations are found in textbooks in auditory physiology and perception.

However, we suggest a numerical experiment: Make a sum of two sine signals with identical amplitudes and the frequencies 100 and 110 Hz. Let the signal last for at least hundred 100 Hz periods. Calculate the Fourier transform. The result is as expected.

Calculate then the signal squared (each element is the square of the same element in the previous signal) and perform the Fourier transformation. Notice the frequencies of the peaks now!

This numeric experiment is of interest since many detectors for oscillatory phenomena in physics do not respond to the momentary amplitude of the signal, but to the *square* of the amplitude (to the intensity instead of the amplitude). Whether or not this is applicable to the beat sensation is just a speculation. Our ears are mainly "square law detectors" since phase information is lost for frequencies higher than ≈ 1 kHz.

7.4.2 Sound Intensity Versus Distance and Time

When sound propagates in air, little energy is lost along the way. This means that the amount of energy crossing a spherical shell of radius of r_1 will be nearly the same as crosses a spherical shell of radius r_2 ($> r_1$). The local sound intensity is the amount of energy per unit area and time. Since the area of a spherical shell of radius r is $4\pi r^2$, the intensity will decrease as $1/r^2$ where r is the distance from the source.

Now, sound rarely spreads out in totally spherical wavefronts. The distance to the ground is usually significantly shorter than the extent of propagation in the horizontal plane. However, the relationship

$$\frac{I(r_2)}{I(r_1)} = \left(\frac{r_1}{r_2}\right)^2$$

applies reasonably well also to limited solid angles (as long as interference phenomena do not play a significant role).

This implies that if, at a concert, we are 10 m from the speakers, the intensity will be 400 times greater than for the audience 200 m away.

However, inside a room an audio pulse will be damped with the passage of time. The pressure waves lead to oscillations in objects, and many objects have a built-in friction where the sound energy is converted to heat. Various materials dampen sound more or less efficiently. A smooth concrete wall is not set into oscillation by sound waves, and sound is reflected from such a surface without much loss of energy. Walls covered with mineral wool or other materials that are more easily set into vibratory motion in response to a sound wave can dampen the sound much more effectively.

Walls and interior in a room can lead to major differences in damping. They affect the so-called reverberation time. In the Trinity Church (in Norwegian, Trefoldighet-skirken) in Oslo, with bare stone walls and few textiles, the reverberation time is so long that music with fast passages becomes fuzzy to listen to, especially when the audience is thin. In a room with a lot of textiles and furniture and people in relation to the overall space, the sound will die out appreciably faster. In an echo-free room, the floor, walls and ceiling are covered with damping materials, and the reverberation time is extremely short. For concert venues and theatre venues, it matters a great deal for a good overall sound experience that the reverberation time is adapted to the sound images that occur. Building acoustics are a separate part of physics, where good professionals are hard to find and therefore much sought after.

7.4.3   Doppler Effect

Most of us know that the sound of an ambulance siren changes pitch when the vehicle passes us. The phenomenon is called *Doppler effect*. We will now derive a mathematical expression for the observed frequency change.

Sound waves travel at a certain speed *in relation to the transporting medium*. No matter what speed the source has, and no matter what speed an observer has, the sound wave passes through, for example, air at rate $v = \sqrt{K/\rho}$ (symbols defined earlier).

To the left in Fig. 7.21, the wavefront is shown to be the maximum in the air pressure waves from a source that is at rest. The sound spreads smoothly in all directions, and as long as the source does not move, all wavefronts will have the same centre. To the right of the same figure, the wavefront is shown when the source of the sound has moved between each time a pressure wave started. Thereafter each pressure wave progresses unabated with the sound speed (e.g. in air).

This means that an observer positioned so that the source of sound approaches her/him will find that the wave peaks are more frequent (more wave peaks per second) than if the source were at rest. For an observer from whom the source of sound is receding, the opposite will be true. This means that the frequency experienced by an observe will differ in the two situations.

When the observer is at rest with respect to the air, the sound waves will approach her/him with the speed v. When the effective wavelength is as shown in the right part of the figure, it follows that the frequency as heard by the observer is f_o:

$$f_o = \frac{v}{\lambda_{\text{eff}}} \ .$$

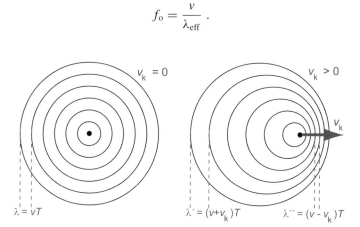

Fig. 7.21 Sound waves spread with the same speed in all directions in the medium through which the sound waves pass. The wavy peaks are equally far apart if the source is at rest in relation to the air. If the source moves relative to the air at the speed v_s, the wave peaks are closer together one side than on the other. The sound speed is set as v

When a source of sound with period T and frequency $f_s = 1/T$ approaches the observer with a speed v_s, one has:

$$f_o = \frac{v}{(v - v_s)T}$$

$$f_o = \frac{1}{1 - v_s/v} f_s \qquad (7.15)$$

where v is the speed of sound in air. For an observer from whom the source is receding, the minus sign is to be changed into plus.

This version of Doppler effect can be described by saying that the wave speed relative to the observer (who is at rest) equals the speed of sound in air, while the effective wavelength is different from a situation where both source and observer are at rest.

A variant of Doppler effect is that when the source is at rest, but the observer is in motion. Then the velocity of the wave peaks relative to the observer is different from the sound velocity in air in general. However, the wavelength is unchanged.

The frequency experienced by the observer will then be proportional to the effective velocity of the wave peaks relative to the observer, compared with the speed with which the waves would have reached the observer if he/she and the source were at rest. For a stationary source, and an observer in motion with the speed v_o towards the source, we have the relation:

$$f_o = (1 + v_o/v) f_s \qquad (7.16)$$

where f_s is again the frequency of the source.

It is perfectly possible to combine the two variants of Doppler effect discussed above, so that we get a more general expression that applies to situations where both the observer and the source are moving in relation to the air where the sound is spreading.

In Eq. (7.16), the frequency f_s can be replaced by the frequency an observer (suffix o) would have experienced if the source (index s) were in motion, i.e. with f_o given by Eq. (7.15). The result will then be:

$$f_o = \frac{v + v_o}{v - v_s} f_s . \qquad (7.17)$$

Here v is the speed of sound in air (e.g. 344 m/s), and v_s and v_o are, respectively, the speeds of the source and the observer relative to the air through which the

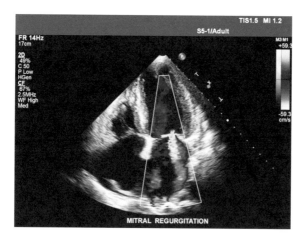

Fig. 7.22 Ultrasound picture of a human heart, superimposed with an ultrasound Doppler image of the blood flow, at a particular point in the heart rhythm. The picture reveals that a heart valve does not close properly during a ventricular compression. The picture is reproduced with a permission from Vingmed. It is difficult to understand a single picture like this. It is recommended to watch a video (search at YouTube with the search words: cardiac ultrasound Doppler heart)

sound is transmitted. In the equation, the following sign convention is observed: If the source moves towards the observer at a rate of v_s *relative to air*, v_s is positive. If the observer moves toward the source at the rate v_o *relative to the air*, v_o is positive.

Note that the sign is based on the relative motion between the source and observer as noted above, while the actual magnitude of the velocity is specified relative to air (or the medium through which the sound waves propagate).

Note that it is *not* irrelevant which is moving, the source or the observer. If the source approaches the observer at a speed close to the speed of sound in air, the denominator will tend to zero and the frequency perceived by the observer will tend to infinity. On the other hand, if the observer approaches the source at a speed equal to the speed of sound in air, he/she will perceive a frequency that is only twice the frequency of the source of sound.

Doppler shift is utilized today in ultrasound diagnostics. In Fig. 7.22, a combined ultrasound and ultrasound Doppler image of a heart is shown. The black and white picture shows the ultrasound picture, while the sector with colours indicates blood flow towards or away from us. The subject has a heart valve that does not close properly when the ventricle compresses.

7.4.4   Doppler Effect for Electromagnetic Waves

Applications of the Doppler effect for sound waves are based on a constant sound speed relative to the medium which the sound passes through. For electromagnetic waves, the situation is completely different. The velocity of light is linked in a not easily comprehensible way to our entire space/time concept, and the velocity of light in vacuum is the same regardless of the speed of the source and how an observer moves. When wavelengths are measured, length contractions are observed due to relativistic effects, and time dilation/contraction take place due to relativistic effects. Therefore, the derivation of Doppler effect for electromagnetic waves becomes a little more complicated than for sound; we will content ourselves by merely reproducing the final expression.

Doppler shift for electromagnetic waves in vacuum is given by the relation:

$$f_0 = \sqrt{\frac{c + v}{c - v}} f_s .$$
(7.18)

Here c is the velocity of light, and v the velocity of the *source relative to observer*, $v > 0$ if the two approach each other. As before, f_s is the frequency of the wave emanating from the source.

This relation shows that light from distant galaxies will be observed to have a lower frequency if the galaxies are moving away from us. The effect is well known and is termed "red shift" in the observed spectra.

Redshift is more pronounced in the light from distant galaxies, as these (in accord with the Big Bang model for the universe) are moving away from us at high speed. The effect is so strong that parts of the visible spectrum are shifted into the infrared region.

This is one reason why the space telescope James Webb is equipped with infrared detectors.

7.4.5   Shock Waves *

From the right part of Fig. 7.21, it appears that the pressure waves lie closer to an sound source moving relative to air than if the source had been at rest. However, the figure was based on an implicit assumption, namely that the source of sound does not catch up with the sound waves generated by it. In other words, the sound source moves at a speed less than the speed of sound in air (or the medium under consideration).

What happens if the audio source moves *faster* than the speed of sound? This state of affairs is depicted in Fig. 7.23. To go from the case at the right of Figs. 7.21, 7.22 and 7.23, we must, however, consider the situation where the source moves at precisely the speed of the sound. In this situation, the pressure waves at the front of the source pile up on each other, and we can have enormous pressure variations within relatively short distances. Such a situation is called a shock wave, a shock front or even a "sound barrier".

Considerable energy is needed to penetrate the sound barrier. The intensity of the shock front can reach 160–170 MW/m². And, perhaps more importantly, the object that is "going through the sound barrier" must be robust enough to withstand the stresses when the pressure variations over the object become very large. The sound intensity of the shock wave is about 200 dB, so that persons aboard a plane passing through the sound wall must be shielded significantly to avoid permanent damage.

Remark: It is not the noise of the engine on the plane that gives rise to the shock wave. It is simply the pressure wave due to the airplane getting through the air. Engine noise comes as an addition to this main component of the pressure wave.

The speed of sound in air is usually taken as 340 or 344 m/s, which comes out to be around 1230 km/h. Fighter planes can fly faster than this, breaking the sound barrier on their way to the highest speeds.

The speed of a supersonic aircraft is given in terms of the Mach number, where:

$$v \text{ measured in Mach} = \frac{v_{\text{plane}}}{v_{\text{sound}}}.$$

The Concorde aircraft had a normal transatlantic cruising speed of about 1.75 Mach, but a peak speed of approximately 2.02 Mach. The space shuttle had a speed of 27 Mach. Remember, in this connection, that the speed of sound in the rarefied air at high altitudes differs from the speed of sound at the ground level.

From Fig. 7.23, one sees that the shock wave forms the surface of a cone after the plane that is the source of the waves. The opening angle of the conical surface is given by:

$$\sin \alpha = \frac{v_{\text{sound}} \times t}{v_{\text{airplane}} \times t} = \frac{v_{\text{sound}}}{v_{\text{airplane}}}.$$

When a supersonic plane is flying at high altitudes, the aircraft will have gone past an observer on the ground several seconds before the observer hears the sound from the plane. Only when the shock wave reaches the earth-based, will he/she hear the plane, and that as a bang, which indicates that the pressure wave on the expanding

Fig. 7.23 Behind a supersonic plane, a shock waveforms (also called a shock front or a "sound barrier") with a conical surface with the plane at the vertex. The angle of the cone depends on how much faster the plane moves with respect to the speed of sound

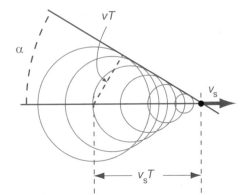

cone surface has reached the ground. The instant at which the bang is heard is not the moment when the plane crosses the sound barrier, but when the shock wave cone hits the observer.

In the case of the Concorde aircraft, the shock wave had a pressure of about 50 Pa at the ground when the plane flew at an altitude of 12,000 m. It was easy to hear the noise from the shock wave shortly after the plane had flown past. Similarly, in the Los Angeles district, we could hear a bang when the spaceship came in for landing on the desert strip a little northeast of the city.

Historically, the American Bell X-1 rocket-powered aircraft was the first vehicle to break the sound barrier. This happened on 14 October 1947; the aircraft then achieved a speed of 1.06 Mach.

7.4.6 An Example: Helicopters *

Few would think of helicopters in the context of supersonic speed, but we must. A Black Hawk helicopter has blades that rotate about 258 times per minute, which corresponds to about 4.3 rotations per second.

The rotor blades have a length of 27 feet, which corresponds to about 9 m.

The speed at the tip of the blade for a stationary helicopter (with the rotor running) is then:

$$\frac{2\pi r}{1/4.3} \text{ m/s} = 243 \text{ m/s}.$$

If the helicopter is flying at a speed of 100 km/h relative to the air, the speed of the blades relative to the air will be 360 m/s on one side of the helicopter. This is about equal to the sound speed!

Manufacturers of helicopters must find a balance between blade speed, the rotation rate and the flight speed in order to avoid problems with the sound barrier. The fact that the speed of the outer edge of the blade does not have the same speed relative to the air through a full rotation makes the task a little easier than with a supersonic plane.

Anyway, it is interesting to calculate the radial acceleration for a point at the extreme end of a helicopter rotor blade. On the basis of the figures above, it follows that:

$$a_r = \frac{v^2}{r} = \frac{243^2}{9} \text{ m/s}^2$$

$$a_r = 6561 \text{ m/s}^2 \approx 670 \text{ g}.$$

In other words, enormous forces work on the rotor, and the material must be flawless to avoid accidents. It is not uncommon for a rotor blade to cost more than 100,000 € per piece.

7.4.7 Sources of Nice Details About Music and Musical Instruments

There is much fun associated with musical instruments. Physicists have contributed to better understanding of many details and continue to do so. Here are some interesting sources you can look at:

Joe Wolfe, Music Acoustics: Basics, The University New South Wales, Australia. http://newt.phys.unsw.edu.au/jw/basics.html (accessed May 2018). Highly recommended!

Alexander Mayer, RIAM (Reed Instrument Artificial Mouth). Institute of Music Acoustics, University of Music and Performing Arts Vienna. http://iwk.mdw.ac.at/?page_id=104&sprache=2 (accessed May 2018).

Seona Bromage, Visualisation of the Lip Motion of Brass Instrument Players, and Investigations of an Artificial Mouth as a Tool for Comparative Studies of Instruments. Ph.D. thesis, University of Edinburgh, 2007.

H. Lloyd Leno, Larry Fulkerson, George Roberts, Stewart Dempster and Bill Watrous: Lip Vibration of Trombone Embrouchures. YouTube video showing lip vibrations when playing trombone: *Lip Vibration of Trombone Enbouchures, Leno,* (accessed May 2018).

Barry Parker, Good Vibrations. The Physics of Music. The John Hopkins University Press, Baltimore, 2009.

7.5 Learning Objectives

After working through this chapter you should be able to:

- Explain general features of reflection and transmission of waves at an interface between two different impedance media.
- Explain conditions for the formation of standing waves, and how such waves are characterized, including the terms nodes and anti-nodes.
- Explain what determines the pitch of some different musical instruments, and how we can achieve different pitches with one and the same instrument.
- Calculate the frequency (approximate) for a vibrating string and for a wind instrument.
- Explain the concept "trapped moving wave" (as opposed to the traditional "standing wave pattern") and explain advantages by this concept.
- Explain what we mean by frequency spectrum, fundamental frequency and harmonics when sound is analysed using, for example, Fourier transformation.
- Explain a tempered scale and calculate the frequency of any tone on a piano.
- Explain what is meant by beats, and derive a mathematical expression that shows that beating has something to do with the sound intensity.
- Calculate (when formulas are given) the amplitude of motion of air molecules and the amplitude of the pressure wave created by a harmonic sound wave with a specified dB value.
- Explain dB, dB(SPL), dB(A) and dBm scales.
- Explain the causes of Doppler shift in different contexts, derive formulas that apply to Doppler shift in air, and perform calculations based on these formulas.
- Explain shock waves, especially the "sound barrier" of supersonic aircraft and the like.

7.6 Exercises

Suggested concepts for student active learning activities: Acoustic impedance, reflective boundaries/interfaces, standing wave, node and anti-node, quantized wave, trapped moving wave, pitch, musical interval, tone scale, octave, sound intensity, difference between physical and phonetic intensity units, sound pressure limit, frequency dependency, decibel scale, dB(SPL), dB(A), audiogram, ultrasound, beating, Doppler effect, shock waves.

Comprehension/Discussion questions

1. For ultrasound examinations of, for example, a foetus, there must be at least as much sound reflected from the interface between the uterine wall and the amniotic fluid as from the interface between the amniotic fluid and the foetus. Why will not reflected sound from the first interface blur the image of the foetus?
2. Some piano tuners base their tuning on a frequency counter alone. Many believe that this is not a good way to tune. Can you give a reasonable explanation for such scepticism?
3. Try to give a verbal description of what is going on physically as we *begin* to blow air into an organ pipe and until the sound becomes stable.
4. We can create a tone by blowing air through a straight tube. By changing the tightening of the lips, we can produce different pitches. How is it related? What is the wave pattern inside the tube made by some of the sounds that can be generated? How do you suppose the spectrum would look like?
5. Can we get a standing wave by adding two waves moving in the opposite direction to each other, one having greater amplitude than the other, but the same frequency? Can we get a standing wave if we add two waves that move in the opposite direction to each other, where one has greater frequency than the other, but the same amplitude?
6. Are standing waves always quantized? Explain.
7. In music, an octave is characterized such that the frequency, for example, of a high C being twice the frequency of a C that is an octave lower. Suppose we have a properly tuned guitar, and we will amuse ourselves by tightening a string so that it will give an octave higher than it normally should be. How much more tightening do you need? [Is this a party game that can be recommended?]
8. A violinist sometimes touches the midpoint of a string while stroking the bow over the string. What does she accomplish with this trick?
9. When sound goes from air to water, which one of the following quantities stays constant: Wavelength, wave speed, frequency, amplitude of displacement of the molecules that propagate sound?
10. On a trumpet we can play different tones by pushing valves that cause air to pass through tubular loops (of different lengths) that extend the effective length of the air string within the instrument. How can we play different tones on a "post horn" or similar instruments where we cannot change the effective length? Can we play the same type of tunes on such an instrument as, for example, on a trumpet?
11. If we inhale helium and talk, we get a "Donald Duck voice" that is light and shrill. What is the reason for that? [Remember that inhaling too much helium can damage health and cause death, so be careful if you want to try this yourself!]
12. When we play an acoustic guitar (see Fig. 7.24), the sound becomes different depending on whether we strum the strings all the way down to near the saddle where the strings end or near the sound hole (or even closer to the middle of the string). What is the reason for the difference in tonal quality? And how would you characterize the difference?

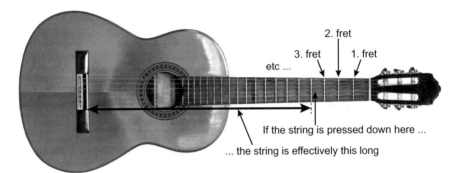

Fig. 7.24 On a classic guitar, a string is shortened if pressed against the first fret. The tone will then be a half-tone higher than with an open string. If the string is clamped at the second fret, the tone becomes two semitones higher, etc.

13. Does it make sense to say: Adding X dB to the sound corresponds to multiplying the intensity of the original sound wave with a definite factor?
14. Explain briefly the difference between dB, dB(SPL), dB(A) and dBm.
15. At an organ concert a listener noticed that after the organist had finished playing, it took a few seconds for the sound to subside totally. What is the reason that the sound dies out slowly? And what happened to the energy that was in the original sound?

Problems

16. An organ pipe is 3.9 m long, and open at the end. What tone do you suppose it emits (compare with Fig. 7.16).
17. The length of the free part of the strings on an acoustic guitar is 65 cm (that is, the part that can vibrate). If we clamp down the G-string on the fifth fret, we get a C (see Fig. 7.24). Where must the fifth fret be located on the guitar neck? The G has a frequency of about 196.1 Hz and the C about 261.7 Hz.
18. Use the information and answers from the previous assignment. For every semitone we go up from where we are, the frequency must increase by a factor of 1.0595. Calculate the position of the first fret, and to the sixth fret. Is the distance between the frets (measured in millimetres) identical along the guitar neck? Show that the distance between the frets is 0.0561 times the length of the string when it was clamped at the previous fret.
19. Check the frequencies indicated in Fig. 7.16. Supposed that we determined the frequency content of the sound data using Fourier transformation. For how long did we have to sample the sound to reach such precision? Is this a realistic way to determine the frequency accurately? Would it be more realistic to report the frequency with five significant digits for the highest frequencies than for the lowest? (Hint: Use the time-bandwidth product from the Fourier transform chapter.)

Fig. 7.25 Example of the frequency spectrum of a trumpet

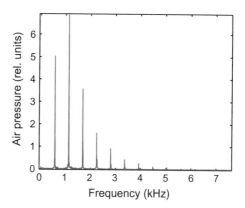

20. Assume (for the time being) that the intensity of the sound that comes from a choir is proportional to the number of singers. How much more powerful, on a decibel scale, will a choir of 100 persons sound compared to a four-person choir (a quartet)?

21. Figure 7.25 shows the frequency spectrum of a trumpet sound.
    (a) Estimate the frequency and relative pressure amplitude of the first five harmonics.
    (b) What is the frequency of the fifth overtone?
    (c) Assume that the intensity of the fundamental tone is 50 dB(SPL). Calculate the sound intensity in dB(SPL) for the entire trumpet sound (enough to include the first four (or five) harmonics).
    (d) Calculate the sound intensity in dB(A) for the entire trumpet sound (enough to include the first four (or five) harmonics).

22. Suppose a person is lying on a beach and listening to a CD player placed 1 m from the head, and that the music has an intensity of 90 dB. How powerful will the music sound to someone who is 4 m away from the speaker? If the neighbour complains about the noise level, what can the first person do to resolve the conflict? Feel free to present a calculation to support your proposal.

23. Two strings on an instrument are both tuned to vibrate at 440 Hz. After a few hours, we notice that they no longer have the same frequency, because we hear a 2 Hz beat when we let both strings vibrate at the same time. Suppose one of the strings still vibrates at 440 Hz. Which frequency or frequencies can the other string have? How much has the tension changed on the string that has lost its tuning?

24. In this assignment, we will compare sound intensities, displacement amplitudes and pressure amplitudes. Remember to comment on the results you get in every part!
    (a) What is the amplitude of air molecules when the sound intensity is 0 dB(SPL) at 1000 Hz? Repeat the same calculation for sound with intensity 100 dB(SPL).

(b) What is the sound pressure amplitude (both in Pascal and in atmospheric pressure) when the sound intensity is 0 dB(SPL) at 1000 Hz? Repeat the calculation for sound with intensity 100 dB(SPL).

(c) What is the displacement amplitude and the pressure amplitude for sound with the frequency 100 Hz and the intensity 100 dB(A)?

(d) There is an upper limit for how large the sound pressure amplitude may be if the sound wave is to be approximately harmonic (sinusoidal). What is this limit? How powerful would the sound be at this limit (specified in dB(SPL))?

25. Suppose you drive a car at 60 km/h and hear that a police car with sirens approaches from behind and drives past. You notice the usual change in sound as the police car passes. Assume that the speed of the police car is 110 km/h and that the upper limit for the frequency of the siren (when heard inside the police car) is 600 Hz. What frequencies do we hear before and after the police car has passed us?

26. Suppose a fighter plane takes off from Bodø airport and reaches 1.75 Mach already at 950 m altitude. What angle does the shockwave have? How long does it take from the moment the plane passes directly above a person on the ground till the moment the person notices the shock wave? Disregard changes in the speed of sound with the height.

27. In an ultrasound examination of a foetus, the Doppler effect is used for measuring the rate of cardiac movement in the foetus. The sound has a frequency of 2.000000 MHz (2 MHz sharp), but the sound back has a frequency of 2.000170 MHz. How much speed had that part of the foster heart where the sound was reflected from, in the short period in which this measurement was made. Sound travels in the foetus with a speed of about 1500 m/s. [Optional additional question: How much time resolution is it possible to achieve for mapping cardiac movement in cases like this?]

28. The Crab Nebula is a gas cloud that can be observed even with small telescopes. It is the remnant of a supernova explosion that was seen on Earth July 4, 1054. Gas in the outermost layers of the cloud has a red colour that comes from hot hydrogen gas. On earth, the hydrogen alpha line H-α has a wavelength of 6562.82 Å. When studying the light from the Crab Nebula, the H-α line has a *width* of 56,942 Å.

(a) Calculate the rate at which the gas in the outer part of the Crab Nebula moves. [Assume that the velocity of light is 3.0×10^8 m/s and that the relativistic Doppler shift for electromagnetic waves can be given approximately as $f_{observ} = (1 - v/c) f_{source}$ if the source moves away from the observer with speed v.]

(b) Assume that the gas in the outer part of the nebula has moved at the same speed ever since the supernova explosion. Estimate the size of the Crab Nebula as it appears now. State the answer both in metres and in light years.

(c) The angular diameter of the Crab Nebula when we see it from Earth is about 5 arc minute. An arc minute is 1/60 of a degree. Estimate the distance (in light years) to the Crab Nebula.

(d) When did the explosion of the star actually take place (approximately).

(e) In reality, the Crab Nebula is not spherical. Viewed from the Earth, it looks more elliptical with the largest and smallest angular diameters of 420 and 290 arc seconds, respectively. Even today, we do not know the distance to the Crab Nebula very accurately. Can you give a good reason for the inaccuracy based on the calculation you have made?

29. Perform a Fourier transform frequency analysis of the sound of two different musical instruments (record sound yourself via microphone and sound card on a PC, on a mobile phone, or use wav-files made available from our Web pages). Determine the frequency of the sound (fundamental tone) and find which tone on the scale it corresponds to. State approximately how many harmonics you find.

30. The left part of Fig. 7.26 shows a time plot of the sound from a tuba. One student used Fourier transform to convert this signal to the frequency spectrum including the harmonics. The student then conducted an inverse Fourier transformation of the frequency spectrum and expected to recover the original time signal. He did not. The result is shown in the right part of the figure. What went wrong?

31. A piano tuner first selects all three C-strings (all of which are activated by one key) to produce the 261.63 Hz frequency. [She actually starts with another frequency, but let's take this starting point here.] She now wishes to tune the F-strings by starting from C and using "re-tuning" where the frequency of F is exactly 4/3 of the frequency of C. This she does for all three F-strings that are struck when we press the key. She then intercepts one of the three F-strings by listening to the beat frequency she gets when she presses the key. By adjusting the beat frequency correctly, she ensures that the string gets the correct frequency on a tempered scale (and can adjust the frequency of the other two F strings after the first). What beat frequency should she choose?

32. Use the numbers for the length of the air column in a trumpet given in Fig. 7.15 to check that:
(a) the fundamental tone is about a B (indicate the frequency).
(b) that the elongation of the air column resulting from the depression of valve 1 corresponds approximately to a complete compared with that when no valves are pressed. Does the frequency go up or down when we press a valve?

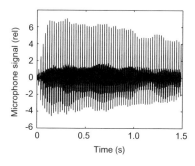 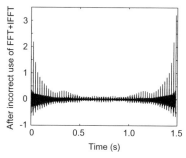

Fig. 7.26 See problem text

33. **Suggested Project**: We invite you to make your own "sound meter". The procedure could be as follows:

- Sample sound from your laptop using, for example the program snippet 2 at the end of Chap. 5.
- Perform a Fourier transformation and get the frequency spectrum of the sound. The intensity for the different frequency components are then proportional to the square of the Fourier coefficients.
- Reduce the relative intensities for various frequency components according to for example the weight function for dB(A).
- Add the weighted intensities for the frequency components.
- Calculate the dB(A) value for your sound, using an arbitrary reference intensity.
- Borrow a sound meter from someone who has one and adjusts the reference intensity in the calculations until you get a similar reading on your own sound meter as for the commercial instrument.

In fact, it is reasonably easy to make your own sound meter in this manner. However, remember that the microphone on the computer as well as the digitizing circuit have their limitations. Especially, it is difficult to get a good determination of weak signals.

For strong signals, it is another serious problem: The sound may produce signals larger than the digitizing circuit can manage. In those cases, "clipping" will occur. It can be discovered if you plot the sampled signal in time domain. Sinus signals will then have a flat top, and no signal can exceed this limit.

For such signals, the sound meter will give wrong readings! However, it is reasonably easy to let your program display a warning in those cases.

References

1. Unknown, http://amath.colorado.edu/outreach/demos/music/MathMusicSlides.pdf. Accessed 18 February 2012
2. Lindosland, http://en.wikipedia.org/wiki/Equal-loudness_contour. Accessed April 2018
3. Lindosland, http://en.wikipedia.org/wiki/A-weighting. Accessed April 2018

Chapter 8
Dispersion and Waves on Water

Abstract This chapter asks how the time development of a wave may be described numerically. The algorithm may offer a better understanding of wave motion than the traditional treatment of the topic. We proceed by discriminating between phase and group velocities and introduce the concept of dispersion. Numerical modelling of dispersion is described in detail, computer programs are provided, and the calculations demonstrate distortion of pulses of waves when they pass through a dispersive medium. Finally, we discuss various phenomena related to gravity-driven surface waves on water, based on a formula for phase velocity of waves on water. As a curiosity, we present at the very end a fun experiment with an oscillating water drop on a hot surface.

8.1 Introduction

Waves on water and sea have fascinated people through the ages. There exists a panoply of waveforms, and the underlying physics is so complex that even today it is almost impossible to make calculations on swirling waves like those illustrated by Katsushika Hokusai almost 200 years ago; see Fig. 8.1.

The waves we treat in this chapter are extremely simple in comparison. Nevertheless, we hope that even our simple descriptions can give you a much deeper understanding of the *phenomenon of waves* than you had prior to reading this chapter, which has three main themes: numerical calculation of the time evolution of a wave, dispersion including differences between phase and group velocities, and a review of gravity-driven waves on water.

Before starting a more thorough analysis, we will undertake a brief recapitulation of oscillations and waves in general. A feature common to all such phenomena is that:

- There is an equilibrium state of the system when oscillations and waves have died out.
- There is a "restoring force" that tries to bring the system back to equilibrium when it is not at equilibrium.

© Springer Nature Switzerland AG 2018
A. I. Vistnes, *Physics of Oscillations and Waves*, Undergraduate Texts in Physics,
https://doi.org/10.1007/978-3-319-72314-3_8

Fig. 8.1 Real waves are extremely complex, like "The Great Wave off Kanagawa". Katsushika Hokusai, Public Domain [1]

- There is an "inertial force" that causes the system to go past the equilibrium state even though the restoring force here is equal to zero.

For a swinging pendulum, the restoring force is a component of gravity; for waves on a string, the tension on the string acts as the restoring force. For sound waves in air or a liquid, pressure differences provide the restoring force through the compression of parts of the volume. The "inertial force" in all these examples is that expressed by Newton's first law. For surface waves on water, there are *two* restoring forces, namely gravity and surface tension.

8.2 Numerical Study of the Time Evolution of a Wave

It is very difficult to understand the mechanisms that lie behind the temporal development of a wave by starting from the wave equation and relying solely on analytical mathematics. If your repertoire consists of only analytical mathematics, you will find it difficult to understand why initial conditions are so crucial to how a wave develops, and how the boundary conditions affect the time development of the wave in detail. Instead, we will use numerical methods to review the mechanisms that govern the time evolution of a wave.

Fig. 8.2 Using numerical methods, a wave is described only at discrete positions in space and at discrete instants in time. Here one and the same wave are indicated at three different times. The first subscript specifies the position index, and the second subscript specifies the time index

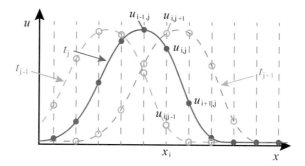

There are several reasons for presenting such a review. The most important is to bring to the fore the underlying algorithm because it can provide a better understanding of wave motion in general.

The starting point is the one-dimensional wave equation for a nondispersing medium (explained later in the chapter):

$$\frac{\partial^2 u}{\partial t^2} = v^2 \frac{\partial^2 u}{\partial x^2} \ .$$

In a numerical calculation, the solution is stated only at discrete instants and positions:

$$u(x, t) \rightarrow u(x_i, t_j) \equiv u_{i,j}$$

where

$$x_i = x_0 + i\Delta x, \quad (i = 0, 1, 2, \ldots, N - 1) \ ,$$

and

$$t_j = t_0 + j\Delta t, \quad (j = 0, 1, 2, \ldots, M - 1) \ .$$

Figure 8.2 illustrates how a wave is described numerically. For each instant of time, a numerical string describes the amplitude at the selected spatial positions. In the figure, parts of the position data points are displayed for three different instants.

In the chapter on numerical methods earlier in the book, it was shown that the second derivative can be expressed in discrete form as follows:

$$\frac{\partial^2 u}{\partial x^2} \equiv u_{xx}(x_i, t_j)$$
$$= \frac{u(x_{i+1}, t_j) - 2u(x_i, t_j) + u(x_{i-1}, t_j)}{\Delta x^2} \ .$$

This can be expressed more succinctly as:

$$u_{xx,i,j} = \frac{u_{i+1,j} - 2u_{i,j} + u_{i-1,j}}{\Delta x^2} . \tag{8.1}$$

In a similar way, the double derivative with respect to time can be expressed as:

$$u_{tt,i,j} = \frac{u_{i,j+1} - 2u_{i,j} + u_{i,j-1}}{\Delta t^2} . \tag{8.2}$$

The discretized version of the whole wave equation takes the form:

$$u_{tt,i,j} = v^2 u_{xx,i,j} . \tag{8.3}$$

Setting Eq. (8.2) in Eq. (8.3) and rearrangement of the terms gives:

$$u_{i,j+1} = u_{i,j} + (u_{i,j} - u_{i,j-1}) + (\Delta t\, v)^2 u_{xx,i,j} .$$

The expression shows that if we know the wave at an instant and at the preceding instant, we can calculate the amplitude of the wave at the next instant by using our prescription. This is an important formula that we should dwell on:

> The algorithm to calculate how a wave evolves in time and space is given by the equation:
>
> $$u_{i,j+1} = u_{i,j} + (u_{i,j} - u_{i,j-1}) + (\Delta t\, v)^2 u_{xx,i,j} . \tag{8.4}$$
>
> These terms are actually quite easy to understand:
> - The first term on the right-hand side states that we must begin with the current amplitude at a point in the wave when we calculate the amplitude for the next instant.
> - The second term corresponds to the assumption that the time derivative of the amplitude at our given point of the wave will be about the same at the next instant as it was in the previous one. This is the "inertial term" corresponding to Newton's first law.
> - The third term states that if the wave in our given point bulges (often bulging away from the equilibrium state), there is a "restoring force" that tries to pull the system back to the equilibrium state. See Fig. 8.2. This restoring force is closely related to the phase velocity of the wave. In the expression, the phase velocity appears in the second power. The phase velocity is therefore determined by how powerfully the *neighbourhood* affects the motion of any selected point in the wave. The algorithm can be visualized as shown in Fig. 8.3.

The algorithm in Eq. (8.4) shows that if we know the wave at all positions at an instant t_j and the wave as it was at the preceding t_{j-1}, then can we calculate the wave

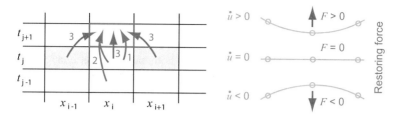

Fig. 8.3 Illustration of the cardinal algorithm that can be used for calculating the time development of a one-dimensional wave when the initial and boundary conditions are given. New amplitude at a particular point is determined by: (1) the amplitude "now" at that point; (2) the approximation that the velocity at the point will be the same at the next instant as in the previous; and (3) the restoring force from the nearest neighbours to the point, which will increase or decrease the change in position according to the sign of the curvature of the restoring force

as it will be at the next instant t_{j+1}. There are hurdles to be jumped over, presented by the initial conditions and boundary conditions, and we will get back to these shortly.

Equation (8.4) is probably the easiest expression to use, if we want to *understand* the rationale behind the algorithm developed below. The expression on the right-hand side of Eq. (8.4) is not suitable for the design of the program code itself. It is advantageous to put Eq. (8.1) into Eq. (8.4), and the result, after some rearrangement, comes out to be:

$$
u_{i,j+1} = 2 \left[1 - \left(\frac{v \Delta t}{\Delta x} \right)^2 \right] u_{i,j} - u_{i,j-1}
$$
$$
+ \left(\frac{v \Delta t}{\Delta x} \right)^2 (u_{i+1,j} + u_{i-1,j}) . \tag{8.5}
$$

Problem at the boundary of the region under consideration

Equation (8.5) is the central expression we use to calculate how a wave evolves in time, but the expression contains some important details that we need to look into. When we start the calculations, we assume that we know the initial conditions along the part of the wave we describe at the start of the calculations. For example, the amplitude at the instant $j = 0$ given by $\{u_{i,0}\}$ for $i = 0, 1, 2, \ldots, N$. But Eq. (8.5) also includes $x_{i+1,0}$ and $x_{i-1,0}$. The points $x_{-1,0}$ and $x_{N+1,0}$ do not exist, so our algorithm must employ some artifice for dealing with these terms. In other words, we must supply so-called boundary conditions for the particular problem at hand. These conditions apply at all instants in the calculations.

In practice, it may be almost impossible to find boundary conditions that are perfect for the calculations we want to make. The most common boundary conditions are "open/free" and "closed/fixed". In the former case, we put $x_{-1,j} = x_{0,j}$ and $x_{N+1,j} =$

$x_{N,j}$, in the latter case we set $x_{-1,j} = x_{N+1,j} = 0$. For a concrete calculation, we must choose how to set the boundary conditions, and in many cases, the response depends strongly on the physical system we try to describe.

For a wave that has zero amplitude at the boundary, we can consider, without incurring any error, the time evolution of the wave until the wave has spread to the edge of the calculation range. By making the calculation region large enough and limiting the time for which we consider the wave evolution, calculations of localized waves can be good even without worrying about boundary effects.

Problem with the starting instant

Another source of difficulty in Eq. (8.5) is the term $u_{i,j-1}$. If we start the calculations at time $t = 0$, there is no $u_{i,-1}$. Therefore, we get trouble already at the start of the calculations.

On the other hand, in all differential equations, we must use the initial conditions to arrive at the particular solution we seek. For a wave, it means that the initial conditions, for example, may be stated as the amplitude at all positions at $t = 0$, along with the time derivative of the amplitude at all positions at the same time. Based on this information, we can calculate positions at the starting instant and approximate positions one time-step earlier.

There are also other ways to specify initial conditions and procedures that can be followed for taking advantage of the initial conditions. We confine ourselves to the amplitude and its time derivative, both as a function of position.

The time derivative of the result at the point i can be specified as follows:

$$\dot{u}_{i,j} \equiv \left(\frac{\partial u}{\partial t}\right)_{i,j} \approx \frac{u_{i,j} - u_{i,j-1}}{\Delta t} \; .$$

Consequently,

$$u_{i,j-1} = u_{i,j} - \Delta t \, \dot{u}_{i,j} \; . \tag{8.6}$$

For $j = 0$ we get:

$$u_{i,-1} = u_{i,0} - \Delta t \, \dot{u}_{i,0} \; . \tag{8.7}$$

Threading together

Assume that the initial conditions are given by the amplitude $\{u_{i,0}\}$ at all positions along the wave and the time derivative of the amplitude $\{\dot{u}_{i,0}\}$ at all positions along the wave at the start time. Then Eq. (8.5) in combination with Eq. (8.7) can be used for the starting instant in the calculations. Equation (8.5) can be used for the remaining instants as many times as we wish. Along the way, one must take account of the boundary conditions.

8.2.1 An Example Wave

As an example, let us calculate how a Gaussian wave moves on a string. The initial conditions are a snapshot of the wave as it is at one point (both position and speed!), and we will follow its development in time.

The displacement as a function of position along the string is given analytically by:

$$u(x, t) = A \exp\left[-\frac{(x - vt)^2}{2\sigma^2}\right] = A \exp[f(x, t)] \tag{8.8}$$

where we have used the notation $\exp[f(x, t)]$ instead of the notation $e^{f(x,t)}$, since the expressions in this chapter are more complex than in previous chapters.

The time derivative of $u(x, t)$ comes out to be:

$$\frac{\partial u}{\partial t} \equiv \dot{u} = A \exp[f(x, t)] \frac{\partial f}{\partial t} = A \exp\left[-\frac{(x - vt)^2}{2\sigma^2}\right](-2)\left(\frac{x - vt}{\sqrt{2}\sigma}\right)\left(-\frac{v}{\sqrt{2}\sigma}\right)$$

$$= \frac{(x - vt)v}{\sigma^2} A \exp\left[-\frac{(x - vt)^2}{2\sigma^2}\right],$$

$$= \frac{v}{\sigma^2}(x - vt)u . \tag{8.9}$$

We choose to describe the wave on a string that is long in relation to the width of the Gaussian function, and we choose to follow the wave only so long that it does not come too close to a boundary. We use in the program a complete adherence to the endpoints along the way in the calculations.

We select the following parameters $A = 1$, $\sigma = 2\sqrt{2}$, $v = 0.3$ and allow x to cover the range from -20 through $+20$ in 400 equal steps. We try with $\Delta t = 0.1$ and follow the movement for 300 time increments. No units are provided, but we assume that all units are SI devices.

A computer program written in Matlab is given below. The code is also available at the "Supplementary material" web page for this book at http://www.physics.uio.no/pow. The program performs the calculations based on the expressions given above.

```
function waveAnimationX

% Generate position array
delta_x = 0.1;
x = -20:delta_x:20;
n = length(x);
nx = 1:1:n;   % Just for plotting purposes

% Generate and plot the wave at t=0
sigma = 2.0*sqrt(2.0);
u = exp(-(x/sigma).*(x/sigma)/2.0); % Gaussian shape
plot(nx,u,'-r');
```

Fig. 8.4 Profiles of the wave
at the start of the calculation
and after a lapse of 300
time-steps for initial
conditions that ensure a
constant wave shape as the
wave evolves in time

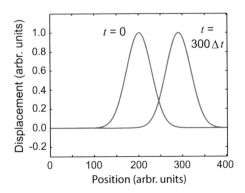

```
axis([1 n+1 -0.3 1.2])  % Ease comparison with animation
figure;

% Generate parameters and time derivative of the wave at t=0
v = 0.5;   delta_t = 0.1;
factor = (delta_t*v/delta_x)^2;
dudt = (v/(sigma*sigma))*x.*u;

% Calculate effective initial conditions:
u_jminus1 = u - delta_t*dudt;
u_j = u;

% The animation (one thousand time steps):
for t = 1:1000
    u_jplus1(2:n-1) = (2*(1-factor))*u_j(2:n-1) - ...
        u_jminus1(2:n-1) + factor.*(u_j(3:n)+u_j(1:n-2));

    % Handle boundary problem (fixed boundary)
    % u_j(-1) = u_j(n+1) = 0
    u_jplus1(1) = ...
        (2*(1-factor)).*u_j(1) - u_jminus1(1) + factor.*u_j(2);
    u_jplus1(n) = ...
        (2*(1-factor)).*u_j(n) - u_jminus1(n) + factor.*u_j(n-1);

    plot(u_j);
    axis([0 n+1 -0.3 1.2])
    drawnow;

    u_jminus1 = u_j;
    u_j = u_jplus1;
end;
```

Figure 8.4 shows the wave at the start and after the passage of 300 time-steps.
We see that the wave moves to the right (positive v) and that the waveform remains
unchanged.

Fig. 8.5 Profiles of the initial position of the string on a guitar just before it is released and made to oscillate

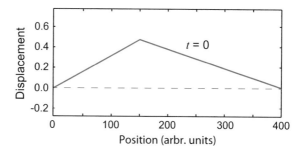

In an exercise at the end of the chapter, you are asked to investigate how the wave evolves if we use a $\ddot{u}$ which is either too small or too large compared to what it should have been. Which term in Eq. (8.9) is now incorrect if we want to preserve the waveform as the wave evolves? Would it be possible to explain the pattern observed in the simulations if you consider the initial condition as a sum of two different waves? It is crucial that you carry out this exercise and try to explain in your own words the mechanisms behind time evolution of a wave.

You are also urged to modify the code so that you can handle a case where the wave hits an interface between two media with different impedances (different phase velocities). It is recommended that you complete the exercise, for that would provide you with a significantly better understanding of waves.

Finally, another highly recommended exercise: When we play the guitar, we pull at the string so that the initial condition is a slanted triangle (with straight edges, see Fig. 8.5) and no motion before we release the string. Use the reasoning and algorithm that lies behind Fig. 8.3 to suggest how the string will move afterwards! It will reveal whether you have understood the algorithm or not!

Then perform a numerical calculation of the motion of the guitar string. It is easy since the string is at rest before you release it, and the endpoints are fixed (no motion is permitted). You may be surprised by the result!

It should be noted that the algorithm we use does not allow for any rigidity in the string. If the string has a certain stiffness, segments a little further than the neighbouring point will also affect the motion. A true guitar string will therefore get a little different motion than our calculations show, at least if we follow the motion over several periods. However, if we use a rubber band as a guitar string, we will observe a pretty close fit with the calculation, because the band has negligible stiffness. There are nice YouTube videos (shot with high-speed camera) showing the motion of a rubber band. Examples are "Motion of Plucked String" by Dan Russell and "Slow motion: Rubber string pulled and released" by Ravel Radzivilovsky. It is fun to compare your own calculations with these videos!

8.3   Dispersion: Phase Velocity and Group Velocity

In the previous section, we studied the mechanisms which govern the time development of a one-dimensional wave. We initially said that the calculations dealt with an idealized situation in which there was no *dispersion*. "No dispersion" means that a wave moves at the same speed no matter what its wavelength. In the calculations, the wave rate v was a constant.

For many physical wave phenomena, the restoring force will vary with the wavelength. In such situations, we say that the medium is *dispersive*. The multicoloured band we get when we send white light through a glass prism is an example of the phenomenon called *dispersion*. The spectrum is a consequence of the fact that light of different wavelengths travels with different speeds through the glass. This is the dispersion property of glass for light.

Let us take a closer look at this. We know that the refractive index of glass varies with the wavelength of light (see Fig. 8.6 for different types of glass). The refractive index increases as the frequency increases (wavelength decreases).

In Chap. 9, we will show at the phase velocity of the electromagnetic waves (light) in glass is given by the relation

$$v_\mathrm{p} = c_\mathrm{glass} = c_0/n(\lambda)$$

where c_0 is the light velocity in vacuum, $c_\mathrm{glass} = v_\mathrm{p}$ is the light velocity in glass, which by definition is the phase velocity of light in glass. $n(\lambda)$ is the refractive index which is wavelength dependent [see Eq. (9.36)].

> Phase velocity is the velocity a constant intensity laser beam (or a perfect harmonic wave) will have when it travels through a medium.

It follows from the data plotted in Fig. 8.6 that the phase velocity decreases as the wavelength decreases. Such a behaviour is called *normal dispersion*.

A slightly different graphic representation is often used to display dispersive behaviour. The alternative is to plot the angular frequency ω as a function of the wavenumber k. For a usual monochromatic wave $A\cos(kx - \omega t)$, the velocity (i.e., phase velocity) is given by:

$$v_\mathrm{p} = \frac{\omega}{k}$$

If we want to send information from one place to another, we cannot just send a constant intensity laser beam. We must make changes in the light output, and the information is conveyed through the changes.

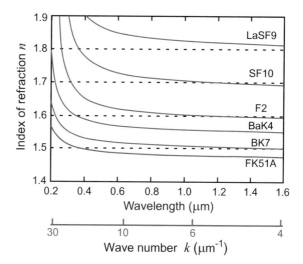

Fig. 8.6 Refractive index of light, from UV to IR, in various types of glass

In an optical fibre, we often send a number of light pulses one after the other. In the case of radio telecommunications, we also make variations in the radio wave that causes the wave to be seen as different "groups" of waves that come one after the other. It is remarkable that these pulses or wave groups propagate through the fibre or the atmosphere at slightly different velocities than a constant intensity laser beam would propagate. The pulses or wave groups propagate with what is called *group velocity*.

There is usually little difference between phase velocity and group velocity for electromagnetic waves. However, when we throw a pebble in a still body of water, the group velocity will only be half the velocity of water waves if a wave-making machine had generated continuous waves of about the same wavelength as we saw in the rings after the stone hit the water. It is therefore important to distinguish between phase velocity and group velocity!

It is dispersion that accounts for the difference between phase and group velocity. The connection between phase velocity v_p, (angular) frequency ω and wavenumber k is:

$$\omega = v_p k$$

When there is no dispersion, v_p is independent of k, and if we plot ω as a function of k, we get a straight line.

With dispersion, however, v_p will be a function of the wavelength and hence k. We can then write:

$$\omega(k) = \mathscr{F}(k) .$$

Fig. 8.7 Relation between the angular frequency ω and the wavenumber k for a given medium is called the *dispersion relation* for the medium. We distinguish between three different classes of media, as indicated in the figure. Note: The three curves represent completely different physical action mechanisms, so the three curves do not have to coincide for low k values. It is the form (curvature) that is important!

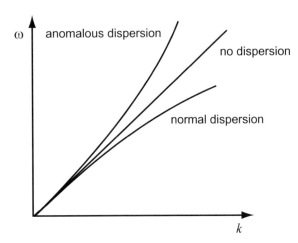

where $\mathscr{F}$ is some function. Then a plot of ω will no longer be a straight line.

> We call $\mathscr{F}$, which gives us the relationship between $\omega(k)$ and k, the *dispersion relation* for the medium in question. For a dispersive medium, an ω versus k plot will be a curved line, as shown in Fig. 8.7. When the curve bends downwards, the phase velocity decreases with the wavenumber (the wavelength becoming smaller). This is called *normal dispersion*. When the curve bends upwards, the phase velocity increases with the wavenumber (the wavelength becoming smaller). This is called *anomalous dispersion*.
>
> It can be shown that the group velocity is determined by the slope of the dispersion relation in the region under consideration:
>
> $$v_{\mathrm{g}} = \frac{\partial \omega}{\partial k} . \tag{8.10}$$

It can be shown that such a definition corresponds to the velocity of the "envelope" of a composite wave packet, at least where the envelope has a Gaussian shape. This corresponds to what we associate with "group velocity". For more complicated "envelopes" it is not always easy to specify a group velocity precisely, since the actual shape of the envelope will change as it moves.

The fact that the group velocity is the derivative of the dispersion relation $\omega(k)$ opens up interesting possibilities. We will use it several times in this book.

For example, let us find an expression for the variation of group velocity with the refractive index. The starting point is then the relationship:

$$\frac{\omega}{k} = \frac{c_0}{n(k)} .$$

Whence follows the relations:

$$v_g = \frac{\partial \omega}{\partial k} = -\frac{c_0}{n^2} \frac{\partial n}{\partial k} + \frac{c_0}{n} ,$$

$$v_g = v_p(\omega) \left(1 - \frac{k}{n} \frac{\partial n}{\partial k} \right) . \tag{8.11}$$

In normal dispersion, $dn/d\omega > 0$, which means that $v_g < v_p$, that is, the group velocity is less than the phase velocity. If we use the data from Fig. 8.6, we see that there is very little difference between the phase velocity and the group velocity when we transmit light through glass—at least as long as the wavelength is greater than 400 nm (visible light and IR). On the other hand, dispersion becomes a major problem if we try to send light of shorter wavelengths through glass.

In modern communication, we use optical fibres and wavelengths in the IR region, where the refractive index is almost completely independent of the wavelength. Then dispersion is very small, and it allows the use of short pulses, which ensures a high data transfer rate.

8.3.1 Why Is the Velocity of Light in Glass Smaller Than That in Vacuum?

This may be an appropriate moment for injecting a small aside, since in practice it has been found that relatively few know why light travels more slowly in glass than in vacuum. A clear indication is obtained by examining the expression for the velocity of light through a medium. This expression, usually given in books of general electromagnetism, is also discussed in detail in Chap. 9 in our book, and it reads:

$$c = c_0/n = \frac{1}{\sqrt{\varepsilon_0 \varepsilon_r \mu_0 \mu_r}}$$

where c_0 is the light velocity in vacuum, n is the refractive index, ε_0 is the permittivity in vacuum, ε_r is the relative permittivity, μ_0 is the magnetic permeability of vacuum and μ_r is the relative magnetic permeability. In glass, which is diamagnetic, μ_r is approximately equal to 1, and we get:

$$c = \frac{1}{\sqrt{\varepsilon_0 \varepsilon_r \mu_0}} = c_0 \frac{1}{\sqrt{\varepsilon_r}} .$$

When we remember that ε_r is a measure of how much polarization (shifting of positive and negative charges in each direction) we can achieve when we put a material into an electric field, we realize that polarization of glass is the reason that light goes slower through glass than in vacuum.

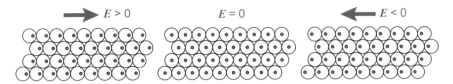

Fig. 8.8 When an electromagnetic wave passes through a slab of glass, the electromagnetic field in the wave will cause the electrons in the electron cloud around each nuclear core to shift as expected from the Coulomb force

Figure 8.8 shows what happens when an electromagnetic wave passes through glass. The electric field will alternate in a harmonic manner, with a value sometimes in one direction (across the direction of motion of the light), sometimes zero, and sometimes in the opposite direction. The electrons in the glass atoms will be affected by the electric field, and the entire "electron cloud" around each nuclear core will shift slightly relative to the core as indicated in the figure. In reality, the displacement is extremely small, since the electromagnetic field from the electromagnetic wave is usually small compared to the electrical forces between the core and electrons.[1] Yet, even in weak light, there will be a collective displacement of the electrons relative to the nuclei, and that is what really matters.

The collective displacement results in the glass being almost regarded as an antenna with oscillating currents. This oscillation in charges leads to the transmission of electromagnetic waves at the same frequency as the wave that started it all. However, we have seen in Chap. 3 (forced fluctuations) that there is generally a phase difference between the movement and the applied force. It is *the combination of the original wave and the phase shifted resonant wave from the oscillating electron clouds*, which ensures that the light velocity in glass is less than in vacuum.

It goes without saying that when the electromagnetic wave has passed the glass and gets into the air (almost like vacuum in our context), there will be no noteworthy polarization of the medium and the wave will not be delayed by the re-emitted wave. The velocity of light speed returns, of course, to (almost) the velocity of light in vacuum.

If we recall what was said in Chap. 3 about forced oscillations, we will also think of the resonance phenomenon. At certain frequencies, the amplitude became particularly large under the influence of the applied force. If you look at Fig. 8.6, you can see clear indications that something special happens to wavelengths just under 200 nm (0.2 μm). Then we are in the UV region. Several different physical processes will take place at the same time, but it may be useful to think that there will be some form of resonance in the electron oscillations around the nuclei. By thinking about the form of the resonance curve in Chap. 3, you can hopefully also imagine

---

[1] In a rapid pulse laser experiment in Germany in 2013, however, the electric field was so powerful that many electrons were stripped away from the core. Then the glass is transformed from being an insulator to a good electric conductor within a few femtoseconds!.

what happens if we go through resonance and reach even shorter wavelengths. Then curves in a diagram similar to Fig. 8.6 will slope the opposite way and we get the so-called anomalous dispersion. For some materials, the supposed resonance frequency will be at much longer wavelengths, and then we can achieve anomalous dispersion even for common visible light. However, it is somewhat strained to compare dispersion unequivocally with resonance in such phenomena, because more physical interactions usually contribute.

This is one of many aspects of physics where the simple laws and patterns discussed in the early chapters of the book appear. Simple principles are often *part* of the explanation even for more complicated phenomena, but seldom the *whole* explanation!

A nice little historical episode in this context:
In Newton's corpuscular model of light, diffraction was explained by the particles being *faster* through glass than in air, but the wave description gave the opposite prediction. Measurement of the velocity of light in glass was therefore regarded, during a certain period, as an important test for seeing whether a wave model or particle model accorded better with experiments. However, we cannot measure the velocity of light velocity in a coherent monochromatic wave. We must have a "structure" in the wave that we can recognize in order to be able to measure the velocity of light. This translates into the measurement of group velocity.

However, no one was able to measure the velocity of light in this way in the eighteenth and early nineteenth centuries. Foucault was the first to carry out the experiment. That happened in 1850, and the result showed that the velocity of light in glass was smaller than that in air, which supported the wave model for light. By this time, however, most physicists had reluctantly abandoned Newton's particle model for light. Experiments of Thomas Young (1801 double-slit experiment) and a work of Fresnel around 1820 (first opposed by Poisson, but corroborated by an experiment conducted by Arago), eventually convinced physicists that the wave model of light gave a better description than particle model. Please read about "Arago spot" in Wikipedia.

8.3.2 Numerical Modelling of Dispersion

Dispersion is a phenomenon that is somewhat difficult to understand. We present here a method which can be used to model dispersion numerically. We hope that, by reading the description of the method and the results furnished by it, you will understand dispersion better. We recommend that you run the computer program and watch how the waves within a group are moving forward, backward or stand still compared to the envelope of the wave group. It is fascinating, and you can easily observe such a pattern in real life when you look, for example, at the wake behind a boat on the sea.

We start this section by pointing out that a frequency analysis of a wave may be carried out both in time domain and in space domain. It is related to Fig. 6.2 in Chap. 6.

We often use the word "wave" rather uncritically, and seldom think that a real physical wave *must* have a limited extent in time and space. This means that when we describe a wave, for example, with the following expressions:

$$y(x, t) = A \cos(k_0 x - \omega_0 t) ,$$

this is at best just an *approximate description* of reality within a limited range of time and space. The velocity such a wave moves with is the phase velocity $v = \omega_0 / k_0$. A Fourier analysis of the time variation of the amplitude (meaning the displacement from the rest position) of such a wave at one fixed position $x = x_0$ would be

$$Y(\omega) = \frac{1}{2\pi} \int_{-\infty}^{\infty} y(x_0, t) e^{-i\omega t} dt \tag{8.12}$$

and Y would give one sharp peak in the frequency domain for $\omega = \omega_0$. This tells us that the amplitude of the wave varies harmonically with time at the fixed position $x = x_0$ and the frequency is $f = \omega_0 / 2\pi$ and the time periodicity is $T = 1/f$.

We could equally well have described the wave as a snapshot at one particular time $t = t_0$. A Fourier analysis can be carried out of the amplitude *as a function of position* for this particular time. We would then have a slightly different expression:

$$Y(k) = \frac{1}{2\pi} \int_{-\infty}^{\infty} y(x, t_0) e^{-ikx} dx . \tag{8.13}$$

The numbers we put into the calculation would be almost identical with the numbers describing the amplitude as a function of time. So mathematically, there will be no difference (in the numbers used). As physicists, however, we need to keep track of the difference and how the analysis should be used. Even in this case Y would give one sharp peak for $k = k_0$. We call k "the wavenumber", but that is the number of wavelengths within 2π metres and can equally well be called 2π times the "*spatial frequency*". This tells us that the amplitude of the wave varies harmonically with position at the particular time $t = t_0$ and that the spatial frequency is $f_s = k/2\pi$ and the spatial periodicity is the wavelength $\lambda = 1/f_s$.

In our numerical simulations of dispersion, we will use a description based on spatial frequencies, as will be apparent in the following.

Our chosen model of a real physical wave

As mentioned above, dispersion will have no influence on the motion of a pure harmonic wave. At the same time, it is also impossible to define a "group" for a pure harmonic wave. Thus, for a simulation of dispersion, we need a different model for a physical wave.

A physical wave changes character (form) from one region in space-time to another, and it can in general be very complicated indeed.

We have chosen to base our discussion on a "wave packet" that is formed by multiplying a harmonic wave with a Gaussian envelope. We describe the wave at a

Fig. 8.9 Frequency analysis of our wave packet (absolute values). Only the small region where the coefficients are clearly different from zero is shown. Along the horizontal axis, only element number has been given, in order to pick out the indices we are interested in

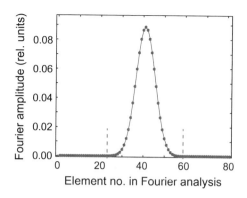

particular time $t = t_0 = 0$

$$y(x, t_0) = \cos(k(x - x_r))\, e^{-\{\frac{x-x_r}{\sigma}\}^2} . \tag{8.14}$$

Here x_r is the position where the wave packet has its maximum and the $1/e$ width of the envelope equals $2\sigma_x$. For our choice of parameters, the wave at the starting instant ($t = 0$) is shown in the upper left of Fig. 8.10.

We may Fourier transform this description of our wave packet as described in Eq. (8.13). From Fig. 5.12 in Chap. 5 and the description in that chapter, we know that the Fourier transform of Eq. (8.14) will have contributions mainly within a band of (spatial) frequencies, which correspond to a band of wavelengths. The band for our chosen model (discrete version) is shown in Fig. 8.9.

If we also bring the inverse Fourier transform into play, we can then state that

Our model of a physical wave $y(x, t_0)$ can be described as a sum (integral) of harmonic spatial waves with different wavelengths, for wavelengths in a limited wavelength band.

The key element in our simulation of dispersion

- We now know that the wave at $t = t_0 = 0$ can be described as a sum of spatial harmonic waves with different wavelengths.
- We also know that a harmonic wave will evolve in time as if dispersion was not present.
- However, dispersion implies that the phase velocity will depend on the wavelength.

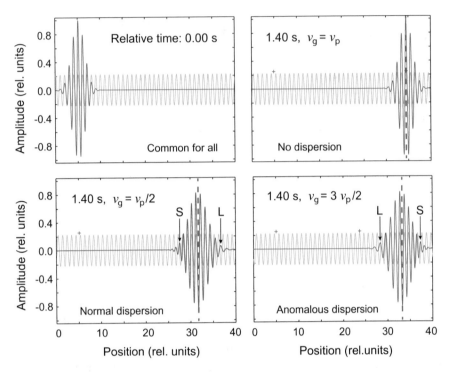

Fig. 8.10 Evolution of the wave packet with different dispersions is shown in blue. The green curve shows a wave with a wavelength corresponding to the maximum in Fig. 8.9 and is only included to show the difference between phase rate and group velocity in an animation. The wave packet at the start of time development is shown at the top left, and the next three graphs show the wave packet after it has been 1.4 s. **S** and **L** indicate shorter or larger wavelengths than the dominant one. See the text for details

- The time evolution of the total wave packet can then be calculated by adding a number of spatial harmonic waves that evolve in time with different phase velocities.

To be more specific, the last step can be implemented by replacing summation of spatial harmonic waves at one instant of time $\cos[k(i)x + \theta(i)]$ followed by a *common* time evolution—with summation of harmonic waves in space and time with *individual* time evolution and arguments $\cos[k(i)x - \omega(i)t + \theta(i)]$. However, the challenge is to determine $\omega(i)$. This is explained in the more detailed description of the actual simulation program at the end of this chapter, and it is illustrated in the Fig. 8.21 there.

Later in this chapter, we will discuss some physical wave phenomena in which the chosen behaviour is manifested.

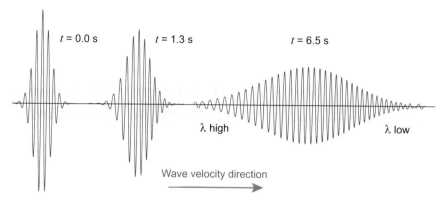

Fig. 8.11 Wave packet at the start, after it has moved (with anomalous dispersion) for a short time, and after a much longer time. Note the difference in wavelength in the front of the wave packet compared to the end. It is a fingerprint of dispersion

The results of our animation are given in three of the four plots in Fig. 8.10. A number of details emerge from the modelling/animation:

Figure 8.10 shows the important characteristics of dispersion:
- The waveform of the envelope curve does not change over time when there is no dispersion.
- When there is dispersion, the wave packet "spreads", its shape changes and the peak amplitude decreases, as shown in Fig. 8.11.
- When there is no dispersion, the group velocity equals the phase velocity, i.e. $v_g = v_p$.
- With normal dispersion, the group velocity is less than the phase velocity, more specifically $v_g = \frac{1}{2}v_p < v_p$ in our case.
- With anomalous dispersion, group velocity is greater than the phase velocity, more specifically $v_g = \frac{3}{2}v_p > v_p$ in our case.
- Group velocities are exactly as expected on the basis of Eq. (8.10).
- Although the wave packet ("the group") moves with a different velocity than the phase velocity, the *individual wave peaks* within the envelope moves approximately with the phase velocity of a wave whose wavelength corresponds to the dominant component in the frequency spectrum (shown green in Fig. 8.10).
- This means that the wave packet under the envelope curve moves forward *relative to the envelope* with normal dispersion and backward with anomalous dispersion.
- This means that at normal dispersion, wave packets will apparently disappear at the front of a wave packet as time passes and appear to grow out of nothing at the rear end of the wave packet. For anomalous dispersion, the opposite holds. See Fig. 8.12.

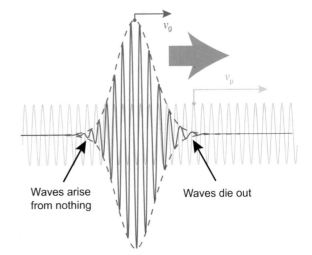

Fig. 8.12 Wave packet after it has moved for a while with normal dispersion. The phase and group velocities, v_p and v_g, respectively, are indicated by arrows

- In normal dispersion, the phase velocity for small wavenumbers, i.e. long wavelengths, is greater than the phase velocity for short wavelengths. Then the longest wavelengths will dominate the fastest part of the group, and the shortest wavelengths will dominate the last (slowest) part of the group. For anomalous dispersion, it is the opposite. In Fig. 8.10, long wavelengths are marked with L and short with S.

8.4   Waves in Water

It is time now to describe waves on the surface of water. However, we will start with qualitative descriptions before we grapple with a mathematical description where it is possible to go into more detail.

In Chap. 6, we derived the wave equation for waves on a string and waves in a medium. It would have been nice to go through a similar derivation for surface waves in water, but this will not be attempted here, since the task is rather demanding. The interested reader is referred instead to books in hydrodynamics or geophysics. We will nevertheless look at some details. In Fig. 8.13 is shown one possible model, which can be used as a basis (the model is the starting point for the derivation in, for example, the book by Persson, reference at the end of this chapter).

Here we consider a vertical volume element parallel to the wavefront has the same volume, regardless of whether it is in a trough (valley) or a crest (top). In the figure, this would mean that $V_1 = V_2$. However, since the pressure is equal to the air pressure

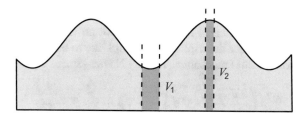

Fig. 8.13 One wave model envisages that vertical volume along the wavefront preserves its value, regardless of whether the column is in a trough (valley) or a crest (peak). The surface area of the cross sections will change, but we ignore that at first

above the water surface (approximately the same above all volume elements) and the pressure increases with the depth inside the water, the pressure at a given height above the bottom is higher in the volume element corresponding to the wave peak compared with that in the wave valley. In this way, we can regard the wave as a longitudinal pressure wave that moves with the velocity of the wave.

In Chap. 6, sound waves in air and water were described as pressure waves. The model in Fig. 8.13 looks similar to that description, but is still quite different!

For sound waves, we considered the gas or liquid as compressible fluids, that is, if we increase the pressure, the volume will decrease. The compressibility modulus was central to the derivation. Gravity, on the other hand, played no role whatsoever.

When surface waves on water are modelled, we completely ignore the compressibility. Regardless of pressure changes, a volume element retains the same volume.

In surface waves, large pressures will mean that the volume element is compressed across the wavefront, that is, in a volume element below a wave peak.

We may wonder whether it is reasonable to operate with completely different models of sound waves and surface waves, and of course there are transition zones where these descriptions will be at variance with each other. However, there are physically good reasons to operate with different models.

In the case of sound waves, we are most interested in frequencies in the audible region (and possibly ultrasound). That is, from about 20 Hz and upwards. The period is 50 ms or less (in part much less). If sound would lead to surface waves as described in this chapter, we must move significant amounts of water up to several metres in 25 ms or less! It would require enormous powers (according to Newton's second law).

On the other hand, we can transfer large amounts of water a few microns within 25 ms as required for sound waves, and still shorter times (higher audio frequencies). The powers that are needed for this task are achievable.

Surface waves on water have a much lower frequency (at least for large wave heights). Then we get time to move large amounts of water from a wave bottom to a wave peak with the available power.

One must also take into account the time scale and Newton's second law, which means that we operate with completely different models of sound waves in water and gravity-driven surface waves on water.

A better model

All in all, the model given in Fig. 8.13 does not provide a good description of surface waves. For a better description, we would like to base ourselves on one of the basic equations in fluid mechanics, namely Navier–Stokes equation:

$$\rho \left(\frac{\partial \vec{v}}{\partial t} + \vec{v} \cdot \nabla \vec{v} \right) = -\nabla p + \nabla \cdot \vec{\mathcal{T}} + \vec{\mathcal{B}}$$

where ρ is mass density, $\vec{v}$ is the flow rate, p is hydrostatic pressure, $\vec{\mathcal{T}}$ is a stress vector (may include surface tension) and $\vec{\mathcal{B}}$ stands for "body forces" that work per unit volume in the fluid. ∇ is the del operator.

It may be useful to look closely at Navier–Stokes equation and recognize that it is largely a further development of Newton's second law for a continuum fluid.

Navier–Stokes equation is nonlinear, which means that solutions of this equation do not necessarily follow the superposition principle. If two functions separately are solutions of the equation, the sum of these functions will not necessarily be a solution of the equation. Another characteristic feature of nonlinear equations is that they can have chaotic solutions, that is, solutions where we cannot predict how the solution will develop in time (in a purely deterministic way). Even the slightest change in initial conditions or boundary conditions could result in the solution after a time being having wildly different values. This has come to be called "the butterfly effect". The flap of a butterfly's wings can cause weather development after a long while to be completely different from what it would have been had the butterfly not flapped its wings.

There are some interesting mathematical challenges associated with Navier–Stokes equation today, but we will not mention it here.

My main concern is to point out that there is a wealth of different phenomena related to motion in fluids, and amazingly many physicists and mathematicians have been interested in water waves. These include Newton, Euler, Bernoulli, Laplace, Lagrange, de la Coudraye, Gerstner, Cauchy, Poisson, Fourier, Navier, Stokes, Airy, Russell, Boussinesq, Koertweg, de Vries, Zabusky, Kruskal, Beaufort, Benjamin, Feir and others. We are talking about monster waves, tsunamis, solitary waves, etc. The field has a rich tradition, also in the Norwegian research milieu, and there is still a lot to be tackled!

In our time, computers have become so powerful and so many numerical methods have been developed for use in mathematics and physics that we can now grab wave descriptions in a completely different way than could be done a few decades earlier. As an example of the development that has taken place, Professor Ron Fedkiw (born 1968), as working with Computer Sciences at Stanford University, received an Oscar award in 2008 for his efforts to animate realistic water waves for use in the film industry (including the film "Poseidon"). For those who are students today and will become familiar with numerical methods for solving mathematical and physical problems, this is extra fun. After completing your studies you will have the skills that would enable you to produce, only with modest effort, realistic animations of similar to those of Ron Fedkiw!

8.4.1 Circle Description

Let's now give a picture-and-word description of the waves themselves. Figure 8.14 shows a vertical cross-section of the wavefront. The solid curve shows the wave at a given moment, and the dotted curve shows the wave a short while later. The wave moves to the right in this case.

In the figure, arrows are drawn to show which direction the water must move in order to let the wave as it is now to become what it will be. The arrows in the upper half are quite easy to understand, while the arrows in the lower half may be harder to get hold of. However, we recall that the wave does not necessarily lead to a net transport of water in the direction of the wave, so water that moves forward in a part of the wave must move backwards into another part of the wave. And water that moves upward in part of the wave must move down in another part. If we keep these facts in mind, the directions of the arrows begin to make sense.

Note that the water must move *both* along the wave propagation direction and across it. This means that the wave is a mixture of a longitudinal and a transverse wave.

If we draw the direction of motion and relative position of the same small volume element at different times while a wave peak passes, we get a chart as in the lower part of the figure. It appears that the water in the surface moves along a vertical circle across the wavefront.

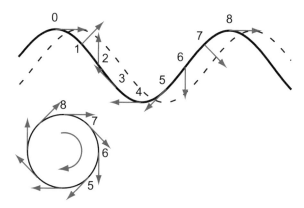

Fig. 8.14 Upper part indicates in which direction the water at the surface moves when the wave rolls to the right. In the lower part, the position and speed of one and the same volume element are drawn as a wave peak passes. The current wave element is that at position 8 at the beginning, but at the next moment it is located on the part of the wave that is in line with point 7 in the upper part. At the next moment, it has a location in the waveform that corresponds to point 6, etc. The result is that the volume element we follow appears to move in the clockwise direction as time passes

Fig. 8.15 When we want to indicate how the water moves between the surface and the bottom, simple sketches like this are used. However, sketches like this give a rather simplistic picture of what is happening

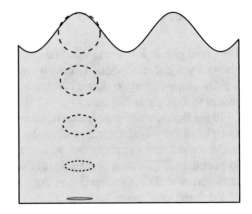

Further down into the water, the circular motion will change from being near-circular (as in the surface) to a more and more flattened ellipse, as shown in Fig. 8.15. At the bottom of the bottom, the movement is almost a pure horizontal movement back and forth. We can notice that when we snorkel at the bottom of a lake, and see aquatic plants and weeds swing slowly back and forth as waves pass on the surface.

This description, however, applies only to shallow water, that is for water that is not much deeper than the wavelength (the distance between the wave crests).

For deeper water, the waves on the surface will only propagate downwards a short distance, but near the bottom, the waves on the surface will not be noticed at all.

It is possible to see the circular motion by spraying small droplets of coloured oil into the water, provided that the mass density of these drops is about the same as that for water. We can then follow the movement of the drops in water waves, as is done in the basement of Abel's House at UiO and in Sintef's wave tanks in Norway. However, I have been told that it is harder to show these circular motions than we might infer from the textbooks.

When we portray water waves by drawing circles and ellipses at different depths, we must recognize that such a description can be easily misunderstood. How should we look at the circles and the ellipses for subsequent volumes in the wave direction? Here there must be some sort of synchronization that does not emerge from the figure and which necessarily has to give a more detailed description than can be conveyed through simple sketches.

The sinusoidal form is by no means the best model for surface waves on water. Often the wave tops are more pointed than the bottoms, as indicated in Fig. 8.16. The larger the amplitude, the steeper the top becomes. However, there is a limit to this tendency. When the wave peak becomes larger than about 1/7 of the wavelength, the wave is often unstable and can, e.g., go over to a breaking wave. At the border, the angle between the upward and downward part of the wave peak is about 120° (an angle that of course does not fully apply to the actual vertex).

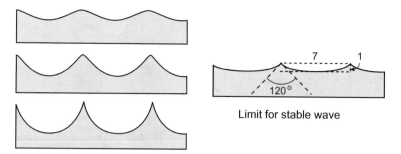

Limit for stable wave

Fig. 8.16 Waveform is usually such that the top is more pointed than the bottom. The effect becomes clearer as amplitude increases. When the peak to peak amplitude is 1/7 of the wavelength, we reach a limiting value after which further increase in amplitude often gives an unstable wave

8.4.2 Phase Velocity of Water Waves

Although we have not shown how the wave equation will actually look for surface waves, we can establish an approximate expression of one characteristic of the solutions, namely the phase velocity of water waves. The expression is:

$$v_p^2(k) = \left[\frac{g}{k} + \frac{Tk}{\rho} \right] \tanh(kh) \qquad (8.15)$$

where k is the wavenumber, g the acceleration due to gravity, T the surface tension, ρ the mass density and h the depth of water. The formula applies to a practically flat bottom (compared to the wavelength).

The first term inside the square brackets indicates the contribution of gravity to the restoring force, while the second indicates the contribution of surface tension. The first term thus corresponds to so-called gravity-driven waves, while the second term corresponds to what we call "capillary waves".

Since the wavenumber k occurs in the denominator of one term and in the numerator of the other, it follows that the gravitational term will dominate for small wavenumbers (long wavelengths), while the surface tension will dominate at large wavenumbers (small wavelengths). It may be interesting to find the wavelength where the two terms are about the same magnitude. We put then:

$$\frac{g}{k_c} = \frac{Tk_c}{\rho} .$$

The subscript c indicates a "critical" wavenumber where the two contributions are equal. The result is:

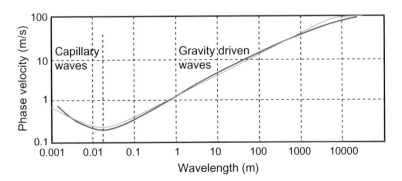

Fig. 8.17 Phase velocity of surface waves on water. The red curve in this figure is taken from R. Nave, (water depth was not specified). The blue curve is calculated using Eq. (8.15) with $h = 1000$ m. Inspired by [2]

$$\frac{1}{k_c^2} = \frac{T}{g\rho} \ .$$

Since $k = 2\pi/\lambda$ we find finally:

$$\lambda_c = 2\pi\sqrt{\frac{T}{g\rho}} \ .$$

For pure water at about 25 °C and 1 atmosphere, $T = 7.197 \times 10^{-2}$ N/m. Then the critical wavelength becomes:

$$\lambda_c \approx 1.7 \text{ cm.}$$

In other words, the surface tension will dominate the phase velocity of waves of wavelength appreciably smaller than 1.7 cm, while gravity will dominate for wavelengths considerably larger than 1.7 cm.

The phase velocity is actually smallest when the wavelength is about 1.7 cm, being only 0.231 m/s. Both shorter and longer wavelengths increase the phase velocity, and at very long wavelengths, the phase velocity can reach more than 100 m/s. Figure 8.17 shows the calculated phase velocity for wavelengths from 1 mm to 10 km. The calculations are essentially based on the fact that the water depth is large relative to the wavelength (something that cannot be attained here on earth for the longest wavelengths!).

We will immediately look at the expression for phase velocity, but first of all remind us of some features of the hyperbolic tangent function. The entire range of hyperbole trigonometric functions can be defined in an analogous manner to normal sine, cosine, etc. (all of which can be described by exponential functions with complex exponents). For the hyperbolic functions, the expressions look like this:

$$\sinh(x) = \frac{e^x - e^{-x}}{2} \; ,$$

$$\cosh(x) = \frac{e^x + e^{-x}}{2} \; ,$$

$$\tanh(x) = \frac{e^x - e^{-x}}{e^x + e^{-x}} \; .$$

In what follows, we focus on how hyperbolic tangent behaves when the argument is much smaller or much larger than 1. Then it applies:

$$\tanh(x) \approx x \text{ for } |x| < 1 \; ,$$

$$\tanh(x) \approx 1 \text{ for } x > 1 \; .$$

In Eq. (8.15), the argument for tanh is equal to hk. The argument can also be written as

$$hk = \frac{2\pi h}{\lambda} \; .$$

It is then natural to distinguish between "shallow water" characterized by $h < \lambda/20$ and "deep water" characterized by $h > \lambda/2$. These limits mean that the shallow water condition corresponds to:

$$hk < \frac{2\pi\lambda}{20\lambda} = \frac{\pi}{10} < 1.0$$

and deep water condition corresponds to:

$$hk > \frac{2\pi\lambda}{2\lambda} = \pi > 1.0 \; .$$

It is time now to discuss some main features of Eq. (8.15). For shallow water first and wavelengths well above 1.7 cm (so that we may ignore the surface tension term) follow:

$$v_p^2(k) = \frac{g}{k} \tanh(kh) \approx \frac{g}{k} kh = gh \; ,$$

$$v_p(k) = \sqrt{gh} \; .$$

We see that the phase velocity is independent of the wavelength (wavenumber). Furthermore, we notice that phase velocity decreases as the depth decreases.

This gives a good effect. When waves come from the ocean towards a longshore beach, waves that are inclined inward will move fastest in the part where the depth is greatest. That is, the part of the longest wave will go faster than the part of the wave that is farther in. Generally, this causes the wavefront to become quite parallel to the shoreline, no matter what direction the waves had before approaching the beach.

For a deep water coast all the way down to the mountain cliffs down to the water, there is no equivalent effect and the waves can come in towards the cliffs in any direction.

For deep water waves, the phase velocity (assuming that the surface tension plays a negligible role):

$$v_f^2(k) = \frac{g}{k} \tanh(kh) \approx \frac{g}{k} 1 = \frac{g\lambda}{2\pi} \,,$$

$$v_f(k) = \sqrt{\frac{g}{2\pi}} \sqrt{\lambda} \approx 1.25\sqrt{\{\lambda\}}\ \text{m/s} \,.$$

where $\{\lambda\}$ means the value of λ (without units) measured in the number of metres.

Thus, in deep water the phase velocity will change with the wavenumber (wavelength). Such a relationship has been called *dispersion* earlier in the chapter.

Increasing the wavelength by two decades in our case, the phase velocity will increase with a decade. This is reflected approximately in Fig. 8.17.

Something to ponder over

It may be interesting to know that the ocean wave with the highest wavelength here on earth has a wavelength of 20,000 km. It rolls and goes all the time. Can you guess what sort of wave it is? Do you want to characterize it as a surface wave that is gravity driven? If so, does it fall under our description above and will it have a wavenumber given by our formulas? You can think about it for a while!

When we treated Eq. (8.15), we said that for wavelengths well over 1.7 cm, gravity dominated the wave motion. For capillary waves with wavelength significantly less than 1.7 cm, the surface tension dominated. These numbers apply at the ground surface.

A water drop will have a shape that is determined by both gravity and the surface tension. When the gravitational force disappears, such as, for example, in the weightless state of Spacelab, it is possible to make water droplets that are almost perfectly spherical, even with a diameter up to

10 cm. Waves on the surface of such water balls will in weightless state be dominated by surface tension even at wavelengths greater than 1.7 cm.

8.4.3 Group Velocity of Water Waves

We have previously given in Eq. (8.15), an expression for the phase velocity of surface waves in water, but reproduce the formula here to refresh the reader's memory.

$$v_p^2(k) = \left[\frac{g}{k} + \frac{Tk}{\rho} \right] \tanh(kh) .$$

As usual, here k is the wavenumber, g the acceleration due to gravity, T the surface tension, ρ the mass density and h is the depth of the water. The expression can be derived if we start by just taking into account gravity and surface tension, and we ignore viscosity, wind and a tiny but final compressibility to the water.

We have previously found expression of the phase velocity for gravity-driven waves for shallow and deep water. Now we will also discuss group velocity and describe three of the four possible simple special cases a little more in depth:

1. Gravity-driven waves with a small depth relative to the wavelength, i.e. the product $hk \ll 1$:

The wavelength is assumed to be large relative to the critical (1.7 cm) and from Eq. (8.15) follows:

$$v_p^2(k) \approx \frac{g}{k} hk ,$$

$$v_p \approx \sqrt{gh} .$$

This has been shown earlier, but let us also look at the group velocity. We then use the relationship $v_p = \omega/k$ and get:

$$\frac{\omega}{k} = \sqrt{gh} ,$$

$$\omega = \sqrt{gh}\, k ,$$

$$v_g = \frac{d\omega}{dk} = \sqrt{gh} = v_p .$$

Therefore,

$$v_g = v_p \, .$$

This means that there is simply no dispersion.

2. Gravity-driven waves in deep water

In this case, we found:

$$v_p^2(k) \approx \frac{g}{k} \, .$$

We set again $v_p = \omega/k$ and get:

$$\frac{\omega^2}{k^2} = \frac{g}{k} \, .$$

This leads to the following dispersion relation:

$$\omega \approx \sqrt{gk} \, .$$

The group velocity is thus seen to be:

$$v_g = \frac{d\omega}{dk} = \frac{1}{2}\sqrt{\frac{g}{k}} \, .$$

$$v_g \approx \frac{1}{2} v_p \, . \tag{8.16}$$

Thus, we see that the group velocity is approximately equal to half the phase velocity.

Wake pattern from ships often fall into this category. The single waves seem to roll faster than the "plow" or "fan" that follows the boat (see Fig. 8.18). As a result, the single waves roll in a way past the "fan" and disappear soon afterwards. We will look into this in some respects.

3. Short ripples in deep water

Here the wavelength of the waves is small relative to the critical wavelength of 1.7 cm. At the same time, the wavelength is much less than the depth of the water. Then we get surface tension-driven waves and

$$v_p^2(k) \approx \frac{Tk}{\rho} \times 1 = \frac{\omega^2}{k^2} .$$

The dispersion relation is easily seen to be:

$$\omega \approx \left(\sqrt{\frac{T}{\rho}} \right) k^{\frac{3}{2}} .$$

The group velocity in this case becomes:

$$v_g = \frac{d\omega}{dk} = \left(\sqrt{\frac{T}{\rho}} \right) \frac{3}{2} k^{\frac{1}{2}} = \frac{3}{2} \sqrt{\frac{Tk}{\rho}} ,$$

$$v_g = \frac{3}{2} v_p .$$

In this case, the group velocity is actually greater than the phase velocity (corresponding to anomalous dispersion). In this case, individual waves seems to appear from nothing at the front of the group of waves, and then move "backwards" through the group and disappear. However, relative to the water, the single waves will always propagate away from the source that created the waves (as long as we do not have reflection), but the illusion of walking backwards is because the group velocity is even greater than the phase velocity.

8.4.4 Wake Pattern for Ships, an Example

Many are not used to identifying what is meant by a group of waves and what is meant by single waves. Left part of Fig. 8.19 attempts to show this. The figure refers to the photograph in Fig. 8.18. The fan with many single waves that extends slightly across the outer edge of the fan forms the group of waves. This fan is expanding at a speed that is the group velocity. However, each single wave will wander in a different direction than the fan as such and with a different wave velocity which is now the phase velocity.

We have previously concluded that for deep water waves, the group velocity is about half the phase velocity [see Eq. (8.16)]. This means that the single waves move

Fig. 8.18 Photograph of a boat with waves forming a V-shaped wake behind it. See further discussion in the next section. Arpingstone, Public Domain, [3]

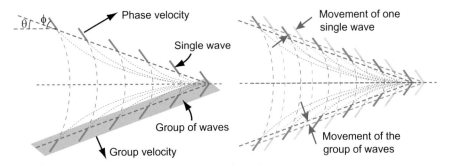

Fig. 8.19 To the left: Identification of the group moving with group velocity and single waves moving at phase velocity in waves from a boat. The figure is a continuation of Fig. 8.18. To the right: Detail showing how far the wave group and how far a single wave has moved across each of its wavefronts over a certain period of time. The figure clearly shows that the group velocity is lower than the phase velocity of these water waves

faster than the group. The single waves therefore appear to arise of almost nothing on the inside of the group, and scroll across the group, and almost disappear when they reach the outer edge of the group.

If you have ever paddled a canoe and watched a bit nervously how fast single waves approach the canoe after a boat has passed, you might have wondered that the waves which looked so scary seem to have vanished on their own before reaching the canoe. Only much later than we first became aware of them do the waves reach the canoe. The waves make it to the canoe only when the *group* reaches there, and the group moves half as fast as the single waves. This time course comes out beautifully in the animation of how a wave packet evolves in time using the computer program discussed earlier in the chapter (listed at the end, just before "Learning objectives").

In the right part of Fig. 8.19, the waves are drawn at one point and a little later. Then it becomes clear that the group has gone a much shorter distance than the single waves in the period we are studying. It always remains true that the wave pattern behind the boat is stationary relative to the boat. When the boat has moved 10 m forward, the entire wave pattern behind the boat has also moved 10 m ahead.

Lord Kelvin (W. Thomson) claimed long time ago (1887) that wakes pattern from ships are fanning out at a constant angle of 19.47°, no matter the speed of the vessel. This was regarded as an established truth. This has lately been shown not to be true. Especially, the wake pattern is quite different from what is seen in Fig. 8.18 for high-speed boats. The subject is described by Ceri Perkins in an easily readable article in Physics World May 30, 2013, and in more details, for example, in the paper "A solution to the Kelvin wake angle controversy" by A. Darmon, M. Benzaquen and E. Paphal, 2013, available at https://www.gulliver.espci.fr/sites/www.gulliver.espci.fr/IMG/pdf/darmonbenzaquen2013.pdf. It is fascinating to see the broad range of pattern a ship wake can take.

Additional comments for the most interested:
There has been a lot of research on phase and group velocities since about 1980. Much of this is related to light.

In February 2015, Giovannini and colleagues published an article in *Science* that shows taking the velocity of light in vacuum is not necessarily c as Einstein's relativity theory indicates. Light velocity equal to c applies only to light in the form of plane waves. For some other wave configurations, which the authors designate as "spatially structured photons", the velocity of light in vacuum is slightly lower than c (not a big difference, but demonstrably smaller).

We have long known that when light goes through glass (or a water drop for that matter), light of different wavelengths travels at different speeds. The refractive index is wavelength dependent, $n(\lambda)$, which is again the expression of dispersion. Light in glass shows normal dispersion.

However, in the last few decades a number of special materials have been developed, and some of these have a highly varying phase velocity for light with different wavelengths. Therefore, we can get widely varying phase and group velocities, and even materials where phase velocity has one direction and group velocity has the opposite direction.

There are also artificial materials and experimental relationships where we can slow down the light enormously, even "stop" it for shorter periods, then start it again (search for Lene Hau at Harvard University to get an insight into an exciting research field. Lene is Danish and is a favourite for the Danish press).

In some materials, a light pulse can travel—according to some people—faster than light in the vacuum, and in principle, we threaten Einstein's relativity theory in this way. When we look at what happens, we see that the claim of "faster than the light speed in vacuum" can be defeated. It all depends on how we define this or that, but Einstein's relativity theory is not exactly threatened by these experiments, on the whole. What the future will bring is something harder to contemplate!

Dispersion is also relevant to matter waves in quantum physics. Group velocity is defined through dispersion relations where $\omega(k)$ is described and we use $v_g = d\omega/dk$. For matter waves, the wavelength through the Broglie relationship is related to the momentum and the frequency of energy. For matter waves, therefore, we have dispersion if the energy does not increase with the momentum in the expected manner.

Dispersion turns up in many other contexts, among them the so-called Kramers–Kronig relation that shows that dispersion is related to the amount of absorption for different wavelengths in the medium. To a certain extent, this is linked to forced oscillations and Q-values, as we have mentioned earlier, but we do not have the time to go further in depth.

8.4.5   Capillary Waves

We all know waves at sea. Less commonly known are oscillations in small water droplets where surface tension is the dominant restoring force. When a drop falls from a tap, it will oscillate while it falls. Examples of this are found in references 1 and 2 at the end of the chapter.

Standing waves in a water drop we can observe when we place some water in the pit of an old-fashioned electric stove, provided that the plate is so hot that the drop floats atop a cushion (steam cushion) that forms. We can get beautiful quantized oscillations with an asterisk shape where an integer number of arms swings back and forth (see Fig. 8.20). Slight variation in heat or size of the drop may cause it to suddenly change the swing pattern from, for example, a five-arm to a four-arm star. The arms are shot out and pulled back in such a way that we will perceive an octagonal star (since we cannot follow the rapid movement with the unaided eye).

The purpose of this description is to recall that classical physics is full of quantified states, in an analogous manner to what is found on the atomic scale described in quantum physics. We have already seen in other chapters other examples of quantization on macroscopic scale, such as oscillations on a string and sound waves in a musical instrument.

The reason for quantization is that we are dealing with waves and the associated boundary conditions. For waves on a guitar string, the quantization is a consequence of the fact that the amplitude at the endpoints must be equal to zero. This is completely analogous to quantization of the wave function in quantum physics (e.g. for a "particle in box").

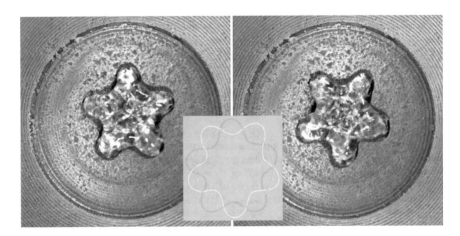

Fig. 8.20 Pictures of an oscillating water drop. The picture to the left is taken a few milliseconds before the image to the right. The two pictures show the extremes of the oscillation of this drop. The movement can be considered as a standing wave in the drop. The images are selected from a video taken by high-speed camera at the Department of Physics, University of Oslo (M. Baziewich and A. I. Vistnes). The video is available at the "Supplementary material" web page for this book at http://www.physics.uio.no/pow

8.5 Program Details and Listing

Given below is a Matlab program that can be used to explore how dispersion affects the time development of a wave packet (a wave group). The program consists of a main program that calls on four functions. One must find out oneself which parameters are to be changed when one wants to switch between no dispersion, normal dispersion and anomalous dispersion. It is natural to change several of the functions if you want to enter the appropriate parameters to model completely specific physical wave phenomena. However, it is imperative to understand the parts of the program that you want to change; otherwise, the result may turn out to be meaningless.

A description of the different parts of the actual program

The function/script $pg3$ is the main program that activates the different modules in the complete program. We have to choose in the program code whether we want normal, anomal or no dispersion.

The wave packet is calculated in function pg_wpack.

Since the wave is limited in extent, a Fourier analysis of $y(x)$ in the function pg_fft will yield a range of spatial frequencies. For our selection of parameters, the result is illustrated in Fig. 8.9. The components with indices 23–59 (marked with blue vertical lines) contain all the components that are notably different from zero (compared to the value of the most powerful component). The different indices correspond to each (spatial) frequency (as explained in detail in the chapter on Fourier analysis). This range of components has to be stated in the code of $fg3$.

Fourier analysis gives us (spatial) frequency, amplitude and phase of each components of interest.

$$k(i) = (i - 1)\frac{2\pi x_{max}}{N} \quad A(i) = 2\,abs(Y(i)) \quad \theta(i) = atan2(imag(Y(i)), real(Y(i)))$$

where $Y(i)$ is the ith element in the Fourier transform of $y(x)$, and x_{max} and N are, respectively, the greatest value of the position and number of points in our description. The factor of 2 in A is because we use only the lower half of the Fourier spectrum (not the folded part). The expressions "abs", "atan2", "imag" and "real" are all Matlab functions.

We know from Chap. 5 on Fourier transformation that we can describe the same function as in Eq. (8.14) by a "reverse Fourier transform":

$$z(x) = \sum_{i=23}^{59} A(i)\cos[k(i)x + \theta(i)] \tag{8.17}$$

where we have included only the components that are worth mentioning for the result. Plotting $z(x)$ calculated with Eq. (8.17) and comparing it to a plot based on Eq. (8.14) will not reveal any difference visually.

Note that in Eq. (8.17), we add cosine functions, each contribution having the same amplitude over the *entire region* for which we calculate. There is no specific information about where the peak of the wave packet should be or how wide it is. All of this is information hidden in amplitudes and phases of the various frequency components that are included.

Both Eqs. (8.14) and (8.17) describe the wave at time $t = 0$. Equation (8.17) is most useful in our context because it is well suited to describe how the wave will *evolve* over time when we have dispersion. Then the algorithm we used first in the chapter will not suffice because there is no simple wave equation when the phase velocity is not constant.

If we use Eq. (8.17), we can get the time evolution simply by replacing $\cos[k(i)x + \theta(i)]$ with $\cos[k(i)x - \omega(i)t + \theta(i)]$. Then each spatial frequency component will evolve with its individual phase velocity. This demonstrates why Fourier analysis sometimes is very useful! However, the challenge is to determine $\omega(i)$. The function *pg_omega* takes care of that challenge.

We have chosen the following three variants:

$$\omega(k) = v_p \frac{k}{k_{\text{dom}}} \qquad \text{No dispersion}$$

$$\omega(k) = \kappa_1 v_p \sqrt{\frac{k}{k_{\text{dom}}}} \qquad \text{Normal dispersion}$$

$$\omega(k) = \kappa_2 v_p \left(\frac{k}{k_{\text{dom}}}\right)^{3/2} \qquad \text{Anomalous dispersion}$$

where v_p is the phase velocity of a contemplated harmonic wave with wavenumber k_{dom} (dominant wavenumber in our Fig. 8.21). Note: We have chosen the parameters so that the group velocities are roughly the same for all three cases. The phase velocities are quite different. κ_1 and κ_2 are small correction factors (1.04 and 1.10)

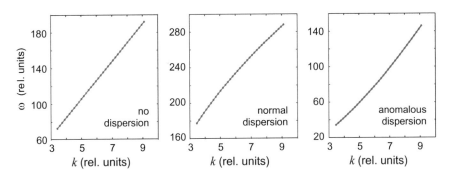

Fig. 8.21 Relation between frequency and wavenumber used in our calculations for the next figure

to optimize Fig. 8.10 and do not matter for the argument that follows. Figure 8.21) shows how ω varies with the wavenumber k.

In the function *fg_omega*, we also defines a parameter "deltat" that is used in the animation of the time evolution of the wave packet. The numbers in the computer code at the end of this chapter are chosen just so that the final position of the wave packet is convenient for plotting. Can be adjusted as wanted to have a different time at the end of the animation.

The animation including plotting is carried out by the function *pg_animer*. This function uses the function *pg_wave* to generate a complete spatial wave for each wavelength component and time in the animation.

Main Program

The code is available at the "Supplementary material" web page for this book at http://www.physics.uio.no/pow.

```
function pg3

% Program to illustrate the difference between phase and group
% velocity. Movement of a wave package is animated (blue). To
% ease the understanding of the difference between phase and
% group velocity, a pure monochromatic wave with the central
% wavelength is animated along with the wave package (in green).
% This will move with the phase velocity.
% Version: 5. October 2017, AIV

% NOTE: Due to the periodicity buried in a FFT, the wave pattern
% will only be valid if the animation stops before the wave
% pattern reaches the right end of the animation plot. If it
% turns up again at the left, the result is not valid.

% NOTE 2: You have to choose several values for the parameters
% in the program listing below!

% Choose first type of dispersion:
disp = -1.0;    % -1,0,+1: normal, no, anomal dispersion

% Create a wave package IN SPACE (!)
N = 4000;
xmax = 40.0;
xlambda = 1.0;        % Spatial wavelength
xsigma = 2.0;         % Width of the package
[x,z] = pg_wpack(N,xmax,xlambda,xsigma);
plot(x,z,'-b');

% Spatial frequency analysis, find amplitude and phase as a
% function of the wavenumber k
 [A,phase,k] = pg_fft(z,N,xmax);

% Pick manually those points in the frequency plot with
% considerable amplitude (use pg_fft, last part).
```

```
imin = 23;
imax = 59;

% Determines omega(k) using the dispersion relation
[omega,deltat] = pg_omega(imin,imax,k,disp);

% Now the movement can be animated
[xavt] = ...
   pg_animer(x,deltat,N,A,phase,k,omega,imin,imax,xmax,disp);
xavt;   % Position to a peak in a monochromatic wave (moving at
        % the phase velocity) after finishing the animation.
        % Start value is xmax/8.Remove the semicolon to have
        % this value written to the screen.
```

Create Wave Package in Space (at $t = 0$)

```
function [x,z] = pg_wpack(N,xmax,xlambda,xsigma)

% Create a wave package in space (!). Version Oct 5 2017 AIV
% Input parameters: N: Points in the description,
% xmax: Defines the interval x is defined ( |0,xmax>),
% xlambda: spatial wavelength for the central wavelengh,
% xsigma: the width in the gaussian shaped wave package.
% Returns: x array as well as the wave package array.

x = linspace(0,xmax*(N-1)/N,N);
xr = xmax/8.0; % Startpoint for the centre of the wave package
xfreq = 1/xlambda;   % Spatial frequency
y = cos((x-xr)*2*pi*xfreq);
convol = exp(-((x-xr)/xsigma).*((x-xr)/xsigma));
z = y.*convol;
return;
```

Frequency Analysis of Wave Package in Space

```
function [A,theta,k] = pg_fft(z,N,xmax)

% Frequency analysis of a wave package in space.
% Version Oct 5 2017 AIV
% Input parameters: z: the array describing the wave package,
% N: number point in this description, xmax: describe the x
% interval |0, xmax>. Returns: Amplitude A and phase (theta)
% in the frequency analysis as a function of the wavenumber k.

Zf = fft(z)/N;
A = 2.0*abs(Zf);   % Ignore the error in Zf(1), don't use it
theta = atan2(imag(Zf),real(Zf));
xsamplf = N/xmax;   % Spatial sampling frequency
xfreq = linspace(0,xsamplf*(N-1)/N,N); % Spatial frequency
k = zeros(1,N);
k = 2.0*pi*xfreq;
```

```
% NOTE: Use the reminder of this function when you need to
% pick frequency components for your wave package. You need
% this in order to choose imin and imax in the program pg3.m
%figure;
%plot(A,'.-r'); % Plot to be able to choose points to be used
%plot(xfreq,A,'.-r');   % Alternative plot
%plot(xfreq,fase,'.-k');
return;
```

Generate the Dispersion Relation omega(k)

```
function [omega,deltat] = pg_omega(imin,imax,k,disp)

% Generate the dispersion relation omega(k).
% Version Oct 5 2017, AIV
% Input parameters: imin, imax: first and last index that
% will be used in the function that creates the animation,
% k: the wavenumber array created by the function pg_fft,
% disp: -1, 0, or +1 represent normal, no and anomalous
% dispersion.
% Returns: omega: the dispersion relation omega(k),
% deltat: a suitable delta_t for the animation in order to
% get useful animation/plots.

if (disp==-1)    % Normal dispersion (here vg = vp/2)
    deltat = 0.015;
    omegafactor = 44.0;
    for i = imin:imax
        omega(i) = omegafactor*sqrt(k(i));
    end;
end;

if (disp==0)    % No dispersion (here vf = const)
    deltat = 0.015;
    omegafactor = 9.5;
    for i = imin:imax
        omega(i) = omegafactor*k(i);
    end;
end;

if (disp==1)    % Anomal dispersion (here vg = 3vp/2)
    deltat = 0.0065;
    omegafactor = 5.5;
    for i = imin:imax
        omega(i) = omegafactor*(k(i)^1.5);
    end;
end;

figure;
plot(k(imin:imax),omega(imin:imax),'.-b');
xlabel('k (rel. units)');
ylabel('omega (rel. units)');
return;
```

Create a Sum of Harmonic Spatial Waves at a Given Time t

```
function [zrecon] = pg_wave(x,t,N,A,phase,k,omega,imin,imax)

% Generate the complete spatial wave using the Fourier
% coefficients. Version Oct 5 2017 AIV
% Input parameters: x: position array, t: current time,
% N: number of points, [A, phase, k]: amplitude, phase and
% wavenumber arrays, respectively, omega: the dispersion
% relation omega(k), [imin, imax]: minimum andmaximum index
% that will be used in the arrays A, phase and k.
% Returns: zrecon: the position of the marker which gives the
% position to where a peak with the central wavelength would
% have ended up (for verification of proper functioning).

zrecon = zeros(1,N);
for i = imin:imax  % Sum over Fourier elements
    arg = k(i)*x - omega(i)*t + phase(i);
    zrecon = zrecon + A(i)*cos(arg);
end;
return;
```

Make an Animation of All Spatial waves

```
function [xavt] = ...
    pg_animer(x,deltat,N,A,phase,k,omega,imin,imax,xmax,disp)

% Animation of a wave package during some time. To ease the
% understanding of the difference between phase and group
% velocity, a pure monochromatic wave with the central
% wavelength is animated along with the wave package.
% Returns how far the monochromatic wave has moved during
% the animation (indicates the phase velocity).
% Input parameters: See the explanations given in the
% functions pg3.m, pd:wpack.m, pg_fft.m, pg_omega.m and
% pg:wave.m  Version: Oct 5 2017 AIV

figure;
count=1;
% The animation loop
for n = 1:200
    % Calculate the wave at time t (manual IFFT)
    t = deltat*n;
    [zrecon] = pg_wave(x,t,N,A,phase,k,omega,imin,imax);
    % Calculate also the wave with central spatial frequency
    % in the distribution
    imean = round((imin+imax)/2.0);
    [zrecon0] = pg_wave(x,t,N,A,phase,k,omega,imean,imean);
    % Calculate marking positions, start and end of movement
    % at phase velocity
    x00 = xmax/8.0;
    xavt = x00 + t*omega(imean)/k(imean);
```

```
% Plots everything
plot(x,2.5*zrecon0,'-g', x,zrecon,'-b', x00,0.25,'+r', ...
   xavt,0.25,'+r');
       xlabel('Position (rel)');
       ylabel('Amplitude (rel)');
       axis([0,xmax,-1.04,1.04])
       title('Movement to a blue wave package');
       S = sprintf('Time: %.2f s',t);
       text(3.0, 0.8,S);
       S = sprintf('Xref: %.2f',xavt);
       text(3.0, 0.65,S);
       S = sprintf('Dispersion code: %.1f',disp);
       text(3.0, -0.8,S);
       M(count)=getframe;
       count=count+1;
       M(count)=getframe;
       count=count+1;
end;
% Animation is played with (1 x 20 frames per sec)
movie(M,1,20);
return;
```

8.6 References

1. R.E. Apfel, Y. Tian et al. : *Free Oscillations and Surfactant Studies of Superde-formed Drops in Microgravity. Phys. Rev. Lett.* 78 (1997) 1912–1915 (Large water drop analysed in the spaceship Columbia.).
2. H. Azuma and S. Yoshihara: *Three-dimensional large-amplitude drop oscilla-tions: Experiments and theoretical analysis. J. Fluid Mech.* 393 (1999) 309–332.
3. For oscillating drops, see an elegant piece of work from: C-T. Chang, S. Daniel, P.H. Steen: *Excited sessile drops dance harmonically*, described in *Phys. Rev. E* 88 (2013) 023015.

    For a system that resembles ours, see for example:

    http://www.youtube.com/watch?v=YcF009w4HEE (Leidenfrost-effect: The dancing druppel (version 2)) or the last half of the video

    http://www.youtube.com/watch?v=b7KpHGgfHkc (JuliusGyula_HotPot 1.3). Both were accessible on 4 October 2017.

    We have also made our own high-speed films of oscillating water droplets, and a few of these will be available at the "Supplementary material" web pages for this book. The films were taken by Michael Baziljevich and Arnt Inge Vistnes 2014.
4. J. Persson: *Vågrörelselära, akustik och optik.* Studentlitteratur 2007.

8.7 Learning Objectives

After going through this chapter, you should be able to:
- Perform numerical calculations of the time course of a one-dimensional wave (with arbitrary shape) when there is no dispersion, based directly on the wave equation.
- Explain the contents of the algorithm for such calculations.
- Explain the difference between phase and group velocity in general and know how each is calculated.
- Explain how we can animate the time development of a wave packet.
- Know typical characteristics of how a wave packet develops over time when there is no dispersion, normal dispersion and anomalous dispersion.
- Provide examples of physical systems with dispersive behaviour, both normal and anomalous.
- Perform numerical calculations of the time course for a one-dimensional wave in dispersive media.
- Explain the contents of the algorithm for such calculations.
- Explain differences in gravity-driven waves in water and sound waves through water.
- Explain the two different "restoring forces" of surface waves on water.
- Enter an approximate criterion for whether it is the surface tension or gravity that dominates in a given case.
- Give examples of surface tension-driven waves and gravity-driven waves.
- Explain a model where we explain/describe waves by (small volumes of) water following a circular motion.
- Find approximate expression of phase velocity and group velocity of waves both in shallow and deep water, starting from the formula

$$v_{\mathrm{p}}^2(k) = \left[\frac{g}{k} + \frac{Tk}{\rho}\right] \tanh(kh) \ .$$

- Recapitulate the main features in Fig. 8.17.

8.8 Exercises

Suggested concepts for student active learning activities: Dispersion, group velocity, phase velocity, anomal/normal/no dispersion, mechanism for wavelength dependence of the speed of light in glass, wave packet, wave envelope, gravity-driven waves, capillary waves, high-speed camera, V-shaped wake.

Comprehension/discussion questions

1. What do we mean by a dispersive medium? How will dispersion affect the wave motion of *a*) a harmonic wave and *b*) a nonharmonic wave?

2. What is the difference between normal and anomalous dispersion?

3. What characterizes dispersion? What is a dispersion relation? Is the dispersion responsible for the phenomenon that waves often come in almost parallel to a sandy beach?

4. Indicate how a guitar string looks (amplitude vs position) just before we release the string. Use the algorithm given in Fig. 8.3 to tell which parts of the string will move in the first time step, second time step and third time step. Perhaps you can guess how the string will actually vibrate?

5. The oscillatory pattern we encounter in the previous task (and in our calculations based on the program *bolgeanimationX*) corresponds to the real motion of a guitar string during the first few oscillations. Eventually, sharp transitions disappear. Can you imagine what physical characteristics of the string, not taken into account here, would affect the fluctuations quite quickly? (Hint: Videos on YouTube, which show a motion entirely in accord with our calculations, use a rubber band rather than a proper guitar string to get the particular vibrating behaviour found in our calculations.)

6. A common misconception about why light goes slower through glass than in vacuum is that the photons are impeded by the glass. Such a view is confronted with a problem when we come to explain that the velocity returns to its value in air as soon as the light has passed the glass. Why is this hard to explain with the aforementioned explanation/model?

7. Why do we use a wave packet in the calculations that give us animation of dispersion?

8. Could you give some kind of explanation that the wavelength is different at the beginning of a wave packet compared to the wavelength at the end of the pack if we have normal or anomalous dispersion?

9. Are surface waves in water transverse or longitudinal waves? Explain.

10. Try to explain why we do not notice any effect of surface waves on water at a depth that is large relative to the wavelength.

11. Explain why waves roll in with the wave peaks parallel to the water's edge on a longshore beach.

12. See the video of Chang and coworkers in Reference 3 above. Find a diagram from the web that shows electron orbitals for the hydrogen atom. Compare the diagrams of Chang et al. with quantum descriptions of atomic orbitals. Comment on similarities and dissimilarities (Hint 1: 2D vs 3D; Hint 2: Quantum Physics operates with *wave* functions; Hint 3: Quantization).

Problems

13. Set up a mathematical expression (based on wavenumber and angular frequency) for a plane, monochromatic harmonic wave. Comment on the phase velocity and group velocity to the extent they are defined.

14. Create your own program to calculate numerical solutions of the wave equation. Feel free to take a look at the program shown under point Sect. 8.2.1 above. Test that a wave described by Eqs.(8.8) and (8.9) appears as shown in Fig. 8.4. Then make the following changes:

(a) Change the time derivative of the amplitude at the starting instant to the negative of what it should have been. Complete the calculations and describe what you observe.

(b) Reduce the time derivative of the amplitude at the starting instant to a half of what it should have been. Complete the calculations and describe what you observe.

(c) Use instead twice the time derivative of the amplitude instead of the correct one at the starting instant. Complete the calculations and see what you observe this time, paying attention to both amplitudes and phases.

(d) How do you want to create the initial conditions to simulate standing waves? [Optional]

(e) What conclusion can you deduce from all the calculations in this task? With a pendulum motion, we can choose position and velocity independently and always get an oscillatory motion that is easy to understand. Does the same apply to waves?

15. Modify the program you used in the previous task so that it can handle the case of a one-dimensional wave along a string meets a material with a different phase velocity. The wave should be able to continue into the new material and may also be reflected at the point where the string changes property (may correspond to the string changing mass per unit length). Attempt both with a 30% increase in phase velocity and a 30% reduction in phase velocity. Describe the results and comment on whether the results are consistent with what is described in Chap. 7 or not.

16. Make some simple sketches that show how you, before you do the calculations (or get to know the results found by fellow students), envisage a guitar string to vibrate. Write *afterwards* a computer program that calculates the motion of a guitar string for at least a couple of vibration periods after the string has been pulled, by means of a plectrum or fingernail, at a point that is at a distance of about 1/3 of the string length from one end, and released from there (after being at rest). Feel free to look at the program shown under point Sect. 8.2.1 above. Describe the motion.

[Check after you have done the calculations, whether there is a match between your calculations and YouTube movies mentioned in the text.]

17. Try to modify the computer program *waveAnimationX* based on the algorithm in Eq. (8.5) early in this chapter so that it can be used to describe the movement of a triangular wave as depicted in Fig. 8.22 for the case where the waveform is conserved during the movement. The wave will eventually hit a fixed/closed boundary and is reflected. Compare the result with the left side of Fig. 7.2.

18. Try to describe how the triangular wave in the previous problem will develop if it runs through a dispersive medium. In this case, a procedure based on Fourier

Fig. 8.22 A triangular wave that moves towards higher x-values. Suppose that the shape is unchanged at first until it hits a wall. See the problem text

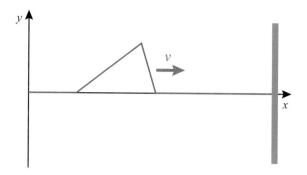

decomposition in spatial components described in the end of this chapter should be used. Do not include the reflecting wall in this case.

19. Check through your own calculation that the wavelength is about 1.7 cm when surface waves on water are controlled as much by surface tension as by gravity. Surface tension for clean water at 25 °C is 7.197×10^{-2} N/m.

20. Determine the phase velocity of surface waves on "deep" water at a wavelength of 1.7 cm (Tip: Use the information from the previous task.).

References

1. Katsushika Hokusai, The Great Wave off Kanagawa (color woodblock print), https://commons.wikimedia.org/wiki/File:The_Great_Wave_off_Kanagawa.jpg. Accessed April 2018
2. R. Nave, http://hyperphysics.phy-astr.gsu.edu/hbase/Waves/watwav2.html. Accessed April 2018
3. Arpingstone, https://commons.wikimedia.org/wiki/File:Wake.avon.gorge.arp.750pix.jpg. Accessed April 2018

Chapter 9
Electromagnetic Waves

Abstract This chapter starts with the integral form of Maxwell's equations—one of the greatest achievements in physics. The equations are transformed to differential form and used, in concert with important assumptions, to derive the wave equation for a plane, linearly polarized electromagnetic wave. The wave velocity is determined by electric and magnetic constants valid also at static/stationary conditions. The electromagnetic spectrum is presented as well as expressions for energy transport and radiation pressure. It is emphasized that the simple plane-wave solution is often invalid due to the effect of boundary conditions; we need to discriminate between near- and far-field conditions. At the end, a brief comment on the photon picture of light is given.

9.1 Introduction

Of all the wave phenomena that are consequential to us humans, sound waves and electromagnetic waves occupy a prominent position. Technologically speaking, the electromagnetic waves rank the highest.

We are going to meet, in many of the remaining chapters, electromagnetic waves in various contexts. It is therefore natural that we go into some depth for the sake of describing these waves; it is not true that all electromagnetism can be reduced to electromagnetic waves. That means, we must be careful to avoid mistakes when we treat this material.

Of all the chapters in the book, this is the most mathematical. We start with Maxwell's equations in integral form and show how their differential versions may be deduced. It will then be shown that Maxwell's equations can lead, under certain conditions, to a simple wave equation. Electromagnetic waves are transverse, which means that the complexity of the treatment is somewhat larger than for longitudinal sound waves.

The chapter takes it for granted that the reader has previously taken a course in electromagnetism, and is familiar with such relevant mathematical concepts as line integrals and surface integrals. It is also a great advantage to know Stokes's theorem, the divergence theorem and those parts of vector calculus which relate to divergence,

© Springer Nature Switzerland AG 2018

A. I. Vistnes, *Physics of Oscillations and Waves*, Undergraduate Texts in Physics,
https://doi.org/10.1007/978-3-319-72314-3_9

gradient and curl, and it is vital that the reader knows the difference between scalar and vector fields before embarking on the chapter.

As already mentioned, mathematics pervades this chapter. Nonetheless, we have tried to point to the physics behind mathematics, and we recommend that you too devote much time for grappling with this part. It is a challenge to grasp the orderliness of Maxwell's equations in its entirety.

Experience has shown that misunderstandings related to electromagnetism arise frequently. A common misconception, incredibly enough, is that an electromagnetic wave is an electron that oscillates up and down in a direction perpendicular to the direction of propagation of the wave. Other misapprehensions are harder to dispel. For example, many believe that the solution of the wave equation is "plane waves" and that the Poynting vector always describes energy transport in the wave. We spend some time discussing such misunderstandings and hope that some readers will find this material useful.

At the end of the chapter is a list of useful mathematical relations and a memorandum of how electrical and magnetic fields and flux densities relate to each other. It may be useful for a quick reference and for refreshing material previously learned.

Let us kick off with Maxwell's stupendous systematization (and extension) of all that was known about electrical and magnetic laws in 1864.

9.2   Maxwell's Equations in Integral Form

Four equations connect electric and magnetic fields:
1. Gauss's law for electric field:

$$\oint \vec{E} \cdot d\vec{A} = \frac{Q_{\text{inside}}}{\varepsilon_r \varepsilon_0} \tag{9.1}$$

2. Gauss's law for magnetic field:

$$\oint \vec{B} \cdot d\vec{A} = 0 \tag{9.2}$$

3. The Faraday–Henry law:

$$\oint \vec{E} \cdot d\vec{l} = -\left(\frac{d\Phi_B}{dt}\right)_{\text{inside}} \tag{9.3}$$

3. The Ampère–Maxwell law:

$$\oint \vec{B} \cdot d\vec{l} = \mu_r \mu_0 \left[i_f + \varepsilon_r \varepsilon_0 \left(\frac{d\Phi_E}{dt}\right)_{\text{inside}} \right] \tag{9.4}$$

We expect that you are familiar with these laws, and therefore do not go into great detail about how to interpret them or what the symbols mean. In the first two equations, the flux is integrated over a closed surface and compared with the source within the enclosed volume (electrical monopole, i.e. charge, and magnetic monopole, which are non-existent). The vector $d\vec{A}$ is positive if it points outward of the enclosed volume.

In the last two equations, the line integral is calculated for electrical or magnetic fields along a curve that limits an open surface. The line integral is compared with the flux of magnetic flux density or electrical flux density as well as flux of electrical currents due to free charges through the open surface. The signs are then determined by the right-hand rule (when the four fingers on the right hand point in the direction of integration along the curve, the thumb points in the direction corresponding to the positive flux).

Prior knowledge of these details is taken for granted.

The symmetry is best revealed if the last equation is written in the following form:

$$\oint \vec{H} \cdot d\vec{l} = \left[i_f + \left(\frac{d\Phi_D}{dt} \right)_{inside} \right]. \tag{9.5}$$

Here, use has been made of the following relationship between the magnetic field strength $\vec{H}$ and the magnetic flux density $\vec{B}$:

$$\vec{H} = \vec{B}/(\mu_r \mu_0)$$

where μ_0 is (magnetic) permeability in vacuum and μ_r is relative permeability.

Use has also been made of the following relationship between the electric field strength $\vec{E}$ and the electric flux density $\vec{D}$ (also called "displacement vector"):

$$\vec{E} = \vec{D}/(\varepsilon_r \varepsilon_0)$$

where ε_0 is the (electrical) permittivity in vacuum and ε_r the relative permittivity.

The left-hand sides of Eqs. (9.3) and (9.5) are line integrals of field strengths ($\vec{E}$ and $\vec{H}$), whereas the right-hand sides are the time derivative of the flux through the enclosed surface plus, for the latter equation, the current density due to free charges. The flux is obtained by integrating the pertinent flux density ($\vec{B}$ or $\vec{D}$) over the surface.

The contents of Maxwell's equations are given an approximate verbal rendering below:

- There are two sources of electric field. One source is the existence of electrical charges (which may be regarded as monopoles). Electric fields due to charges are radially directed away from or towards the charge, depending on the signs of the charges. (This is the content of Gauss's law for electric field.)
- The second source of electric field is a time-varying magnetic field. Electrical fields that arise in this way have a curl (or circulation); that is, the field lines tend

Fig. 9.1 James Clerk
Maxwell (1831–1879).
Public domain [1]

to form circles across the direction along which the magnetic field changes in time.
Whether there are circles or some other shape in space depends on the boundary
conditions. (This is the content of Faraday's law.)

- There are two contributions to magnetic fields as well, but there are no magnetic
  monopoles. Therefore, magnetic fields will never flow out radially from a source
  point similar to electric field lines near an electrical charge. (This is the content of
  Gauss's law for magnetic fields.)

- On the other hand, magnetic fields can arise, as in the case of electric fields, because
  an electric field varies over time. An alternative way of generating a magnetic field
  is to have free charges in motion that form a net electric current. Both these sources
  provide magnetic fields that tend to form closed curves across the direction of the
  time variation of the electric field or the direction of the net electrical current.
  However, the shape of these closed curves in practice is entirely dependent on the
  boundary conditions. (This is the content of the Ampère–Maxwell law.)

It was the physicist and mathematician James Clerk Maxwell (1831–1879,
Fig. 9.1)) who distilled all knowledge of electrical and magnetic laws then available in one comprehensive formalism. His publication "A Dynamical Theory of
Electromagnetic Field", published in 1865, shows that it is possible to generate electromagnetic waves and that they travel with the speed of light. His electromagnetic
theory is considered to be on a par with Newton's laws and Einstein's relativity theory. The original 54-page long article (https://doi.org/10.1098/rstl.1865.0008 Phil.
Trans. R. Soc. Lond. 1865 vol. 155, pp. 459–512) can be downloaded for free from:
The Royal Society.

Maxwell–Heaviside–Hertz: However, the four Maxwell's equations, as we know them today,
are far from the equations Maxwell himself presented in "A Treatise on Electricity and Magnetism"

Fig. 9.2 Michael Faraday (1791–1897). Parts of an old 10-pound banknote from Great Britain

in 1865. His original article consisted of 20 equations with 20 unknowns. Maxwell did not use the vector field formalism familiar to us.

Oliver Heaviside (1850–1925) gave in 1884 the equations in about the form we use today. Heaviside, who was from a poor home, left school when he was 16 and receive no formal education subsequently. Nevertheless, he made a number of important contributions to physics. It is fascinating to read about him, for example, in Wikipedia. (There are certain similarities between Heaviside and Faraday. Faraday's story is also fascinating, and highly recommended to read, and is even honoured on a British banknote; see Fig. 9.2. Heaviside did not receive similar recognition.)

The German physicist Heinrich Hertz (1857–1894) was the first to demonstrate how we can send and receive electromagnetic waves. It happened in 1887 when Hertz was 30 years old.

It is interesting that Hertz is honoured by a number of stamps from many different countries, while Maxwell is far from getting the same honour.

Recapitulation from electromagnetism: It might be appropriate to begin with a little repetition of some details here. We will later see that magnetic permeability and, in particular, electrical permittivity play an important role in electromagnetic waves. The values in vacuum μ_0 and ε_0 are rather uninteresting. They are primarily related to the choice of units for electrical and magnetic fields.

The relative values, however, are of far more interest. The relative (magnetic) permeability is related to how much magnetic field is generated in a material when it is exposed to an external magnetic field. In a diamagnetic material, a tiny magnetic field is generated in the material, and the field is directed opposite the external magnetic field. Even in a paramagnetic material, only a tiny magnetic field is generated, but it is in the same direction as the extraneous field. The magnetic field generated in the material is only of the order of 10^{-5} times the external magnetic field in each of these cases. In a ferromagnetic material, a significant magnetic field is generated inside the material, and it is in the same direction as the applied field. There are many details related to these processes, and we do not deal with these here.

Since most of the substances we come in contact with are either diamagnetic or paramagnetic, we can simply set the relative permeability equal to 1.0 and ignore the interaction of the magnetic field with materials in the processes we are going to discuss.

For the electrical field, it is different. The relative (electrical) permittivity tells us something about the amount of electrical field that occurs inside a material when subjected to an external electric field. In Fig. 9.3, a schematic representation of what is happening is given. An external electric field will cause the electron cloud around an atomic core to shift slightly. However, since there are so

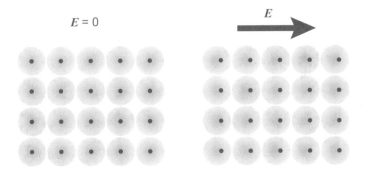

Fig. 9.3 In a material such as glass, an external electric field can easily cause a polarization of the charge distribution in each individual atom of the material. This polarization leads to an electric field inside the material directed opposite to the applied electric field

many atoms, even an almost negligible shift in position of the negatively charged electron clouds relative to the positively charged nuclei, the generated electric field inside the material can easily reach the same magnitude as the outer electric field (e.g. half the size).

There is no talk of moving free charges here, only of a local distortion of the charge distribution of each individual atom, which imparts, nonetheless, a polarization to the entire material. Note that we are talking about "polarization" in a certain sense. We will soon talk about polarization in a completely different context, so it is imperative that you do not mix different terms with the same name!.

9.3   Differential Form

We will now show how we can go from the integral form of Maxwell's equations to the differential form. The integral form can be applied to macroscopic geometries, for example to find the magnetic field at a distance of 5 m from a straight conductor where there is a net electrical current. The differential form applies to a small region of space. How "small" this might be is a matter for discussion. Maxwell's equations were developed before we had a proper knowledge of the structure of atoms and substances on the microscopic level. Maxwell's equations in differential form are often used in practice *on an intermediate length scale that is small in relation to the macroscopic world and yet large compared to atomic dimensions.*

In going over from the integral to differential form, two mathematical relationships are invoked that apply to an arbitrary vector field $\vec{G}$ in general:

Stokes's theorem (more correctly the Kelvin–Stokes theorem, since the theorem first became known through a letter from Lord Kelvin. George Stokes (1819–1903) was a British mathematician/physicist. Lord Kelvin (1824–1907), whose real name was William Thomson, was a physicist/mathematician contemporary of Stokes.)

Stokes's theorem:

$$\oint \vec{G} \cdot d\vec{l} = \int_A (\nabla \times \vec{G}) \cdot d\vec{A} . \tag{9.6}$$

The theorem provides a relation between the line integral of a vector field and the flux of the curl of the vector field through the plane limited by the line.

The second relationship we use is the *divergence theorem* (discovered by Lagrange and rediscovered by several others). Joseph Louis Lagrange (1736–1813) was an Italian/French mathematician and astronomer:

Divergence theorem:

$$\int \nabla \cdot \vec{G} dv = \oint_A \vec{G} \cdot d\vec{A} . \tag{9.7}$$

The divergence theorem provides the connection between divergence to a vector field in a volume and the flux of the vector field through the surface which bounds the volume.

Gauss's law for electric field:

We start with Gauss's law for electric field.

$$\varepsilon_r \varepsilon_0 \oint \vec{E} \cdot d\vec{A} = Q_{\text{inside}} .$$

By using the divergence theorem, we get:

$$\oint \varepsilon_r \varepsilon_0 \vec{E} \cdot d\vec{A} = \int \nabla \cdot (\varepsilon_r \varepsilon_0 \vec{E}) \, dv = Q_{\text{inside}} .$$

We now choose such a small volume that $\nabla \cdot (\varepsilon_r \varepsilon_0 \vec{E})$ is approximately constant over the entire volume. This constant can then be taken outside the integral sign, and integration over the volume element simply gives the small volume Δv under consideration. Accordingly:

$$\int \nabla \cdot (\varepsilon_r \varepsilon_0 \vec{E}) \, dv \approx (\nabla \cdot \vec{D}) \Delta v = Q_{\text{inside}}$$

$$\nabla \cdot \vec{D} = \frac{Q_{\text{inside}}}{\Delta v} = \rho$$

where ρ is the local charge density. We are led thereby to Gauss's law for electric fields in differential form:

$$\nabla \cdot \vec{D} = \rho \ .$$

$$(9.8)$$

Gauss's law for magnetic field:

The same approach leads us to the differential form of Gauss's law for magnetic field:

$$\nabla \cdot \vec{B} = 0 \ .$$

$$(9.9)$$

The Faraday–Henry law:

We will now rephrase Faraday's law. The starting point is thus:

$$\oint \vec{E} \cdot d\vec{l} = - \left(\frac{d\Phi_B}{dt} \right)_{inside} \ .$$

The application of Stokes's theorem now gives:

$$\oint \vec{E} \cdot d\vec{l} = \int_A (\nabla \times \vec{E}) \cdot d\vec{A} = - \left(\frac{d\Phi_B}{dt} \right)_{inside} \ .$$

The magnetic flux through the surface can be expressed as:

$$\Phi_B = \int_A \vec{B} \cdot d\vec{A} \ .$$

Hence,

$$\int_A (\nabla \times \vec{E}) \cdot d\vec{A} = - \frac{d}{dt} \int_A \vec{B} \cdot d\vec{A}$$

$$= - \int_A \frac{\partial \vec{B}}{\partial t} \cdot d\vec{A} \ .$$

In taking the last step, we have assumed that the area element dA does not change with time. In addition, we have changed the ordinary derivative to partial derivative since the magnetic flux density $\vec{B}$ depends on both time and spatial relationships, but we assume that spatial conditions do not change in time. Again, for a small enough area A, the functions to be integrated can be regarded as constants and placed outside the integral signs, which leads to the result:

$$\nabla \times \vec{E} = - \frac{\partial \vec{B}}{\partial t} \ .$$

$$(9.10)$$

This is Faraday's law in differential form.

The Ampère–Maxwell law:

The same procedure can be used to show the last of Maxwell's equations in a differential form, namely the Ampère–Maxwell law. The result is:

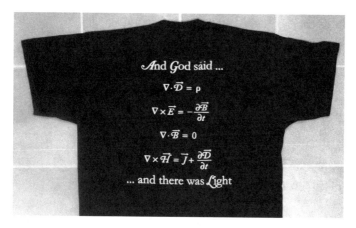

Fig. 9.4 Maxwell's equations on a T-shirt

$$\nabla \times \vec{H} = \vec{j}_f + \frac{\partial \vec{D}}{\partial t} \tag{9.11}$$

where $\vec{j}_f$ is the electric current density of the free charges.

Einstein had pictures of Newton, Maxwell and Faraday in his office, indicating how important he thought their works to be. It is therefore not surprising that the Physics Association at UofO has chosen Maxwell's equations on their T-shirts (see picture 9.4) as a symbol of a high point in physics, a high point as regards both how powerful equations are and how mathematically elegant they are! (It should be noted, however, that mathematical elegance did not seem to be as polished at Maxwell's time as it is today.)

Collected:

Let us assemble all of Maxwell's equations in differential form:

$$\nabla \cdot \vec{D} = \rho \tag{9.12}$$

$$\nabla \cdot \vec{B} = 0 \tag{9.13}$$

$$\nabla \times \vec{E} = -\frac{\partial \vec{B}}{\partial t} \tag{9.14}$$

$$\nabla \times \vec{H} = \vec{j}_f + \frac{\partial \vec{D}}{\partial t} \tag{9.15}$$

Maxwell's equations in the presence of the Lorentz force

$$\mathbf{F} = q(\vec{\mathbf{E}} + \vec{\mathbf{v}} \times \vec{\mathbf{B}})$$

form the full basis for classical electrodynamic theory.

9.4 Derivation of the Wave Equation

The wave equation can be derived from Maxwell's equations using primarily the last two equations along with a general relation that applies to an arbitrary vector field **G**:

$$\nabla \times (\nabla \times \vec{\mathbf{G}}) = -\nabla^2 \vec{\mathbf{G}} + \nabla(\nabla \cdot \vec{\mathbf{G}}) . \tag{9.16}$$

In words, the relation states that "the curl of the curl of a vector field is equal to the negative of the Laplacian applied to the vector field plus the gradient of the divergence of the vector field" (pause for breath).

Application of this relation to the electric field yields:

$$\nabla \times (\nabla \times \vec{\mathbf{E}}) = -\nabla^2 \vec{\mathbf{E}} + \nabla(\nabla \cdot \vec{\mathbf{E}}) .$$

We recognize the curl of the electric field in the expression on the left-hand side. Replacing it by Faraday's law, interchanging the right and left side of the equation, and changing the sign, we get:

$$\nabla^2 \vec{\mathbf{E}} - \nabla(\nabla \cdot \vec{\mathbf{E}}) = -\nabla \times \left(-\frac{\partial \vec{\mathbf{B}}}{\partial t} \right) . \tag{9.17}$$

On the right-hand side, we change the order of differentiation to find:

$$= \frac{\partial}{\partial t}(\nabla \times \vec{\mathbf{B}}) .$$

Applying now the Ampère–Maxwell law, and using the relation

$$\vec{\mathbf{B}} = \mu_r \mu_0 \vec{\mathbf{H}}$$

we get:

$$= \frac{\partial}{\partial t}\left[\mu_r \mu_0 \left(\frac{\partial \vec{\mathbf{D}}}{\partial t} + \vec{\mathbf{j}}_f \right) \right] . \tag{9.18}$$

For the left-hand side of Eq. (9.17), Gauss's law is used for electric field to replace the divergence of electric field in the second link on the left side with charge density ρ divided by total permittivity.

$$\nabla^2 \vec{E} - \frac{\nabla \rho}{\varepsilon_r \varepsilon_0} .$$
(9.19)

By equating the right-hand side of (9.19) to the left-hand side of (9.18), and transposing some terms, we end up with:

$$\nabla^2 \vec{E} - \varepsilon_r \varepsilon_0 \mu_r \mu_0 \frac{\partial^2 \vec{E}}{\partial t^2} = \frac{\nabla \rho}{\varepsilon_r \varepsilon_0} + \mu_r \mu_0 \frac{\partial \vec{j}_f}{\partial t} .$$
(9.20)

This is a nonhomogeneous wave equation for electric fields. The source terms are on the right side of the equality sign.

In areas where the gradient of charge density ρ is equal to zero (i.e. no change in electrical charge density), while there is no time variation in electrical current density $\vec{j}_f$ of free charges, the inhomogeneous equation is reduced to one simple wave equation:

$$\nabla^2 \vec{E} - \varepsilon_r \varepsilon_0 \mu_r \mu_0 \frac{\partial^2 \vec{E}}{\partial t^2} = 0$$

or in the more familiar form:

$$\frac{\partial^2 \vec{E}}{\partial t^2} = \frac{1}{\varepsilon_r \varepsilon_0 \mu_r \mu_0} \nabla^2 \vec{E} .$$
(9.21)

Well, to be honest, this is not an ordinary wave equation, as we have seen before, since we have used the Laplacian on the right-hand field. In detail, we have:

$$\nabla^2 \vec{E} = \left(\frac{\partial^2 E_x}{\partial x^2} + \frac{\partial^2 E_x}{\partial y^2} + \frac{\partial^2 E_x}{\partial z^2} \right) \vec{i}$$
$$+ \left(\frac{\partial^2 E_y}{\partial x^2} + \frac{\partial^2 E_y}{\partial y^2} + \frac{\partial^2 E_y}{\partial z^2} \right) \vec{j}$$
$$+ \left(\frac{\partial^2 E_z}{\partial x^2} + \frac{\partial^2 E_z}{\partial y^2} + \frac{\partial^2 E_z}{\partial z^2} \right) \vec{k} .$$
(9.22)

We search now for the simplest possible solution and investigate if there is a solution where $\vec{E}$ is independent of both x and y. In that case, all terms of the type $\partial^2 E_u / \partial v^2$ will vanish, where $u = x, y, z$ and $v = x, y$. If such a solution is possible, it will involve a planar wave that moves in the z-direction, since a plane wave is just unchanged in an entire plane perpendicular to the wave direction of motion.

Equation (9.22) then reduces to the following simplified form:

$$\nabla^2 \vec{E} = \left(\frac{\partial^2 E_x}{\partial z^2} \right) \vec{i} + \left(\frac{\partial^2 E_y}{\partial z^2} \right) \vec{j} + \left(\frac{\partial^2 E_z}{\partial z^2} \right) \vec{k} = \frac{\partial^2 \vec{E}}{\partial z^2} \qquad (9.23)$$

and Eq. (9.21) along with Eq. (9.23) finally gives us a common wave equation:

$$\frac{\partial^2 \vec{E}}{\partial t^2} = c^2 \frac{\partial^2 \vec{E}}{\partial z^2} \qquad (9.24)$$

where

$$c = \frac{1}{\sqrt{\varepsilon_r \varepsilon_0 \mu_r \mu_0}} . \qquad (9.25)$$

is the phase velocity of the wave.

We already know that one solution of the wave equation, Eq. (9.24), is

$$\vec{E} = \vec{E}_0 \cos(kz - \omega t) \qquad (9.26)$$

where $\vec{E}_0$ is a constant vector whose direction can be chosen arbitrarily within the $x - y$-plane.

Let us now see whether we are able to derive a wave equation for the magnetic field. To this end, we start with Eq. (9.16), but apply it to the magnetic flux density and write

$$-\nabla^2 \vec{B} + \nabla(\nabla \cdot \vec{B}) = \nabla \times (\nabla \times \vec{B}) .$$

We use next the Ampère–Maxwell law in order to replace the curl of $\vec{B}$ with the time derivative of the electric flux density $\vec{D}$ plus the current density of free charges. As in the corresponding derivation for the electric field, we interchange the order of time and space derivatives, obtaining thereby a term containing the curl of $\vec{E}$. We invoke Faraday's law and the vanishing of the divergence of $\vec{B}$ (Gauss's law for magnetic fields), to arrive finally at the following equation for $\vec{B}$:

$$\nabla^2 \vec{B} - \varepsilon_r \varepsilon_0 \mu_r \mu_0 \frac{\partial^2 \vec{B}}{\partial t^2} = -\mu_r \mu_0 \nabla \times \vec{j}_f . \qquad (9.27)$$

We observe that the magnetic flux density also satisfies an inhomogeneous wave equation, in which the source term is the curl of the current density of free charges. For regions of space which are source-free, we obtain a homogeneous wave equation. We seek the simplest solution to the equation, and ask, as we did in the case of the electric field, whether a plane-wave solution exists for a wave propagating in the z-direction. With the same simplifications as those used earlier, we obtain

$$\frac{\partial^2 \vec{B}}{\partial t^2} = c^2 \frac{\partial^2 \vec{B}}{\partial z^2} \tag{9.28}$$

where the wave velocity c is precisely that given in Eq. (9.25), applicable to the electric field.

We already know that, in this case as well, the equation does have a solution, which can be written in the form:

$$\vec{B} = \vec{B}_0 \cos(kz - \omega t) \tag{9.29}$$

where $\vec{B}_0$ is a constant vector whose direction is essentially arbitrary.

9.5 A Solution of the Wave Equation

Equations (9.26) and (9.29) are valid solutions of the two wave Eqs. (9.24) and (9.28), respectively. But the solutions must also satisfy Maxwell's equations individually, in practice the Ampère–Maxwell law and Faraday's law.

We start with Faraday's law

$$\nabla \times \vec{E} = -\frac{\partial \vec{B}}{\partial t}$$

and substitute the solution for the electric field (9.26) (in determinant form):

$$\begin{vmatrix} \vec{i} & \vec{j} & \vec{k} \\ \dfrac{\partial}{\partial x} & \dfrac{\partial}{\partial y} & \dfrac{\partial}{\partial z} \\ E_x & E_y & E_z \end{vmatrix} = -\frac{\partial \vec{B}}{\partial t}$$

$$\left\{ \left(\frac{\partial E_z}{\partial y} - \frac{\partial E_y}{\partial z} \right) \vec{i} - \left(\frac{\partial E_z}{\partial x} - \frac{\partial E_x}{\partial z} \right) \vec{j} + \left(\frac{\partial E_y}{\partial x} - \frac{\partial E_x}{\partial y} \right) \vec{k} \right\} = -\frac{\partial \vec{B}}{\partial t}.$$

For the plane-wave solution sought by us, partial derivatives with respect to x or y will vanish, and the expression takes the simpler form shown below:

$$-\frac{\partial E_y}{\partial z} \vec{i} + \frac{\partial E_x}{\partial z} \vec{j} = -\frac{\partial \vec{B}}{\partial t}.$$

We already notice that $\vec{B}$ cannot have any component in the z-direction, that is, the direction of propagation of the wave. A similar analysis using Ampère–Maxwell's law shows that nor can $\vec{E}$ have a z-component (except for a static homogeneous field, which is of no interest in the present context).

We choose now the following solution for $\vec{E}$:

$$\vec{E} = E_0 \cos(kz - \omega t)\, \vec{i} \tag{9.30}$$

which means that $E_y = 0$, hence also $E_y = 0$ and also $\partial E_y / \partial z = 0$

Consequently,

$$\frac{\partial E_x}{\partial z}\, \vec{j} = k E_0 \sin(kz - \omega t)\, \vec{j} = -\frac{\partial \vec{B}}{\partial t}\ .$$

This equation will be satisfied if

$$\vec{B} = B_0 \cos(kz - \omega t)\, \vec{j}\ . \tag{9.31}$$

Furthermore, since

$$-\frac{\partial \vec{B}}{\partial t} = \omega B_0 \sin(kz - \omega t)\, \vec{j}$$

and the (phase) velocity of this plane wave is ω / k which must be equal to c from Eq. (9.25), we also get an important connection between electric and magnetic field in an electromagnetic wave:

$$E_0 = c B_0\ . \tag{9.32}$$

We have then shown that *one* possible solution to Maxwell's equations for a space where no charges are present is a planar electromagnetic wave

$$\vec{E} = E_0 \cos(kz - \omega t)\, \vec{i} \tag{9.33a}$$
$$\vec{B} = B_0 \cos(kz - \omega t)\, \vec{j} \tag{9.33b}$$

where

$$E_0 = c B_0$$

You are reminded that solutions of wave equations generally depend on a great extent on boundary conditions. In our derivation, we have searched for a solution that gives a planar wave. In practice, this amounts to assuming that the area under consideration is located *far* from the source of the wave, as well as free charges and currents generated by such charges. The plane-wave solution is therefore, in principle, *never a perfect solution* of Maxwell's equations, but an exact solution may in some cases be quite close to a planar wave solution. It is our task as physicists to decide whether or not we can model real field distribution with a plane-wave description in each case. See the description of the near field and far field given below.

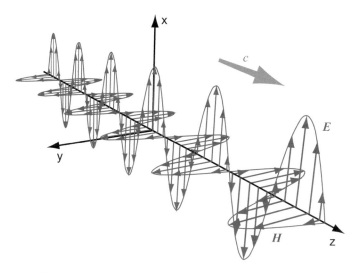

Fig. 9.5 A snapshot of the simplest form of electromagnetic wave, namely a plane wave. Such a wave can be realized sufficiently far from the source of the wave and from materials that can perturb the wave. Experience has shown that figures of this type cause many misunderstandings, which are discussed in the last part of this chapter

Since Maxwell's equations are linear, we can have plane electromagnetic waves *in addition* to other electrical or magnetic fields with completely different characteristics. *The sum* of all contributions will then not follow the relationships given in the blue box above!

Figure 9.5 shows a snapshot of an electromagnetic wave with properties as given in Eq. (9.33). Such a static figure does not give a good picture of the wave. It is therefore advisable to consider an animation to get an understanding of time development. There are several animations of a simple electromagnetic wave on the Web (but some of them have wrong directions of the vectors!).

9.6   Interesting Details

What determines the speed of light?
We saw in the derivation of the wave equation that electromagnetic waves have a phase velocity

$$c = \frac{1}{\sqrt{\varepsilon_r \varepsilon_0 \mu_r \mu_0}} \ . \tag{9.34}$$

In vacuum, $\varepsilon_r = 1$ and $\mu_r = 1$, and the velocity of the wave becomes

$$c_0 = \frac{1}{\sqrt{\varepsilon_0 \mu_0}} . \tag{9.35}$$

This is simply an expression for the velocity of light in vacuum.

It must have been a wonderful experience for Maxwell when he first understood this. The speed of light was known, but not associated with any other physical parameters. Then, Maxwell derives the wave equation and finds that the equations allow for the existence of electromagnetic waves, and—as it happens—these waves will propagate with the familiar speed of light! The surprise must have been particularly high because the speed of light closely follows the electrical and magnetic constants ε_0 and μ_0, which were determined by *static* electrical and static magnetic phenomena.

In glass, the velocity of light is given by Eq. (9.34), but for glass μ_r is practically equal to 1. That means that it is simply the relative electrical permittivity of the glass, which causes light to be slower in glass than in vacuum. This too is surprising since the relative permittivity can be determined by putting a glass plate between two metal plates and measuring change in capacitance between the metal plates. Even this measurement can be made by using static fields, and equally this quantity plays an important role for light oscillating with a frequency of the order of 10^{15} times per second!

There is no dispersion in vacuum, but in a dielectric material dispersion may occur because ε_r (and/or μ_r) is wavelength dependent. We discussed this in Chap. 8 when treated dispersion and the difference between phase velocity and group velocity.

It will be noted that when we discuss the passage of light through glass, we are dealing with a material constant called *refractive index n* which varies slightly from one glass to another. The phase velocity of light is lower in glass than in vacuum. The refractive index n can be defined as the ratio of the light velocity in vacuum to the light velocity in glass: $n = c_0/c$. The word *refractive* index will be explained in detail when we in Chap. 10 describe how light rays change direction when the beam is inclined towards an interface between air and glass or the other way round (Snel's law).

Glass is diamagnetic and $\mu_r \approx 1.0$. From the above expressions, then the refractive index is approximately equal to the square root of the relative permittivity:

$$n \approx \sqrt{\varepsilon_r} . \tag{9.36}$$

Relative permittivity is also called dielectric constant.

Plane wave

The wave we have described is plane because the electric field at a given instant is identical everywhere in an infinite plane normal to the wave propagation direction z. Another way of expressing this is to say that the "wavefront" is plane. The wavefront of a plane wave is a surface of constant phase (i.e. the argument of the sine or cosine function is identical at all points of the surface at a given time).

The fact that the electric field everywhere is directed in the $\pm x$-direction is a characteristic feature of the solution we have found. We say that the wave is *linearly polarized* in the x-direction. We return to polarization in Chap. 10, but already mention here that another solution to Maxwell's equations is a so-called circularly polarized wave. For such a solution, the electric field vectors in a snapshot corresponding to Fig. 9.5 will look like the steps in a spiral staircase, and the arrows themselves will form a "spiral" whose axis coincides with the z-axis. The magnetic field will also form a spiral, and in this case too the electric and magnetic fields will be perpendicular to each other and perpendicular to the direction of propagation. You can find nice animations of circularly polarized electromagnetic waves on the Web.

In addition, we will return later to an important discussion of the validity of the simple electromagnetic waves we have described so far.

9.7 The Electromagnetic Spectrum

In deriving the wave equation for electromagnetic waves, we placed (initially) no restrictions on the frequencies and wavelengths. In principle, more or less "all" frequencies (with the corresponding wavelengths) were eligible for consideration.

It turns out also in practice that we can generate electromagnetic waves for a wide range of frequencies (and wavelengths). Figure 9.6 shows an approximate overview of the frequency ranges/wavelength ranges we operate in, what we call the waves at different frequencies and what such waves are used for. We say that figures like 9.6 present "the electromagnetic spectrum".

Figures of this type must be taken with a large pinch of salt. They seduce many people into thinking that there exist tidy plane waves at each of the specified frequencies, but that is not the case. The spreading of waves in time and space, energy transport (or its absence) and several other factors vary widely from one frequency to another. We will come back to this a little later in this chapter.

9.8 Energy Transport

When we discussed sound, we saw that a sound wave carry energy many metres away from the source, although the molecules that contributed to the transmission through oscillatory motion only moved back and forth over distances of the order of $1\,\mu\text{m}$ or less (when we ignore the diffusive motion of the molecules).

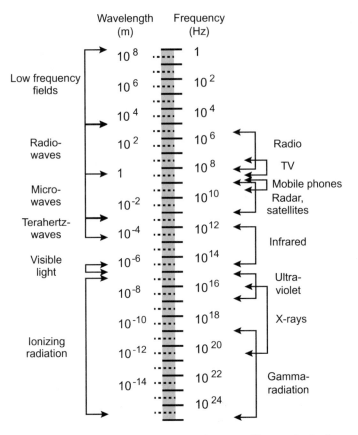

Fig. 9.6 Electromagnetic waves can exist in an impressive range of frequencies (and corresponding wavelengths). Surveys such as this may, however, give an impression of a greater degree of similarity between different phenomena than it is in practice. We will come back to this, for example, when we discuss the difference between near field and far field later in the chapter

In a similar manner, an electromagnetic wave can carry energy, something we all experience when we bask in the Easter sun on a mountain or when we lie on a sunny beach in summer.

An electric field has an energy density given by:

$$u_E(z,t) = \tfrac{1}{2} E(z,t) D(z,t) \, .$$

Similarly, the energy density of a magnetic field is given by:

$$u_H(z,t) = \tfrac{1}{2} H(z,t) B(z,t) \, .$$

When a plane electromagnetic wave (as described above) passes, the instantaneous energy density will be:

$$u_{tot}(z, t) = \tfrac{1}{2}E(z, t)D(z, t) + \tfrac{1}{2}H(z, t)B(z, t)$$

$$= \frac{1}{2}E_0 \cos() \, \varepsilon E_0 \cos() + \frac{1}{2}B_0 \cos() \frac{B_0}{\mu} \cos() .$$

The arguments of the cosine function have been omitted in order to avoid clutter.

But we know that $E_0 = cB_0$. In addition, we want to look at the *time-averaged* energy density, and we know that the mean value of $\cos^2()$ is equal to half. Consequently, we find for the time-averaged energy density:

$$\bar{u}_{tot} = \frac{1}{4}\varepsilon E_0^2 + \frac{1}{4\mu}\left(\frac{E_0}{c}\right)^2 .$$

Now, energy density is energy per unit volume. How much energy will cross a hypothetical surface A perpendicular to the direction of wave motion over a time Δt? Such a quantity defines the (time-averaged) wave *intensity*:

$$I = \text{intensity} = \frac{\text{Energy passed by}}{\text{Area} \times \text{Time}} = u_{tot} \times c .$$

The expression is relevant only when we consider a long time compared to the time a wavelength needs to pass our surface. Intended for the energy density we found in town, we get:

$$I = \frac{1}{4}\left(c\varepsilon E_0^2 + c\frac{1}{c^2\mu}E_0^2\right) .$$

But we know that

$$c = \frac{1}{\sqrt{\varepsilon\mu}}$$

from which follows the relation

$$\frac{1}{c^2\mu} = \varepsilon$$

and we see that the energy contributions from the electric field and the from the magnetic field are precisely equal!

Consequently, the intensity of an electromagnetic wave is given by the expression:

$$I = \tfrac{1}{2}c\varepsilon E_0^2 = \tfrac{1}{2}cE_0D_0 . \tag{9.37}$$

By using the familiar ratio between electric and magnetic fields, the result can also be written as follows:

$$I = \frac{1}{2}c\frac{1}{\mu}B_0^2 = \frac{1}{2}cH_0B_0 \ . \tag{9.38}$$

If we choose to specify the strength of the electric and magnetic fields in terms of the effective values instead of the amplitudes, Eqs. (9.37) and (9.38) can be recast as:

$$I = c\varepsilon E_{\text{eff}}^2 = cE_{\text{eff}}D_{\text{eff}} \tag{9.39}$$

and

$$I = \frac{c}{\mu}B_{\text{eff}}^2 = cH_{\text{eff}}B_{\text{eff}} \ . \tag{9.40}$$

A small digression: The term "effective value" can be traced to alternating current terminology. We can then state the amplitudes of harmonically varying current and voltage, but we can also specify the equivalent value of direct current and direct voltage that supply the same power to a given load; these direct current/voltage values are called effective values. In our case of electromagnetic waves, it is rather artificial to speak of direct currents and suchlike, yet we speak of effective values in the same way as for alternating currents and voltages in a wire.

We can also derive another expression that connects electrical and magnetic fields to an electromagnetic wave in the remote field. Going back to Eqs. (9.39) and (9.40), and using the relationship $B = \mu H$, we get:

$$c\varepsilon E_{\text{eff}}^2 = \frac{c}{\mu}B_{\text{eff}}^2 = c\mu H_{\text{eff}}^2$$

and are led thereby to the relation:

$$\frac{E_{\text{eff}}}{H_{\text{eff}}} = \sqrt{\mu/\varepsilon} \ .$$

For vacuum, we obtain:

$$\frac{E_{\text{eff}}}{H_{\text{eff}}} = \sqrt{\mu_0/\varepsilon_0} \equiv Z_0 = 376.7\ \Omega \tag{9.41}$$

where Z_0 is called *the (intrinsic) impedance of free space.*

The expressions have a greater scope than that warranted by our derivation. However, we must be careful about using the terms of electromagnetic waves in regions near sources and near materials that can interfere with the waves. We refer to the so-called near field and far field a little later in this chapter.

9.8.1 Poynting Vector

There is a more elegant way to specify energy density (equivalent to intensity) than the expressions presented in the previous section. The elegance is a consequence of the fact that plane electromagnetic waves are transverse, with the electrical and magnetic vectors perpendicular to each other and to the direction of propagation of the wave.

We saw that if the electric field was directed in x-direction and magnetic field in y-direction, the wave moved in z-direction. We know that for the cross-product, the relation $\vec{i} \times \vec{j} = \vec{k}$ holds, which suggests that we may be able to utilize this relationship in a smart way.

We try to calculate:

$$
\begin{aligned}
\vec{E} \times \vec{B} &= E_0 \cos() \, \vec{i} \times \frac{E_0}{c} \cos() \, \vec{j} \\
&= \frac{c E_0^2}{c^2} \cos^2() \, \vec{k} \\
&= \mu (c \varepsilon E_0^2) \cos^2() \, \vec{k} \, .
\end{aligned}
$$

The time-averaged values are (using Eq. (9.37) in the last part):

$$
\overline{\vec{E} \times \vec{B}} = \mu (\tfrac{1}{2} c \varepsilon E_0^2) \, \vec{k} = \mu I \, \vec{k} \, .
$$

Since $B = \mu H$, it follows that:

$$
\vec{i} = \overline{\vec{E} \times \vec{H}} \, . \tag{9.42}
$$

Here, we have introduced an intensity vector that points in the same direction as the energy flow.

More often, we operate with the *instantaneous intensity* in the form of a "Poynting vector". This is usually designated by the symbol S or P. We choose the first variant and write:

$$
\vec{S} = \vec{E} \times \vec{H} \, . \tag{9.43}
$$

Poynting vector provides us with a nice expression of energy flow in an electromagnetic wave.

However, the Poynting vector can be used only in the trouble-free cases where we have a simple plane electromagnetic wave far from the source and far away from disturbing elements. Put in another way: it can only be used in the far-field region (see below) where the electromagnetic fields are totally dominated by pure electrodynamics.

The English physicist John Henry Poynting (1852–1914) deduced this expression in 1884, 20 years after Maxwell wrote his most famous work.

9.9 Radiation Pressure

The electric and magnetic fields will exert a force on particles/objects struck by an electromagnetic wave. It is possible to argue that the electric field in the wave causes "forced oscillations" of charges, and that moving charge, in turn, experiences a force $\vec{F} = q\vec{v} \times \vec{B}$. This force works in the same direction as that in which the electromagnetic wave moves.

It can be shown that an electromagnetic wave causes a radiation pressure given by:
$$p_{\text{radiation}} = S_{\text{time-avg}}/c = I/c$$

if the wave is completely absorbed by the body being taken. If the body reflects the waves completely, the radiation pressure becomes twice as large, i.e.

$$p_{\text{radiation}} = 2S_{\text{time-avg}}/c = 2I/c \ .$$

In both of these terms, $S_{\text{time-avg}}$ is the absolute value of the time-averaged Poynting vector. The direction of the radiation pressure is usually identical to the direction of the Poynting vector.

It is the radiation pressure that causes the dust in a comet to always turn away from the sun. The gravitational pull exerted by sun on the dust is proportional to the mass, which in turn is proportional to the cube of the radius. The force due to the radiation pressure is proportional to the *surface* (cross section) that can absorb or reflect the wave, and the cross section goes as the square of the radius. This results in gravity dominating over radiation pressure for large particles, while the converse happens for small particles.

It is possible to regard radiation pressure as a flow rate of electromagnetic momentum. In such a picture, it can be said that the momentum per time and per unit surface moving with the wave is equal
$$S_{\text{time-avg}}/c$$

which is the same expression as for radiation pressure when the body absorbs the wave completely.

The description above applies in the event that light is either absorbed or totally reflected on the surface of a material. The situation is different for light passing through a transparent medium.

There are two different descriptions of how the momentum of light changes when light enters a transparent medium. In one description, it is claimed that the momentum increases, and in another description, the opposite is claimed. This is an optical dilemma that partly depends on whether light is regarded as waves or as particles. In this way, there is a clear parallel between the dilemma we have today and the dilemma that existed from the seventeenth century to about 1850 mentioned in the previous chapter, where we wondered whether the group velocity of light in glass was larger or smaller than the phase velocity.

If you want to learn a little more about today's dilemma, start by reading a popular scientific article by Edwin Cartlidge in Physics World.

9.10 Misconceptions

First: A small reminder ...

Note that nothing actually protrudes from an electromagnetic wave. For any arbitrary point in space, the field itself changes the value. The field has a direction in space, but no arrows shoot out to the side and no sinusoidal curves are found along the wave. It is therefore a totally different situation than when, for example, we pluck a guitar string where the string actually moves across the longitudinal direction.

9.10.1 Near Field and Far Field

We have repeatedly reminded the reader of this chapter that the electromagnetic waves we have derived in Eq. (9.33) and illustrated in Fig. 9.5 are the *simplest* wave solutions of Maxwell's equations. Usually, these relationships *do not* apply to time-dependent electromagnetic phenomena in general! To understand this, we need to look more closely at the details in our derivation.

First, we ended up with inhomogeneous differential equations in Eqs. (9.20) and (9.27) as a result of combining Maxwell's equations. Only by ignoring the source terms did we arrive at the simple homogeneous wave equations that became the starting point for the plane-wave solution.

Even if there are no charges and currents in the region of our interest, fields from nearby regions can have a big influence. For example, will electric fields from charge distributions in an antenna and magnetic fields from electric currents in an antenna dominate the electromagnetic fields pattern nearby the antenna, even if it is placed in vacuum. This pattern is *not* what we find in an electromagnetic wave.

A rule of thumb in this context is that we use the word "nearby" for distances up to a few times the calculated wavelength ($\lambda = c/f$), *and/or* up to several times

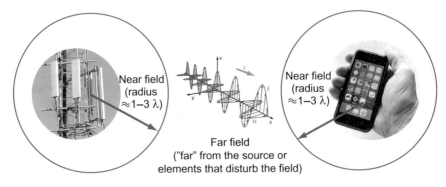

Fig. 9.7 "Near fields" dominate the electromagnetic field pattern at a distance up to the order of a calculated wavelength $\lambda = c/f$ away from charges/electrical currents. In the near-field zone, the solution of Maxwell's equations is often very different from the solution in the far-field zone (far from the source of the fields and far from disturbing elements)

the extent of the object in space away from a region where there are free charges or currents. In regions that are influenced by boundary conditions in the broadest sense, we find "*near fields*", as opposed to "*far fields*", which we find in areas where boundary conditions have almost no influence (Fig. 9.7).

It may be useful to think about how far the near-field region extends from different sources. For a light source, the wavelength is about 500 nm. The near-field range extends a few times this distance away from the source, i.e. of the order of a few microns (thousands of millimetres) away from the source.

For a mobile phone that operates at 1800 MHz, the calculated wavelength is about 16 cm. A few times this distance takes one over to the far-field zone.

To sum up:

For the *far-field region*, the following relationships we have established for simple plane electromagnetic waves are:
1. The electric and magnetic fields are perpendicular to each other.
2. There is a fixed ratio between electric and magnetic fields.
3. The Poynting vector provides a measure of transport of electromagnetic energy.
4. The energy that passes a cross section has left the source once and for all and does not (normally) return.
5. It may therefore be natural to use the word "radiation" for energy transport.

For the *near-field zone*, however, the following applies:

1. The electric and magnetic fields are normally *not* perpendicular to each other.
2. There is *no* a fixed ratio between electrical and magnetic fields.

3. The Poynting vector does *not* provide a measure of transport of electromagnetic energy.

4. Energy can build up in the vicinity of the source for some periods of time,
   but retracts again in other periods of time. Only, a tiny part of the energy
   that goes back and forth to the vicinity will leave the source like waves
   (and this energy transport is generally not apparent before we get into the
   far-field zone).
5. It is therefore not natural to use the word "radiation". We describe the
   situation more like "fields".

9.10.2 The Concept of the Photon

I would like to append a few comments concerning the term "photon".

The majority of today's physicists believe that light is best described as elementary
particles, called photons.

A photon is perceived as an "indivisible wave packet or energy packet" with a
limited extension in time and space. The word photon was originally used for visible
light where the wavelength is of the order of 500 nm (the Greek word "phos" means
"light"). This means that even a wave packet containing quite a few wavelengths will
be tiny compared to macroscopic dimensions. In this case, then, it is not particularly
odd that we perceive this as a "particle". The notion of the indivisible energy packet is
assigned the energy $E = h\nu$ where h denotes Planck's constant and ν is the frequency.

Problems soon arise with "photons" in the realm of mobile telephony (and radio
waves). In that case, a wave packet consisting of a few wavelengths will inevitably
occupy a spatial extent of several metres (up to kilometres). Does it make sense to
regard such a packet as "indivisible" and to think that energy is exchanged instanta-
neously from the antenna to the packet and from the latter to surrounding space?

For power lines and 50 Hz fields, the problem is even worse. For 50 Hz, a wave
packet of several times the wavelength would soon extend to dimensions comparable
to the perimeter of the earth! We then get serious problems imagining a photon that
extends several times the wavelength. And if we consider the photon as small particles
instead of an extended wave packet, it will be problematic to explain wavelengths
and a variety of other properties. Furthermore, the distribution of electrical and
magnetic fields near power lines is significantly different from that of light. This
can be grafted into a quantum mechanical description, but then one has to resort
to strange special variants where the quantum mechanical description really only
mimics classic electromagnetism.

A description based on Maxwell's equations gives us a formalism that scales
smoothly from STATIC electric and magnetic fields to electromagnetic waves with
frequencies even larger than the frequency of visible light. Electromagnetism also

provides a powerful explanation of the speed of light and what affects it, and the transformations of the special theory of relativity also come naturally out of electromagnetism.

Nevertheless, problems arise with the description of the interaction between a classical electromagnetic wave (or an electromagnetic field) and atomic processes. This is because classical electromagnetism cannot be used to describe atomic transitions.

In spite of this, I am among the physicists who believe that Maxwell's equations and electromagnetism are by far preferable to the photon concept for describing the vast majority of currently known phenomena, but not those involving atomic transitions. In my opinion, we have so far not reviewed the interaction between electromagnetic waves and atomic transitions with sufficient thoroughness. I represent a minority, but this minority is not getting smaller—quite the contrary, in fact. I mean that the last word has not been written about how physicists will think in the coming 50 years.

In a separate tract, I will explore this knotty issue and will not delve into it here.

9.10.3   A Challenge

Hitherto, we have seen that for both oscillations and mechanical waves there is an alternation between two energy forms as the oscillation/wave evolves. For example, for the mass–spring oscillator the energy changed in time between potential and kinetic energy, and the sum was always constant. In a travelling sound wave, at every point in space where the wave is passing, the energy density changes between potential energy (pressure) and kinetic energy, and the sum is always constant. For a travelling wave along a string, it is likewise.

For a travelling electromagnetic wave, it is not easy to see the same pattern. The electric field has the maximum at the same time and place as the magnetic field, at least for a plane. Have we overlooked something?

I suspect something is missing in our standard descriptions of electromagnetic waves. I have an idea I will follow up in the coming years. Perhaps this is a gauntlet you too want to take up?

9.11   Helpful Material

9.11.1   Useful Mathematical Relations

Here, we list some useful relationships from the mathematics you have hopefully met earlier:

Common to all expressions is that we operate with a scale field:

$$\phi = \phi(x, y, z)$$

and a vector field

$$\vec{a} = a_x \vec{i} + a_y \vec{j} + a_z \vec{k} \, .$$

A gradient is defined as:

$$\text{grad}\phi \equiv \nabla\phi \equiv \frac{\partial\phi}{\partial x}\vec{i} + \frac{\partial\phi}{\partial y}\vec{j} + \frac{\partial\phi}{\partial z}\vec{k} \, .$$

The divergence is defined as:

$$\text{div } \vec{a} \equiv \nabla \cdot \vec{a} \equiv \frac{\partial a_x}{\partial x} + \frac{\partial a_y}{\partial y} + \frac{\partial a_z}{\partial z} \, .$$

The divergence of a gradient is:

$$\text{div grad } \phi \equiv \nabla \cdot (\nabla\phi) \equiv \frac{\partial^2\phi}{\partial x^2} + \frac{\partial^2\phi}{\partial y^2} + \frac{\partial^2\phi}{\partial z^2} \equiv \Delta\phi \, .$$

The curl is defined as:

$$\text{curl } \vec{a} \equiv \nabla \times \vec{a} \equiv$$

$$\begin{vmatrix} \vec{i} & \vec{j} & \vec{k} \\ \frac{\partial}{\partial x} & \frac{\partial}{\partial y} & \frac{\partial}{\partial z} \\ a_x & a_y & a_z \end{vmatrix} =$$

$$\left(\frac{\partial a_z}{\partial y} - \frac{\partial a_y}{\partial z} \right)\vec{i} + \left(\frac{\partial a_x}{\partial z} - \frac{\partial a_z}{\partial x} \right)\vec{j} + \left(\frac{\partial a_y}{\partial x} - \frac{\partial a_x}{\partial y} \right)\vec{k} \, .$$

Notice what are vector fields and what are scalar fields. In general:

- A gradient converts a scalar field into a vector field.
- A divergence works the other way.
- Div-grad starts with a scalar field, passes through a vector field and ends with a scalar field again.
- A curl, in contrast, starts with a vector field and ends with a vector field.

The symbol ∇ is involved in different operations depending on whether it works on a scalar field or a vector field, and it is especially challenging to use ∇^2 on a vector since we must then use the Laplacian on each of the components in the vector separately:

$$\nabla^2\mathbf{a} = \left(\frac{\partial^2 a_x}{\partial x^2} + \frac{\partial^2 a_x}{\partial y^2} + \frac{\partial^2 a_x}{\partial z^2} \right)\vec{i}$$

$$+ \left(\frac{\partial^2 a_y}{\partial x^2} + \frac{\partial^2 a_y}{\partial y^2} + \frac{\partial^2 a_y}{\partial z^2} \right)\vec{j}$$

$$+\left(\frac{\partial^2 a_z}{\partial x^2} + \frac{\partial^2 a_z}{\partial y^2} + \frac{\partial^2 a_z}{\partial z^2}\right)\vec{k}\ .$$

Some other useful relations appear below:

$$\text{curl grad}\phi = \nabla \times (\nabla\phi) = 0\ ,$$

$$\text{div curl }\mathbf{a} = \nabla(\nabla \times \mathbf{a}) = 0\ ,$$

$$\text{curl(curl}\mathbf{a}) = \text{grad(div}\mathbf{a}) - \triangle\mathbf{a} = \nabla \times (\nabla \times \mathbf{a}) = \nabla(\nabla \cdot \mathbf{a}) - \nabla^2\mathbf{a}\ .$$

9.11.2   Useful Relations and Quantities in Electromagnetism

Here are some relationships from electromagnetism as a refresher of prior knowledge:

- Electric field strength $\vec{E}$ is measured in V/m.
- Electric flux density $\vec{D}$ is measured in C/m^2.
- Magnetic field strength $\vec{H}$ is measured in A/m.
- Magnetic flux density $\vec{B}$ is measured in T.
- E-flux density is also often referred to as electric displacement.
- Free space electrical permittivity ε_0 is measured in F/m = (As)/(Vm) and defined as

$$\varepsilon_0 \equiv \frac{1}{\mu_0 c_0^2} \approx 8.854188 \times 10^{-12}\ \text{F/m}$$

- The relative permittivity ε_r is usually a number larger than 1.0.
- Free space magnetic permeability μ_0 is measured in H/m and defined as:

$$\mu_0 \equiv 4\pi \times 10^{-7}\ \text{H/m} \approx 1.256637 \times 10^{-6}\ \text{H/m}$$

- The relative permeability μ_r is close to 1.0 for most materials. Ferromagnetic materials are an exception.
- The speed of light in vacuum is given exactly as:

$$c_0 \equiv 299{,}792{,}458\ \text{m/s}$$

The SI basic units are now the speed of light in vacuum and the second. The length 1 metre is no longer one of the basic units!

- The relation between field strengths and flux densities is as follows:

$$\vec{D} = \varepsilon_r \varepsilon_0 \vec{E}$$

$$\vec{B} = \mu_r \mu_0 \vec{H}$$

9.12   Learning Objectives

After working through this chapter, you should be able to:
- Convert Maxwell's equations from integral to differential form (assuming Stokes's theorem and divergence theorem are given).
- Derive the wave equation for electromagnetic field in vacuum provided that Eq. (9.16) is given.
- Explain what simplifications are introduced in the derivation of the wave equation for electromagnetic fields in vacuum.
- Explain which part of Maxwell's equations is responsible for an electromagnetic wave to travel through free space.
- Explain carefully the difference between "plane wave" and polarization.
- Specify the amount of energy transport in a plane electromagnetic wave.
- Apply the Poynting vector and know the limitations of this concept.
- State and apply expression of radiation pressure in an electromagnetic field in a plane wave.
- Explain what we mean by near field and far field and why these sometimes are very different.
- Explain the characteristics of electromagnetic fields that differ in the two zones.
- Explain some problems using the photon term for all electromagnetic fields/waves.

9.13   Exercises

Suggested concepts for student active learning activities: Electromagnetic wave, line integral, surface integral, vector field, near field, far field, pure electrodynamics, polarization, dielectric, index of refraction, relative permittivity, electromagnetic spectrum, energy density, energy transport, radiation pressure.

Comprehension/discussion questions

1. It is not easy to comprehend Fig. 9.5 correctly. It is so easy to think of waves in a material way, similar to surface waves on water. However, an electromagnetic wave is much more abstract, since it is just the abstract quantities of electric and magnetic fields that changes with position and time. Discuss if it becomes easier to comprehend Fig. 9.5 if we state that an electric and magnetic field actually *change the property of the space* locally (even in vacuum) and that it is this *changed property of space* that moves as the electromagnetic wave passes by.

2. Explain briefly how to characterize a region in space where the divergence of the electric field is different from zero. Similarly, explain briefly how to characterize a region in space where the curl of the electric field is different from zero.

3. In going from the integral form of Maxwell's equations to the differential form, we use an argument based on the "intermediate" scale of length/volume. What do we mean by this?

4. Suppose we measure the electric and magnetic fields in an electromagnetic wave in the far-field zone. Can we determine the direction of the waves from these measurements?

5. We apply an alternating voltage across a capacitor, or we send an alternating current through a solenoid. Attempt to find the direction of the electric and magnetic fields and relative magnitudes. Will these fields follow the well-known laws that apply to the electric and magnetic fields for plane electromagnetic waves?

6. It is sometimes said that for an electromagnetic wave in vacuum, the electric and magnetic fields are perpendicular to each other. Magnetic fields and electric fields do not have this relationship to one another a short distance from a solenoid ("coil"), even if it is in vacuum and high-frequency electric and magnetic fields are present. What causes this?

7. Is polarization a property of all electromagnetic waves, not just light waves? Can sound waves have a polarization? By the way: What do we mean by "polarization"?

8. An electromagnetic wave (e.g. strong light) may have an electric field of about 1000 V/m. Could it lead to electric shock if one is exposed to this powerful light?

9. The magnetic field in intense laser light can be up to 100 times as powerful as the earth's magnet field. What will happen if we shine with this laser light on the needle of a compass?

10. Poynting vector indicates the power in an electromagnetic wave. Can we use the Poynting vector to calculate the power that springs from a power line to residents nearby? Explain your answer.

11. If you flash with the light from an electric torch, would you experience a recoil similar to that one gets on firing a gun? Discuss your answer.

12. In any physical system/phenomenon, one may identify a length scale and a timescale. What is meant by such a statement when we consider electromagnetic waves?

13. A person measures the electric field E and the magnetic field B in vacuum for the same frequency f and position, but finds that $E/c \gg B$. Is this an indication of malfunction for one of the two instruments used in the measurements?

14. In several equations in this chapter, the relative electrical permittivity ε_r is included.

(a) The speed of light is linked to this quantity. How?

(b) The relative permittivity tells us something about what physical processes take place when light travels through glass. What processes are we thinking about?

(c) Many think that it makes sense that light slows down on going from air or vacuum to glass, but they find it hard to understand that light regains the original speed upon leaving the glass. What, in your opinion, accounts for their difficulty?

Problems

15. Show that a plane electromagnetic wave in vacuum satisfies all four Maxwell's equations.

16. Write down Maxwell's equations in integral form, and state the correct names for them. Give a detailed derivation of Ampére's law in differential form.

17. The derivation of the wave equation from Maxwell's equations follows about the same tricks whether one uses them to arrive at the wave equation for the electric field or for the magnetic field. Make a list showing which steps/tricks are used (a relatively short account based on essential points without going into detail will suffice).

18. Find the frequency of yellow light of wavelength 580 nm. Do the same with wavelength of about 1 nm. The fastest oscilloscopes available now have a sampling rate in the range of 10–100 GHz. Can we use this kind of oscilloscope to see the oscillations in electric fields in the X-ray waves? What about yellow light?

19. An electromagnetic wave has an electric field given by $\vec{E}(y, t) = E_0 \cos(ky - \omega t)\vec{k}$. $E_0 = 6.3 \times 10^4$ V/m, and $\omega = 4.33 \times 10^{13}$ rad/s. Determine the wavelength of the wave. In which direction does the wave move? Determine $\vec{B}$ (vector). If you make any particular assumptions in the calculations, these must be stated.

20. An electromagnetic wave of frequency 65.0 Hz passes through an insulating material with a relative permittivity of 3.64 and relative permeability of 5.18 for this frequency. The electric field has an amplitude of 7.20×10^{-3} V/m. What is the wave speed in this medium? What is the wavelength in the medium? What is the amplitude of the magnetic field? What is the intensity of the wave? Are the calculations you have made really valid? Explain your answer.

21. An intense light source radiates light equally in all directions. At a distance of 5.0 m from the source, the radiation pressure on a surface that absorbs the light is approximately 9.0×10^{-9} Pa. What is the power of the emitted light?

22. A ground surface measurement shows that the intensity of sunlight is 0.78 kW/m². Estimate the power the radiation pressure will exert on a 1 m² large solar panel? State the assumptions you make. As a matter of interest, we may mention that the atmospheric pressure is about 101,325 Pa (about 10^5 Pa).

23. For an electromagnetic wave, it is assumed that the electric field at one point is aligned in the x-direction and magnetic field in the $-z$-direction. What is the direction of propagation of the wave? What if the fields were in the $-z$- and y-direction, respectively? Did you make any assumption for finding the answers?

24. An ordinary helium–neon laser in the laboratory has a power of 12 mW, and the beam has a diameter of 2.0 mm. Suppose the intensity is uniform over the cross section (which is completely wrong, but it can simplify the calculations). What are the amplitudes of the electric and magnetic fields in the beam? What is the average energy density of the electric field in the beam? What about the energy density in the magnetic field? How much energy do we have in a 1.0 m long section of the beam?

25. Measurements made a few hundred metres from a base station indicated an electric field of 1.9 V/m and a magnetic field of 1.2 mA/m (both at about 900 MHz). A knowledgeable person concluded that the measurements were not mutually consistent. What do you think was the reason for this conclusion?

26. Measurements at the ground just a few tens of metres from a power line registered an electric field of 1.2 kV/m and a "magnetic field" of 2.6 μT (microtesla) (both at 50 Hz). In practice, it is often magnetic flux density reported at low frequencies, but we can convert from B to H, and then find that 2.6 μT corresponds to the magnetic field value 2.1 A/m. Is there correspondence between electric field and magnetic field in this case? Comment on similarities/differences between the situations in the previous task and in this task.

27. One day, the electric and magnetic fields are measured at the same location near the power line as in the previous task, and the values are found to be 1.2 kV/m and 0.04 A/m. Can we conclude that there is something wrong with one of the measuring instruments in this case?

28. According to Radiation Protection Info 10–11: Radio Frequency Fields in our Environment (Norwegian Radiation Protection Agency) (http://www.nrpa.no/filer/5c7f10ca06.pdf, available 10 May 2018), the "radiation" from base stations, wireless networks, radio, etc., is less than 0.01 W/m^2 across our country. Calculate the electric field and magnetic field equivalent to 0.01 W/m^2 if we think that the radiation is dominated by mobile phone communications from a base station at 1800 MHz.

29. When we use a mobile phone somewhere where the coverage is poor so that the phone gives maximum power, the mobile phone supplies about 0.7–1.0 W power while communicating. Calculate the intensity 5 cm from the mobile phone if you assume an isotropic intensity around the phone. Compare the value with measured intensities from base stations, wireless networks, etc., given in the previous task.

30. It is not customary to report the "radiation" from a mobile phone in terms of power density (intensity) measured in W/m^2, but in Specific Absorption Rate (SAR).
    (a) Search the Web to find out about SAR. State the URL for the source you are using.
    (b) Explain what SAR implies and what is the SAR unit?
    (c) What do you think is the reason why such a unit has been adopted in this case, even though we use power density from base stations and suchlike, with about the same frequency as the mobile phone?

31. Let us consider interplanetary dust in our solar system. Suppose the dust is spherical and has a radius of r and a density of ρ. Suppose all radiation that hits the dust grain is absorbed. The sun has a total radiated power of P_0 and a mass M. The gravity constant is G. The distance from the sun is R. Derive an expression that indicates the relationship between the power exerted by the radiation pressure from the sun rays to the dust grain and the gravitational force between the sun and the dust grain. Determine the radius of the dust when the two forces are equal as we insert realistic values for the quantities that are

involved. ($\rho = 2.5 \times 10^3$ kg/m, $P_0 = 3.9 \times 10^{26}$ W, $M = 1.99 \times 10^{30}$ kg, $G = 6.67 \times 10^{-11}$ Nm^2/kg^2).

32. Relate the gravitational force between the earth and the sun, and the force on the earth due to the radiation pressure from the sun. The earth's mass is 5.98×10^{24} kg. You can estimate the radius of the earth by recalling that the distance between a pole and the equator is about 10,000 km.

Reference

1. PD, https://commons.wikimedia.org/wiki/File:James_Clerk_Maxwell_big.jpg. Accessed April 2018

Chapter 10
Reflection, Transmission and Polarization

Abstract In this chapter, Maxwell's equations are used for deducing laws of reflection/transmission of an electromagnetic wave entering an idealized plane boundary between two insulators, e.g. air (or vacuum) and glass. The expression for the Brewster angle is derived and Fresnel's equations are presented. Snel's law is derived using the principle of minimum time. Emphasis in the last part of the chapter is put on polarization and how it may be changed by the use of birefringent material like calcite or polarization filters. Use of polarization in polariometry as well as in stereoscopy is mentioned, and a brief comment on evanescent waves is given.

10.1 Introduction

In Chap. 9, we found that a plane electromagnetic wave with the phase velocity

$$c = \frac{1}{\sqrt{\varepsilon_0 \varepsilon_r \mu_0 \mu_r}}$$

$$= \frac{1}{\sqrt{\varepsilon_0 \mu_0}} \frac{1}{\sqrt{\varepsilon_r \mu_r}} = \frac{c_0}{\sqrt{\varepsilon_r \mu_r}}$$

is a possible solution of Maxwell's equations in an infinite homogeneous medium containing no "free charges". The symbols have their usual meanings.

The speed of light in a medium (without free charges) is the quotient of the speed of light in vacuum c_0 and the refractive index n for the medium:

$$c \equiv \frac{c_0}{n} .$$

The vast majority of media we are going to consider are diamagnetic or paramagnetic. This applies, for example, to optical glass, for which $\mu_r \approx 1.00$. As a result, we may write:

$$n \approx \sqrt{\varepsilon_r} .$$

A. I. Vistnes, *Physics of Oscillations and Waves*, Undergraduate Texts in Physics,
https://doi.org/10.1007/978-3-319-72314-3_10

In other words, the index of refraction is, in a manner of speaking, directly related to the "polarization susceptibility" of the medium, and the relative permittivity is a measure of this. The more easily an external electric field can distort the electron cloud around the atoms from their equilibrium positions, the slower is the speed of light in that medium.

For substances whose atoms are arranged in a regular and special way, as in a calcite crystal, it is easier to displace the electron clouds away from equilibrium when the electric field has one particular direction relative to the crystal than other directions. This causes light to travel more slowly (through the crystal) for one orientation of the crystal relative to the direction of light polarization than for other orientations. Calcite crystals, which have this property, are said to be birefringent. Doubly refracting materials are widely used in modern optics.

Other substances have the property that they only transmit light with the electric field in a particular orientation. Such substances can be used as so-called polarization filters, which are used in photography, material characterization, viewing 3D movies, and in astronomy.

We will treat birefringence and polarization filters in this chapter, but we *start* by analyzing how waves are partially reflected and partially transmitted when they strike an interface between two different media (in contact with each other). Again, Maxwell's equations are central to the calculations.

A running topic throughout the chapter is polarization, but polarization appears in two quite different contexts. Be careful not to confuse them!

10.2   Electromagnetic Wave Normally Incident on An Interface

Generally, there are infinitely many different geometries and as many different solutions of Maxwell's equations when an electromagnetic wave reaches an interface between two media. We need to simplify enormously in order to extract regularities that can be described in a mathematically closed form.

> In this section, we will use the Faraday–Henry law together with an energy balance sheet to find out what fraction of an electromagnetic wave is reflected and what is transmitted when the wave enters, for example, from air into glass. We assume that the electromagnetic wave is approximately plane and strikes normally a plane interface between two different homogeneous media without free charges. We make the following assumptions for the second medium and the interface:
> 1. Assume that the medium itself is homogeneous within a volume of λ^3 where λ is the wavelength.
> 2. Assume that the interface is flat over an area much greater than λ^2.
> 3. Assume that the thickness of the interface is much less than the wavelength λ.

As long as we consider light of wavelength in the 400–800 nm range travelling through glass, where the atoms are a few tenths of a nanometre apart, these three assumptions are reasonably well fulfilled. But the conditions are certainly not met in all common cases. When light goes through raindrops, the drops are often so large that we can almost use the formalism that will be derived presently. But when the drops are so small that the above conditions are not met, Maxwell's equations must be used directly. For drops that are small, we get the so-called Mie scattering, which produces not a regular rainbow but an almost colourless arc.

Also for electromagnetic waves in completely different wavelength ranges than light, it is difficult to satisfy the three assumptions. Take for example X-rays with wavelength around 0.1 nm. Then, the wavelength is about the same as the distance between the atoms. For radio waves as well, the assumptions cannot be easily satisfied. This means that the laws to be deduced in this chapter are often limited in practice to electromagnetic waves in the form of visible light, or in any case nearby wavelengths.

The purpose of the following mathematics in this chapter is to derive useful expressions, but also to point out clearly the assumptions we base the calculation on. This is important so that we can judge the validity of the formulas in different contexts. Within the rapid growing field of nanotechnology, it becomes clear that common expressions are not applicable everywhere.

So, let us study what happens when an electromagnetic wave meets an interface head on. Let us suppose that the three above assumptions are satisfied and that we send electromagnetic waves normally to the interface. Part of the wave will be reflected at the interface and travel back in the original medium, while the rest of the wave is transmitted into the next medium and continues there. In Fig. 10.1, the three waves are drawn in a manner that brings out their main features. The waves that are drawn in can be considered, for example, as one component of the electric field (in a given direction perpendicular to the normal to the interface). The index of refraction on the left side of the figure is n_1 and that on the right side n_2, and we have not yet said anything about which of these is the larger. For the same reason, we have not considered whether the reflected wave would have the opposite sign to the incoming

Fig. 10.1 An electromagnetic wave travelling perpendicular to another medium is partially reflected and partially transmitted. The waves are depicted separately in order to indicate instantaneous electric fields for each of them

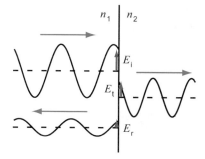

Fig. 10.2 Integration path (*blue arrows*) and electric field (*red arrows*) defining positive directions when applying Faraday's law to find relations between electric fields from different components. See the text for details

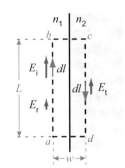

wave at the interface itself. We proceed tentatively, and calculate the signs shown in the figure, and we will discuss the details later.

First step: Faraday's law

We choose the rectangular integration path shown in Fig. 10.2 with a length L and width w. The integration path is oriented so that the long sides are parallel to the electric fields of the electromagnetic wave. We are ready to apply Faraday's law:

$$\oint \vec{E} \cdot d\vec{l} = -\left(\frac{d\Phi_B}{dt}\right)_{inside} . \tag{10.1}$$

We deal with the line integral first:

$$\oint \vec{E} \cdot d\vec{l} = \int_{ab} + \underbrace{\int_{bc}}_{=0} + \int_{cd} + \underbrace{\int_{da}}_{=0}$$

$$= (E_i + E_r)L - E_t L .$$

The integrals along bc and da contribute nothing because the paths are perpendicular to the electric fields; the first integral is positive, and the last negative, because, as shown in Fig. 10.2, the field is oppositely directed to the line element in the latter case.

As for the right-hand side of Eq. (10.1), our assumption that the interface is infinitely thin makes it permissible to choose w, and therefore the area $A = Lw$, to be arbitrarily small. Next, we express Φ_B as a surface integral, and get the simple result:

$$-\left(\frac{d\Phi_B}{dt}\right)_{inside} = -\frac{d}{dt}\int_A \vec{B} \cdot d\vec{A} \approx 0$$

the last step being a consequence of the smallness of the area A.

The foregoing manipulations of Eq. (10.1) lead us to the result:

$$E_i L + E_r L - E_t L = 0$$

which implies that

$$E_i + E_r = E_t \ . \tag{10.2}$$

We can apply a similar reasoning to the Ampére-Maxwell law to get

$$H_i + H_r = H_t \ .$$

Second step: Energy conservation

We can also set up an *energy balance sheet*: All energy incident per unit time on the interface must be equal to the energy that leaves the interface per unit time. We know from Chap. 9, that the intensity of an electromagnetic wave is given by:

$$I = c u_{\text{tot}} = \tfrac{1}{2} c \vec{E} \cdot \vec{D} = \tfrac{1}{2} c \varepsilon_0 \varepsilon_r E^2$$

where u_{tot} is the energy density in the wave, and c is the speed of light in the medium under consideration. The energy balance sheet comes out to be:

$$\tfrac{1}{2} c_1 \varepsilon_0 \varepsilon_{r1} E_i^2 = \tfrac{1}{2} c_1 \varepsilon_0 \varepsilon_{r1} E_r^2 + \tfrac{1}{2} c_2 \varepsilon_0 \varepsilon_{r2} E_t^2 \ ,$$

$$c_1 \varepsilon_{r1} (E_i^2 - E_r^2) = c_2 \varepsilon_{r2} E_t^2 \ ,$$

$$c_1 \varepsilon_{r1} (E_i + E_r)(E_i - E_r) = c_2 \varepsilon_{r2} E_t^2 \ .$$

But, since $E_i + E_r = E_t$:

$$c_1 \varepsilon_{r1} (E_i - E_r) = c_2 \varepsilon_{r2} E_t \ .$$

Let us examine the constants appearing above. To this end, we recall the expression, given earlier in this chapter, for the speed of light:

$$c_1 = \frac{c_0}{n_1} \approx \frac{c_0}{\sqrt{\varepsilon_{r1}}} \ .$$

Multiplying by ε_{r1} and replacing the $\approx$ sign with equality, we get

$$c_1 \varepsilon_{r1} = \frac{c_0}{\sqrt{\varepsilon_{r1}}} \varepsilon_{r1}$$

$$= c_0 \sqrt{\varepsilon_{r1}} = c_0 n_1 \ .$$

Substituting this expression (and its counterpart for medium 2) in Eq. (10.2), we obtain

$$n_1 (E_i - E_r) = n_2 E_t \ . \tag{10.3}$$

Third step: Combine

We combine now Eqs. (10.2) and (10.3) and eliminate, to begin with, E_t in order to find a relation between E_i and E_r:

$$n_1 E_i - n_1 E_r = n_2 E_i + n_2 E_r$$

$$(n_1 - n_2) E_i = (n_1 + n_2) E_r \ .$$

The ratio between the amplitudes of the reflected and transmitted waves is found to be:

$$\frac{E_r}{E_i} = \frac{n_1 - n_2}{n_1 + n_2} \ . \tag{10.4}$$

We see that the right-hand side can be positive ($n_1 > n_2$), negative ($n_1 < n_2$) or zero ($n_1 = n_2$).

For $n_2 > n_1$, the ratio is negative, which means that E_r has a sign opposite to that of E_i (i.e. to say, E_r is in the opposite direction to that indicated in Fig. 10.1).

For $n_2 < n_1$, the expression in Eq. (10.4) is positive, which means that E_r has the same sign as E_i (i.e. E_r has the direction shown in Fig. 10.1).

Let us conclude by combining Eqs. (10.2) and (10.3) by eliminating E_r in order to find a relation between E_i and E_t. This gives:

$$n_1 E_i - n_1 E_t + n_1 E_i = n_2 E_t \ .$$

The ratio between the amplitudes of transmitted and incident waves is easily found:

$$\frac{E_t}{E_i} = \frac{2 n_1}{n_1 + n_2} \ . \tag{10.5}$$

We see that the electric field of transmitted wave has always the same sign as that of the incident wave.

Equations (10.4) and (10.5) provide the relationship between electric fields on both sides of the interface. When we judge how much of the light is reflected and transmitted, we want to look at the intensities. We have already seen that the intensities are given by expressions of the type:

$$I_i = \frac{1}{2}c_1\varepsilon_0\varepsilon_{r,1}E_i^2 \approx \frac{1}{2}c_0\varepsilon_0 n_1 E_i^2 \, .$$

We are led to the following relation between the intensities:

$$\frac{I_r}{I_i} = \frac{n_1 E_r^2}{n_1 E_i^2} = \left(\frac{n_1 - n_2}{n_1 + n_2}\right)^2 \tag{10.6}$$

and

$$\frac{I_t}{I_i} = \frac{n_2 E_t^2}{n_1 E_i^2} = \left(\frac{2n_1}{n_1 + n_2}\right)^2 \times \frac{n_2}{n_1} \, . \tag{10.7}$$

If we choose to look at what happens at the interface between air and glass (refractive index 1.00 and 1.54, respectively), we get:

Reflected:

$$\frac{I_r}{I_i} = \left(\frac{0.54}{2.54}\right)^2 \approx 0.045 \, .$$

Transmitted:

$$\frac{I_t}{I_i} = \left(\frac{2}{2.54}\right)^2 \times 1.54 \approx 0.955 \, .$$

Thus, we see that about 4.5% of the intensity of light normally incident on an air–glass surface is reflected, while about 95.5% is transmitted. This is the case when the glass surface has not received any special treatment ("anti-reflection coating").

Finally, it may be noted that the reflection at the surface leads to the creation of some standing waves in the area in front of the interface.

10.3   Obliquely Incident Waves

10.3.1   Snel's Law of Refraction

Willebrord Snel of Royen was born in the Netherlands in 1580. He later changed his name to Willebrord Snellius and died in 1626. His name should be written either as Snel or Snellius, but it is most commonly spelled as Snell. We have chosen the original name Snel.

Snel's law of refraction gives us the relation between the inclination of a light ray before it strikes an interface between two materials and its inclination after the interface.

The law of refraction can be derived in several ways. We will use "Fermat's principle" which is also called *principle of minimum time*. Fermat's principle is expressed in our times by saying that *the optical path length must be stationary*. Speaking a little imprecisely, this means that for the route along which light transports energy ("where light actually goes"), optical path length is the same (in the first approximation) for an array of optical paths that are close to one another. This means that the optical path length must be a maximum, minimum or stationary for small variations in the selected path. When we deduce Snel's law of refraction, we use the minimum as the criterion.

We refer to Fig. 10.3. A beam of light is sent from the point P in a medium with refractive index n_1 to P' in a medium with refractive index n_2. We assume in the figure that $n_2 > n_1$. Since light travels faster in medium 1 than in medium 2, the shortest time to cover the distance between the two points will be achieved by travelling a little longer in medium 1, instead of travelling along the straight line connecting the two points. If we use the symbols in the figure, it follows that the time for travel is:

Fig. 10.3 In the derivation of Snel's law of refraction, we use the coordinates given in this figure. The angles θ_1 and θ_2 are called the "incident angle" and the "refraction angle", respectively. See also the text

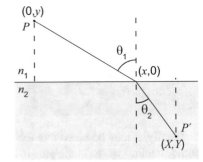

$$t = \frac{\sqrt{x^2 + y^2}}{c_0/n_1} + \frac{\sqrt{(X-x)^2 + Y^2}}{c_0/n_2}$$

$$= \frac{1}{c_0}\left(n_1\sqrt{x^2 + y^2} + n_2\sqrt{(X-x)^2 + Y^2}\right) .$$

The independent variable here is x and the minimum time can be determined by setting $dt/dx = 0$, and this gives:

$$\frac{dt}{dx} = \frac{1}{c_0}\left(n_1\frac{\frac{1}{2} \times 2x}{\sqrt{x^2 + y^2}} + n_2\frac{\frac{1}{2}(X-x) \times 2 \times (-1)}{\sqrt{(X-x)^2 + Y^2}}\right) = 0 ,$$

$$n_1 x\sqrt{(X-x)^2 + Y^2} - n_2(X-x)\sqrt{x^2 + y^2} = 0 ,$$

$$\frac{n_1}{n_2} = \frac{(X-x)\sqrt{x^2 + y^2}}{x\sqrt{(X-x)^2 + Y^2}} = \underbrace{\frac{X-x}{\sqrt{(X-x)^2 + Y^2}}}_{\sin\theta_2} \times \underbrace{\frac{\sqrt{x^2 + y^2}}{x}}_{1/\sin\theta_1} .$$

We arrive finally at the refraction law commonly attributed to Snel:

$$\frac{n_1}{n_2} = \frac{\sin\theta_2}{\sin\theta_1}$$

or

$$n_1\sin\theta_1 = n_2\sin\theta_2 . \tag{10.8}$$

Fermat's principle has clear links to Huygen's principle and also the thinking behind quantum electrodynamics (CED). The waves follow all possible ways, but in some cases the waves reinforce each other, and in other cases, they will oppose each other. In other words, it is interference that is actually running the show, and the central idea behind this phenomenon is the role played by the relative phases of the different contributions. The "minimum time" criterion achieves the desired result automatically, since minimum time means that many waves, which we can imagine to have been sent from P, will take close to the minimum travel time and all these waves will automatically have the same phase and therefore interfere constructively.

10.3.2 Total Reflection

Total reflection is of course an important effect that anyone who likes to dive underwater knows well. The point is that if light goes from a medium with refractive
index n_1 to a medium with index n_2 and $n_1 > n_2$, the "incidence angle" θ_1 will be
smaller than the "refraction angle" θ_2 for the transmitted beam. We can first send the
beam normally to the interface and then gradually increase the angle of incidence.
The refraction angle will then gradually increase and will always be greater than the
incidence angle.

Sooner or later, we will have an angle of incidence that leads to a refraction angle
of almost 90°. If we increase the angle of incidence further, we will not be able to
satisfy Snel's law, because the sine of an angle cannot exceed unity.

The incidence angle (θ_c) for which the angle of refraction is 90°, called the "critical
angle", is found by setting $\theta_1 = \theta_c$ and $\theta_2 = 90°$ in Snel's law:

$$n_1 \sin \theta_c = n_2 \sin \theta_2 = n_2 \sin 90° = n_2 .$$

The critical angle of incidence can be expressed as:

$$\sin \theta_c = \frac{n_2}{n_1} . \qquad (10.9)$$

If the angle of incidence is increased beyond the critical angle, there will no
longer be a transmitted beam. Everything will be reflected from the interface
back into the original medium, leading to a phenomenon called *total reflection*.

If we are underwater and look up at the surface, the critical angle will be given
by:

$$\sin \theta_c = \frac{1.00}{1.33}$$
$$\theta_c = 48.8° .$$

If we try to look at the surface along a greater angle than this (relative to the
vertical), the water surface will merely act as a mirror.

Total reflection is used to a large extent in today's society. Signal cables for the
Internet and telephony and almost all information transfer now largely take place via
optical fibres. For optical fibres having a diameter that is many times the wavelength
(so-called multimode fibres), it is permissible to say that total reflection is at work
here.

An optical fibre consists of a thin core of super-clean glass. Outside this core is
a layer of glass whose refractive index is very close to, but slightly less than that of

the core, the difference being about 1%. The consequence is that the critical angle becomes very close to 90°. This means that only the light that moves very nearly parallel to the fibre axis is reflected at the interface between the inner core and the next layer of glass outside. It is important that the waves are as parallel as possible to the axis so that pulses transmitted into the fibre should retain their shape before being relayed.

In many optical fibres, the diameter of the inner glass core is only a few times the wavelength. Such fibres are called single-mode fibres, and most are used in telecommunications and similar applications. For single-mode fibres, it is really misleading to explain the waveform in the fibre with total reflection. Instead, we must use Maxwell's equations directly with the given geometry. The wave image inside the fibre can no longer be considered a plane wave as we find it in vacuum far from the source and from disturbing boundary conditions. The boundary conditions imply a completely different solution. We will come back to this when we deal with waveguides in Chap. 16.

Single-mode fibres are challenging to work with because the cross section of the fibre is very small and the light entering the fibre must have a direction very close to the fibre direction. It is therefore difficult to get light *into* the fibre without too much loss. Standardization of coupling devices, however, makes it easy for telecommunication equipment, but it is quite a challenge to connect light into a fibre from a beam in air in a laboratory.

It is much easier to get light into multimode fibres because they have larger cross sections and the direction of the incoming light is not as critical. Multimode fibres, however, are not suitable for long-distance communications since pulses "fade out" after travelling relatively short distances.

10.3.3 More Thorough Analysis of Reflection

We will now look more closely at reflection and transmission when a (more or less) plane electromagnetic wave strikes obliquely an interface between two media. We make the same assumptions as mentioned in the beginning of the chapter that the interface is plane, "infinitely large and infinitely thin".

A major challenge in the derivation that will follow consists of keeping track of geometry. Waves are inclined towards the interface, and the outcome depends on whether the electric field that meets the interface is parallel to the interface or inclined obliquely with respect to it. You may want to spend some time to understand the decomposition of the electric field vector E in Fig. 10.4 before reading further.

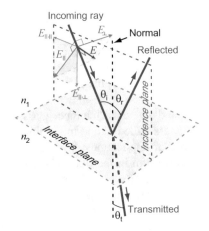

Fig. 10.4 Geometrical details for discussing the propagation of an electromagnetic ray inclined obliquely towards a plane interface between two media. The electrical field vector of the ray is resolved into a component normal to the incidence plane and a component parallel to this plane (lying in the plane of incidence). The latter component is further resolved into a component that is parallel to the interface and one that is normal to the interface. See the text for details

We draw a "ray" travelling obliquely towards the interface. We draw a normal to the interface at the point where the ray meets the interface. The plane containing the incident ray and the normal will be called the *plane of incidence*. The angle between the incident beam and the normal will be denoted by θ_i. See Fig. 10.4.

The reflected beam will lie in the input plane and have the same angle with the incident ray as the incident beam, i.e. $\theta_i = \theta_r$. The transmitted beam will also be in the same plane as the other rays, but it makes an angle θ_t with the normal (extended into medium 2).

We shall not go into any detailed proof that the three rays are in the same plane, but Maxwell's equations are symmetrical with regard to time. It is believed that if one solution of Maxwell's equations is an incident beam that divides into a reflected and a transmitted beams, then another solution is that where the reflected and transmitted waves can be considered as two incident rays coming *against* the interface and combining into a single output ray (similar to the original incident ray, but with the opposite direction of motion).

Since we can reverse, at least hypothetically, the time course for what is happening, it means that the solution must have a certain degree of symmetry. One consequence is that the three rays must lie in the incidence plane.

We *start* by assuming that all three rays lie in the incidence plane and $\theta_i = \theta_r$ in Fig. 10.4, and then use Maxwell's equations to determine how much of the incoming energy that is reflected and transmitted at the interface.

However, the wave has an arbitrary polarization. This means that the electric field $\vec{E}$, which is perpendicular to the incident beam, may have any angle relative to the incidence plane. The result for the component of electric field which lies in the incidence plane $E_\parallel$ is slightly different from that for the component perpendicular to the incidence plane $E_\perp$.

First step: $E_\perp$

We start by treating the component of electric field perpendicular to the incidence plane. This component will at the same time be parallel to the interface, which was also the case for the wave incident at the interface (discussed in the previous section). Faraday's law used as in Fig. 10.2 gives as before:

$$E_{i,\perp} + E_{r,\perp} = E_{t,\perp}$$

where i, r and t again represents incoming, reflected and transmitted. $\perp$ indicates the component that is perpendicular to the incident plane, which in turn is parallel to the interface. However, we do not pursue this component in detail.

It is more interesting to look at the component that lies in the incidence plane, but the treatment here is a little more complicated. The component *lying in the incidence plane* can be resolved into a component that is *normal to the interface* and one that is *parallel to the interface*.

In Fig. 10.4, we have tried to indicate that the electric field of the incoming wave has components both normal and parallel to the *incidence plane*, and that the latter component, $E_\parallel$, can in turn be resolved into a component $E_{\parallel,\parallel}$ parallel to and a component $E_{\parallel,\perp}$ perpendicular to the *interface/boundary surface* (Note: For simplicity reasons, we drop the vector notation for all components of the electric fields.).

In Fig. 10.5, *only* the component of the electric field parallel to the incident plane is drawn. Decomposition of this component is, respectively, $E_{\parallel,\parallel}$ and $E_{\parallel,\perp}$. The first

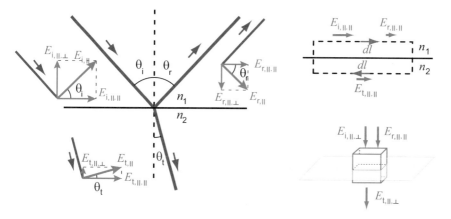

Fig. 10.5 Components of the electric field in the incoming plane for incoming, reflected and transmitted rays. The field decomposition on the left is drawn separately for the incoming, reflected and transmitted rays so as to avoid clutter. The diagrams on the right specify the positive directions of the components in the mathematical treatment. See also the text for details

part of the subscript indicates component relative to the incidence plane, and the second part indicates the component with respect to the interface.

From Fig. 10.4, we see that $E_{\parallel,\parallel}$ is perpendicular to $E_\perp$ (the component of electric field normal to the incidence plane), although both are parallel to the interface. Also note that $E_{\parallel,\perp}$ is perpendicular to the interface and thus parallel to the normal (defining the plane of incidence).

Second step: $E_{\parallel,\parallel}$

We can apply Faraday's law to the $E_{\parallel,\parallel}$ components of incident, reflected and transmitted waves, and we find, just as for waves incident normally on the interface:

$$E_{i,\parallel,\parallel} + E_{r,\parallel,\parallel} = E_{t,\parallel,\parallel} \; .$$

The positive direction is defined in the right part of the figure. It follows then that:

$$E_{i,\parallel} \cos \theta_i + E_{r,\parallel} \cos \theta_r = E_{t,\parallel} \cos \theta_t \; .$$

Since $\theta_i = \theta_r$, we are finally led to state:

$$E_{i,\parallel} + E_{r,\parallel} = \frac{\cos \theta_t}{\cos \theta_i} E_{t,\parallel} \; . \tag{10.10}$$

Third step: Gauss' law

We need yet another equation to eliminate one of the three quantities in order to find a relation between the other two. For the case where the ray was normal to the interface, we used an energy balance sheet to get an equation. In the present case of oblique incidence, it will be not so easy, since we have to take into account many components at the same time. Instead, we choose to use Gauss's law for electric fields on a small cube with surfaces parallel to the interface and the incidence plane. The cube has sides with area A and normal to the $d\vec{A}$, and we write:

$$\oint \vec{D} \cdot d\vec{A} = Q_{\text{free,enclosed}} \; .$$

The advantage of this choice is that all components of the electric field that are parallel to the interface will give zero net contribution to the integral. They enter and leave the side surfaces in the same medium, and these field components are approximately constant along the surface as long as we allow the cube to have a side length small compared with the wavelength. On the other hand, we get contributions from the component that is normal to the end faces of the cube that are parallel to the interface; see Fig. 10.5. By specifying how we define positive field directions in the right part of the same figure, follow:

$$D_{i,\|,\perp} + D_{r,\|,\perp} = D_{t,\|,\perp} \, ,$$

$$\varepsilon_0 \varepsilon_{r1} E_{i,\|,\perp} + \varepsilon_0 \varepsilon_{r1} E_{r,\|,\perp} = \varepsilon_0 \varepsilon_{r2} E_{t,\|,\perp} \, .$$

We use now the relation $n \approx \sqrt{\varepsilon_r}$, and get:

$$n_1^2 E_{i,\|,\perp} + n_1^2 E_{r,\|,\perp} = n_2^2 E_{t,\|,\perp} \, .$$

Using the definition of positive directions for the vectors in the right part of Fig. 10.5, it follows that:

$$-n_1^2 E_{i,\|} \sin \theta_i + n_1^2 E_{r,\|} \sin \theta_r = -n_2^2 E_{t,\|} \sin \theta_t \, .$$

We invoked Snel's law of refraction (derived above):

$$n_1 \sin \theta_i = n_2 \sin \theta_t$$

and moreover $\theta_i = \theta_r$. We then eliminate θ_t and get:

$$-n_1^2 E_{i,\|} \sin \theta_i + n_1^2 E_{r,\|} \sin \theta_i = -n_2 E_{t,\|} n_1 \sin \theta_i \, .$$

Dividing throughout by $n_1^2 \sin \theta_i$, we get:

$$E_{i,\|} - E_{r,\|} = \frac{n_2}{n_1} E_{t,\|} \, . \tag{10.11}$$

Fourth step: Combining

We now have two equations that connect $E_\|$ for incoming, reflected and transmitted waves. One equation can be used for eliminating one of the three quantities and obtaining the relationship between the two others. For example, if we subtract Eq. (10.10) from Eq. (10.11), we get:

$$2E_{r,\|} = \left(\frac{\cos \theta_t}{\cos \theta_i} - \frac{n_2}{n_1} \right) E_{t,\|} \, . \tag{10.12}$$

Details, Brewster angle

Equation (10.12) is in fact interesting in itself, because it shows that the contents of the parenthesis can be made to vanish. When this happens, no part of the incident wave will be reflected if E lies in the incidence plane (because then $E_\perp = 0$). The incident angle θ_i where this happens is called *the Brewster angle*. Let us explore this special case in some detail. The condition is that:

$$\frac{\cos \theta_t}{\cos \theta_i} = \frac{n_2}{n_1} \; .$$

Using Snel's law once again, we get:

$$\frac{\cos \theta_t}{\cos \theta_i} = \frac{\sin \theta_i}{\sin \theta_t}$$

$$\sin \theta_i \cos \theta_i = \sin \theta_t \cos \theta_t \; .$$

We know that $\sin(2x) = 2 \sin x \cos x$, thus

$$\sin(2\theta_i) = \sin(2\theta_t) \; .$$

We also know that $\sin x = \sin(\pi - x)$, which implies

$$\sin(2\theta_i) = \sin(\pi - 2\theta_t) \; .$$

This relation will be satisfied if

$$2\theta_i = \pi - 2\theta_t \quad \text{or} \quad \theta_i = \pi/2 - \theta_t \; .$$

With $\theta_i = \theta_r$, we are finally led to the result:

If

$$\theta_r + \theta_t = \pi/2 \tag{10.13}$$

there will be no reflected light with polarization parallel to the incidence plane. Then, the angle between the reflected and transmitted rays equals $\pi/2$ as indicated in Fig. 10.6.

Since the angles of incidence and reflection are equal, it is easy to show that the angle where we have no reflected light with polarization in the incidence plane is characterized by the angle between reflected and transmitted rays being 90°.

We wish to find an expression for the angle ($\theta_i \equiv \theta_B$) for which this holds, and start with:

$$\frac{n_2}{n_1} = \frac{\cos \theta_t}{\cos \theta_i}$$

Fig. 10.6 When the angle between the reflected and transmitted rays is 90°, there is no electric field parallel to the incident plane in the reflected ray

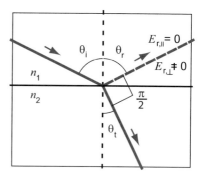

and combine this with $\cos \theta_t = \cos(\pi/2 - \theta_i) = \sin \theta_i$ to get:

$$\tan \theta_i = \frac{n_2}{n_1} \equiv \tan \theta_B . \tag{10.14}$$

The angle θ_B is called *Brewster's angle*. At the interface between air and glass with refractive index 1.54, we find:

$$\tan \theta_B = \frac{1.54}{1.00}$$
$$\theta_B \approx 57° .$$

Since $\theta_r + \theta_t = \pi/2$, we can easily determine θ_t. The result is about 33°.

It may be worth noting that there will also be no reflection (for light with electric vector parallel to the incidence plane) if the light goes from glass to air. For this case, we have:

$$\tan \theta_B = \frac{1.00}{1.54}$$
$$\theta_B \approx 33° .$$

In other words, the Brewster effect can occur when light enters a new medium, regardless of whether the refractive index becomes higher or lower! By comparison, total reflection (which we will return to if a little) occurs only when the light hits a medium with a lower refractive index.

Fig. 10.7 Unpolarized light reflected at an air–glass interface can be fully polarized when the angle of incidence is equal to the Brewster angle. These photographs show this. The picture on the left is taken without a polarization filter. The picture on the right is taken with a polarization filter oriented so as to let only light polarized parallel to the incidence plane. All reflection is removed at the Brewster angle, and we look directly at the curtains behind the glass window practically without any reflection. This means that, at the Brewster angle, all the reflected light is fully polarized in a direction perpendicular to the incidence plane (parallel to the air–glass interface). Note that reflections on the painted surface are affected similarly to reflections from the glass. NB: Many modern windows are now surface treated in different ways. Then, we do not get any direct interface between air and glass, and the Brewster effect as described disappears totally or in part

10.3.4 Brewster Angle Phenomenon in Practice

It is actually relatively easy to observe that light reflected from a surface at some angles is fully polarized.

The essential point is that ordinary unpolarized light can be decomposed into light with polarization parallel to the incidence plane and perpendicular to it. For the component parallel to the incidence plane, we can achieve zero reflection if the light comes in at the Brewster angle. In that case, the reflected light will be completely polarized normal to the incidence plane. We can observe this by using a polarization filter that only lets through light polarized in a certain direction. Figure 10.7 shows an example of this effect.

10.3.5 Fresnel's Equations

In order to arrive at relations involving reflection and transmission, we used Maxwell's equations, but these laws were derived long before Maxwell systematized electromagnetic phenomena in his equations. Fresnel derived equations which describe reflection and transmission already in the first half of the nineteenth century. You can read more about this e.g. in Wikipedia under the keyword "Fresnel equations". Here we will present only two formulas and a graph. In Eqs. (10.15) and (10.16), and in Fig. 10.8, the reflection coefficient is given for light fully polarized perpendicular to the incidence plane (R_s) and fully polarized parallel to the incidence plane (R_p) [The suffixes s and p are from German: *Senkrecht* (vertical) and *parallel*, respectively.]. The reflection coefficient refers to intensities, so in our language use, for example,

Fig. 10.8 Reflection and transmission coefficients of electromagnetic waves directed obliquely at an interface between two media with refractive index $n_1 = 1.0$ and $n_2 = 2.0$. The subscript s indicates that the electric field component of the wave is normal to the incidence plane, and the index p that the component is parallel to the incidence plane

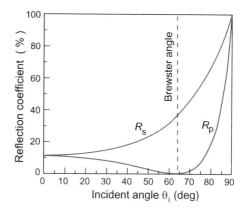

$$R_s = \left(\frac{E_{r,\perp}}{E_{i,\perp}}\right)^2.$$

The complete expressions can be written as follows:

$$R_s = \left(\frac{n_1 \cos\theta_i - n_2\sqrt{1 - \left(\frac{n_1}{n_2}\sin\theta_i\right)^2}}{n_1 \cos\theta_i + n_2\sqrt{1 - \left(\frac{n_1}{n_2}\sin\theta_i\right)^2}}\right)^2, \qquad (10.15)$$

and

$$R_p = \left(\frac{n_2 \cos\theta_i - n_1\sqrt{1 - \left(\frac{n_1}{n_2}\sin\theta_i\right)^2}}{n_2 \cos\theta_i + n_1\sqrt{1 - \left(\frac{n_1}{n_2}\sin\theta_i\right)^2}}\right)^2. \qquad (10.16)$$

The transmission can be found by using the relations $T_s = 1 - R_s$ and $T_p = 1 - R_p$.

If the light falling on the surface is totally unpolarized (with all polarizations equally present), the total reflection is given by $R = (R_s + R_p)/2$.

Figure 10.8 gives the reflection as a percentage for different angles of incidence. The figure applies to $n_1 = 1.0$ and $n_2 = 2.0$. For a wave that approaches the interface normally, the reflection is about 11% and of course independent of the polarization direction. The Brewster angle for these refractive indices is about 63°, and for this angle, the reflection is about 36% for waves polarized normally to the incidence plane.

Note further that the reflection coefficient goes to 1.0 (100%) when the angle of incidence goes to 90°. This applies to both components of the electric field.

10.4   Polarization

We have already mentioned polarization a great deal in this chapter, meaning the direction of the electric field vector when an electromagnetic wave travels through space.

> However, polarization is not always in a particular *plane*. The electric field of an electromagnetic wave may change direction in a systematic manner as the wave moves. If we draw an electric field vector at closely spaced points along the line of propagation, the tip of all the field vectors may describe, for example, a helix with one turn per wavelength. In that case, the wave is said to be circularly polarized.

Figure 10.9 shows four different forms for polarization, where elliptical polarization is intermediate between linear polarization (polarization in a plane) and circular polarization.

It might seem that linear polarization is very different from circular, but the fact is that it is quite simple to switch from one to the other. Start by considering a plane

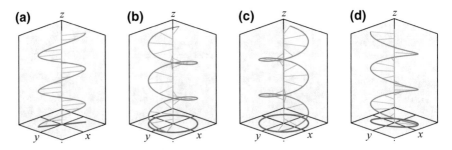

Fig. 10.9 Four different polarizations of a plane electromagnetic wave travelling in the z-direction. The green bars perpendicular to the z-axis indicate the size and the direction of the electric field at some z-values, all at the same time. The blue curves mark the tip of the electric field vector drawn from every point along the z-axis. The red curve shows the projection of the blue curve onto the xy-plane. **a** Plane polarized wave with the polarization plane −60° relative to the xz-plane. **b** Left-handed circular polarized wave. **c** Right-handed circular polarized wave. **d** Elliptically polarized wave, in this case 45% circular, 55% linear, the plane for the linear polarization is −30° relative to the xz-plane

linearly polarized electromagnetic wave moving in the z-direction. The polarization lies in a plane between the xz-plane and the yz-plane (similar orientation as in part a of Fig. 10.9). We can say that $E_x(t)$ and $E_y(t)$ vary "in step" or "in phase".

Mathematically, we can describe the wave on the left of Fig. 10.9 in the following way:

$$\vec{E} = E_x \cos(kz - \omega t)\,\vec{i} + E_y \cos(kz - \omega t)\,\vec{j}$$

where $E_x < E_y$.

If we delay the x-component by a quarter period compared to y-component (e.g. by using a quarter wave plate), and the amplitudes are equally large, polarization is circular (similar to c in Fig. 10.9), and the polarization follows a spiral as on a normal screw. We say that we have a right-handed circular polarization because the polarization direction follows our fingers on the right hand if we grasp the axis that indicates the direction of propagation, with the thumb pointing in this direction.

However, if we advance x-component by a quarter of a period compared to y-component, the polarization is left-handed circular (as for b in Fig. 10.9).

Mathematically, we can describe a left-handed circularly polarized wave (as **b** in Fig. 10.9) as follows:

$$\vec{E} = E_x \cos(kz - \omega t)\,\vec{i} + E_y \sin(kz - \omega t)\,\vec{j}$$

where $E_x = E_y$. The electric field in the x-direction is, as we see, shifted a quarter period (or a quarter wavelength) relative to the electric field in the y-direction.

The polarization of a plane electromagnetic wave can be specified either in terms of two plane polarized waves with orthogonal polarizations as basis vectors, or with a right-handed and a left-handed circularly polarized wave as basis vectors.

Be sure that you understand what is meant by a "plane, electromagnetic wave with (e.g. right-handed) circular polarization".

10.4.1 Birefringence

In the previous section, we claimed that it is easy to change from linear polarization to circular or vice versa. All that is needed is to change the phase of the time variation of one component of the electric field with respect to the other. But how do we achieve such a phase change in practice? Change in phase corresponds to a time delay, and a delay can be achieved if the wave moves more slowly when the electric field vector has one direction in space compared to when the field vector has a direction perpendicular to the first.

> There exist materials in which waves polarized in one direction have a different velocity than waves polarized in a direction perpendicular to the first. This means that the refractive index is different for the two polarizations. Such materials are called birefringent (meaning doubly refracting).

A glass cannot be birefringent because it is matter in a disordered state, where bonds between atoms have all possible directions in space. To get a birefringent material, there must be a systematic difference between one direction and another, and this difference must be constant within macroscopic parts of the material (preferably an entire piece of the material). A birefringent material is therefore most often a crystal. Calcite crystals are a well-known example of birefringent material and will be described in some detail in the next sub-chapter.

It is interesting to note that birefringence was first described by Danish scientist Rasmus Bartholin in 1669.

It is possible to make a thin slice of a calcite crystal that has just the thickness required to delay the waveform by a quarter period in one component of electric field vector as compared to the perpendicular component perpendicular. Such a disc is called a "quarter wave plate". A quarter wave plate will ensure that linearly polarized light is transformed into circularly polarized or vice versa. A quarter wave plate will only work optimally for a relatively narrow wavelength range. When such a plate is bought, the wavelength for which it is to be used must be specified.

Two different refractive indices in one and the same material give rise to a peculiar phenomenon. The upper part of Fig. 10.10 shows how a straight line looks when we see it through a calcite crystal oriented in a special way. The orientation is such that we see *two* lines instead of one. It is easy to understand the term "birefringent material" when we see such a splitting of an image.

We can imagine that the light from the line (surrounding area) has all possible linear polarization directions. Light with a particular polarization travels at a different speed compared to light polarized along a perpendicular direction. That is, the refractive indices for light with these two polarizations are different, which is why we see two lines through the crystal.

The last two pictures in the figure show how the line looks when we interpose a polarization filter between the crystal and our eyes. For a specific orientation of the filter, we allow the passage of light with only one polarization direction. By rotating the filter in one direction, we only see one line through the crystal. If we rotate the filter 90°, we only see the other line through the crystal. This is a good indication that the two refractive indices are linked to the polarization of the light through the crystal.

Fig. 10.10 Upper part of the figure shows a straight line viewed through a birefringent substance (oriented in a well-chosen manner). We see *two* lines! These are due to the fact that light with different polarization has different refractive indexes through the crystal. This can be demonstrated by holding a linear polarization filter in front of the crystal. If we orient the polarization filter in one way, we only see one of the two lines, but if we rotate the polarization filter by 90°, we see only the other line. A mark is made on the filter to show the rotation made between the two lower pictures

Remark: So far, we have set the relationship between electric field strength $\vec{E}$ and electric flux density (or the displacement vector) $\vec{D}$ as follows:

$$\vec{D} = \varepsilon_0 \varepsilon_r \vec{E}$$

where ε_0 is the permittivity in empty space, and ε_r is the relative permittivity (also called the dielectric constant). Both of these quantities have been simple scalars, and therefore, the vectors $\vec{D}$ and $\vec{E}$ have been parallel.

In terms of the components, the equation can be written as:

$$D_i = \varepsilon_0 \varepsilon_r E_i \tag{10.17}$$

where $i = x$, y, or z.

In birefringent materials, this simple description no longer holds. Electric field directed in one direction could provide the polarization of a material (e.g. calcite) also in a different direction. To incorporate this behaviour into mathematical formalism, the scalar ε_r must be replaced with a tensor with elements $\varepsilon_{r,i,j}$ where i and j correspond to x, y and z. Then, Eq. (10.17) is replaced by:

$$D_j = \varepsilon_0 \varepsilon_{r,i,j} E_i \; . \tag{10.18}$$

This is just one example of how a simple description needs refinement when a physical system displays properties that lie beyond the realms of the most elementary.

We mention these details to remind you that one of the tasks of physics is to provide mathematical modelling of the processes we observe. When the processes in nature are complicated, correspondingly complicated mathematical formalism is needed.

10.4.2 The Interaction of Light with a Calcite Crystal

All light originates in some process involving matter. When it is created, light acquires a polarization determined by the geometric constraints that are a part of the process whereby light is created. When light passes through vacuum, its polarization does not change, but as soon as it interacts with matter again, polarization can change. There are many different mechanisms that affect the polarization of light. This means that by studying change of polarization that accompanies the passage of light through matter, we can gain more knowledge of the material. A collective name for all such studies is "polariometry".

To get an idea of the mechanism responsible for the change in the state of polarization, let us discuss what happens when light is sent through a piece of mineral calcite. The chemical formula of calcite is $CaCO_3$, and we will consider calcite crystals. These are "birefringent"; that is, when we consider an object through a clear calcite crystal, the object looks double. The unit cell in a calcite crystal is relatively complicated.[1] Figure 10.11 is a perspective sketch of four $CaCO_3$ as some of the molecules are located within the unit cell. All the $CaCO_3$ molecules in the crystal are oriented so that the carbonate groups (CO_3^{2-}) are approximately in a plane perpendicular to a preferred direction called the optic axis. The orientation of the carbonate groups is such that there is a significant degree of rotational symmetry around the optic axis.

In Fig. 10.11b, we have indicated what happens when light passes the crystal with a polarization parallel to the carbonate plane. When the electric field is aligned as shown, the electron clouds around each atomic core will undergo a slight displacement relative to the core. Each atom then acquires a polarization (redistribution of electrical charge). Energy is stolen from the electromagnetic field of the light and temporarily stored in the polarization of the crystal. When the electric field then goes to zero and increases again in the opposite direction, it will induce polarization of the crystal again, but now with the opposite displacements of the electron clouds relative to the atomic nuclei.

However, we do not build more and more polarization as time passes. The stored energy in the polarization of the material will in some way act as "antennas" and generate electromagnetic waves. These waves have the same frequency as those which created the polarization originally. It is this polarization of the material and re-emitting of electromagnetic waves from the small induced dipoles in the material which causes light to move at a slower speed in the crystal compared with vacuum. As soon as the wave goes out of the crystal, there is no matter to polarize (when we ignore air) and the light velocity becomes the same as in vacuum.

Now comes something exciting! If we send light into the calcite crystal so that the electric field in the light wave has a direction *perpendicular* to the carbonate planes as in Fig. 10.11c, we will, as before, have displacement of the electron clouds relative to the atomic nuclei. But now the electron cloud is shifted across the carbonate planes.

---

[1] See for example Wikipedia for "calcite".

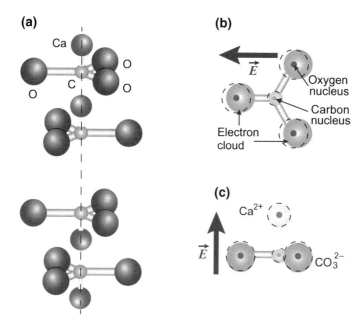

Fig. 10.11 Calcite is built up by the atomic groups CaCO₃. Part **a** gives a perspective drawing that indicates how these groups are oriented relative to each other. There is a large degree of symmetry around the direction marked with the dashed line, the so-called optic axis. In **b** and **c**, a snapshot of how an external electric field from passing light will polarize the atoms. Dashed circles indicate the location of the electron clouds when there is no external electric field. In **b**, the polarization is across the optic axis, whereas in **c** the polarization is along the optic axis

Due to the special symmetry of the crystal, light polarized in the direction of the symmetry axis will have a smaller charge polarization than when the polarization of light is perpendicular to the symmetry axis.

The result is that when light is transmitted through the crystal, the beam will be split into two, an "ordinary ray" with polarization normal to the optic axis and an "extraordinary ray" with a polarization perpendicular to the former. Snel's law does not apply. At about 590 nm, the refractive index for the ordinary ray is $n_o = 1.658$ and for the extraordinary ray $n_e = 1.486$.

It is quite natural that effects similar to that seen in calcite are observed only in crystalline materials, or at least materials with different properties in one direction compared to another (anisotropic media). However, we can have similar effects also for an initially isotropic material if it has been exposed to stress in a certain direction on account of which it is no longer isotropic. An isotropic plastic material can be made slightly anisotropic by, for example, bending or stretching it. By the way, some types of plastics are often slightly anisotropic if they are made by moulding where the molecules have been given a certain uniformity locally as the plastic was pressed into the mould from a particular point of feeding.

A Challenge

In this chapter, we have used the word "polarization" for two widely different conditions. We used the word when we mentioned different electric permittivities (which are connected with the difference between the electric field $\vec{E}$ and the electric field strength $\vec{D}$). This reflects how much we can deform the electron clouds relative to the nuclei of the atoms and generate a polarization (asymmetry) in charge distributions. We also used the word when we distinguished between e.g. linear and circular polarization. Make sure you fully understand the difference between these two different (but still related) terms with the same name. Otherwise, you should discuss with fellow students and/or teaching-assistant/lecturer.

10.4.3   Polarization Filters

Linear polarization filters

When we discussed the Brewster angle, we saw an example of a linear polarization filter. Roughly speaking, we can say that such a filter (if it is thick enough) peels off one component of the electric field vector in the electromagnetic waves (visible light). If the light is totally unpolarized initially, the intensity will be halved after the light has passes through a linear polarization filter.

What does the term "unpolarized light" mean? It is actually a little difficult to explain. We have seen that Maxwell's equations can give us plane and polarized (or circularly polarized) waves. Is that not true for all electromagnetic waves? Well, it is true, but light is usually generated from a very large number of sources that are independent of each other. When we turn on the light in a room, for example, light is generated from every fraction of a millimetre of filament in an old-fashioned light bulb and the light emitted from each part is independent of the other parts. All the waves pass through the room and the contributions will, to a large degree, overlap in time and space. The result is that if we follow the polarization at a tiny point in the room, polarization will still change during a fraction of a millisecond. There is also quite a different time development at a small point in the room and at another point just a few millimetres away from the first.

We shall describe such more or less chaotic waves in Chap. 15 when we refer to coherence. Unlike chaotic light (unpolarized light), for example, we have laser light and it is exciting to see how waves are added to each other, but that will come later.

Let us assume *for now* that we have a horizontal light ray with unpolarized light. Using a linear polarizing filter, we can make sure all the transmitted light has electric field that is aligned horizontally.

If we insert another such filter, and orient it just like the previous one, any light transmitted by filter 1 will be transmitted also by filter 2.

If filter 2 is rotated 90° so that it can only let through vertically polarized light, there will be no such light coming through filter 1. Then, *no* light will emerge from filter 2 (left part of Fig. 10.12).

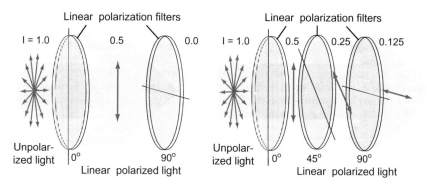

Fig. 10.12 Two sequential linear polarization filters aligned 90° apart, transmit no light (*left part*). If a third filter is placed between the first two, with the polarization direction different from the other two, some light will in fact go through the filters (*right part*)

However, if we, for example, rotate filter 2 by 45° relative to filter 1, light with horizontal polarization after filter 1 will actually have a component also in the direction of filter 2. Light that now passes filter 2 acquires polarization 45° relative to the polarization it had after filter 1. We *change* polarization, but the amplitude of the electric field is now less than what it was before filter 2 (Only the E field component in the direction of filter 2 is transmitted).

The intensity of the light passing through filter 2 is given by Malus's law:

$$I = I_0 \cos^2(\theta_2 - \theta_1) . \tag{10.19}$$

Here, I_0 is the intensity of the light after it has passed filter 1. The argument for the cosine function is the difference in the angle of rotation between filters 1 and 2. Malus's law only applies to linear polarization filters!

Now let us start with two polarizers with polarization axes perpendicular to each other, and place a third polarization filter between the first two. If we choose an orientation other than 90° relative to the first, we get light through all three filters (right part of Fig. 10.12). This is because the middle filter has changed the polarization of the light before it hits the last filter.

It is important to note that a polarization filter of this type actually plays an active role as it *changes* the polarization of light passing through it.

Remarks:
We will now present a picture that might serve as a useful analogue to what is happening in a linear polarization filter. Imagine that the filter consists of pendulums that can swing in only one plane. If we attempt to push the pendulums in a direction along which they can actually swing, the

pendulums will swing. A swinging pendulum can propagate its motion to a neighbouring pendulum of the same type, and so a wave can propagate through the material.

However, if we try to push the pendulums in a direction in which they *cannot* swing, there will be no oscillations. No wave can then propagate through the medium. If we push obliquely, the pendulums will swing, but only in the direction along which they can actually swing. This means that the swing direction in the wave will change when the wave propagates through the medium, but we get a reduction in the wave because only the component of our push that is along the swinging plane will be utilized in the oscillations.

10.4.3.1   Circular Polarization Filters in Photography *

A circular polarization filter is basically a filter that only lets through circularly polarized light. There are two variants of such filters, one type that allows right-handed circularly polarized light to pass through, and another type that passes through the left-handed circularly polarized light. A purely circular polarization filter has the same effect even if it is rotated around the optic axis.

Today, however, there is a completely different type of filter called the circular polarization filter. We are thinking of polarization filters used in photography. In many photographic devices, autofocus is based on circularly polarized light. If we want a polarization filter in front of the lens, the filter must be made such that circularly polarized light reaches a detector inside the device.

Such a circular polarization filter is assembled in a very special way. When light enters the filter, it first meets an ordinary linear polarization filter. Just behind this filter is a so-called quarter wave plate with a special orientation. As a result, the light is first converted into pure linearly polarized light, and subsequently converted into almost completely circularly polarized light. The light that enters the camera is therefore circularly polarized and the autofocus works.

We will look closely at the details in this context.

A quartz wave plate is made of a birefringent material, for example calcite. We have already seen that in a birefringent substance the phase velocity of light polarized in a certain direction differs from the phase velocity of light polarized perpendicular to the aforementioned orientation.

In polarization filter used in photography, the orientation of the birefringent substance is chosen so that the electric vector after the linear polarization filter forms 45° with each of the two special directions in the birefringent substance. We decompose the electrical vector as shown in Fig. 10.13. The E_o component will go through the substance with a certain phase velocity (i.e. a certain wavelength), while the E_e component passes through the substance at a different phase velocity (and wavelength).

By choosing a certain thickness of the birefringent substance, we can arrange E_o to have just one-quarter wavelength difference from what E_e has when it leaves the filter. In that case, we achieve exactly what we want, namely that linearly polarized light is transformed into circularly polarized light.

Fig. 10.13 Schematic drawing of a so-called circular polarization filter used in photography. The light passes through an ordinary linear polarization filter and then through a quarter wave plate. The two parts are close to each other. The orientation of the plate is chosen so that mean wavelengths are converted from linear to circular polarized light

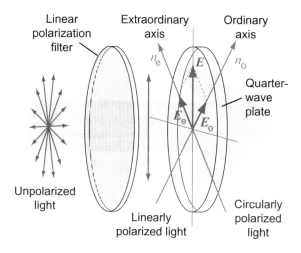

Looking further at this argument, we discover that we cannot get a perfect transformation from linear to circular light for all wavelengths in the visible range at the same time. In practice, therefore, the thickness of the birefringent substance will be chosen such that the centre wavelengths (around spectral green) get optimal conversion while other wavelengths get a less perfect transformation. It does not matter because autofocus must have some light that is circularly polarized and does not need perfect circular polarization for all wavelengths.

Remarks:
A linearly polarized wave can be considered as a sum of a right-handed and a left-handed circularly polarized wave, and a circular polarized wave can be considered as a sum of two linearly polarized waves with polarization perpendicular to each other (and phase shifted). This means that we can combine circular polarization filters and linear polarization filters in different ways.

However, filter combinations in which the photographic circular polarization filters are included give a lot of surprises just because these filters are composed of two elements.

If we place two photographic circular polarizing filters with the inner surfaces facing each other, the light will pass through both filters with approximately the same intensity as after the first filter. Intensity is almost independent of the angle of rotation of one filter relative to the other. This is due to the fact that the light after passing the first filter is approximately circularly polarized and thus has a circularly symmetrical E-field distribution (when considering intensity).

If, on the other hand, we place two such filters with the outer surfaces against each other, two linear polarization filters will follow in succession in the light path. The pair of filters will then behave like two ordinary linear polarization filters, and the intensity of the emerging light is given by Malus's law (Eq. 10.19).

The special construction of the filters means that in photography, we achieve the same effect as with linear polarization filters, from the photographic point of view. Polarizing filters are used to remove reflection (as shown in Fig. 10.7) and to remove the effect of haze in the atmosphere (since light from the haze is partially linearly polarized). With the help of polarizing filters, we can achieve a big contrast between blue sky and white clouds, which adds extra life to the pictures (see Fig. 10.22).

10.4.4 Polariometry

We have previously seen that a polarization filter sandwiched between two crossed polarization filters causes light to escape through the combination of three filters. This gives us an excellent starting point for studying certain material properties. Any material that changes the polarization of light will ensure, when interposed between crossed polarizers, that some light passes through the arrangement. For example, many plastic objects will have differences in optical properties in different parts of the article depending on how the plastic material flowed into a mould prior to and during curing. Anisotropy in different parts of the material causes the polarization direction of light to rotate slightly or the conversion of some plane polarized light to circularly polarized light. The effect is often dependent on the wavelength. As a result, we can get beautiful coloured images through the crossed polarization filters if we send white light through the arrangement.

Figure 10.14 shows the image of a plastic box for small video cassettes in the crossed configuration. I could have used white light (e.g. from the sun or from a filament lamp), two crossed linear polarization filters and the plastic cassette holder between the filters. However, since the filters available to me were not as big as the cassette holder, I chose instead to use a computer screen as the source of plane polarized light. Many data monitors, cellular phone monitors and some other displays based on liquid crystal technology, give rise to plane polarized light. I placed the plastic holder directly against the computer screen and used a polarization filter just in front of the camera lens.

As shown in Fig. 10.14, anisotropies in the plastic are revealed well by polariometry. Variants of this method are used for many different materials and in many different contexts in industry and research. You can purchase specialized equipment for this type of analysis.

It may be useful to remember that light from, for example, mobile phones is usually linearly polarized. If you wear polarized glasses, you may experience a black screen and think something is wrong with your mobile phone, while the picture is completely normal when you do not wear the glasses!

10.4.5 Polarization in Astronomy

In recent years, several studies have been conducted on the polarization of the light from the sun and from distant light sources in the universe. Admittedly, it is not exactly what astronomers are primarily occupied with. Usually, the challenge is to gather enough light to get good pictures or spectroscopic data. If we insert a polarization filter, we lose half the intensity of light. And if we want information about the polarization of the light, we would like to have at least two photographs taken with light polarized along perpendicular directions. This means that a study based on a straightforward procedure will take at least four times as long as one image taken without paying any regard to polarization.

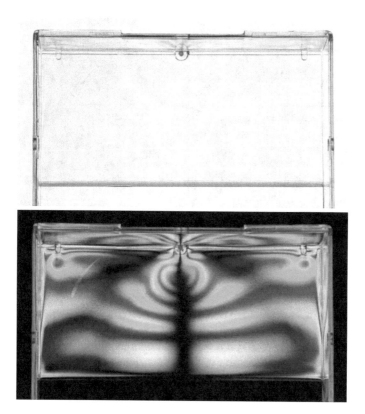

Fig. 10.14 Photography of a plastic box for small video cassettes in polarized light (*top*), and when sandwiched between two crossed linear polarizers (*bottom*). Light can be transmitted only if the polarization of the light transmitted by the first polarizer changes as it passes the plastic

The reason why polarization is still interesting in astronomy is much the same as for the polariometry of different materials. For example, let us consider light from the sun. The light may be emitted as unpolarized light in processes known as "blackbody radiation" (radiation from a hot body). However, the light will interact with plasma and atoms along its way to the earth. If the electrons in a plasma are influenced by a strong "quasi-static magnetic field", the movement of the electrons will not take place equally easily in all directions (remember the cross-product in the expression of the Lorentz force).

When for example the light from a part of the sun passes the electrons in a plasma, the electromagnetic field of the electromagnetic wave (the light) will set the electrons in the plasma into motion. Without any quasi-static magnetic field, the electrons will oscillate in the same direction as the electric field in the light, and the polarization will not change. But if there is a powerful quasi-static magnetic field present, the electron oscillation could get a different direction than the electric field of the light. The result is that a polarization is imparted to the light that is conveyed to us. The

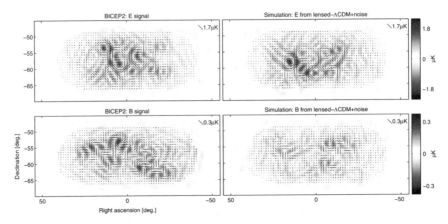

Fig. 10.15 Polarization in the electromagnetic waves originating from the Big Bang has recently been mapped. Figure taken from [1] under a CC BY-SA 3.0 license

direction of polarization will tell us something about the magnitude and direction of the quasi-static magnetic field in that part of the sun where the light came from.

A number of other factors can also affect the polarization of light from astronomical objects, and there is much that needs to be mastered! In the coming years, polariometry will undoubtedly give us information about astronomical processes that until recently was unavailable by any other means. Please read the article "Polarization in astronomy" in Wikipedia. There are also many postings and YouTube videos available on the Web that mention the so-called POLARBEAR Consortium project that uses the polarization of electromagnetic waves in space exploration (Fig. 10.15).

10.5   Evanescent Waves

In Chap. 9, we distinguished between near field and far field and pointed out that many known relationships between the electric and magnetic field of electromagnetic waves apply only in the far field. We mentioned that the near field extends no further than a few calculated wavelengths beyond the sources or structures that cause the near field.

This recognition has grown in the last decade and has had a big impact in e.g. optics. When we derived the expression for total reflection above, we relied entirely on Snel's law and on manipulating a few mathematical expression.

However, if we use Maxwell's equations in a more thorough analysis of total reflection, we will realize that *an electric field on the inside of the glass at*

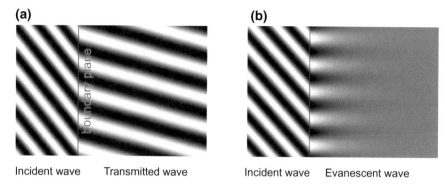

(a) Incident wave   Transmitted wave   **(b)** Incident wave   Evanescent wave

Fig. 10.16 Evanescent waves at total reflection (*right*). The boundary line between two media with different refractive indices is marked with a red line. Only incoming and transmitted/evanescent waves are displayed; that is, the reflected wave is not shown. Aluminum, CC BY-SA 3.0 GNU Free Documentation License 1.2. Modified from original [2]

> *total reflection cannot end abruptly at the interface between glass and air. The electric field must decrease gradually.*
>
> A detailed solution of Maxwell's equations for the region near the interface shows some kind of standing wave where the amplitude (and intensity) decreases exponentially when we move away from the interface (see Fig. 10.16). This standing wave is called an evanescent wave, and it fades over a distance of the order of a wavelength.

Evanescent waves are found in many situations, not only when total reflection takes place. A very important example is the interface between metal and air or metal and another dielectric. In the metal, however, electrons will move along the interface in a special way. We call this phenomenon "plasmons" ("surface plasmon polariton waves"). Plasmons (collective electron motion) are a powerful contributor to how the electromagnetic field will change in the region near the interface between the two materials, and consequently also the evanescent waves outside the metal.

Evanescent waves are now very popular in physics research, not least because we have also had a significant development in nanotechnology recently. Today we can make structures much smaller than the wavelength of light. The result is, among other things, that smart ways have been found to improve resolution, for example, in microscopy. In Chap. 13, we will describe diffraction, and according to the classical analysis of diffraction, we could never achieve a better resolution than the so-called diffraction-limited resolution. Today, however, for special geometries, we can surpass this limit.

The evanescent waves are primarily significant in distances less than about $\lambda/3$ away from the interface. There is room for much creativity over the coming years in the field of evanescent wave research and utilization of these!

10.6   Stereoscopy

People have a well-developed "depth perception". The two pictures captured by our eyes are slightly different because the two eyes are 6–7 cm apart. This means that the viewing angle for nearby objects is different for each eye, but almost identical for remote objects.

Ever since the infancy of photography, people have experimented with taking pairs of photographs that correspond to the pictures formed on our retinas, the so-called stereoscopic pair. We could look at the pictures individually through special stereo- scopes (see Fig. 10.17). This technique was also used in commercial "ViewMaster" binoculars that were highly popular in the 1970s. The technique is still employed in modern Virtual Reality technology.

Three-dimensional images can also be created by placing a stereoscopic picture pair on top of each other, where the image to the left eye is transmitted through a blue-green colour filter while the image to the right eye is through a reddish colour filter. The resulting image looks somewhat strange, with reddish and blue-green objects side by side (see Fig. 10.18). When such an image is viewed through so- called anaglyph glasses which are red on the left side and blue-green on the right, the blue-green parts of the image will be visible through the red filter (little colour drops and the object looks dark). The red parts of the image will pass through the red filter as well as white light and will only look white and become "invisible".

The use of a crude form of colour coding serves its purpose in anaglyph glasses, but colour reproduction is not satisfactory for many purposes.

That is where polarization filters come in, and polarization of light is just perfect for this purpose. We have two eyes and need to "code" two pictures so that one picture

Fig. 10.17 Stereoscopic picture pairs with accompanying lenses were developed already over 100 years ago. The picture shows a 1908 stereo viewer used for looking at a pair of stereoscopic photographs

Fig. 10.18   Stereoscopic images can be made by adding two colour-coded images on top of each other. The images are traversed through coloured glasses ("anaglyph glasses") to ensure that a stereoscopic pair of photographs is perceived only by the eye each image is made for

reaches one eye and another picture the other eye. If we allow the light from one image to be horizontally polarized and the light from the other image is vertically polarized (we assume that we now look horizontally against the pictures), by using a horizontal axis polarization filter on one glass and vertical axis on the second glass, we will achieve exactly what we want.

The use of linearly polarized light works well as long as we see a movie and the head is held straight up (so that the interocular axis is horizontal). But if we cock our head at $45°$, each of the eyes will get as much light from each of the two pictures. In that case, we will see double images on both eyes.

However, if we use circularly polarized light, we will forego this disadvantage. The two projectors needed for stereoscopy must provide right-handed and left-handed circularly polarized light. The glasses must have the corresponding polarization filters.

Several hundred movies have been recorded with stereoscopic techniques so far (see list on Wikipedia under "List of 3D movies"). Many stereoscopic televisions have been on the market. They are based on spectacles that makes sure that every alternate picture is presented to only one eye. Commercial photographic appliances and camcorders intended for the consumer market already exist. Only the future will show how large a share will stereoscopic images/movies get.

For the sake of curiosity, we conclude with a stereoscopic image that can be viewed without any aids (see next page). It is formed by many point pairs that are positioned so that one point in each pair fits one eye and the other point in the pair fits the other eye. A variety of books based on this principle have been made in many variants and with different effects (Fig. 10.19).

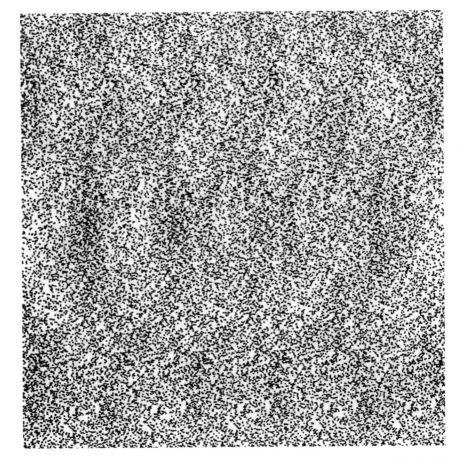

Fig. 10.19 Stereoscopic image made using dots. The illustration was made by Niklas En [3]. Reproduced with permissions. Bring the book (or screen) all the way up to the nose and let the book (screen) recede slowly, very slowly from the face. Do not try to focus on the dots in the image, but let the actual stereoscopic overall image come into focus (perhaps focusing on "infinite" at the start). This stereoscopic image will, when you finally notice it, appear to be almost as far back from the paper (screen) as your eyes are in front of the paper

10.7 Learning Objectives

After working through this chapter, you should be able to:

- Using Maxwell's equations to deduce on your own, the relationships between incident, reflected and transmitted waves when a planar electromagnetic wave strike normally a flat interface between two different dielectric materials.

- Explain "Fermat's principle" (also called the principle that optical distance must be stationary). Apply this principle to derive Snel's law of refraction and the law according to which "the angle of incidence equals the angle of emergence" when light is reflected from a flat surface.
- Explain the phenomenon "total reflection" and be able to give an example of the use of total reflection in modern technology.
- Explain the calculation of reflection and transmission when a planar electromagnetic wave is incident obliquely at a flat interface between two media (especially keeping track of the two components of the electric field in the calculations).
- Explain the phenomenon associated with the Brewster angle, and set up a mathematical expression for this angle.
- Define the reflection coefficient and transmission coefficient.
- Explain the difference between a linearly and a circularly polarized plane electromagnetic wave, and state mathematical expressions for the two examples.
- Explain what characterizes a birefringent material and explain how we can use such a material to transform a linearly polarized wave into a circularly polarized wave.
- Explain what happens when light is sent through several subsequent polarization filters, and be able to state Malus's law.

10.8 Exercises

Suggested concepts for student active learning activities: Homogeneous, thin interface, Fermat's principle, total reflection, optical fibre, incidence angle, refraction angle, Brewster phenomena, birefringence, quarter wave plate, polariometry, rotational symmetry, plane/circular/elliptical polarization, right-handed circular, linear polarizing filter, crossed polarizers, circular polarizing filter in photography, evanescent waves, stereoscopy, anaglyph glasses.

Comprehension/discussion questions

1. Can water waves and/or sound waves in the air be reflected and transmitted (as we have seen for transverse waves)?
2. When we see a reflection in a window, we often see two images slightly displaced in relation to each other. What causes this? Have you even seen more than two pictures every now and then?
3. You direct a laser beam to a glass plate. Can you achieve total reflection? Explain.

4. How can you decide if your sunglasses are of the polaroid type or not?
5. How can you determine the polarization axis of a single linear polarization filter?
6. Name two significant differences between total reflection and the Brewster angle phenomenon.
7. The speed of sound waves in air increases with temperature, and the air temperature can vary appreciably with height. During the day, the ground often becomes hotter than air, so that the temperature of the air near the ground is higher than slightly further up. At night, the ground is cooled (by radiation) and we can end up with the temperature in the air being lowest near the ground and rising slightly (before it gets cooler still further up). Can you use Fermat's principle to explain that we often hear sounds from distant sources better at night than in the day?
8. Why does the sea look bright and shiny when we look at a sunset in the ocean?
9. Is it possible to create a *plane* electromagnetic wave which is simultaneously *circularly* polarized. As usual: Justify the answer!
10. When referring to Fig. 10.4, it was said that Maxwell's equations are symmetrical with regard to time. If one solution is as given in Fig. 10.4, another solution will be the one where all the rays go in the opposite direction. Would it be possible to demonstrate this in practice? (Look especially at the light that comes from the bottom towards a interface. Would there not be a reflected beam down in this case?) Do you have any examples of experimental situations similar to this case?

Problems

11. A light source has a wavelength of 650 nm in vacuum. What is the velocity of light in a liquid with refractive index 1.47? What is the wavelength in the fluid?
12. Light passes through a glass that is completely immersed in water. The angle of incidence of a light beam that strikes the glass–water interface is 48.7°. This corresponds to the critical angle where the transition from some transmission to pure total reflection occurs. Determine the refractive index of the glass. The refractive index of water at 20 °C at 582 nm is 1.333.
13. Assume that one (multimode) optical fibre has a difference in the refractive index of 1% between the glass in the inner core where the light is and the surrounding layer of glass. Determine the maximum angle (relative to the fibre axis) the light may have and yet get total reflection. Determine the minimum and maximum time a short pulse will use to travel 1.0 km along the fibre (due to different angles of the light beam in the fibre). What will be the largest bit rate (pulses per second) that can be transmitted along the fibre (if we just consider this difference in efficient path length only)?
14. When a parallel unpolarized light beam hits a glass surface with an angle of 54.5°, the reflected beam is fully polarized. What is the value of the refractive index for the glass? What angle does the transmitted beam have?
15. A horizontal unpolarized ray of light passes through a linear polarization filter with polarization axis turned 25.0° from the vertical. The light ray continues through a new, identical polarization filter whose axis is turned 62.0° from the

vertical. What is the intensity of the light after it has gone through both filters compared to the intensity before the first filter?

16. A horizontal unpolarized ray of light passes through a linear polarization filter with polarization axis turned $+15.0°$ from the vertical. The light ray continues through a new, identical polarization filter whose axis is rotated $−70.0°$ from the vertical.

(a) What is the intensity of the light after it has gone through both filters compared to the intensity before the first filter?

(b) A third polarization filter, identical with the other two, is then put in, but now with the axis turned $−32.0°$ from the vertical. The third filter is placed *between* the other two. What is the intensity of the light now going through all three filters?

(c) Would the result be different if the third filter was located *after* the other two instead of between them?

17. Show that if we send a thin beam of light through a flat glass plate of uniform thickness, the beam passing through the glass will have the same direction as the incoming beam, but with a parallel displacement. Show that the parallel displacement d is given by:

$$d = t \sin(\theta_a − \theta_b)/\cos(\theta_b)$$

where t is the thickness of the glass plate, θ_a is the angle of incidence, and θ_b is the angle between the normal and the refracted ray in the glass.

18. Show mathematically that the angle of incidence is equal to "angle of reflection" (angle between the reflected beam and the normal to the reflecting surface at the point of incidence) using Fermat's principle.

19. A birefringent material has a refractive index of n_1 for light with a certain linear polarization direction and n_2 for light with polarization perpendicular to the first. If this material is to be used as a quarter wave plate for light wavelengths of 590 nm, light with one polarization must travel a quarter wavelength more within the plate than light with perpendicular polarization. Show that the plate must have a (minimum) thickness given by:

$$d = \lambda_0/[4(n_1 − n_2)]$$

where λ_0 is the wavelength in vacuum (air). Find the minimum thickness of a quarter wave plate made of calcite ($n_o = 1.658$ and $n_e = 1.486$, where the suffix o in n_o stands for "ordinary" and can correspond to our n_2, while the suffix e stands for "extraordinary" and can correspond to our n_1). What is the next thickness that will provide quartz wave plate function? What is the function of a quarter wave plate?

20. Determine how much a beam of light is deviated if it passes through an equilateral triangular glass prism in such a way that the light beam inside the prism is parallel to a side surface. The glass has a refractive index of n.

Fig. 10.20 A "polarization filter" for radio waves

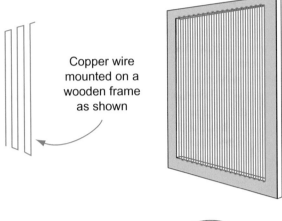

Copper wire mounted on a wooden frame as shown

Fig. 10.21 Light that gives us the rainbow goes through the water droplets as shown in this figure at an angle $\phi \approx 60°$

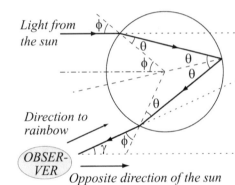

Light from the sun

Direction to rainbow

OBSER-VER

Opposite direction of the sun

21. Show that Eq. (10.6) can be derived from Eqs. (10.15) and (10.16) in the event that all are valid at the same time.

22. We can create a polarization filter for radio waves using a copper wire stretched over a frame as shown in Fig. 10.20. Explain the mode of action and explain which direction of polarization the radio waves must have to be stopped by the filter and which direction has minimal influence on the waves. It may be that the filter would be even more effective if the filter was made *slightly* differently. Do you have any good ideas in this way?

23. The light paths in raindrops when we see a rainbow are described in the article " The rainbow as a student project involving numerical calculations" written by David S. Amundsen, Camilla N. Kirkemo, Andreas Nakkerud, Jørgen Trømborg and Arnt Inge Vistnes (*Am. J. Phys.* 77 (2009) 795–798).

    Figure 10.21 shows the path of the ray. As stated in the figure, the angle $\phi \approx 60°$ for the light rays gives us the rainbow (for details, read the original article).

    (a) Calculate the angle θ, given the refractive index of water is approximately 1.333.

    (b) Is there total reflection for the light that hits the rear boundary of the water

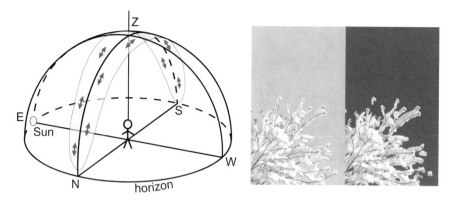

Fig. 10.22 Scattered light from the sun is polarized in a band over the sky 90° away from the direction of the sun. This can be used to remove a good part of the scattered light when we take a photograph. The right part shows images without and with polarization filter

drop for this value of θ?

(c) Calculate the Brewster angle both for the light that goes from air to water and at the interface water to air.
(d) Would you think that the light from a rainbow is highly polarized or not?
(e) How could we experiment experimentally if the light from the rainbow is highly polarized or not? (See Fig. 1.1.)

24. Light from the sun is scattered in the atmosphere, and it is this scattering that makes the sky blue during the day and red in the evening sky before the sun sets. The scattering of light is called Rayleigh scattering, and when the atmosphere is very clean and dry, the scattered light is polarized in a special way. Looking towards the sky in a direction 90° from the direction to the sun, the light will be significantly linearly polarized. The left part of Fig. 10.22 shows the direction of polarization for the region where polarization is most pronounced for a situation where the sun is on the horizon.

This polarization uses photographers occasionally to get a deep blue sky. In the right part of Fig. 10.22, it is shown how the sky looks without polarization filter.

(a) How should the filter be set for maximum effect?

The right part of the Fig. 10.23 shows the photograph of a shining water surface. Reflected sky light is strong for the water surface far away, but there is a pronounced dark area near us (marked with a dashed oval), where almost no light is reflected. We look straight down to the bottom of the water. The image is taken *without* polarization filter (just as we see it by our eyes). The phenomenon can be observed when the sun is near the horizon, and we look in a direction about 90° away from the sun (marked with dashed line in the image).

(b) Can you explain in detail the physical effects that conspire to create a "black hole" in the reflected sky light?
(c) The phenomenon is easiest to see when we stand at least a few metres above

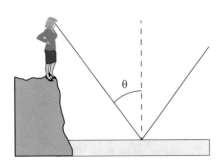

Fig. 10.23 Wide angle image of a shining water surface. There is hardly any light from the dark area (*marked with a dashed oval with a centre*). The sun is on the horizon in a direction 90° to the right from the direction marked with a vertical dashed line. The sky was clear everywhere, but brighter to the right (closer to the sun on the horizon, outside the field of view.)

the surface of the water, as shown in the left part of Fig. 10.23. Can you determine the θ angle in the figure that corresponds to the direction of the darkest part of reflected sky light?

References

1. P.A.R. Ade et al., Detection of *B*-mode polarization at degree angular scales by BICEP2 (2014). https://doi.org/10.1103/PhysRevLett.112.241101
2. Aluminum, GNU Free Documentation License 1.2, https://en.wikipedia.org/wiki/Evanescent_ field. Accessed April 2018
3. N. En, Forskning och framsteg, no. 4 (Sweden, 1992)

Chapter 11
Measurements of Light, Dispersion, Colours

Abstract Light and sound are the two most important wave phenomena people experience in daily life. We start this chapter by defining various concepts and units used for characterizing light and light sources, both in a purely physical setting and as judged by human vision. We then give important characteristics of the wavelength ranges of the various receptors in the human eye and how our colour vision makes the foundation of additive colour mixing in modern colour picture technology. We also point out how dispersion leads to colour spectra and boundary colours for specific boundary conditions. A few curious phenomena found in human colour vision are also mentioned.

11.1 Photometry

When discussing sound, we mentioned the dB scales used for specifying sound intensity and the like. We saw then that the sound volume could be given in terms of a purely physical unit, such as the amplitude of the average oscillations of air molecules, or as the local pressure amplitude while a sound wave passes, or as dB(SPL). Such measurements have a limited utility since oscillations with a frequency of less than about 20 Hz or higher than about 20 kHz do not matter, no matter how powerful such pressure oscillations in air is. Therefore, we had to introduce our own volume scale related to how sensitive the human ear is to sound with different frequencies. We weighed contributions at different frequencies according to the sensitivity curves for our hearing and devised scales such as dB(A) and dB(C).

Similarly, two parallel measuring systems are employed when we want to specify light levels/light intensities. A "radiometric" measurement system is based on physical energy measurements in units related to watt in one way or another. On the other hand, the "photometric" measuring system is based on visual impression, i.e. on the sensitivity curve for the human eye. Here the basic unit of measurement is lumen (which is itself based on the SI unit candela).

There is no simple proportionality between radiometric values and brightness as perceived by the eye. A pure infrared source may have a high radiometric intensity in

© Springer Nature Switzerland AG 2018
A. I. Vistnes, *Physics of Oscillations and Waves*, Undergraduate Texts in Physics,
https://doi.org/10.1007/978-3-319-72314-3_11

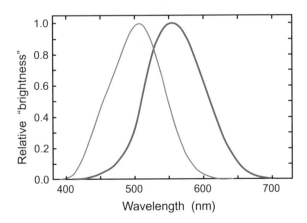

Fig. 11.1 Curves showing how "bright" the human eye perceives different spectral colours to be when the radiance to the eye is kept constant. The red curve (*on the right*) gives the "brightness curve" for colour vision (cones), while the blue curve gives the corresponding information for night vision (rods). The first has a peak at 555 nm (green), while the other has a peak at about 505 nm. Curves of this type vary slightly from person to person, and a standard curve can be made from many measurements. Dickyon, Public Domain, Modified from original [1]

W/m^2, yet the eye will not perceive almost any light from such a source. Hence, we need also units of measurement that are linked to human perception of light levels.

Figure 11.1 shows the sensitivity curve of the eye, for both colour vision (the cone cells) and night vision (the rod cells). It is only the shape of the curve shown, not absolute sensitivity (the peaks are normalized to 1.0).

The changeover between radiometry and photometry is completely analogous to the changeover between sound intensity in W/m^2 or dB(SPL) on one side and dB(A) (or dB(C)) on the other (see Chap. 7 on sound and hearing). We must also integrate physical measurements in W/m^2 and weigh the contributions for different wavelengths with the eye's sensitivity to brightness (Fig. 11.1) for the same wavelength. For colour vision, this means, to take an example, that it takes twice the intensity of light at 510 nm (or 620 nm) to contribute as much to the perceived brightness as light at 555 nm (see Fig. 11.1).

Absolute scale in photometric contexts is determined by setting a correlation between radiometry and photometry for the peak of the curve. The context exists in the definition of the base unit of photometry in the SI system, namely *candela* (abbreviated *cd*):

Monochromatic light with frequency 540×10^{12} *Hz (wavelength about 555 nm) and radiant intensity 1/683 W/sr, by definition, has a brightness of 1.0 candela (cd).*

We will define more precisely what we mean by radiation intensity. At the moment, it is enough to note that the definition connects a radiometric quantity with a photometric quantity for just a single wavelength.

The number 1/683 seems strange, but is related to the fact that the unit used previously was "normal light", which corresponded approximately to the light from a candle. Candela is chosen to match the old unit.

It may be noted that in the SI system there are only seven basic units, so candela really plays a central role. It is basically rather odd when we notice that the unit is so closely linked to human perception. On the other hand, the definition of candela given in the SI system is approximately the same as if it were a radiometric quantity, since the light source has only one wavelength.

Why do we see wavelengths around 500 nm?

As we saw in Chap. 9, visual light is restricted to a very narrow wavelength interval compared to the very impressive range of electromagnetic waves. One can wonder why this is so.

From an evolution perspective for life on earth, the explanation is rather obvious. Electromagnetic waves are strongly absorbed in water for a broad range of wavelengths (Fig. 11.2). Only for a relative narrow range around 500 nm, the absorption is much less than for electromagnetic waves in the infrared and ultraviolet range. The absorption minimum is actually near 500 nm. This number should be compared with the sensitivity curves in Fig. 11.1.

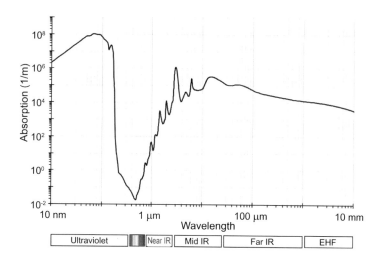

Fig. 11.2 Electromagnetic waves are strongly absorbed in water for a broad range of wavelengths except for the relatively narrow range 200–800 nm. The range of visual light is marked with the spectrum box at the bottom of this figure. See text for details. Kebes, CC BY-SA 3.0 GNU Free Documentation License 1.2. Modified from original [2]

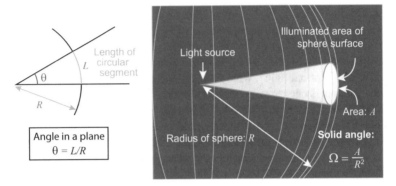

Fig. 11.3 An angle between two lines in a plane is defined as the length of L to an arc (of a circle with centre at the intersection of the lines) divided by the radius R to the circle. The unit of measurement is *radian*. A *solid angle* is defined as an area of A of a spherical segment (part of a sphere with a centre at the vertex of a conical surface that delimits the solid angle) divided by the square of the radius R^2. The unit for the measurement of a solid angle is *steradian* (sr)

Since all life on earth is based on aqueous media, vision and reflection of colour pigments from within plants and animals would be severely hampered if our vision was based on electromagnetic waves outside the 200–800 nm.

The solid angle

We will shortly go through the most common concepts related to the measurement of light, but we start by defining the concept called "solid angle", since this is included in several quantities in terms of which light is measured.

Figure 11.3 illustrates that the definition of the solid angle is a natural extension of the definition of the plane angle. A plane angle is given as the length of a circular arc divided by the radius of the circle. The angle is then independent of the chosen radius.

A solid angle is the area of a spherical segment limited by the conical surface, divided by the square of the radius of the shell.

$$\Omega = \frac{A}{R^2} \ .$$

With this definition, the solid angle turns out to be independent of the choice of the radius, which is what we want. A solid angle that includes a hemispherical shell will have the value

$$\Omega_{\text{hemispherical shell}} = \frac{4\pi R^2/2}{R^2} = 2\pi \ .$$

A solid angle that includes the *entire* space (all possible directions from a point) will then be 4π. The SI unit for the solid angle is a *steradian* (abbreviated *sr*).

Fig. 11.4 A pocket torch emits light rays in the form of a divergent beam with a certain solid angle. The beam hits a surface, the illuminated part of which then acts as a secondary light source emitting light more or less in all directions in front of the surface (but only a matt or Lambertian surface will become an isotropic source). The figure forms a starting point for defining different photometric quantities

Measurements based on solid angles are usually interesting only when they are conducted far from the source compared with the physical size of the source.

The simple units of measurements

Measurement of light involves many more concepts than measurement of sound does, possibly because the output of a light source can be everything from a narrow beam to light travelling along almost all directions. A laser can produce a very narrow beam of light, while an old-fashioned Edison-type filament lamp emits with approximately equal brightness in any direction. It is not easy to generate narrow sound beams, unlike the case of light beams, simply because the wavelength of sound waves is much greater than the wavelength of light.

The Wikipedia article on "SI radiometry units" lists altogether 36 different radiometric quantities that are in use. We will content ourselves with the six most common quantities used. We choose to provide both the radiometric and photometric quantities at the same time for each type of measurement, using the abbreviation "RAD" and "PHOT" to distinguish what is what.

We choose to refer to Fig. 11.4 that shows a light beam from a pocket torch (flashlight) that strikes a surface a short distance away.

We often wish to send a powerful beam of light from a torch. When companies advertise that a flashlight reaches for example 800 m, the beam must have high power and it must not be too wide. In order for us to see what the beam hits, the light must have not only a high power, but a significant part of this power must lie in the wavelength range that is perceivable to the eye.

Light sources, light intensities, illuminations, etc., are described through a number of concepts and units. Let us describe the most important ones:

How powerful light do we have within a defined solid angle?

- RAD: *Radiant intensity* $I_{e,\Omega}$ is measured in W/sr. Characterizes radiation per unit solid angle in a given direction.
- FOT: *Luminous intensity* I_v (brightness/light intensity) is measured in candela: cd = lm/sr. Characterizes visible light intensity from a light source (per unit solid angle) in a given direction.

On a dark winter night, we wish to light our living room to make it a cosy place. We would evidently use lamps that spread the light a lot. Lights with narrow beams will not provide a pleasant lighting. We are then interested in how much light we get from a light source, and in this case, we do not care how effective the lamp is (how much current it draws). In the past, we switched from a 40 W to a 60 W bulb when we wanted more light. Today, we look at how many lumens the lamp gives since different types of lamps have different efficiencies in generating visible light. To get more light, for example, we change from 500 to 800 lm, and then, following consideration becomes important:

Total amount of light, summed over all directions:

- RAD: *Radiant flux* Φ_e is measured in W. Characterizes the power of *light* from sources (note: not how much power is drawn from the supply!).
- FOT: *Luminous flux/luminous current* Φ_v is measured in lumen: lm ≡ cd sr. Characterizes the flow of visible light coming from the source.
  Here it is worth noting that the watt in radiometry corresponds to the lumen within photometry!

The relationship between lumen and candela is that the lumen integrates light in all directions, while candela indicates how much light goes within a chosen solid angle. For a light source that emits approximately the same light in all directions, the number of lumen will be 4π times the number of candela. For light sources with a very narrow light beam, the candela value will be much greater than the lumen value.

A light bulb does not last forever. For example, some last for 500 h, and others are claimed to last for several thousand hours. The cost of using different types of light bulbs, of course, depends on the power consumption, but it also depends on how much light we can get out of a bulb before it is exhausted. Therefore, the total amount of light energy may be of interest in certain contexts:

Total amount of energy/light, summed up over time:

- RAD: *Radiant energy* Q_e is measured in J. Characterizes the source.
- FOT: *Luminous energy* Q_v is measured in cd sr s = lm s (candela steradian second = lumen second). Characterizes the source.

You may have seen that the old Edison light bulbs often had a milky coating on the inner surface of the outer glass envelope. The same is found on many modern LED bulbs as well. There are several reasons for such a coating, but one of the reasons is that the light bulb should not be too uncomfortable to look at directly. It is more comfortable if a light source emits light from a large surface than a small one.

We cannot look directly at the sun since the intensity is so large compared to its size. But the moon can be viewed directly without hurting the eyes. However, if all the light from the moon had come from a tiny area in the sky, our eyes would feel uncomfortable when looking directly at the moon.

In other words, it is not just the amount of light that comes from a light source that matters, but also how large the area (seen by the observer) the light comes from. This takes into account the following criterion:

How powerful is the light within a given solid angle, per unit surface of the source:

- RAD: *Radiance* $L_{e,\Omega}$ is measured in $W/(sr\,m^2)$. Characterizes radiated power per square metre of projected surface per steradian in a given direction.
- FOT: *Luminance* L_v is measured in cd/m^2 (equivalent to $lm/(sr\,m^2)$.) Characterizes radiated visible light intensity from each (projected) square metre of the source.

When we use a flashlight for, say, reading a map in the dark or when we have a reading lamp and want to read a book, we need to ask for the amount of light that illuminates the map or book. This is one of the simpler light measurement quantities, and here are the definitions:

Light intensity that falls upon a surface:

- RAD: *Irradiance* E_e is measured in W/m^2. Characterizes radiation intensity towards a surface.
- FOT: *Illuminance* E_v is measured in $lm/m^2 \equiv lux$. Characterizes luminous flux towards a surface.

A book usually consists of black letters on a white paper. This is not accidental, because the paper itself does not emit light, but it "reflects" light that falls on the paper. In order for us to read a book, there is a requirement for how much light per unit surface comes from the paper to our eyes.

There is a similar requirement for a mobile phone screen or a computer screen or a TV. The quality of the image and how comfortable it is to view the screen depend on how much light per unit area comes from the screen:

Light intensity radiating from a surface:

- RAD: *Radiant exitance* M_e is measured in W/m². Characterizes radiation intensity from a surface.
- FOT: *Luminous emittance* M_v is measured in lm/m² $\equiv$ lux. Characterizes Luminous flux that radiates from a surface.

Now we are done with reviewing the quantities that are commonly used in light measuring. In addition, there is another item that falls in the list of the photometric quantities and is frequently used in connection with energy efficiency. This last quantity is:

Efficiency to convert electric power into visible light:

- FOT: *Luminous efficiency* η is measured in lm/W. Characterizes how effective a light source is to convert physical power into visible brightness. In this context, the number of watts is simply electrical power drawn from the mains.

Some additional comments

Experience has shown that it is difficult for many people to grasp the difference between radiation intensity and radiance. We will therefore try to concretize. Radiation intensity tells us how much intensity emanates from a light source in a certain direction, as compared to other directions. A spotlight bulb will provide significantly higher intensity in the direction of the spot than in a direction beyond the cone where most of the light is confined. We can describe this distribution by, for example, a graph that shows the radiation intensity as a function of angle from the axis of the luminous cone. The intensity (measured, e.g. as irradiance) at different distances from the cone will vary with the distance, but the radiation intensity, which is the power per unit solid angle, will be independent of the distance from the source.

Radiance, a quantity that closely resembles radiation intensity, is used in cases where the light source is extended so that it is meaningful to specify the amount of radiation intensity that comes from different parts of the source. How much radiation intensity there is, for example, from areas approximately in the middle of the sun (seen from the earth) and how much radiation intensity do we get from areas further out to the edge of sun (seen from us)? We normalize the intensity of radiation from the *projected* surface and arrive at the radiance. Areas near the edge of the sun (seen from the earth) are inclined in relation to the view from the earth. Then we divide the radiation intensity with an apparent area seen from us, that is, an imaginary area that is perpendicular to the direction of view from us that covers the area we calculate the radiation intensity from. For areas at the centre of the solar disc, we do not need to make such a correction since the solar surface in the middle of the solar panel is perpendicular to the direction of vision for us. Then projected surface will be identical to real surface, which simplifies the calculation of the radiance (radiation intensity per (projected) surface of the light source).

Lumen (lm) is a derived unit: candela multiplied by the solid angle. Lumen indicates how much visible luminous intensity a source emits (integrated across all directions). A source that provides many lumens will provide more visible light than a source with few lumens (see Fig. 11.6).

Let us try to put place of the quantities in context:

Suppose that we have a spotlight source that has a light flux of 4π lumen in all. This is an integrated brightness for the source (including all directions.)

The light intensity from this source is 1 candela in all directions (assuming a dot light source). The brightness is 1 candela no matter how far away from the light source we are, because the brightness characterizes the source and not the light at a given location.

However, if we set up a screen across the light direction, the illumination density/illuminance on the screen will decrease with the distance from the source. Our spotlight source with light flux 4π lumen will have an illuminance of 1 lux on the inside of a spherical shell with radius 1 m centred on the light source. If we double the radius of the spherical shell, the illuminance decreases to 1/4 lux.

Our eye can adapt to an enormous range of light intensities. The ratio of the largest and smallest intensity it can handle is about 10^{10}–10^{12} (depending on what source we consult). However, the eye cannot handle such a large light intensity range at one and the same time. The eye adapts to the average level to which it is exposed. After such adaptation, the ratio of the largest and smallest intensity which the eye can handle *at the same time* is about 10^4. Within this range, the eye can distinguish between about 50–100 brightness levels (grey shades).

Figure 11.5 shows an overview of luminance in our surroundings under different conditions. For many years, it has been argued that the faintest light the human eye can perceive is a few photons per accumulation time in the rod cells (i.e. five to ten photons per second within a tiny area of the retina). However, this is an opinion based on photons as indivisible particles, a view about which there is no consensus (see

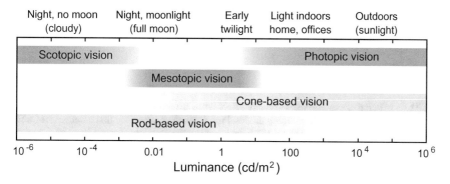

Fig. 11.5 Approximate light level (luminance) in the environment where our visual sense works. The figure is inspired by Fig. 16.2 in [3]

the reference list at the end of the chapter). Regardless of the exact level, it is an incredibly low intensity compared with full sunlight.

Modern light-sensitive detectors (so-called single-photon detectors) are somewhat more sensitive than the eye, but can suffer permanent damage if the light intensity becomes more than about 10^6 times the lowest light intensity they can detect. However, there are many factors that easily make a comparison with the eye misleading. Sensitive detectors are expensive!

11.1.1   Lumen Versus Watt

In recent years, much attention has been paid to the efficiency of light sources. An Edison light bulb, to coin a phrase, converts about 80% of the energy to radiated energy. The rest is spent hot in heating the filament and the terminals, etc. Of the radiated energy, most is in the infrared region, which is not visible. When using such lamps in environments where we need heating, the low efficiency of making visible light is not a disadvantage. In environments where, on the contrary, we need to remove heat (in warm climates), such lamps are clearly undesirable. We get too little visible light out of the energy used.

In recent years, there have been a lot of different light bulbs on the market with many different sockets and operating voltages (see Fig. 11.6). There are classical filament bulb, halogen bulb, fluorescent lamp and the so-called energy-saving bulb based on the fluorescent lamp, and a number of different LED (light-emitting diodes)-based bulbs.

Fig. 11.6   In part **a**, some examples of today's light bulbs are shown. A particular interesting source of illumination is organic light-emitting diodes. Picture **b** shows one of the first commercial OLED lamps available in the market. The OLED light panel is extremely thin, as seen in panel **c**, and the light is radiating evenly from the total area of the film (**e**) providing a pleasant illumination. The colour spectrum reveals that this OLED radiates light relatively evenly from the visible spectrum (**d**). It is not as good as for a classical filament bulb, but far better than some fluorescent lamps

There are big differences in how efficient the light bulbs are, even when they are of the same type. Fortunately, the manufacturer often indicates the light flux in the number of lumens, but even this is not sufficient. Some bulbs collect most of the light into a relatively narrow beam while others provide light in almost any direction. It has become quite difficult to find one's bearings.

On the Web, we can find different overviews of the light output for different types of light sources. These overviews change several times a year due to developments that take place. From my own observations in shops, and a table on Wikipedia, April 2018, under the term " luminous efficacy ", the following overview for different light sources can be given (light source, light output given in lm/W):

- Stearin candle: 0.3
- Classical incandescent lamp: 5–18
- Halogen bulb: 17–24
- Fluorescent tube (including energy savers): 46–104
- White LED (light-emitting diodes): 45–170
- Theoretical limit for white LED: ca 260–300
- Theoretical limit for an optimal blackbody radiating source (5800 K): 251.

It is worth noting that the light output from, for example, LED lamps differs considerably from one type of LED to another (a factor of four in difference according to actual observations and the overview in Wikipedia). If we want to save energy and think about the environment, we should take a close look at the specifications before choosing new lamps and bulbs!

There is currently intense research and development on new types of light sources. The fact that the Nobel Prize in Physics in 2014 went to the pioneers who found out how we can make blue LED sources shows how important lighting is for the world community. That UNESCO pointed out 2015 as "The International Year of Light" reinforces the sense of importance attached to light.

11.2 Dispersion

We have previously found that dispersion causes light with different wavelength to have different velocities through glass. This means that the refractive index varies with wavelength.

In Fig. 11.7, a diagram shows how the refractive index of light changes with wavelength of a regular type of optical glass (Schott BK7). The curve varies considerably for different types of glass, so if we want to demonstrate the different colour phenomena mentioned in this chapter, a type of glass should be used which gives a large dispersion (often associated with high refractive index). In binoculars, we prefer to search for materials with the least possible dispersion in order to minimize the so-called chromatic aberration (more about chromatic aberrations in Chap. 12). This applies primarily to wavelengths in the visible area.

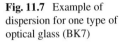

Fig. 11.7 Example of dispersion for one type of optical glass (BK7)

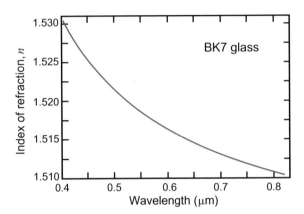

Newton's classic colour experiment is usually described in more or less the following words: "When light is sent through a glass prism, we get a spectrum". In practice, more is needed to ensure that the spectrum has the quality we expect and Fig. 11.8 indicates this. We must send light through a *narrow* slit, and need a lens to make a sharp image of the slit on a screen (as described in the next chapter). We may mimic this situation if we, for example, send sunlight through a narrow slit, without a lens, to make a somewhat blurred image of the slit on the screen. The quality of the spectrum is then rather poor.

Only when these conditions are satisfied, can we put the prism in the light path with a side edge parallel to the slit. The light beam will then deflect, but will form a shifted image of the slit on the screen. We may need to rearrange the lens's location so that the image on the screen becomes as sharp as possible.

The resulting spectrum can be described as a *multitude* of images of the slit, slightly offset from one another. If the light source contains a continuous range of wavelength components throughout the visible region, red light would appear in one place, green somewhere else, and blue on yet another place. The constellation of all these images constitutes a visible "spectrum" on the screen.

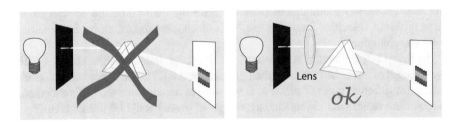

Fig. 11.8 Newton obtained a coloured spectrum when he imaged a slit on a screen and let the light on its way pass through a glass prism. The geometry of the layout is crucial to the result

11.3 "Colour". What Is It?

A number of details do not appear in such a simple description of Newton's spectrum as the one given above. Firstly, what do we mean by "colour"? Many people have a very inadequate impression of colours.

Colour is something we *experience*, a sense impression. The colour sensation has a complicated connection with the physical stimuli that affect our eye. The light is partially absorbed into special proteins in the retinal rods and cones, cells called "photoreceptors". The rods are the most photosensitive receptors and are responsible for sight in the dark. The rods cannot provide colour information and will no longer be discussed here.

The cones, on the other hand, provide colour information. There are three types of cones that can initially be called blue-sensitive, green-sensitive and red-sensitive. These are also referred to as S-, M- and L-cones where the letters stand for "short", "medium" and "long" wavelength for the peak in their sensitivity curves.

In short, the sensitivity and the most sensitive area of the three cones are as follows:

- S-cones, 380–560 nm, peak 440 nm
- M-cones, 420–660 nm, peak 540 nm
- L-cones, 460–700 nm, peak 580 nm.

Quite different numbers are reported in different studies because of individual differences from person to person, and partly because determination of a sensitivity curve is not a trivial task so the values are somewhat dependent on the measurement method used. CIE (see reference list) has adopted a standard that gives the average value for each of the peaks.

Figure 11.9 shows the sensitivity curves for the three types of cones. The figure must be so understood that if we send *monochromatic light* (light with only one wavelength), the curves show the sensitivity of each of the three cone types. At

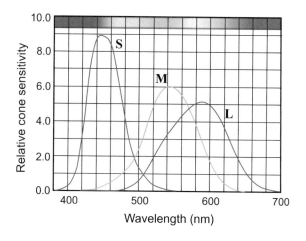

Fig. 11.9 Relative sensitivity curves for the three types of cones in our eye

about 444 nm, the "blue-sensitive" cones (S-cones in the figure) are about twice as sensitive as at about 480 nm (8.8 and 4.2 along the vertical axis, respectively, in the plot). This means that one must double the intensely of light at 480 nm to get the same response from this cone as light at 444 nm. Put in another way: if a person is born with defective green-sensitive (M) and red-sensitive (L) cones, he or she will have exactly the same viewing experience for monochromatic light at 480 nm as for monochromatic light at 444 nm, but at half the intensity. For monochromatic light with wavelength less than 380 nm and greater than 560 nm, this type of cone gives a negligible response.

The special thing is that the three curves in Fig. 11.9 overlap one another, in part very substantially! This means that monochromatic light with a wavelength of approximately 570 nm will stimulate (excite) both the red-sensitive and the green-sensitive cones about equally! We may gather then that the expressions "red-sensitive" and "green-sensitive" are really misleading, and we therefore proceed to mention only cones of types S, M and L (shortcuts for, respectively, "short", "medium" and "long" wavelengths).

In summary: The only signal that a visual cell (one cone) can send is a train of *identical neural pulses*, no matter what light stimulus excites the cell. However, the amount of light required to excite a cone is wavelength-dependent and the sensitivity curve differs for S-, M- and L-cones. The number of pulses per second coming from a cone is approximately proportional to the intensity of the light (assuming that the wavelength distribution of the light is kept constant when the intensity increases). These characteristic features apply to all three types of cones. There is no difference in the shape of the nerve pulses from S-, M- and L-cones.

The response is in a way "digital"; either we have a pulse or there is no pulse. The brain deduces colour information because it keeps track of what type of visual cell (cone) each nerve fibre comes from. If the brain receives, for example, about the same number of neural pulses per second sent by M- and L-cones from a region of the retina, and little from the S-cones in the same area, the brain gives us a greenish-coloured colour experience of the light that reaches this part of the retina (compare with Fig. 11.9).

A small digression:
When you see curves such as those in Fig. 11.9, you hopefully associate them with a fundamental phenomenon from the book's first chapters. You encountered curves that looked very similar to each of the curves in Fig. 11.9 when we discussed forced oscillations and resonance. Then, the resonance curves had frequency along the *x*-axis and some amplitude along the *y*-axis, while in Fig. 11.9 we

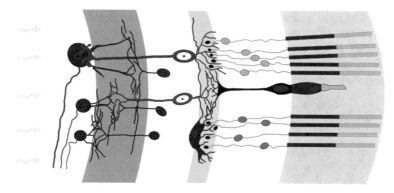

Fig. 11.10 Signals from the cones (to the right) are processed by many types of cells in our eye and on our way to and in the brain itself. The vision process is therefore very complicated. (Light is actually coming in from the left.) Cajal/Chris, Wikimedia Commons, CC BY-SA 3.0. The figure is based on [4]

have wavelength along the x-axis. However, since frequency and wavelength are linked via $f\lambda = c$, the wavelength axis can easily be converted to a frequency axis. According to semi-classical theory, there is a correlation between forced oscillations and absorption in the visual receptors.

Our vision is very advanced and the raw image we get from the actual visual cells (cones) is further processed in many different ways. Already in the retina we have four types of cells that process the response from the photoreceptors (cones) (Fig. 11.10). These are so-called horizontal cells, bipolar cells, amacrine cells and ganglion cells. Each cell has its own function, including contrast enhancement, or reacting specifically to temporal changes in brightness (e.g. when the subject is moving). The cells are also involved in the signal processing related to colour discrimination. There is also extensive processing of the signals from the retina in certain relay hubs in the visual field, and even more in the brain's visual centre. What an impressive machinery lies behind our visual perception!

We will first focus on the simplest principles of colour (chromatic) experience, and the *main rule* in this context is that the colour is determined by the interrelation between the amount of light absorbed in the three types of cones. Then we will mention other factors that also affect our colour experience.

11.3.1 Colourimetry

If we consider only monochromatic waves in the visible region, we find that light absorption in the three types of cones will change in a unique manner as the wavelength of light is varied. Monochromatic light provokes visual sensations known as "prismatic colours". These colours are in a class by themselves and are perceived as "saturated" colours. We cannot make a spectral red any redder than it already is (at least not with the colour index assigned to it).

If we let in light with *many* wavelengths, the response from the cones will equal (almost) the *sum* of the responses due to the individual monochromatic contributions. The summation rule can be traced to the superposition principle. Needless to say, this description holds only within a limited intensity range, but it will be adopted here for the sake of simplicity.

Energy absorption in M-cones can be expressed mathematically as follows:

$$M = \int \phi(\lambda)M(\lambda)d\lambda \qquad (11.1)$$

where $\phi(\lambda)$ is the spectral intensity distribution of the incident light (formally called colour stimulus function). $M(\lambda)$ is the spectral energy sensitivity for M-cones corresponding to the middle curve in Fig. 11.9.

Energy absorption in the two other types of cones can be described in an analogous manner. The three expressions so obtained describe only relative absorption (because the expressions do not take calibration into account).

It is not hard to see that monochromatic light of ca 570 nm plus monochromatic light of ca 420 nm will provide the same stimulation of the three types of cones as will a mixture of monochromatic light at 660, 500 and 410 nm. The only proviso that must be met is:

$$M = \phi_1(570)M(570) + \phi_1(420)M(420) =$$

$$\phi_2(660)M(660) + \phi_2(500)M(500) + \phi_2(410)M(410)$$

and likewise for L and S. We obtain three equations with three unknowns (assuming that the two ϕ_1-values are known).

The upshot of the foregoing analysis is that *we can get the same colour sensation from widely different physical stimuli.* By "stimulus" we mean specific distribution of intensity for various wavelengths—in other words, the spectral distribution of the stimulating light. The spectral distribution of the light that appears to us as a special green colour can, as a matter of fact, be rather different from the spectral distribution of another light source, although we perceive the latter as exactly the same green colour as the first (termed "metamerism"). It is by no means to be supposed that "colour" is synonymous with the spectral colour defined above!

How fortunate for us that this happens to be so! We exploit it extensively today, especially in photography and coloured visual displays on a TV screen or computer monitor. In all these cases, we usually start out with three colours and mix them with each other in different proportions to form "all other colours". But there are some limitations.

Fig. 11.11 "Colour Horseshoe" defined by CIE in 1931. Detailed description of this "xy colour space diagram" is in the text. NRC, Public Domain [5]

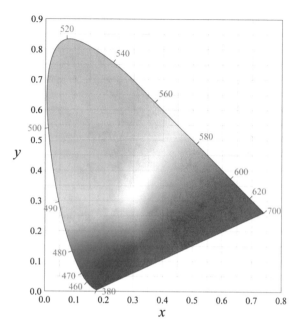

CIE 1931

Figure 11.11 shows a so-called colour horseshoe, which is devised according to a special scheme. Along its curved edge lie the "pure" spectral colours from red to violet. On the straight line between red and violet, called the "line of purples", lie the nonspectral magenta and purple-red mixtures. The centre of the horseshoe corresponds to light that is perceived as white.

Along the axes, so-called x and y coordinates are determined from a transformation of (S, M, L). How may a colour stimulus be determined by *three* parameters: (S, M, L) while the colour horseshoe reproduces colours only in a *two*-dimensional plot?

The three stimuli indicate *both* colour and light intensity information. For a given light intensity (or rather *luminance*), the three parameters will not be independent of each other. Only two can be chosen freely. By using an appropriate transformation of the cone absorbances, we can transform into two independent parameters x and y which indicate the colour irrespective of the light intensity (luminance). The inverse transformation is *not* unambiguous! The colour horseshoe in principle indicates all the colours we can experience at a given luminance and can therefore be regarded as a general "colour chart". It is therefore called a *chromaticity chart*.

The mathematics behind the current transformations has been developed over many years. The colour horseshoe was adopted in 1931 as a standard for colour measurement by CIE (Commission International de l'Eclairage in French, The International Commission on Illumination in English). The transformations used are still

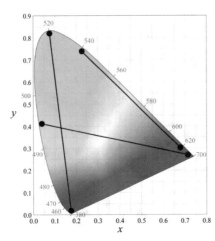 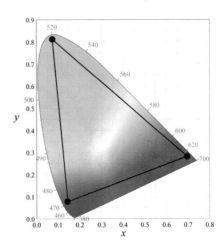

Fig. 11.12 Finding the colour for additive colour mixing corresponds to the "centre of gravity" for the colour coordinates that are included in the mixture. To the left are given three examples of the colours that can be obtained by mixing two colours, marked with straight lines. Further information is in the text. *Right part:* The colour scope of a computer screen using three types of coloured pixels lies within the triangle spanned by the colour coordinates of the pixels. The amount of colour within the resulting triangle is significantly less than the amount of colour that the entire colour horseshoe represents

being discussed, and several Norwegian physicists (such as Arne Valberg, Knut Kvaal and Jan Henrik Wold) have worked on this issue for many years.

The colour horseshoe is useful in many ways. If we start with two colours (two coordinates in the colour horseshoe) and mix them with equal weight (defined appropriately), our colour scheme will correspond approximately to the point in the colour horseshoe that is in the middle of the two points we started with. This is indicated in the left part of Fig. 11.12. Starting with equal amounts of near-spectral stimuli at 540 and 620 nm, the colour we perceive will be quite similar to the colour of a wavelength of 572 nm wavelength (most would denote it as yellow). However, if we mix in approximately equal amounts of near-spectral stimuli at 495 and 680 nm, we will perceive the colour mixture as approximately "white" (a light grey without colour).

When we consider a computer screen, a mobile phone screen, an iPad, a TV screen or the like, there are three types of light that build up the image: "red", "green" and "blue". These stimuli each have their coordinates (chromatic points) (x, y) in the colour horseshoe. *The colours we can form with these three primary colours are within the triangle that the three coordinate points form in the colour horseshoe.* The amount of all colours we can form with the three primary colours is called *colour scope*, for example, the computer screen.

Fig. 11.13 Spectral lines show beautiful saturated colours when viewed directly in the laboratory. After photography and reproduction (as here), the amount of colour becomes much smaller

We can try to select three points and draw lines between them to identify which colours can be produced by the three primary colours, and we will then discover that a variety of colours lie *outside* the triangle as the points expand. An example of such a triangle is given in the right part of Fig. 11.12. Since an inner triangle can never cover the entire colour horseshoe, it means that the colours we can see on a computer screen, etc., are a rather pale image of the amount of colour we can experience in nature. A variety of colours on flowers, for example, are far more saturated when you see the flower in reality than what we can reproduce on a computer screen (or photograph for that matter).

An example of the lack of colour achievable with three-colour reproduction is shown in Fig. 11.13. In the figure, there are two spectra of gases: one with a few spectral lines and one with many more. Spectral lines are in fact the most saturated colours we can have, and when the lines are observed directly in a laboratory, we notice this. The same spectral lines shown in a photograph are just a pale copy of reality (as the figure shows).

In industrial context, different colour systems have been developed than CIE. Colours are included in far more parts of a modern society than we usually think about. For example, food quality is assessed using colours. Two colour systems used in industrial context are NCS and Munsell.

For example, Natural Colour System (NCS) is used when we buy paint in a store. The paint is mixed in the store. There are four basic colours in NCS (in addition to white and black), namely red (R), yellow (Y), green (G) and blue (B). When a colour code is 1020-Y90R, the first two digits give the mixture of white and black. In our case, these numbers are 10, which means that it must be mixed 10% black and 90% white. The two next digits tell us how much we should have of this grey colour in relation to the colour we want. The numbers 20 means that 20% of the grey colour we mixed will be used and 80% colour/colour. Y90R means that it will use 10% yellow and 90% red, and this colour mixture will be mixed with the grey colour in the ratio of 20% grey and 80% red/yellow.

11.3.2 Colours on a Mobile Phone or Computer Display

Let us now make a practical check on how colours are generated on a TV, mobile phone or computer screen. Figure 11.14 shows in the centre a small part of a computer screen with Windows icons. We have taken a picture closer to the screen to see details. A clip from the Google icon has the colours green, white and dark blue. Another snippet has the colours red, yellow, white and light blue.

To the right and to the left, there are selected representative "pixels" that the image is built with. Each pixel on this screen has three vertical fields. These fields can give the colours red, green and blue, respectively, and only these. These three colours correspond to the points shown in the right part of Fig. 11.12. We see here quite clearly that, for example, the colour yellow on the computer screen really is generated only by means of red and green light. The pixels are so small that the light from the red and the green field in a pixel hits the same eye cells.

Additionally note that dark blue or blue-black is generated virtually by using zero red and green light and only weak blue light. Light blue (slightly light blue-green), however, is generated with almost maximum of blue, some green and a little red. White is generated with powerful red, strong green and powerful blue at the same time. It is fascinating that, using only three primary colours, we can generate as many colours as we actually can!

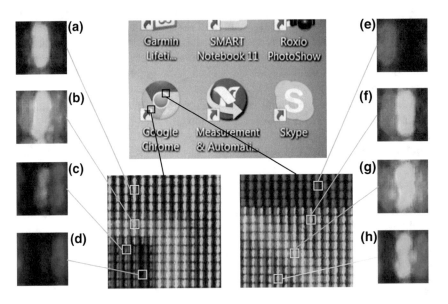

Fig. 11.14 Photographs of a computer screen. A section of icons on the "desktop" on a Windows computer is shown at the top middle. Two small clips from the Google icon are shown underneath. Pixels from different coloured areas are shown to the far left and right of the figure. A particular pixel *can only* emit red, green or blue light (over its entire area). The colours the pixels give us are **a** green, **b** white, **c** dark blue (blue-green), **d** blue-black, **e** red, **f** yellow, **g** white and **h** light blue (blue-green)

11.3.3 Additive Versus Subtractive Colour Mixing

Art posters use, for example, 7-colour, 9-colour, 13-colour printing. One of the reasons for this is that the colour gamut in the final image should be as large as possible. It is natural to draw parallels to the triangle on the right in Fig. 11.12 in this context. However, we must remember that when colours are mixed using pigments that are illuminated by an external light source, all colour blends are far more complicated than those we have mentioned above. We have so far just referred to *additive* colour mixture that occurs when mixing (superposition) of light. In an art poster (or in a painting or print of a photograph), *subtractive* colour mixing is at work. Pigments absorb some of the light that falls on them, and the light sent back to us will cause the pigmented surface to appear with a certain colour when illuminated, for example, with sunlight. If we add more pigments together, e.g. by mixing yellow and blue pigments, the surface will often look green. However, if the pigments are illuminated by light with only a few wavelengths (e.g. light from some diodes (LED) or from fluorescent tubes), there is no guarantee that the mixture of yellow and blue pigments will look green!

The best colour rendering is achieved with light sources that have a continuous spectral distribution, i.e. common old-fashioned light bulbs or halogen bulbs. In art exhibitions and the like, it is therefore important to use such lighting instead of fluorescent lamps, energy-saving light bulbs or simple types of LED lamps.

Incidentally, the word subtractive colour mixture is a little misleading. To find the cone absorption when the stimulus corresponds to light reflected from a mixture of two pigments, the spectral reflection coefficients of the pigments must be *multiplied* with each other.

By the way, Helmholtz was the first to describe the difference between additive and subtractive colour mixing. This happened about 200 years after Newton's colour mixing model based on the superposition of light (additive colour blend).

It is not trivial to make pigments from scratch. Often one uses natural pigments from, for example, plants or minerals. Only a limited number of pigments are available, and when we print an artwork, it may sometimes be useful to use more than three "colours" (pigments) to get as good reproduction as possible, even if the original is only found as RGB (three target figures) from a digital camera. We cannot expand the colour gamut relative to the image (colour gamut spun out of the RGB values), but we can *reproduce on paper* the colour gamut better than by using fewer pigments.

In order to achieve a larger colour gamut, we need to start with more than three stimuli already in the data recording. It helps little to start with a digital camera with only three detectors per pixel and believe that if only we had a good printer, then the overall result would be almost perfect! This is analogous to audio recording: if we are going to process sound at a sampling rate several times that used in CD audio, then there is no point in starting with a low resolution in the processing of sound and raise it later on. We must have the highest sampling rate already at the very first digitization of sound. In studio recording of sound, it is now fairly common to use higher sampling rates than the CD standard. For capturing images, a trend has started to employ cameras with more than three detectors per pixel, and as such, screens with more than red, green and blue luminous points. It is not inconceivable that in the future photography apparatus and computer monitors will be based on technology with more than three basic stimuli.

11.4 Colour Temperature, Adaptation

The formalism given in Eq. (11.1) and the description of mathematical transformations leading to the CIE chart can leave an impression that a certain spectral distribution of light will always give us the same visual impression (colour). That is wrong. Human vision is highly advanced and has incorporated a form of adaptation that is very useful. In short, the colours in a Norwegian flag will be judged to be red, white and blue, whether we view the flag under lamplight in the evening or in the sunshine under a brilliant blue sky during the daytime. This happens despite the fact that the spectral distribution of the light from the flag is quite different in the two cases. The ability to adapt is a consequence of natural selection: it was useful also for the Neanderthals to be able to make out the colours in the light from the fire in the evening as well as in direct sunlight.

The big difference in spectral distribution in these examples is well presented in Planck's description of "black body radiation", which is close to what is emitted by a hot body. In an incandescent lamp, the temperature of the filament is in the range of 2300–2700 K. The surface of the sun has a temperature of about 5500 K. This causes the continuous spectrum of these light sources to be quite different, as shown in Fig. 11.15. In the light bulb, the intensity of the blue spectral colours is much less than for the red ones, while in the spectrum of the sun, blue spectral colours are more prominent than the red ones.

Figure 11.16 illustrates the importance of different lighting if we detect the appearance of an object without adaptation. A digital camera is used to capture images of one and the same object in lamp and sunlight. The colour temperature correction in the camera could be set manually. Photos are taken at three different settings on the camera for each of the light sources. The settings are: "colour temperature 2700, 4000 and 5880 K".

Fig. 11.15 Spectral distribution of electromagnetic waves emitted by hot bodies (blackbody radiation). The curves are values calculated from Planck's radiation law for different temperature of the body

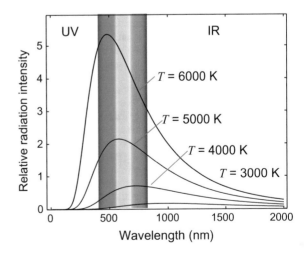

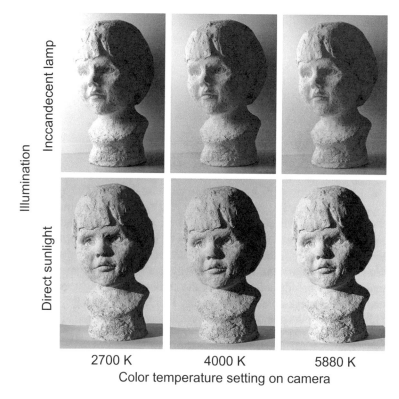

Fig. 11.16 Photographs of one and the same head sculpture in plaster in two different lights and three different manual colour temperature settings on the camera. The figure indicates the actual difference in spectral colour distribution from the object in lamplight and in sunlight. Nevertheless, with our visual sense, we perceive that the head looks almost white regardless of whether it is viewed under lamplight or sunlight. This is due to the visual adaptation capacity

Figure 11.16 shows that when we take pictures of objects in lamplight after setting the camera for approximately 2700 K, the pictures look as we expect them to, but they have a ghastly red appearance for a colour temperature setting of 5880 K. Similarly, images in sunlight often produce correct colour in the pictures if the camera is set to about 5880 K while the image looks very blue if we chose a colour temperature setting of 2700 K.

Most modern digital camera has a built-in form of adaptation similar to that of human vision. If we shoot the pictures in Fig. 11.16 with such a camera, the result would have been relatively "correct" in most cases, but if the object and its surroundings by themselves are very reddish or bluish, the camera could perform unintended corrections that can not easily be avoided.

We have now seen that human vision (the eye plus all further processing even in the brain) has a fabulous capacity for adapting to different spectral distribution of the dominant light source. What we call a red, green or blue surface depends not only on the absorption by the cone (S, M, L) of the light from the surface, but also to a large extent on the surroundings. It has some important implications: if we apply colour correction of digital images, for example, in Photoshop or similar software, it is important to have a grey surface along with the image where the colours are to be assessed. If we still find that the grey surface looks grey, we have a reasonable guarantee that our eyes have not adapted to the image to be corrected. If we do not check the eye's adaptation mode in relation to a grey surface, we may deceive ourselves, and the final result may become unsightly.

11.4.1   Other Comments

There are also other forms of adaptation in our visual sense. The eye also adapts with respect to intensity. In sunlight, a "grey" flat surface reflects much more light than a "white" surface will reflect at dusk. Nevertheless, we call the first one grey and the other white. What we call white, grey or black surfaces depends not so on the intensity of light from the surface as on *the relative intensity from the surface in relation to the environment*. In a similar way, our colour assessment of a field in the visual image is greatly influenced by the colours in the neighbouring fields of the visual image. An example is given in Fig. 11.17. There are many other delights related to the eye and the rest of the visual system, not least related to contrasts (e.g. read about Mach band in Wikipedia) but we cannot take the time to talk more about this matter than we have already done.

Finally, a short comment about the colour horseshoe: If we look at Fig. 11.11 on several different computers, we will find that the colours look quite different from screen to screen. This is partly due to the fact that the three pixel colours are slightly different from one type of screen to another. Graphic artists often perform a screen calibration and transform colour information using matrices before viewing images on screen or before printing the images. Such a transformation is usually called a "colour profile". In the process of achieving a good colour profile, a standard card (an image) is used, for example, in the subject when shooting. The colour profile can then be designed so that the end result gets as close to the original default as possible. The X-rite colour checker is an example of such a chart. It can be obtained from Edmund Optics (see reference list).

Colour management is one of the issues we have to grapple with when handling today's technology. Colour correction is a profession!

11.5   Prismatic Spectra

Now that we know a little more about how we perceive colours, we are ready to return to Newton's colour spectrum from a prism. Many think of spectral colours as the colours they see in the rainbow: ROYGBIV (or Roy G. Biv): red, orange, yellow, green, blue, indigo and violet. But what do we really see when we examine a Newtonian spectrum produced by a narrow opening? Well, the spectrum looks like the top of Fig. 11.18. The curious thing is that, in fact, we see only red, green, blue

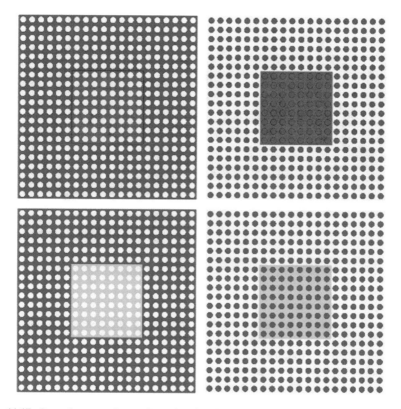

Fig. 11.17 Our colour experience of a surface is affected by colours close by. The red squares in the upper part of the figure have exactly the same spectral distribution, but look different. Similarly, the green squares in the lower part are identical, but look different since the neighbouring areas have a completely different colour

Fig. 11.18 Spectrum from a narrow slit (top) and from increasing slit width below

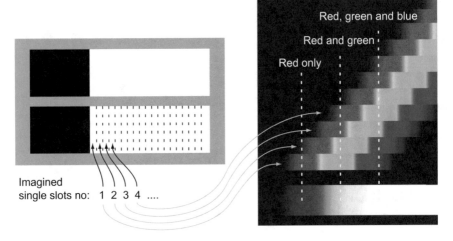

Fig. 11.19 Imaging of an edge can be considered as the sum of images of many slots next to each other. See the text

and partly violet. There is very little yellow and orange, and clearly less yellow than in the rainbow! How can it be explained?

The explanation is found by increasing the width of the opening a little. In the next two examples in Fig. 11.18, we have simulated spectra from slots with increasing width. *Now* we see a yellow area more easily! What is this due to?

This is because when we consider spectral colours, it is only a very narrow wavelength range that gives us the colour impression "yellow". Most of the yellow we perceive is due to the blending of red and green spectral colours (additive colour mixing). We will then get a coordinate point in the colour horseshoe that lies a bit within the edges.

To understand how we think of the colour scheme, refer to Fig. 11.19 which shows how the image would look if we did not image a hole on the screen but instead an edge between a surface without light and a surface with homogeneous "white" light. The light also passes here a glass prism. We can imagine that the bright area is a sum of many single slots adjacent to one another. Each of the slots (if sufficiently narrow) will provide a spectrum that looks red, green and blue. Each column is slightly offset in relation to the neighbouring area, so that the spectra are also slightly shifted in relation to one another.

If we add up the light striking different parts of the screen, we see that at the left end only the red light comes in. The sum is then red. Just to the right of this position, we get a mixture of red and green light, but no blue. The sum of these two lights will be perceived as yellow. Just to the right of this field again, we will get a mixture of red, green and blue. The sum is perceived as white. This is how it continues, and the result will be what we see at the bottom of the figure.

We see then that when we image an edge on a screen but make the light go through a prism, the edge will be coloured with a red and a yellow stripe.

Fig. 11.20 There are two types of border colours, depending on which page is black and which is white in relation to the prism orientation. The left part of the figure shows the black and white distribution of light we start with. The right part shows how the image of the original light distribution would look if the light went through a prism. The colour effect is a simulation customized visual impression from a computer screen. Real reds look prettier, but they must be experienced in vivo!

If we image an edge where the light and dark have exchanged places, the violet/blue area will not blend with the other colours. Next to this, we get an area of cyan (a mixture of green and violet, crudely speaking). Next to this, we get a mix of all colours and we experience this as white.

> An edge of this type will have two coloured stripes in the borderland between white and black: blue violet and cyan. The red-yellow and the violet-cyan stripes we call *boundary spectra* (sometimes also called boundary colours). An example of the boundary colours is shown in Fig. 11.20. When you see edge colours in practice, the stripes are much narrower (and nicer) than we can get an impression of in this figure, but it depends of course on distances and many other details.

If you look through a pair of binoculars and choose to look at a sharp borderline between a white and a black area in the periphery of the field of view, you will almost always see boundary colours at the edge. On good binoculars, where attempts have been made for reduction of the dispersion of light through glass (using combinations of different glass types in the lenses), the edge colours are not very clear. In cheaper binoculars, the border colours are significant and destroy the contrast and sharpness of the image we see.

Coming back to the rainbow: why do we see yellow much more clearly in the rainbow than in a Newtonian spectrum where we use a narrow slit (opening)? In the rainbow, the "opening" is in practice raindrops, and the angular extent of each raindrop is compatible with a small opening. But the sun itself has an extent of about half a degree (angular diameter) in the sky! The rainbow, therefore, becomes a summation of many rainbows that lie a little apart (which originates from different zones on the sun surface). It is *this* summation (corresponding to using a wide opening) that gives us a bright yellow in the rainbow!

11.5.1   A Digression: Goethe's Colour Theory

In Newton's spectrum, we have a highly specialized geometry that gives us the usual spectrum. Historically, Goethe responded to Newton's explanation, not least because Newton did not clearly realize that his spectrum only emerged by *imaging* a slit (after the light from the slit passed through a prism). Goethe showed that other geometries gave completely different colours. Among other things, the "spectrum" from the inverse geometry of Newton, namely a narrow black stripe on a white background, has a different colour gradient than the Newtonian spectrum as indicated in Fig. 11.21. Goethe thought that Newton's explanation was much too simple, and we have to draw the chromatic boundary conditions to understand the colours that are perceived in different geometries.

Goethe explored many different geometries and found many symmetries in the phenomena and introduced certain "colour harmonies", but we should not go into detail.

In Norway, the poet André Bjerke was an important disciple of Goethe. He led a discussion group over a number of years, among which the physicists Torger Holtsmark and Sven Oluf Sørensen were keen participants. A book on the subject: "Goethe's Colour Theory. The Selection and Comments of Torger Holtsmark" was published by Ad Notam Gyldendal's publishing house in 1994.

Personally, I have not discovered that Goethe's has a greater power of explanation than our usual physics model (based on light as waves) in terms of colour phenomena. Figure 11.19 shows the main principle of how we can proceed to build how the colours will come out of a variety of geometries.

On the other hand, Goethe's colour theory has had an important historical function because it focused on Newton's spectrum, not merely "the light that went through a prism". It focused on symmetries and geometries in a great way as until then was not as well known as now. In my language suit, I would like to say that the Goethe fans' points remind us that calculations based on Maxwell's equations depend to a large extent on the boundary conditions! Occasionally, we are physicists too sloppy when we describe phenomena and when we provide explanations. In that case, we will lose valuable details and may be left with wrong beliefs.

Fig. 11.21 Colour spectrum we get from a Newtonian column and the colour spectrum we get from a "reverse slit", i.e. a black narrow stripe on a light background. Also this picture is the result of a simulation. Real spectra (without going through photography or computer monitors) are far more beautiful to look at!

11.6 References

In Norway, it is perhaps the University College in Gjøvik that has the best competence in colour concentrated in one place. They are gathered under the umbrella The Norwegian colour Research Laboratory (https://www.ntnu.edu/colourlab).

Norsk Lysteknisk Komite is the Norwegian national body of the global lighting organization CIE. More information on the websites of "Lyskultur, Norsk kunnskapscentrum for lys" (https://lysveileder.no/).

Here are a few other links that may be of interest if you are interested in reading more about colours:

- http://www.brucelindbloom.com/ and http://www.efg2.com/ (accessed May 2018)
- International Commission of Illumination, Commission Internationale de L'Eclairage (CIE) is an organization that deals with lighting and visual perception of light. Their website is at http://www.cie.co.at/. In particular, details of the transition between physics and perception are given in the Photometry report. The CIE System of Physical Photometry ISO 23539:2005(E)/CIE S 010/E:2004.
- X-rite colour checker to facilitate colour correction can be purchased including at Edmund Optics: https://www.edmundoptics.com/test-targets/color-gray-level-test-targets/large-x-rite-colorchecker (accessed May 2018).
- A book on the topic: Richard J. D. Tilley: *Colour and Optical Properties of Materials*. John Wiley, 2000.
- On the limit of the absolute sensitivity of the eye: See G. D. Field, A. P. Sampath, F. Rieke. *Annu. Rev. Physiol.* 67 (2005) 491–514.

Note:
Matlab programs used to generate the border colours, and the reverse spectrum (the last four figures in this chapter) are available from the author Arnt Inge Vistnes.

11.7 Learning Objectives

After working through this chapter, you should be able to:

- Explain the need for different quantities such as radiant power, radiation intensity, radiance, irradiance, brightness, light output within radiometry and photometry, and make some calculations where you change from one quantity to another.
- Explain the connection between colour perception (chromatic colour) and physical spectral distribution (which can be obtained from a spectroscope, e.g. by dispersing light with a grating or a prism). For example, you should be able to explain that "yellow light" actually does not need to have any spectral yellow in it whatsoever, and yet to be perceived as "yellow".

- Explain the CIE colour horseshoe and the theory of additive colour mixing, and explain what colours we can and cannot reproduce fully, using, for example, digital pictures on a TV or computer screen.
- Explain the term colour temperature and what it implies for human colour experience and for photography.
- Provide relatively detailed qualitative explanations of how we can achieve a "Newtonian" spectrum, boundary spectrum and reverse spectrum, and point out the importance of boundary conditions.
- Reflect a little over a detector, for example, the eye's visual receptor (cone cell in the retina) has only a limited sensitivity range, and usually associate it with so-called forced oscillations earlier in the book.

11.8 Exercises

Suggested concepts for student active learning activities: Radiometric and photometric quantities, radiant energy, luminous energy, solid angle, radiation intensity, difference between lumen and lux, luminous efficiency, photopic, scotopic, photoreceptors, rods and cones, sensitivity curves, dispersion, colourimetry, chromaticity chart, colour horseshoe, CIE diagram, spectral colour, additive colour mixing, subtractive colour mixing, colour temperature, adaptation in intensity and colour, colour spectrum, border colours, Goethe's colour theory.

Comprehension/discussion questions

1. From Fig. 11.9, you can find a few spectral regions where only one of the cones absorbs light. What does it mean for the colour experience we can have for different spectral colours within each of these ranges?
2. Consider the colour horseshoe in Fig. 11.11, and especially the wavelengths that lie along the edge of the horseshoe. Attempt to estimate the lengths of the rim that correspond to spectral colours in the range 400–500 nm, 500–600 nm and 600–700 nm, respectively. This is related to how easily we can detect *changes* in chromaticity when we assess colours. What would you expect this relationship to be (qualitatively)?
3. Do you find any clues in Fig. 11.9 for the relation you found in the previous task? Attempt to write down your argument in as precise and easily intelligible form as possible! (This is an exercise in being able to argue clearly within physics.)
4. Why are wavelengths not listed along the straight edge of the colour horseshoe?
5. Try to see boundary colours by looking, through binoculars or a lens, at a sharp edge between a bright and a dark area. Make a sketch that shows approximately what you see. Point out how the edge should lie in order to see red-yellow

Fig. 11.22 Photograph of various structures seen through Fresnel lenses in glass (objects are out of focus). Boundary colours are visible

and violet-cyan boundary colours, respectively. Alternatively, you can point out boundary colours in Fig. 11.22.

6. Digitalization circuits like those used for digitizing sound usually have 12–24 bit resolution. How many decades of sound intensity can we cover with such equipment? For photographic apparatus, 8–16-bit resolution is commonly used for specifying light intensity. How many decades of light intensity can we cover with such a camera? Compare with the intensity range that human hearing and vision can handle. Why do sound recordings and photographs still work satisfactorily despite the limited bit resolution?

7. How much variation in intensity, in powers of tens, can be handled by our auditory and visual senses (see Fig. 11.5)? Explain that senses that function over such a large "dynamic range" must actually be based on providing a logarithmic response (at least not a linear response).

8. In tests where we want to determine how sensitive the eye is to light (absolute threshold), it has been found that the responses from many visual cells within a certain area of the retina are summed. These visual cells are believed to be linked to the same nerve fibre ("spatial summation area"). Such a summation area corresponds to the light source having an extent of about 10 arcs of minute ($1/6°$) seen from the eye's position. Discuss which of the photometric quantities must be under our control in this type of experiment. Also discuss why some of the other quantities are not relevant.

9. White glass beads are often used in light bulbs (domes) to get an even and pleasant lighting without sharp shadows. Despite the fact that the dome is spherical, the radiance from the edge of the dome seems to be approximately equal to the radiance of the central areas (see Fig. 11.23). Discuss the reason for this is and

Fig. 11.23 Photograph of
the light from a spherical
dome of white glass

explain why the radiometric unit *radiance* is suitable for coping with this property.
Can you see other dome constructions or shapes that *not* would give the same
result? (Hint: Read about "Lambert's cosine law" on Wikipedia.)

Problems

10. Suppose, for the sake of simplicity, that light can be described as a stream of
    photons with an energy equal to $E = hf$, where h is Planck's constant and f
    is the frequency of the photons. An ordinary laser pointer usually has a power
    of a few milliwatts. What attenuation is needed to bring the intensity of such a
    laser beam to a level that corresponds to the limit of what our eye can perceive
    (assuming the limit is equivalent to about "500 photons per second") when the
    laser has a wavelength of 532 nm?

11. A table in Wikipedia with physical data about the sun states that the sun's "lumi-
    nosity" is 3.846×10^{26} W.
    (a) The word "luminosity" (within astronomy) indicates that this is a photopic
    unit (a unit related to the human visual sense). What quantity is it *really*, judging
    from the tables of radiometric and photometric quantities given in Sect. 11.1?
    (b) The same table contains the following information: "Mean intensity" =
    2.009×10^7 W m^{-2} sr^{-1}. The sun has a diameter of 1.392×10^9 m. Describe in
    words what "mean intensity" tells us and what term should actually be used for
    this quantity. Show by calculation that you actually obtain the expected relation-
    ship between the "mean intensity" and "luminosity" given here in the task text.
    (c) We will calculate theoretically how much power can be captured from sun-
    light at the surface of the earth, e.g. in a solar collector or solar panel. What

radiometric or photometric quantities are then of interest? Find the value of this quantity, given that the distance between the earth and the sun is 1.496×10^{11} m and that about 30% of the sunlight coming into the outer atmosphere is reflected or absorbed there. [Note: In addition to the straightforward calculations, it is in addition important in this problem to give the result with an appropriate number significant figures, taking *all* the given information into consideration.]

12. Eigerøy's Lighthouse occupies a pride of place among the coastal lighthouses of Norway (see Fig. 11.24). Built in 1854, it was the first lighthouse with a cast iron tower in Norway. The lighthouse is 33 m high, and the light comes out 46.5 m above the sea level. The right part of Fig. 11.24 shows a part of the impressive lens system along with the light bulb (plus a spare bulb). The lighthouse emits light with brightness 3.9×10^6 cd in three light beams 90° apart (three lens sets 90° apart and the entire lens set rotates around the light bulb with a revolution time of 4 s). The lighthouse is one of the strongest along our coast, and it is claimed that "the light reaches 18.8 nautical miles beyond the shore".

   (a) How could it be that a light bulb with a few hundred watts can be seen 18.8 nautical miles away?

   (b) Would the light from the lighthouse be seen even further than 18.8 nautical miles if, for example, we doubled the power? (Hint: Check if the distance of 18.8 nautical miles has any bearing on the curvature of the earth.)

   (c) Assume that the light bulb is 500 W and has about the same luminous efficiency as incandescent lamps. Estimate the solid angle of the rays.

13. In this exercise, we will compare two very different light sources: (1) an ordinary, old-fashioned 60 W light bulb (incandescent lamp) that shines freely in the room (no screen) and (2) a 4 mW green laser pointer wavelength 532 nm. The

Fig. 11.24 Eigerøy Lighthouse is a powerful monument from a time before the GPS made its entry. *To the right*: the lens system that surrounds the light bulb is impressive. There are three equivalent lens systems that are located 90° apart. The last "wall" is open to permit entry. The lens systems are over two metres tall

laser beam is circular and has a diameter of 9.0 mm at a distance of 10.38 m from the pointer (the beam is actually the most intense in the middle, but we can use an approximation that the light intensity is equally large within the entire specified diameter of the beam). Remember, "60 W" for the incandescent lamp is the power that the lamp draws from our mains, while "4 mW" indicates the power of the actual light emitted by the laser pointer.

(a) Specify the radiation flux and radiation intensity of each light sources.

(b) Estimate luminous flux and luminous intensity for both light sources for the direction the light is most powerful. (Hint: Use Fig. 11.1 and the information given in Sect. 11.1.)

(c) Compare the brightness (luminous intensity) of the light beam from the Eigerøy Lighthouse (previous task) with the brightness of the laser pointer specified in this task. Do you think it might be appropriate to replace the old light sources in the coastal lighthouse with a revolving laser?

14. Find out about how wide the rainbow is (angular width from red to violet). Compare this with the angular diameter of the sum. Does it seem to be in accord with the assertion that the extent of sun may have some bearing on the colours we observe in the rainbow (compared to a common Newtonian spectrum obtained by using a very narrow slit)?

15. Pick up a few colour cards from a paint shop and try to determine which colour system the colours are listed under; if it is the NCS coding, you might be able to decipher the codes?

16. Figure 11.25 shows parts of the wrapping of an LED bulb. Make sure you understand all the information provided on the package, and answer specifically the following questions:

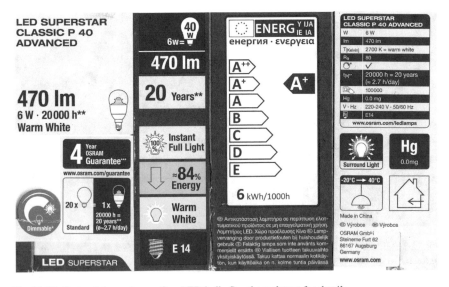

Fig. 11.25 Parts of the package of an LED bulb. See the task text for details

(a) What is the luminous flux and about how much luminous energy can this light bulb give?

(b) Determine the approximate brightness (luminous intensity) of the bulb and the approximate illuminance we can expect on a white A4 sheet held at a distance of 1 m from the light bulb. What assumptions did you make to reach the numbers you specified?

(c) What is the luminous efficiency of this LED bulb?

Acknowledgements I would like to extend my heartfelt thanks to Jan Henrik Wold (at the University of South-Eastern Norway, Drammen), and Knut Kvaal (at the Norwegian University of Life Sciences, Ås for useful comments on this chapter. Any errors and omissions in this part of the chapter are entirely my responsibility (AIV) and not theirs.

References

1. Dickyon, https://en.wikipedia.org/wiki/Luminosity_function. Accessed April 2018
2. Kebes, https://commons.wikimedia.org/wiki/File:Absorption_spectrum_of_liquid_water.png. Accessed April 2018
3. E.F. Schubert, *Light-emitting diodes*, 2nd edn, Figure 16.2 page 276. (Cambridge University Press, 2006). ISBN-13:978-0-511-34476-3
4. S.R.Y. Cajal, Histologie Du Système Nerveux de l'Homme et Des Vertébrés, Maloine, Paris (1911), https://en.wikipedia.org/wiki/Retina#/media/File:Retina-diagram.svg. Accessed April 2018
5. BenRG, https://en.wikipedia.org/wiki/CIE_1931_color_space#/media/File:CIE1931xy_blank. svg. Accessed April 2018

Chapter 12
Geometric Optics

Abstract In this chapter, light is considered as rectilinear rays radiated in all directions from a point at an object. The rays are refracted at air/glass interfaces according to the laws of refraction discussed in a previous chapter. We show that all rays emerging from a point object entering a spherical interface will collect (and use equal time) to a focal point to a good approximation for common conditions. From this treatment, the "lens makers' formula" and "lens formula" are derived, and simple but very useful "ray optics" rules are established. We show how convex and concave lenses can be combined for many different purposes, e.g. for making telescopes, microscopes and loupe (magnifying glass). Light-collecting efficiency, aperture, f-number, depth of view, as well as image quality are discussed. The chapter concludes with a description of the optics of the human eye, defines "near point" and "far point", and shows how spectacles can be used to improve vision in common cases.

12.1 Light Rays

In this chapter, we will see how lenses can be used to make a camera, telescope and microscope, even glasses (spectacles). We will use the term "rays" in the sense of a thin bundle of light that proceeds along a straight line in the air and other homogeneous media. Figure 12.1 illustrates our view of how light rays behave.

The concept of light rays is at variance with the notion that light may be regarded as energy transported by plane electromagnetic waves. Plane waves, in principle, have infinite extent across the direction of propagation of wave. We can reasonably expect that when we make a narrow light beam, it would behave much like a plane wave within the expanse of the beam. A laser beam apparently behaves exactly as we expect light rays to behave.

However, we will see in Chap. 13 that a laser beam does not travel in straight lines all the time. Diffraction (will be treated in a later chapter) can lead to unexpected results. In addition, there is a huge difference between electromagnetic waves in a laser beam and the light from sunlight or lamplight. The laser light usually has a fixed polarization that we find everywhere along the beam, and the intensity is quite stable

© Springer Nature Switzerland AG 2018

A. I. Vistnes, *Physics of Oscillations and Waves*, Undergraduate Texts in Physics,
https://doi.org/10.1007/978-3-319-72314-3_12

Fig. 12.1 "Light ray" is a useful term in geometric optics, although it is useless in certain other contexts

and varies in a well-defined way across the beam of light. The light from the sun or lamps is far more chaotic in all respects and is much more difficult to describe than laser light. We will return to this in Chap. 15 when we come to treat the phenomenon of "coherence".

Despite the fact that light from the sun and from lamps is very complicated, it appears that the phrase "ray of light" is serviceable and useful as long as we work with lenses and mirrors which are almost flat surfaces over areas which are "several times" (at least ten?) the wavelength. Furthermore, the thickness of the interface between, e.g. air and glass, must be very thin in comparison with the wavelength. You may recognize the criteria we set up in Chap. 10 when we looked at the rules for reflection and transmission of electromagnetic waves at interfaces between two media.

This means, among other things, that Snel's refraction law will apply locally to any area on ordinary lenses. Even for a 5 mm Ø lens, the diameter is 10,000 wavelengths, and most lenses are even larger.

When it comes to water drops, the situation is totally different. For ordinary large water drops, we can apply reflection and refraction laws to calculate, for example, the appearance of the rainbow, but for very small water drops the approximation totally breaks. Then we have to return to Maxwell's equations with curved interfaces, and the calculations will become extremely extensive. The spread of light from such small drops is called Mie scattering. Only after the arrival of powerful computers have we been able to make good calculations for Mie scattering. Prior to that, Mie scattering was seen only as an academic curiosity.

Before we begin to discuss the concept of light rays, I want to point out a challenge. We have previously seen that Maxwell's equations, together with energy conservation, provide the magnitude and direction of reflected and transmitted electric fields after plane electromagnetic waves reach a plane interface between two different dielectric media. From Maxwell's equations, both reflection law and Snel's law of refraction follow. However, we have also seen that both these laws can be derived from the principle of minimum time (Fermat's principle), or more correctly, the principle that the time the light uses along its path has an extreme value. In other words, we have two different explanations. What should we count as more fundamental? It is not particularly helpful to say this, but here is something for you to ponder over.

If we have a light source that emits light in all directions, we can see that the light "chooses" to follow the path that takes the shortest time if we select beforehand the position of the light source and the endpoint. However, if we send a well-defined laser beam in a given direction towards the interface, the beam will be bent in a direction given by Snel's law (derived from Maxwell's equations). If we have chosen an endpoint that does not lie along the refracted ray, the light will not reach the endpoint chosen by us. We must change the incoming beam until the refracted beam reaches the endpoint.

With such a description, the criterion of the shortest possible time from the initial to the final point is rather meaningless. The direction of the refracted beam is fully determined by that of the incident beam. Nevertheless, it remains true that if we choose an endpoint somewhere along the refracted beam, the light path represents the route by following which light uses the least time, but that is somehow an extra bonus, not the primary gain. In geometric optics, in other words, we do not invoke Fermat's principle, but the time aspect still appears in a somewhat related way. [We will come back to contemplation of this type when we discuss diffraction, because then we will also learn about Richard Feynman's thoughts on the foundations of quantum electrodynamics (QED).]

12.2 Light Through a Curved Surface

Imagine a glass sphere in the air and, at a short distance, a luminous point that emits light in all directions (at least the light falls on the sphere). We will now investigate how different light rays envisaged to be emanating from the luminous spot will proceed when they hit different points on the surface of the glass sphere.

In Fig. 12.2 P is the luminous point, and we have chosen a section where both this point and the centre of the sphere C lie. A light ray from P following the line between P and C will hit the spherical surface normally. The part of the light transmitted will travel straight on and continue along the extension of the line PC.

We then choose a ray of light that hits the spherical surface at a point A in the plane we consider. The line CA and its extension will then define the incidence normal, and the incidence plane and the emergence plane are in the plane we consider. The beam will *locally* appear to hit a flat surface, and the usual Snel refraction law applies. The

Fig. 12.2 Light rays from a
luminous point (object) at P
will form an "image" at the
point P'. See the text for
details

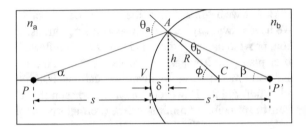

refracted ray is given a certain direction, it will cross the first light beam (which went
through the centre) at the point P'.

We need some geometry to determine where the intersection point P' is located.
Let R be the radius of the sphere and C its centre. The line that goes through P, C
and P' will be called the optical axis. The point V, where the optical axis intersects
the spherical surface, is called the vertex. The distance from P to V is denoted by s
and that from V to P' by s'. The vertical distance from the point A and the optical
axis is denoted by h, and δ denotes the distance between the vertex V and the point
where the normal from A meets the optical axis. The symbols for various angles are
indicated in the figure.

To make the calculations as general as possible, we will denote by n_a the refractive
index of light in the medium where the light source is located (left in the figure), and
n_b will signify the refractive index of the sphere (to the right of the figure). We will
assume that $n_b > n_a$.

Snel's law gives:

$$n_a \sin \theta_a = n_b \sin \theta_b .$$

We also have:

$$\tan \alpha = \frac{h}{s + \delta}, \qquad \tan \beta = \frac{h}{s' - \delta}, \qquad \tan \phi = \frac{h}{R - \delta} .$$

Furthermore, we know that an exterior angle of a triangle is equal to the sum of the
opposite interior angles:

$$\theta_a = \alpha + \phi, \qquad \phi = \beta + \theta_b . \tag{12.1}$$

We will avail ourselves of a simplification that is frequently made in dealing with
geometrical optics, namely the so-called *paraxial approximation*. This means that
we confine ourselves to situations wherein the angles α and β are so small that both
sine and tangent can be replaced by the angle itself (in radians). Under the same
approximation, δ will be small compared to s, s' and R. The equations above then
take the simplified forms shown below:

$$n_a \theta_a = n_b \theta_b \tag{12.2}$$

and

$$\alpha = \frac{h}{s}, \qquad \beta = \frac{h}{s'}, \qquad \phi = \frac{h}{R} . \qquad (12.3)$$

Upon combining the first equality in Eq. (12.1) with Eq. (12.2), we obtain:

$$n_a \alpha + n_a \phi = n_b \theta_b .$$

Using the second equality in Eq. (12.1), one finds:

$$n_a \alpha + n_a \phi = n_b \phi - n_b \beta$$

which can be transformed into:

$$n_a \alpha + n_b \beta = \phi(n_b - n_a)$$

Upon inserting the expressions for α, β and ϕ from Eq. (12.3) and cancelling the common factor h, one finally gets:

$$\frac{n_a}{s} + \frac{n_b}{s'} = \frac{n_b - n_a}{R} . \qquad (12.4)$$

This formula is quite important. It should be observed that the relationship applies independently of the angle α so long as the paraxial approximation holds (small angles). *All* light rays from a light source that makes a small angle with the optical axis will cross the optical axis at the point P'. The luminous point P is called the *object point*, and the intersection point P' is called the *image point*.

So far so good. But what of it? Is it something special that light rays cross each other? Will the different light bundles together give a special result, or might a light beam extinguish another, or that nothing extraordinary will happen at the crossing point anyway?

Figure 12.3 shows the same two light beams as in the previous figure, but we have now turned our attention to something else, namely the *time* light uses in going from the object point to the image point. For the straight ray that goes along the optical axis, the velocity of light will be c/n_a up to the vertex, and c/n_b afterwards. The velocity in glass is smaller. Similar reasoning applies to the ray that goes in the other direction (via A). We see that the distance from the object point to A is longer by γ than that from the object point to the vertex. Thus, light will take a longer time to cover the path PA than PV. On the other hand, we see that the distance AP' is shorter than VP' by an amount equalling ε. We notice that $\varepsilon < \gamma$. If we carry out a thorough analysis (not attempted here), we will find that the *time* light uses for

Fig. 12.3 How long will
two light beams take in
going from a luminous point
at P to an "image" point at
P'? Note the two dashed
circle sectors with centre in
P and P', respectively. See
the text for details

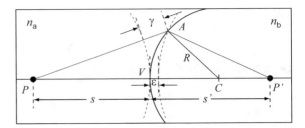

covering the distance γ in air equals the time it takes for travelling the distance ε in glass (but this is true only if the paraxial approximation holds).

> In other words, the light uses the same time from the light source (the object) to the intersection point (image) regardless of the direction of the light ray (within the paraxial approach). Since the light has the same frequency, regardless of whether it is air or glass, this means that there are exactly as many wavelengths along a refracted ray as along the straight one. Consequently, the light coming to the intersection will always be in phase with each other. Consequently, their amplitudes are added. If we could place a screen across the optical axis at P', we would be able to confirm this by seeing a bright spot just there. The word "image" can be used because we can form a real picture of the light source at this place.

12.3 Lens Makers' Formula

In the previous section, we saw how the light rays from a light source (object point) outside a glass sphere converged at the image point inside the sphere. However, such a system is of rather limited interest. We will now see how we can put together two curved surfaces, for example, from air to glass, and then from glass back to air, to obtain rules applicable to lenses. Suppose we have an arrangement as indicated in Fig. 12.4. To find out how Eq. (12.4) is used, we choose to operate with three different refractive indices and let the lens be "thin"; that is, the thickness of the lens is small compared to the distances from the object and image as well the radii of the two interfaces. Under these conditions (and still within the paraxial approach), we get:

$$\frac{n_a}{s_1} + \frac{n_b}{s_1'} = \frac{n_b - n_a}{R_1} \, ,$$

$$\frac{n_b}{s_2} + \frac{n_c}{s_2'} = \frac{n_c - n_b}{R_2} \, .$$

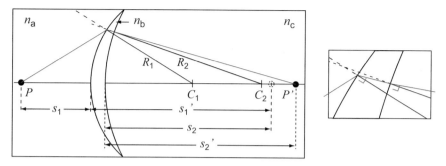

Fig. 12.4 A lens may be composed of two curved interfaces between air and glass. The image of P, if we had only the first interface, is marked with a dashed circle. *To the right*, detailed ray paths through the lens, first with refraction towards the normal (air to glass) and then away from the normal (glass to air). See the text for details

For a glass lens in air, $n_a = n_c = 1$ and $n_b = n$. Furthermore, the image point of other boundaries will be opposite to what we used in deriving Eq. (12.4). It can be shown that we can implement this in our equations by setting $s_2 = -s'_1$. We then make an approximation by ignoring the thickness of the lens; i.e. the lens is supposed to be "thin". Consequently, the equation pair above can be written as:

$$\frac{1}{s_1} + \frac{n}{s'_1} = \frac{n-1}{R_1}$$

$$-\frac{n}{s'_1} + \frac{1}{s'_2} = \frac{1-n}{R_2} \ .$$

Addition of the two equations gives:

$$\frac{1}{s_1} + \frac{1}{s'_2} = (n-1)\left(\frac{1}{R_1} - \frac{1}{R_2}\right) \ .$$

If the lens is regarded as an excessively thin element, it is natural to talk about object distance and image distance relative to the (centre of the) lens itself, instead of working with distances from the surfaces. We are then led to the equation that is called the *lens makers' formula*

$$\frac{1}{s} + \frac{1}{s'} = (n-1)\left(\frac{1}{R_1} - \frac{1}{R_2}\right) \ . \tag{12.5}$$

A special case is that when the object point is "infinitely far away" (s' very much larger than the radii R_1 and R_2). In this case, $1/s \approx 0$, and

Fig. 12.5 Section through a variety of lens shapes. From left to right: positive meniscus, planoconvex and biconvex; negative meniscus, planoconcave and biconcave

$$\frac{1}{s'} = (n-1)\left(\frac{1}{R_1} - \frac{1}{R_2}\right).$$

The image distance for this special case (when the object point is "infinitely far away") is called the "focal distance" or "focal length" of the lens and denoted by the symbol f. The image is then located at the *focal point* of the lens (one focal length from the centre of the lens). With the given definition of focal length f, we end up with:

$$\frac{1}{s} + \frac{1}{s'} = \frac{1}{f} \qquad\qquad (12.6)$$

where the focal distance f is defined as

$$\frac{1}{f} = (n-1)\left(\frac{1}{R_1} - \frac{1}{R_2}\right). \qquad\qquad (12.7)$$

The first of these formulas, called the **lens formula**, will be used in the rest of this chapter.

Before going further, let us see how lenses will look for different choices of R_1 and R_2 in the lens makers' formula. These radii can be positive or negative, finite or infinite. Different variants are given in Fig. 12.5.

Lenses with the largest thickness at the optical axis are called *convex*, whereas those with smallest thickness at the optical axis are called *concave*.

We pause to remind ourselves of what has been done so far.

The above derivations involved a number of approximations, and the ensuing formulas are necessarily approximate. This is typical of geometrical optics. The simple formulas are only approximate, and all calculations based on them are so elementary that they could easily have been done in high school. One might feel that, at university level, one would be able to deal with more complicated problems

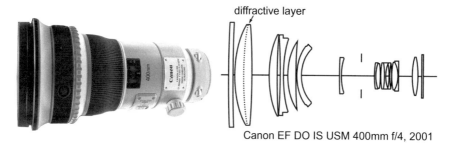

Canon EF DO IS USM 400mm f/4, 2001

Fig. 12.6 Example of a modern lens for photography: Canon EF 400 mm f/4 DO IS USM objective. Instead of a simple thin lens, as we think of an objective in our treatment of geometrical optics, the Canon lens has 12 groups with a total of 18 elements (lenses) that function together as one. In modern camera lenses, most elements have spherical surfaces, but some have aspherical surfaces. Some elements are made of extra dispersive glass; that is, the refractive index has a different variation with the wavelength than in other most common glass types. Photograph: GodeNehler, Wikimedia Commons, CC BY-SA 4.0, Modified from original [1]. Drawing: Paul Chin (paul1513), GNU Free Documentation License, CC BY-SA 3.0, [2].

and reach more accurate descriptions, but the advanced problems are in fact too complicated to be appropriate for a general book like this. Today, numerical methods are used for the more advanced calculations. It has been found that making a perfect lens is not possible. We have to make a trade-off, and a lens to be used mostly at short distances will have to be designed in a different way than a lens that is meant primarily for long distances.

We have based our treatment on spherical interfaces. This is because until recently it was much easier to fabricate lenses with spherical surfaces than with other shapes. In recent years, it has become more common to fabricate lenses with aspherical shapes, and then, the problem arising from the paraxial approximation is made less severe. We can reduce the so-called spherical aberration by designing aspherical surfaces.

Looking back, we see that Eq. (12.4) contains the refractive indices. Now, we know that the refractive index depends on the wavelength, and this means that the image point P' will have a different position for red light than for blue light. Using multiple lenses with different types of glass (different refractive indices), we can partly compensate for this type of error (called chromatic aberration). Overall, however, it is a very challenging task to make a good lens (Fig. 12.6). It is not hard to understand why enthusiasts are examining new objectives from Nikon, Canon, Leitz, etc., with great interest just after they appear on the market. Have the specialists been able to make something special this time, and if so, in what respect? The perfect lens does not exist!

12.4   Light Ray Optics

We are now going to jump into that part of optics which deals with glasses (spectacles), cameras, loupes (jeweller's magnifying glasses), binoculars, microscopes, etc. The term "ray optics" is distinct from "beam optics", which focuses more on how a laser beam changes with distance (where diffraction is absolutely essential).

There are three main rules to which we will appeal continually.

1. For a *convex* lens, incoming light parallel to the optical axis will go through the focal point after the light has gone through the lens. For a *concave* lens, incoming light parallel to the optical axis will be refracted away from the optical axis after the light has gone through the lens. The direction is such that the light beam appears to come from the focal point on the opposite side of the lens (focal point on the same side as the object).
2. Light passing through the centre of the lens (where the optical axis intersects the lens) will move in the direction along which it came in.
3. Rays passing the front focal point of a *convex* lens will go parallel to the optical axis *after* the lens. For *concave* lenses, rays entering the lens along a direction that passes the rear (or back) focal point will continue parallel to the optical axis *after* the lens.
4. Rule 3 is identical to Rule 1 if we imagine that the ray is travelling in a direction opposite to the actual direction, for both convex and concave lenses.

Light rays drawn according to rule 1, 2 and 3 are coloured red, blue and green, respectively, in a number of the remaining figures in this chapter.

The rules originate in part from the lens formula. We saw that if the object distance s was made infinitely large, the image would be at a distance equal to the focal length of the lens. If multiple light rays are drawn from the object in this case, the rays will enter (approximately) parallel to the optical axis, and all such rays shall pass through the focal point. Whence follows the first rule.

However, the lens formula can be "read" in either direction, so to say. If we place a very small light source on the optical axis at a distance equal to the focal length in front of the lens, the light from the source will go to the lens in many different directions, but the image will then be at a distance of $s' = \infty$. This means that, irrespective of where they go through the lens, the rays will continue almost parallel to the optical axis after the lens.

The middle rule may be even easier to understand. At the centre of the lens (where the optical axis cuts through the lens), the two surfaces are approximately parallel. If a beam of light is transmitted through a piece of plane glass, the light beam will be refracted at the first interface, but refracted back to the original direction as it passes through the other interface. The emergent ray will be slightly offset relative to the

Fig. 12.7 A luminous object point not on the optical axis will be imaged at a point opposite to a convex lens. The image is not on the optical axis. Three guides are used to find the location of the pixel

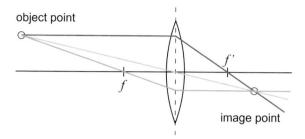

incoming light beam, but if the angle is not too large and the lens thin, the parallel offset will be so small that we can neglect it in our calculations.

Object beyond the focal point

If these rules are applied to a luminous object point that is not on the optical axis, we get the result illustrated in Fig. 12.7. The three special light rays, specified by our general rules, meet exactly in one image point. The image point is on the opposite side of the optical axis in relation to the object point.

If the object is no longer a luminous point, but an extended body, such as an arrow, we find something interesting (see Fig. 12.8). From each point of the body, light is emitted, and for each point in the object, there is a corresponding image point. Our simplified rules of ray optics imply that, for all points in the object lying in a plane perpendicular to the optical axis, the corresponding image points will lie in a plane perpendicular to the optical axis on the opposite side of the lens (under conditions indicated in the figure). That is, we can image an object (such as the front page of a newspaper) into an image that can be captured on a screen. The image will then be a true copy of the object (newspaper page), except that it will have a magnification or reduction compared with the original, and the image will be upside down (but not mirrored).

The magnification is simply dependent on s and s'. If $s = s'$, the object and image will be of the same size. If $s' > s$, the image will be larger than the object (original), and vice versa. The *linear magnification* or *real magnification* is simply given by:

$$M = -\frac{s'}{s} \, .$$

The minus sign is included only to indicate that the image is upside down in relation to the object.

It is also possible to define a magnification in area. In that case, the square of the expression will be given. (The minus sign is then often irrelevant.)

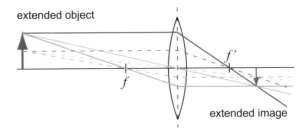

extended object

f'

f

extended image

Fig. 12.8 An extended object may be thought of as a plurality of object points, and each point is imaged in a corresponding point on the opposite side of a convex lens. As a result, the object as such is depicted as an image. The image is upside down and has a different size than the object

Object within the focal distance?

So far, there have been relatively easily comprehensible connections between object and image, and we have been able to capture the image on a screen and see it there. But what would happen if we place the object closer to the lens than the focal length? Figure 12.9 shows how the three reference rays now go. They diverge after traversing the lens! There is no point where the light rays meets and where we can collect the light and look at it. On the other hand, the rays of light appear to come from one and the same point, a point *on the same side of the lens as the object*, but at a different place.

> In cases like this, we still talk about an image, but refer to it as a *"virtual image"* as opposed to a "real image" like that discussed earlier. A virtual image cannot be collected on a screen. On the other hand, we can look at the virtual image if we bring in another lens in such a way that the whole thing, after going through the new lens, creates a real image.

For example, if we look at the light coming through the lens in Fig. 12.9, using our eyes, our eye lens will gather the light to project a real image on the retina. Then we see the image. The image is formed on the retina as a result of the light from the object passing through the free-standing lens and subsequently through the eye lens.

However, we can get exactly the same image on the retina if we remove the outer lens and replace the real object with an imaginary magnified object placed as indicated by the "virtual image" in the figure. This is why we speak of a "virtual" image.

Concave lens

Using a concave lens alone, we cannot form a real image for any position of the object (see Fig. 12.10). Concave lenses on their own always provide virtual images, and it is somewhat unusual and demanding to work with ray diagrams for concave lenses. If we decide to use the lens formula, we say that the focal length is negative for concave lenses. We also need to operate with negative object distances and negative image

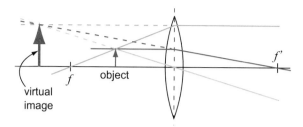

Fig. 12.9 When an extended object is placed within the focal length of a convex lens, no image is formed on the opposite side of the lens. On the contrary, the dashed lines indicate that the object and lens appear to be replaced by an enlarged object on the same side of the lens as the real object. This apparently enlarged object is called a virtual image

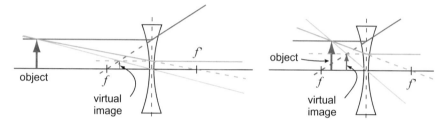

Fig. 12.10 A concave lens alone can never form a real image. If we consider an object through a concave lens, the virtual image looks smaller than the real object

distances depending on whether or not the object and/or image are on the "regular" side of the lens. There are a set of rules for how to treat s, s' and f in the formula for all combinations of cases.

Note that in Fig. 12.10 the object is outside (inside) the focal point in the figure on the left (right). There is no significant difference in the image formation by a concave lens when the object distance is equal to the focal length. This is contrary to what happens with a convex lens.

It is strongly recommended that you study *all* details in how the light rays are drawn in Figs. 12.9 and 12.10 and compare them with the rules in the first part of Sect. 12.4. You will need this for understanding later figures.

12.4.1 Sign Rules for the Lens Formula

The lens makers' formula and lens formula can be used for both convex and concave lenses and mirrors, but sometimes we have to operate with negative values for positions, radii of curvature radii and focal lengths for the formulas to function.

The rules for light coming against lenses or mirrors are as follows:

- Object distance $s > 0$ if the object is a real object; otherwise, $s < 0$.
- Image distance $s' > 0$ if the image is real (real rays meet in the image), $s' < 0$ otherwise.
- Focal distance $f > 0$ for convex lenses, $f < 0$ for concave lenses.
- Focal distance $f > 0$ for concave mirror, $f < 0$ for convex mirror.

In addition, the following convention applies:

- Magnification m is taken to be positive when the image has the same direction as the object, $m < 0$ when the image is upside down.

It is nice to have these rules for signs, but experience shows that they sometimes are more confusing than useful. For that reason, some people choose to decide the signs by drawing the ray diagram, obtaining an approximate value for the image distance relative to object distance, and checking whether the image is real or imaginary. Thereby the sign comes out on its own. The procedure nevertheless means that we know the rules for the focal distance for convex and concave lenses and mirrors.

Slavish use of the lens formula and sign rules without simultaneous drawings based on ray optics will almost certainly lead to silly errors sooner or later!

12.5 Description of Wavefront

Ray optics is useful for finding the size and position of the image of an object after light has passed through a lens. However, we initially mentioned in this chapter that we usually describe light as electromagnetic waves and that the concept of "light rays" does not square with a wave description.

However, it is relatively easy to go from a wave description to a ray description. The link between them is the fact that when a wave propagates, it moves perpendicular to any wavefront. It may therefore be interesting to see how wavefronts of light from a source develop as they pass through a lens.

In Fig. 12.11, we have drawn a light source as a luminous point that emits spherical waves of a definite wavelength. The wavelength of light is very small in relation to the lens size, so the distance between the drawn wavefronts is in the range of a few hundred wavelengths.

In Fig. 12.11a, we have chosen to place the object point on the optical axis at a distance of $2f$ from the lens centre plane, where f is the focal length of the lens. The wavefront hits the lens and continues through the lens and continues after the light has passed the lens. The wavefront must be continuous, and since the light goes at a lower speed through the glass than in air, the wavelength inside the glass is less than in air. The wavefronts lie closer together.

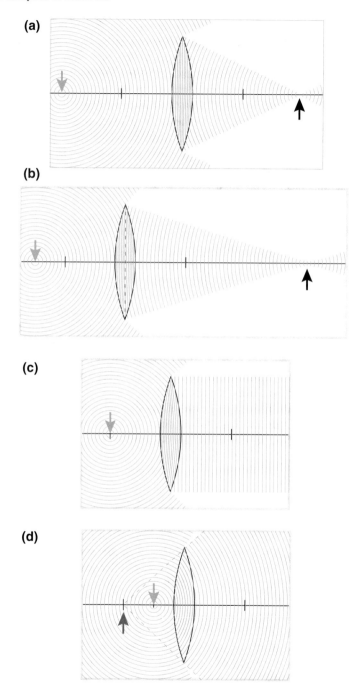

Fig. 12.11 An object point (vertical orange arrow) emits light in all directions as spherical wavefronts. The figure shows how the wavefront changes as they pass through a biconvex lens when the object distance changes. See text for details

In this case, the image point according to the lens formula will be a distance of $2f$ behind the lens. Then the wavefront will appear as shown in the figure, and we also see that the actual light beam narrows and gets a minimum value just at the image point (marked with black arrow). In this case, wavefront within the lens itself is flat, but it is also the only region of the wavefront that is flat in this case.

In Fig. 12.11b, the object point is located at a distance of $\frac{3}{2}f$ from the lens. Now the curvature of the wavefront is larger (the radius of curvature is small) than in the previous case, and the curved glass surface does not manage to align the wavefront so that they get flat inside the glass. After the light has passed the lens, the wavefront has too little curvature (too large radius of curvature) to form the image at a distance of $2f$ behind the lens. According to the formula, the image is now at a distance of $3f$.

In Fig. 12.11c, the object point is located at a distance of f from the lens. The curvature of the wavefront that hits the lens is now so great that the lens as a whole does not manage to gather the light at an image point on the opposite side. In this case, the wavefront becomes flat after the light has passed the lens. The light continues as a cylindrical light beam with unchanged diameter.

In Fig. 12.11d, the object point is located at a distance of $f/2$ from the lens. Now the wavefront of the light after it has passed the lens will be curved the opposite of what we had in the first two cases. There is no image point at which the light is gathered. On the other hand, we see that the wavefront after the light has passed the lens has a curvature that corresponds to the lens removed and that we had put the object at a distance f in *front* of the lens (marked with red arrow). It is this geometry we previously referred to as the "virtual" image.

Since waves in principle can go as well as backwards, Fig. 12.11 can be used to some extent also for light moving the opposite way. However, there are significant differences in the regions in which the light is located.

12.6 Optical Instruments

Several lens combinations were used for making optical instruments in the early 1600s. The telescope opened the heavens to Galilei, and he could see the four largest moons of Jupiter, an observation which had a decisive influence on the development of our world view. The microscope (only a single-lens microscope at that time) enabled Leeuwenhoek to see bacteria and cells and paved the way for a rapid development and increased understanding of biological systems. Optical instruments have played an important role and still have a huge impact on our exploration of nature.

We will presently look at how we can build a telescope and microscope using two lenses. First of all, however, we will take on a simple lens used as loupe, since this construction is classically included in both telescope and microscope.

12.6.1 Loupe

The simplest version of a loupe ("magnifying glass") is a simple convex lens. The ray path of a loupe is somewhat different from what we have indicated in the figures so far.

An object is placed in Fig. 12.12a one focal length away from a convex lens (loupe). The light from an object on the optical axis (red) will appear, after passing the lens, as rays parallel to the optical axis (plane wavefront perpendicular to the optical axis). Light from a point in the object, which is at a distance d from optical axis (green), will also appear almost like parallel rays, but now at an angle θ with the optical axis.

If we place the eye somewhere behind the loupe, the eye lens will form an image of the object on the retina. Since the light rays coming into the eye (from each object point) are almost parallel (the wavefront is approximately plane), the eye must adjust the eye lens as if one is looking at an object far away. The focal length of the eye lens is then the distance between the eye lens and the retina (see later).

We often use a loupe to get an enlarged image of an item we can get close to. The best image we can achieve without a loupe is obtained when the object is as close to the eye as possible without sacrificing the sharpness of the image (see Fig. 12.12b). This distance to the eye is a limiting value s_{min}. A "normal eye" (see Sect. 12.8) cannot

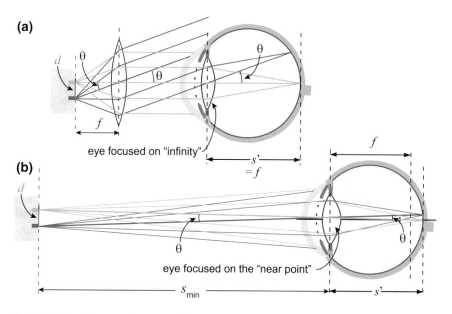

Fig. 12.12 Ray diagram when we consider an object with and without a loupe. In part **a**, an object is placed in the focal plane of a convex lens used as a loupe. The rays are caught by the eye and form a real image on the retina. Part **b** shows the ray path when we keep the object as close to the eye as possible, while at the same time maintaining a sharp image. See text for further information

focus on objects closer than about 25 cm. Therefore, we choose to set $s_{min} = 25$ cm. The curvature of the eye lens is maximized and the focal length f in this case is slightly shorter than s'. Thus, also in this case s' fits with the size of the eye.

We see from Fig. 12.12 that the image *on the retina* becomes larger when we use a loupe as compared to looking at the object with the unaided eye. The loupe gives us a magnification.

Since the size of the eye itself is unchanged in the two cases, the size of the exposed part of retina will be proportional to the *angle* θ between the incident red and green light rays in Fig. 12.12.

Let θ be the angle subtended by light rays from two object points (as judged from the centre of the eye lens). In Fig. 12.12, these points are shown in red and green. The magnification of a loupe is defined as the ratio of the tangent of the subtended angle when the light reaches the eye via a loupe and the tangent of the angle when the light comes directly (without the loupe) from the object when it is at a distance $s_{min} = 25$ cm. We often refer to this as "*angular magnification*", as opposed to magnification we have mentioned earlier, where we found the ratio of the magnitudes of an image to the object. For a loupe with a focal length f, the angular magnification becomes:

$$M = \frac{d/f}{d/s_{min}} = \frac{s_{min}}{f} .$$

Here d is the distance between the two selected points on the object perpendicular to the viewing direction (the distance between red and green object points in Fig. 12.12). The focal length of the loupe is f.

A loupe with a focal length of 5 cm will then have an magnification of 25 cm/5 cm = 5. We usually write 5 X (fivefold magnification). Note that the magnification here is positive because the image we see through the loupe is upright in relation to the object (the image does not turn upside down).

In short, we can say that the loupe has only the function that the object can be moved closer to our eye than can be achieved without the loupe. The effective distance is simply the focal length. If we have a loupe with a focal length of 2.5 cm, we will examine a butterfly wing at an effective distance of 2.5 cm instead of having to move the butterfly 25 cm away from the eye to get a sharp image. The result is a real image on the retina that is about ten times as large as that seen without the loupe.

In a microscope or telescope, a loupe is used with another lens (an objective). Loupes may have focal lengths down to about 3 mm. It automatically gives an almost 100-fold magnification compared to whether we had not used the loupe.

Note that the distance between the loupe and the eye has no bearing on the distance between the red and the green image points on the retina in Fig. 12.12a. However, if the eye is drawn too far away from the loupe, we will not be able to see both the red and the green points at the same time. The field of vision is thus greatest when the eye is closest to the loupe, but the magnification is independent of the distance between the loupe and the eye.

12.6.2 The Telescope

A telescope consists of at least two lenses (or at least one curved mirror and one lens). The lens (or mirror) closest to the object (along the light path) is called *objective*, whereas the lens closest to the eye is called *eyepiece* or *ocular*. The purpose of the objective is to create a *local image* of the object (in a way moving the object much closer to us than it actually is). The eyepiece is used as a loupe to view the local image.

Although the local image is almost always *much* smaller than the object, it is also much closer to the eye than the object itself. Once we can use a loupe when viewing the local image, we can get an (angular) magnification of up to several hundred times. However, a regular prismatic binocular has a limited magnitude of about 5–10 X. Bigger binoculars require a steady stand so that the image does not flitter annoyingly in the field of view.

Figure 12.13 shows a schematic illustration of the optical arrangement for a telescope. We use the default selection of light rays from the object's largest angular distance from the optical axis (from the top of the arrow). Points in the object on the optical axis will be imaged on optical axis, and usually these lines are not shown.

We notice that the objective provides a real inverted image a little further away from the lens than the focal plane. Objects that are very far away will be depicted

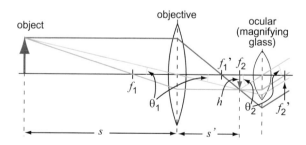

Fig. 12.13 For a telescope, the object is far away compared to the focal length. The lens creates a real, diminished "local" image just behind the focal length. This picture is then considered with a loupe. Total magnification is measured as angular magnification to the object viewed through binoculars compared to no binoculars. The picture in this type of telescope appears inverted (upside down)

fairly close to the focal plane. Objects that are closer fall further and beyond the focal point of the lens.

The eyepiece is positioned so that the image from the lens falls into the eyepiece's focal plane. Then all the light rays from a selected point in the object, after they have gone through the eyepiece, will appear parallel. The eye will focus on infinity and a real image will be formed on the retina.

Magnification

The magnification of the telescope is given as the ratio of the tangents of the angles between the optical axis and the light rays passing through the centre of the lenses.

From Fig. 12.13, we see that the angular magnification can be defined as:

$$M = -\frac{\tan \theta_2}{\tan \theta_1} = -\frac{h/f_2}{h/s'} \ .$$

In other words, the magnification varies with the distance from the objective to the real local image (between the objective and ocular). This distance will vary depending on how close the object is to the objective. It is more appropriate to specify the magnification as a number. It is achieved by selecting the magnification when the object is infinitely far away. Then s is infinite and s' becomes equal to the focal length of the lens f_1. The magnification can then be written as:

$$M = -\frac{h/f_2}{h/f_1} = -\frac{f_1}{f_2} \ .$$

In other words, the angle magnification equals the ratio between the focal lengths of the lens and the eyepiece.

For a telescope with focal length $f_1 = 820\,\text{mm}$ and eyepiece with focal length $f_2 = 15\,\text{mm}$, the angular magnification becomes:

$$M = \frac{820}{15} = 54.7 \approx 55\,\text{X} \ .$$

Note that since there are so many approximations made in the simple variant of geometrical optics that there is no point in specifying magnification with more than two significant figures.

Eyepiece projection *

Before leaving the telescope, we will mention a useful small detail. It is good to look through a telescope or a microscope, but today we often want to record the observed details in such a way that others may also see them. Since the eyepiece usually works like a loupe, we cannot capture a real image by placing, for example, a CMOS image

Fig. 12.14 In the case of ocular projection, the eyepiece is not used as a loupe, but as a second imaging lens. See text for details

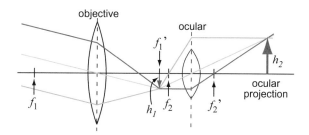

sensor (photo chip) or an old-fashioned photographic film somewhere behind the loupe. One possibility is to remove the entire eyepiece and place the CMOS sensor exactly where the real image is formed. It is quite common today. The CMOS chip may be the internal chip in a digital single-lens reflex camera (DSLR camera) with interchangeable lenses. We then remove the regular camera lens and use the telescope lens instead.

Let us take an example: we have a telescope with an 820 mm focal length objective. Suppose we want to take pictures of the moon. The angular diameter of the moon is about half a degree. The size of the real image formed by a telescope lens will then be:

$$h = 820 \text{ mm} \times \tan(0.5°) = 7.16 \text{ mm} .$$

If the CMOS chip is 24 mm wide, the moon will cover $7.2/24 = 0.3$ of this dimension. The image of the entire moon has a diameter of only 30% of the smallest dimension of the image ("height"). It will be impossible to get nice details of the moon surface even though the exposure of the image is optimal.

Is there a possibility of enlarging (blowing up) the moon image on the CMOS chip? Yes, it is possible by use of "ocular projection".

The principle is quite simple. Normally the eyepiece is used as a loupe, and then, the image from the objective is placed at the focus of the eyepiece. If we push the eyepiece further away from the objective, the real image will be outside the focal plane, and then, the eyepiece can actually create a new real image using the first real image as its own object. Figure 12.14 shows the principle. In order for the new real image to be larger than the first real image, the eyepiece must be pushed only *slightly* farther from the objective than its normal position. In principle, we can get as large a real image as we want, but the distance from the eyepiece to this last real image is in proportion to the size of the image. We must then have a suitable holder to keep the CMOS piece a proper distance behind the eyepiece. With such a technique, we can easily take pictures of details on the surface of the moon with the 820 mm focal length telescope.

However, there is a catch in the method. The lens captures as much light regardless of whether or not we use eyepiece projection. When the light is spread over a larger surface, it means that the brightness per pixel on the CMOS chip decreases. The exposure must then take place over a longer period of time to get a useful image.

It should also be added that eyepieces are usually optimized for normal use. Lens defects may show up in eyepiece projections that would otherwise go unnoticed.

Ocular projection can also be used in microscopy, and the method is very useful in special cases.

12.6.3   Reflecting Telescope

Large astronomical telescopes usually use curved mirrors as objectives. The main reason for this is that reflection laws for a mirror are not wavelength-dependent. Long wavelength light behaves approximately the same as short wavelength light, which eliminates the chromatic deviation due to the wavelength dependence of the refractive index of glass.

As with lenses, it is easiest to make curved mirrors when the surface is spherical. However, this shape is not good because parallel light rays will be focused at different locations depending on how far from the axis the light rays come in. Mathematics shows that it would be far better to choose a surface that has the shape of a paraboloid. In the left part of Fig. 12.15, there are examples of three different light rays coming against a parabolic mirror parallel to the optical axis. The rays are reflected according to the reflection laws and end up in exactly the same point (focal point). A telescope with such a parabolic mirror as an objective can get very sharp images and at the same time very high brightness.

Unfortunately, it is complicated to make telescopes with parabolic surface with the precision needed for light waves (since the wavelength is so small). It is therefore common for low-cost telescopes to use mirrors with spherical surface, but with an *opening* small relative to the *radius* (see the right part of Fig. 12.15). The difference

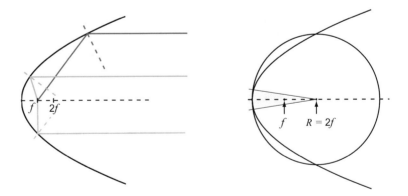

Fig. 12.15 *Left part*: A parabolic mirror ensures that all light rays coming in parallel to the optical axis are focused at the same spot, irrespective of whether the rays are near or farther away from the optical axis. *Right part*: A parabolic mirror and a spherical mirror have the same shape provided that the "opening angle" is small, i.e. the mirror diameter is small compared with the focal length.

Fig. 12.16 Example of construction of image formation for a concave mirror

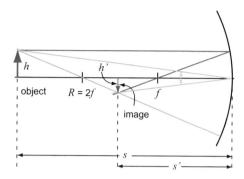

between parabolic and spherical shape is then not so great. Alternatively, we can combine a spherical mirror with a glass correction lens ("Schmidt corrector plate") to get a good overall result at a lower price than if we were to make a near-perfect parabolic mirror.

Construction rules for mirror

We can study image formation by a curved mirror in a manner rather similar to that used for thin lenses. We combine the properties of spherical and parabolic shapes to make the rules as simple as possible and get:
  1. Any incident ray travelling parallel to the optical axis on the way to the mirror will pass through the focal point upon reflection.
  2. For any incident ray that hits the centre of the mirror (where the optical axis intersects the mirror), the ray's incident and reflected angles are identical.
  3. Any incident ray passing through the focal point on the way to the mirror will travel parallel to the principal axis upon reflection.

In Fig. 12.16, the image formation of a concave mirror is shown where the object is slightly beyond twice the focal length. Take note of all the details concerning how the three ray-tracing lines are drawn.

We can use the lens formula also for a mirror, but then be extra careful to consider the sign to get it right.

A concave mirror (concave mirror) will form a real image of the object, provided that the object is placed farther away from the mirror than one focal length.

A problem with a mirror is that the image forms in the same area as the incident light passes through. If we set up a screen to capture the image, it will firstly remove the light that reaches the mirror, and secondly, we get diffraction effects due to the edge between light and shadow (see Chap. 13). There are several tricks to mitigate these drawbacks. One of the classic tricks is to insert an oblique flat mirror for reflecting part of the beam away from the area the light enters (see Fig. 12.17). A telescope of this type is called a *Newtonian reflector*.

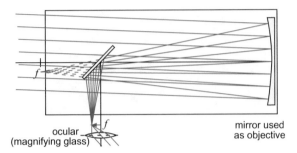

Fig. 12.17 In a Newtonian reflector, a slanting mirror is used to bend the light beam from the main mirror so that we can use an eyepiece and look at the stars without significantly blocking the incoming light. The slanting mirror does remove a small part of this light

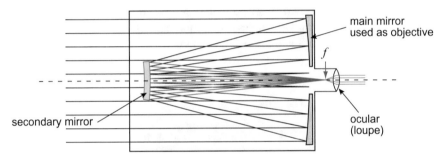

Fig. 12.18 In a Cassegrain telescope, a secondary mirror is used to reflect the light beam from the main mirror back through a circular hole in the main mirror. The secondary mirror does remove a small part of the incoming light. Light rays are drawn in black before the secondary mirror and in red thereafter

Another choice of construction is to use a curved mirror to reflect the light from the main mirror back through a circular hole in the main mirror (see Fig. 12.18). This design, called a *Cassegrain telescope*, makes it possible to make compact telescopes (short compared to their focal length). Schmidt correction plate is often used also for Cassegrain telescopes.

12.6.4 The Microscope

In the telescope, we used the objective to create a local image of the object, and this image was viewed through a loupe. The strategy works well when the object is so far away that we cannot get close to it. It is precisely in such situations that we need a telescope.

When we look at, for example, the cells in a plant stem, we have the object right in front of us. We do not need to create any local image, because we have the original. Then we use another strategy to see an enlarged image. The strategy is really exactly

Fig. 12.19 Ray path in a microscope. The object can be placed arbitrarily close to the focal point of the objective, which, consequently, forms a real enlarged image of the object. This image is then viewed with the eyepiece that acts as a loupe

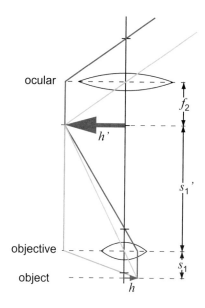

the same as for ocular projection. We place the object just outside the focal point of the lens (which now has a small focal length) to form a real-life image well behind the lens. This enlarged image of the object is then viewed by a loupe. The ray diagram of a microscope is illustrated in Fig. 12.19.

The magnification due to the objective alone is:

$$M_1 = \frac{s_1'}{s_1}.$$

This magnification can, in principle, be arbitrarily large, but then the actual image will move far from the lens, and the microscope would become unmanageable. By using a very short focal length lens, preferably only a few mm, we can achieve a significant magnification even for a tube length (distance between the objective and eyepiece) of 20–30 cm.

In addition, the loupe gives, as always, an (angular) magnification of:

$$M_2 = \frac{25 \text{ cm}}{f_2}.$$

The total magnification of the microscope comes out to be:

$$M_{\text{tot}} = \frac{25 \text{ (cm) } s_1'}{f_2 s_1}.$$

In digital microscopes, a CMOS chip can be put directly in the plane where the real image from the objective is formed. The picture is then viewed on a computer screen and not through an eyepiece (ocular). It is difficult to define a magnification in this case since the digital picture can be displayed in any size.

For an 8 mm objective and a 30 cm tube length, and an eyepiece with focal length 10 mm, the total magnification (expressed in mm in the middle equality) is:

$$M = \frac{25 \text{ (cm) } s_1'}{f_2 s_1} \approx \frac{250 \cdot (300 - 10)}{10 \cdot 8} = 906 \approx 900 \text{ X} \ .$$

Note:
Lately, it has been an extreme revolution within microscopy. By use of advanced optics, different kinds of illumination of the object, use of fluorescent probes, optical filters, scanning techniques, extensive digital image processing, and more, it is today possible to take still pictures and videos of real cells moving around, showing details we a few years ago thought would be impossible to catch by a microscope.

The physical principles behind these methods are exciting and take the wave properties of light into account to the extreme. For the interested reader, we recommend to start with the Wikipedia article on microscopy.

12.7   Optical Quality

12.7.1   Image Quality

Here is a word of warning. If we want to buy a microscope (or telescope for that matter), we can in fact get inexpensive microscopes with the same magnification as that provided by costlier instruments. The magnification itself is really far less important than the image quality. Heretofore, there has been no well-established system for specifying the image quality. Accordingly, there has been room for considerable trickery, and many have bought both microscopes and binoculars, which was no better than throwing money out of the window, because the image quality was too bad. Hitherto, one could rely only going to an optician for buying binoculars and getting a vague subjective sales talk on quality. So frustrating!

Fortunately, this is about to change. The practice of specifying optical quality in terms of measurements based on a "Modulation Transfer Function" (MTF) appears to have gained a firm foothold now. This is primarily a method of determining how sharp and contrasting images we can get. Colour reproduction does not come within the purview of this method.

To put it briefly, the MTF values tell us how close the lines in a black- and white-striped pattern can be, before the differently coloured stripes begin to blend into each

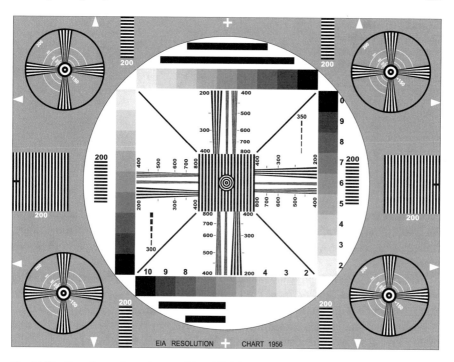

Fig. 12.20 One of several popular test objects "EIA Resolution Chart 1956" to measure the quality of an optical system. The striped patterns are used when testing the resolution, while the grey-toned steps at the edges can be used to test whether optics and CMOS chips provide a good reproduction of different brightness levels. BPK, Public Domain, [3]. A newer ISO 12233 Chart may eventually replace the EIA chart as a high-resolution test pattern for testing imaging systems

other. With increasing density, the stripes at first become more and more grey at the edges, but eventually the stripes disappear altogether.

Several test images have been developed that can be used to determine MTF values, thus telling a little about the optical system's quality in terms of contrast and resolution. In Fig. 12.20, there is given an example of a widely used test plate with various striped patterns with gradual change in the number of stripes per mm. For example, if we have such a high-quality test circuit board (high resolution), we can, for example, look at the pattern through a telescope, camcorder or camera and see how nice stripe details we can detect in the final images/pictures. We will return to this issue later in the book, but with a different test object. (See also problems at the end of this chapter.)

The quality of an optical system can be impaired because of many reasons. Diffraction, due to the fact that the light has a wave nature, will always play a role. Diffraction, however, will only provide a limitation for very good optical systems. Most systems have more serious sources of degradation of optical quality than diffraction.

To avoid spherical and chromatic aberrations in lenses, modern objectives and eyepieces are often composed of several (many) lens elements (see Fig. 12.6). We

Fig. 12.21 A schematic
figure showing the principles
of the structures underlying
the new type of anti-reflex
treatment based on
nanotechnology

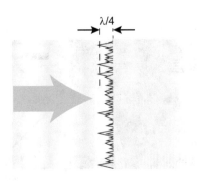

know from previous chapters that when light goes from air to glass, about 5% of the intensity at the surface is reflected. If the light is inclined towards the surface, the reflection may be even greater (for some polarization, as we saw in Fresnel's equations).

With 5% reflection at every outer and inner glass surface in an objective consisting of say eight elements, quite a bit of light will go back and forth several times between elements, which will tend to diminish sharpness and contrast.

For many years, we have been tackling this problem by applying anti-reflection coatings on glass surfaces (see Chap. 13). Reflection can be reduced substantially by this remedy. The problem is, however, that such treatment depends both on the wavelength and on the angle with which the light hits the surface. Anti-reflection treatment of this type significantly improves image quality, but the treatment is not as good as we would like for systems such as cameras and binoculars where light with many wavelengths is admitted at the same time.

Since about 2008, the situation has changed dramatically for the better, and there is some fun physics behind it! Nikon calls their version the "Nano Crystal Coating", whereas the competitor Canon calls it "Subwavelength Structure Coating". Figure 12.21 shows the main principle.

When in Chap. 10 we calculated how much light would be reflected and transmitted at an interface between air and glass, our use of Maxwell's equations was based on some assumptions. To put it plainly, we said that the interface had to be "infinitely smooth, flat and wide" and "infinitely thin" in relation to the wavelength. Then integration was easy to implement and we got the answers we received. We claimed that the conditions could be fulfilled quite well, for example, on a glass surface, since the atoms are so small in relation to the wavelength.

The new concept that is now being used is based on "nanotechnology", which in our context means that we create and use structures that are slightly smaller than the wavelength of light.

The surface of the glass is covered with a layer that has an uneven topography with elements whose size along the layer is less than the wavelength, and the thickness of the layer is about one-quarter wavelength (not as critical as the traditional anti-reflection coating). From a traditional viewpoint, such a layer seems an absurd idea. One might think that the light would be splintered all over when it meets the randomly slanting surfaces, but this is not right. The light as we treat it is an electromagnetic wave that is extensive in time and space. In a manner of speaking, we can say that the wave sees many of the tiny structures *at the same time*, and details smaller than the wavelength will not be followed separately when the wave has propagated several wavelengths further.

Another way to describe the physics of these new anti-reflection coatings is to say that the transition from air to glass gradually occurs over a distance of about a quarter wavelength. Then the reflection is greatly reduced.

It should be added that since the atoms are small compared to the nanocrystals used, we can still use Maxwell's equations to see what happens when an electromagnetic wave hits a lens with nanocrystals on the surface. For example, if the structures are 100–200 nm large, there are about 1000 atoms in the longitudinal direction of these crystals. All calculations on such systems require the use of advanced numerical methods.

The nanocrystalline coating is so successful that about 99.95% of the light is transmitted and only 0.05% is reflected. This new technology is used nowadays on all expensive lenses from Canon and Nikon and has improved the image quality significantly.

12.7.2 Angle of View

So far, we have drawn the three ray-tracing lines from the object to lens plane to image plane without worrying about whether the lines are going outside or within the lens elements themselves. This is all right as long as we are only interested in finding out where the image is formed and what magnification it has. The light from an object follows all possible angles, and as soon as we have established where the image is placed and how large it is, we can fill in as many extra light rays as we wish. We have enough information about how the additional lines must be drawn.

At this point, it is meaningful to consider which light rays will actually contribute to the final image we see when we look, for example, through a telescope. Based on this type of consideration, we can determine what angle of view a telescope or microscope will provide.

Figure 12.22 gives an indication of how this works in practice. The dashed lines indicate the extremities of which light rays from the arrow's tip are captured by the lens and how they continue onwards. In this case, we see that only half of the light that the lens catches will go through the eyepiece. If the object had an even greater angular extent, we could risk that no light from the outermost parts of the object would reach the eyepiece, although in fact some light passes through the lens. By doing this type of analysis, the maximum image angle of, for example, a telescope can be determined, assuming that we actually know the diameter of the objective as well as the ocular.

In practice, it is not quite so simple, because the lenses and eyepieces that are used are composed of several lenses to reduce spherical and chromatic aberrations,

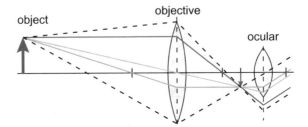

Fig. 12.22 Once we have established the three supporting lines to display image formation, we can fill in with all imaginable light rays that actually pass through a lens (for simple optical layout with a few elements). When multiple lenses are combined, not necessarily all light coming through the first lens will go through the next. This type of consideration can provide an approximate target for what angle of view a telescope or microscope will have

etc. Nevertheless, considerations of this type can provide a rough and ready measure of the field of vision.

It should also be added that different constructions of eyepieces provide quite different experiences when we look through, for example, a telescope. In the old days, we had to keep the eye at a certain distance from the nearest element in the eyepiece to see anything, and what we saw was generally black except a small round field where the subject was. Today, good eyepieces give much more latitude (up to 10 mm) in choosing the position of the eye (in relation to the eyepiece) for viewing an image, and the image we see fills more or less the effective angle of view of the eye. No black area outside is noticed, unless one actively looks for it. As we look through such eyepieces, we get the impression that we are not looking through a telescope at all, but simply *are* at the place shown in the picture. In astronomy, we speak of a sense of "space walk" when such eyepieces are used.

12.7.3 *Image Brightness, Aperture, f-Stop*

Everyone has handled a binocular where the image is bright and nice, and other binoculars where the image is much darker than expected. What determines the brightness of the image we see through binoculars?

In Fig. 12.23, a single lens with object far away has been drawn, along with the image that forms approximately in the focal plane. When the total image angle that the object spans is θ, the focal length of the lens f, and the extent of the image in the image plane is h_1, we have:

$$\tan(\theta/2) = \frac{h_1/2}{f}$$
$$h_1 = 2f \tan(\theta/2) . \tag{12.8}$$

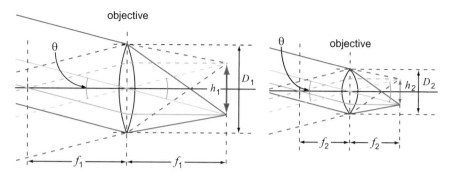

Fig. 12.23 A lens collects a limited amount of light from an object, and this amount of light is spread out over the area of the image being formed. In this figure, the object is assumed to be very far away so that the image is formed in the focal plane. *Comparison between left and right part:* If the lens diameter is reduced to half, while focal length is reduced to half, the light intensity (irradiance) of the image that is formed will remain unchanged. See text for details

For example, if we view the moon, the angular diameter will be about half a degree. If the focal length is 1 m, the image lens will have a diameter of 8.7 mm. How much light is gathered from the image of the moon in the focal plane? It depends on how much light we actually capture from the light emitted from the moon. When the light reaches the lens, it has an irradiance S given, e.g. in the number of microwatt per square metre. Total radiant power collected by a lens of diameter D is $S\pi(D/2)^2$ (in microwatts). The total radiant power will be distributed over the image of the moon in the focal plane, so that:

$$S\pi(D/2)^2 = S_i\pi(h_1/2)^2 .$$

The irradiance S_i in the image plane becomes:

$$S_i = \frac{\pi(D/2)^2}{\pi(h_1/2)^2} S .$$

If we use Eq. (12.8) and rearrange the terms, we get:

$$S_i = \frac{S}{4\tan^2(\theta/2)} \left(\frac{D}{f}\right)^2 \qquad (12.9)$$

where θ is the angular diameter of the moon and f and D are, respectively, the focal length and the diameter of the lens.

The first factor is determined by the light source alone; the second, by the lens alone. The greater the ratio D/f, the brighter is the image formed by the lens (and this is the image that may be viewed by an eyepiece or detected by a CMOS chip or a similar device).

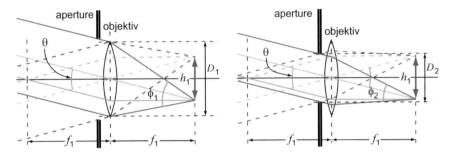

Fig. 12.24 If an aperture is used to reduce the light intensity in the image of an object, the size of the image will always be unchanged—only the brightness goes down. The ratio between focal length and diameter of the light beam that goes to the lens indicates the so-called aperture stop, or f-stop

In Fig. 12.23, two different lenses are drawn, with different radii and different focal lengths. If the focal length goes down to half (right part of the figure), the size of the image (diameter) will be half the size of the left part. But if the lens diameter also drops to half, the area that can catch, for example, light from the moon, will go down to a quarter. However, when the *area* of the image also goes down by a factor of four, it means that the irradiance in the image plane is identical to what we have in the left part of the figure. The ratio D/f is the same in both cases. It has therefore been found that the ratio D/f is a measure of the brightness of the image formed by a lens.

If we insert a photographic film or a CMOS chip into the focal plane and capture the image of, for example, the moon, we will have to collect light for a certain amount of time in order to get a proper exposure. If the lens is referred to as "fast", we will need less time than if the lens is referred to as "slow". A telescope with a large-diameter objective lens (or mirror) will capture much more faint light from distant galaxies than a small-diameter telescope. However, it is not the diameter alone, but the ratio D/f that determines the brightness of the image.

Aperture and f-stops

Cameras use an aperture to change the amount of light that will fall on the film or the CMOS chip. An aperture is simply an almost circular opening whose diameter can be changed by the operator. This is indicated in Fig. 12.24. The image size does not change if we reduce the opening and bring less light onto the CMOS chip, but the irradiance in the image plane will decrease.

The light gathering power of a camera objective is usually indicated by a so-called f-number (also called f-stop or aperture) defined by:

$$\text{f-number} \equiv f/D$$

where f is the lens's focal length and D is the effective diameter of the light beam let through by the aperture and objective. If we compare with the expression in Eq. (12.9), we see that the irradiance of the light that hits the film or the detector chip is inversely proportional to the square of the f-number.

The f-number is usually written in a special way. The most commonly used notation is "f:5.6" or "f/5.6" (f-number = 5.6). Typical f-numbers are 1.4, 2, 2.8, 4, 5.6, 8, 11, 16, 22 corresponding to relative irradiance of about 1/2, 1/4, 1/8, 1/16, 1/32, 1/64, 1/128, 1/256 and 1/512. The higher the f-number, the less light reaches the image plane.

These steps are called "stops", and one stop wider will admit twice as much light by increasing the diameter by a factor of $\sqrt{2} \approx 1.4$. We see that if we change the f-number with one stop, it corresponds to the irradiance in the image plane by a factor of two up or down. Increasing f-numbers correspond to less effective diameter of the lens and thus less light intensity on the light-sensing device. To get the same amount of energy collected per pixel in the CMOS chip, then the exposure time must either be doubled (for each increment in the f-number) or halved (for decreasing f-numbers).

Depth of view

From Fig. 12.24, we can notice another detail. If we move the CMOS bit slightly back or forth in relation to where the image is, a luminous point will be replaced by a luminous disc. The light beam towards the focal plane has a solid angle which decreases when the lens opening is reduced, indicated as ϕ_1 and ϕ_2 in the figure. This means that if we move the sensor chip a little way away from the focal plane, the pixels on the chip will be larger when the aperture is completely open (maximum light intensity in) than when the aperture is smaller (less light emits).

Fig. 12.25 When a luminous point in an object is imaged by a lens, its image will be a point located approximately in the prevailing focal plane. If we are to depict more objects that do not lie at the same distance from the lens, there is no focal plane for the various images that are formed. Then, luminous points in the objects that are at different distances from the lens than the one we have focused on will be depicted as circular discs, and the image will be "blurred". By reducing the aperture, blurring will be reduced (angle ϕ_2 is less than angle ϕ_1 in Fig. 12.24) and we say that we have gained greater depth of field. In the photograph on the left, a lens with f-number 3.5 is used with a shutter speed 1/20 s. In the photograph on the right, the same lens and focus are used, but now with f-number 22 and shutter speed 1.6 s

If we take a picture of a subject where not all objects are the same distance from the lens, we will not be able to get the image of all the objects in focus at the same time. If we have a wide aperture, the blur of the objects that are not in the focal plane will be greater than if the aperture is narrower. The smaller the opening (higher the f-stop number), the less blurred will be the picture. In photography, we say that we have greater "depth of field" (DOF) at small aperture (large f-number) compared to large aperture (small f-number). Figure 12.25 shows an example of this effect.

12.8 Optics of the Eye

Figure 12.26 is a schematic representation of the structure of a human eye. Optically, it is a powerful composite lens that forms a real image on the retina in the back of the eye. The amount of light that reaches the retina can be regulated by adjusting the size of the dark circular opening (pupil) in the iris. The retina has a very large resolution (which allows us to see details), but only in a small area around the point where the optical axis of the eye meets the retina. This area is called *macula lutea*, meaning the "yellow spot" (see Fig. 12.27), and the photoreceptor cells in this region are the so-called cones, which enable us to discriminate between different colours. In other parts of the retina, the density of photoreceptor cells is not so large, and the majority of these cells, called rods, are more sensitive than the cones, but they do not provide any colour information (see Chap. 11). Curiously, the light must pass through several cell layers before it reaches the photoreceptor cells. This may have an evolutionary origin since humans stay outdoors in daytime, when the sunlight is quite powerful. There are species that live in the deep dark layers of the ocean depths where the light reaches the optic cells (only rods) directly without going through other cell layers first.

The major focusing action of the eye comes from the curved surface between the air ($n = 1$) and the cornea. The contribution of the eye lens consists merely in modifying the optical power of the combined refractive system. The refractive indices of aqueous humour and vitreous humour are approximately 1.336 (almost

Fig. 12.26 Schematic representation of the human eye

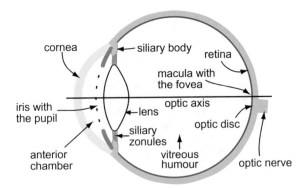

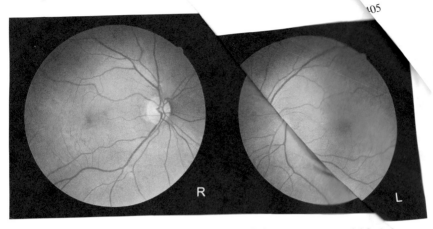

Fig. 12.27 Pictures of retinas taken by a camera through the pupils with a special flash dev. e. The image was taken in a routine check at c)optikk, Lørenskog, 2018. The blind spots are the yellow areas on the image where blood vessels and nerves originate from the eyeball. The so-called yellow spot (*macula lutea* in Latin) is an approximately 1.5 mm diameter area roughly in the middle of the slightly dark oval area somewhat to the left of the centre in the **R** (*right*) eye picture and somewhat to the right of the centre in the **L** (*left*) eye picture. In the macula lutea, the cones are most abundant and the area is responsible for our colour vision and sharp-sightedness. The light must pass through several cell layers before reaching the light-sensitive visual cells

the same as for water), while the eye lens has a refractive index of 1.41. The difference between these refractive indices is much smaller than that between air ($n = 1$) and the cornea ($n = 1.38$), which explains why the total optical power (or refractive power or focusing power) is largely due to the refraction at the surface of the cornea.

The size of the eye remains almost unchanged during use, as indicated in Fig. 12.28. When focusing on objects that are close to us or far from us, it is the shape of the eye lens that is adjusted. The shape is controlled by the ciliary muscle and zonular fibres (which are attached to the lens near its equatorial line). When the ciliary muscle is relaxed, the zonular fibres are stretched, which makes the lens become thinner and reduces its curvature; the converse takes place when the ciliary muscle contracts. For a normal eye, the focal point of the lens will generally be on the retina, and objects that are "infinitely far away" will then form a real upside-down image on the retina. When the muscle contracts, the zonular fibres slacken, allowing the lens to become more spheroidal and gain greater optical power. Then the focal point falls somewhere in the vitreous chamber, and objects that are not so far from the eye could form a real image on the retina. The image spacing s' in the lens formula always stays constant (about 20 mm), but the focal length of the overall lens changes; this process is called *accommodation*.

With advancing age, the eye lens hardens, the tension of the zonular fibres deteriorates, and the activity of the ciliary muscle declines. As a result, ageing leads to what is called *presbyopia*—the continuous loss of the ability of the eye to focus on nearby objects. Specifically, the nearest point on which middle-aged person is able

In other words, the eye lens can only change the total optical power by about ten per cent. The cornea takes care of the rest of the optical power.

Not all eyes are "normal". Some have corneas with excessive curvature, on account of which the optical power becomes too large in relation to the distance between the lens and the retina. When such a person tries to see an object far away, the real image will not hit the retina but lie somewhere inside the vitreous chamber. The person will therefore not have a sharp vision when he/she looks at something far away. Such a person we call *nearsighted*. Nearsightedness can be remedied using glasses or contact lenses. The eye's natural dioptric power must be contraindicated since it is too large, and the glasses or outer eye lenses must be concave (negative eyepiece).

If someone has a cornea with an abnormally small curvature, the optical power of the eye becomes too small. When such a person tries to look at an object 25 cm from the eye (nominal near point), the real image will not fall on the retina but will theoretically fall behind it. The image on the retina will be blurred. Such a person is called *long-sighted*. Again, we can compensate for the error by inserting an extra lens in the form of glasses or external eye lenses until we look sharply also for bodies 25 cm away. In this case, the lens must be convex (positive eyepiece).

When young, a person with an imperfect vision can get, with the help of spectacles, an almost normal vision, with a near point of 25 cm and a far point at infinity. As the age increases, the capacity for *accommodation* decreases, and one pair of spectacles will no longer be able to give a near point at 25 cm and at the same time the far point at infinity. It becomes necessary to continually wear glasses and take them off, maybe even to switch between two sets, to see satisfactorily both at short and long distances. There are also so-called "progressive glasses" where the upper part of the glass has one focal length and lower part another (with a continuous gradation between them).

One type of lens imperfection can be described by saying that the cornea has an asymmetrical shape, almost like an ellipsoid with one meridian being significantly more curved than the meridian perpendicular to it. In such a case, there are different optical powers for the two directions. This lenticular error is called *astigmatism* and can be corrected using lenses that have cylindrical rather than spherical surfaces. Such lenses are called cylinder lenses. They can be combined with spherical surfaces if desired.

Today it is quite common to undertake laser surgery for changing the corneal surface if the lens has significant inborn flaws. In that case, parts of the cornea can be burned and shaped so that the person gets a normal vision and does not have to wear glasses (until age-related defects make it necessary).

Examples

It is rather easy on our own to form a rough impression of what glasses we need in case we are slightly nearsighted or long-sighted. Here are a few examples:

Suppose we can focus sharply only up to 2.0 m, that is, the far point without glasses is 2.0 m. This means that the optical strength of the lens is:

$$\frac{1}{f} = \frac{1}{2.0 \text{ m}} + \frac{1}{0.02 \text{ m}} = 50.5 \text{ dioptre} .$$

The optical strength is thus too large by 0.5 dioptre, since it should have been 50.0 dioptre for the far point. The remedy is to use glasses with optical strengths of −0.5 dioptre, at any case for viewing distant objects. The elegance of these calculations is that we need to only add or subtract optical strengths to get the optical strength of the combination.

In the next example, we take a person who is unable to focus on distances closer than 50 cm. This person's optical strength is then:

$$\frac{1}{f} = \frac{1}{0.5 \text{ m}} + \frac{1}{0.02 \text{ m}} = 52.0 \text{ dioptres} .$$

In this case, when considering the near point, the optical strength should be 54.0 dioptres, which means that there is a deficit of 2 dioptres. The person thus needs spectacles of optical strength +2.0 dioptres to move the near point from 50 to 25 cm.

12.9 Summary

Geometric optics is based on the thinking that the light from different objects propagates like "light rays" in different directions, where each ray behaves approximately as (limited) plane electromagnetic waves. These light rays will be reflected and refracted at interfaces from one medium to another, and their behaviour is determined by Maxwell's equations and satisfies the laws of reflection and refraction in common materials.

When an interface between two media is curved, light rays incident at the interface will have different inclinations compared with the refracted rays.

For thin lenses, we can define two focal points, one on each side of the lens. Light rays will have infinitely many different directions in practice, but it suffices to use two or three guides to construct how an object is imaged by a lens so that we get an image. The guides are characterized by the fact that light parallel to the optical axis is broken through the focal point on the opposite side of a convex lens, but away from the focal point on the same side as incoming light beam for concave lenses. Light rays through the centre of the lens are not broken. We normally draw only lines to the centre planes of the lenses instead of incorporating detailed refraction on each surface. Guides may go beyond the physical extent of the lens, but only the light rays that actually pass through the lenses contribute to the light intensity of the image.

All light rays that go along different paths from one object point to one image point spend the same time on the trip, whether they go through the central or peripheral parts of the lens. This ensures that different light rays interfere constructively when they meet. When real light rays meet in this way, light can be intercepted by a screen, and we speak of the formation of a real image at the screen. If the real beams diverge from each other after passing through a lens, but all seem to come from a point behind the lens, we say that we have a virtual image at this point. If we consider light rays that diverge from each other with the help of our eye, the light rays will again be collected on the retina and form a real image there. Therefore, we can "see" a virtual image, even though we cannot collect this image on a screen.

The lens formula is a simplification of lens makers' formula, and only object distance, image distance and focal length are included. In the lens formula, the focal length is considered positive for a convex lens and negative for a concave lens. The signs for the object distance and image distance change with how the light rays reach the lens relative to where they exit. We must consider the sign in each case in order not to commit an error. A drawing that shows the beam angle is essential to avoiding mistakes in such cases (must check that the result looks reasonable).

A lens can be used as a loupe. Magnification is then an angle magnification, because placing the object at the lens's focal point, the virtual image will apparently be infinitely far away and be infinite. Different angles that indicate maximum propagation of an object lead to a similar physical extent to the actual image on the retina when we consider the object through the louse. The primary function of the loupe is that we can effectively keep the object much closer to the eye than the eye's near point. That is, we can effectively position the object much closer to the eye and still look sharp, compared to looking at the object as close as possible (sharp) without aids.

Lenses can be assembled for optical instruments such as telescopes and microscopes. For the telescope, an objective is used to create a local image of the object that we can consider with a loupe. The result can be a significant magnification. For the microscope, the object is placed outside, but very close to the focus of the lens. The real image that the lens then makes is significantly larger than the object. Again, a loupe is used to view the real image that the lens makes.

Alternatively, a CMOS chip is inserted into the image plane and the eyepiece (loupe) is removed. This is often the case with many of today's "digital microscopes". In such cases, "magnification" is a very poorly defined term since it will in practice depend on which computer screen size the image is finally displayed.

The human eye has a fixed image distance of approximately 20 mm. The focal length of the optical system is mainly determined by the cornea, but can be slightly adjusted since the eye lens strength can be varied within an interval. A normal eye can focus sharply on objects at a distance of approximately 25 cm to infinity. This corresponds to a total lens strength from 54 to 50 dioptres. If the cornea has a too large or too small curvature, the lens strength is too large or too small. Then we will not be able to focus sharply over the entire range from 25 cm to infinity, and we need glasses to compensate for deficiencies in the optical strength of the eye lens.

12.10 Learning Objectives

After going through this chapter, you should be able to:

- Explain why light from objects can be regarded as "light rays" when the light hits for example a lens.
- Calculate where the image point of a point source is after the light from the source has met the surface of a glass sphere.
- Explain the terms object, image, focal point, object distance, image distance, focal length, radius of curvature, concave, convex, real and virtual image.
- Derive (possibly with some help) the lens makers' formula of a positive meniscus lens and specify the simplifications usually introduced.
- Explain main steps in the derivation of the lens formula under the same conditions as in the previous paragraph.
- State the three main rules used in the construction of the ray path through lenses and mirrors (ray optics) and apply these rules in practice.
- Explain why sometimes we need to change the sign for some quantities when the lens formula is used.
- Explain two different ways to specify the magnification of optical instruments.
- Explain how a loupe is routinely used and what magnification it has.
- Describe how a telescope and microscope are constructed and what magnification they have.
- Describe how a reflecting telescope works and how it avoids undue obstruction of the incoming light.
- Calculate how large an image of a given subject (at a given distance) we can obtain in the image plane for different camera lenses.
- Calculate approximately the angle of view of a camera or binoculars when the relevant geometrical data are specified.
- Explain briefly how nanotechnology has led to better photographic lenses.
- Explain the optical power of a lens/objective and know what the f-numbers tell us.
- Explain the concept "depth of field" and how this changes with the f-number.
- Explain the optics of the eye, and explain what the terms near point, far point and accommodation mean.
- Know the optical strength of the eye and how the optical strength is varied.
- Calculate, from simple measurements of near point and far point, the approximate optical strength of the spectacles a person may need.

12.11   Exercises

Suggested concepts for student active learning activities: Light ray, ray optics, beam optics, wavefront, paraxial approximation, focal point, object/image, lens makers' formula, lens formula, convex, concave, (real) magnification, angular magnification, loupe, magnifying glass, normal eye, ocular, objective, telescope, microscope, reflecting telescope, optical quality, angle of view, image brightness, aperture, f-stop, depth of field. near point, far point, optical strength, dioptre.

Comprehension/discussions questions

1. The laws of reflection and refraction mentioned in this chapter apply to visible light. Light is regarded as electromagnetic waves. Will the same laws apply to electromagnetic waves in general? (As usual, the answer must be justified!)
2. An antenna for satellite TV is shaped like a curved mirror. Where does the antenna element itself be placed? Is this analogous to the use of mirrors in optics? Which wavelength does satellite TV signals have? And how big is the wavelength relative to the size of the antenna disc?
3. An antenna for satellite TV with a diameter of 1 m costs about hundred USD, while a mirror for an optical telescope with a diameter of 1 m would cost an estimated thousand times more. Why is there such a big difference in price?
4. A "burning glass" is a convex lens. If we send sunlight through the lens and hold a piece of paper in the focal plane, the paper can catch fire. If the lens is almost perfect, would we expect, solely on the basis of geometric optics, all the light to be collected at a point with almost no extent?
5. If you have attempted to use a burning glass, you may have discovered that the paper is easier to light if the sunspot hits a black area on the paper compared with a white one. Can you explain why?
6. Cases have been reported that fishbowls with water and spherical vases with water have acted as a burning glass, causing things to catch fire. Is it theoretically possible from the laws we have derived in this chapter? What about "makeup mirrors" with a concave mirror, may such a mirror pose any threat?
7. Based on the lens makers' formula, we see that the effective focal length depends on the wavelength since the refractive index varies with the wavelength of light. Is it possible for a biconvex lens to have a positive focal length for one wavelength and negative focal length for another wavelength?
8. How can you quickly find the approximate focal length of a convex lens (converging lens)? Do you also have a quick test for a concave lens (diverging lens)?
9. Does the focal length change when you immerse a convex lens into water?
10. Does the focal length change when you immerse a concave mirror into water?
11. If you look underwater, things look blurred, but if you are wearing diving goggles, you experience no blurring. Explain! Could you get rid of extra spectacles with no layer of air anywhere? In that case, should the spectacles have concave or convex lenses?
12. A real image (created, e.g. by an objective) can be detected by placing a paper, a photographic film or CMOS chip in the image plane. Is it possible to record a virtual image in one way or another?
13. The laws of reflection and refraction, lens maker's formula and lens formula are all symmetrical with respect to which way the light goes. In other words, we can interchange the object and image. Can you point out mathematically how this reversibility is expressed in the relevant laws? Are there any exceptions to the rule?

14. (a) We have a vertical mirror on a wall. A luminous incandescent lamp is held in front of the mirror so that the light reflected by the mirror hits the floor. However, it is not possible to form an image of the incandescent lamp on the floor. Why? (b) We have a laser pointer and we use it in a similar way to the incandescent lamp, so that the light from the laser pointer is reflected by the mirror and reaches the floor. *Now* it appears that we have formed a picture of the laser pointer (the opening of this) on the floor. Can you explain what is going on?

15. How long must a mirror be and how high must it be placed on a vertical wall so that we can see all of ourselves in the mirror at once? Will the distance to the mirror be important?

16. The two cameras in an iPhone model have, according to Apple, objectives with focal lengths of 28 and 56 mm, respectively. Is it possible that the lenses actually have these focal lengths? How do you think the numbers should be understood? Why does Apple choose to give the numbers this way? (Hint: Prior to the digital revolutions, the picture size on the film was usually 24×36 mm. See also problems 23–24.)

Problems

17. Draw a ray diagram for a convex lens for the following object distances: $3f$, $1.5f$, $1.0f$ and $0.5f$. For one of these distances, only two of the usual three standard rays can be used in the construction of the image. Which? State whether we have magnification or demagnification of the image, whether the image is up or down, and whether the image is real or virtual.

18. Determine, by starting from the lens formula and one of the rules for drawing ray diagrams, the smallest and largest magnification (in numerical value) a convex lens may have. Determine the condition that the magnification will be 1.0.

19. Repeat the calculation in the previous task, but now for a concave lens. Determine again the condition that the magnification will be (approximately equal to) 1.0.

20. When we find the image, formed by a convex lens, of an object "infinitely far away", we cannot use the three standard light rays for the construction of the image. How do we proceed in such a case to find the location of the image in the image plane?

21. We have a convex meniscus lens with faces that correspond to spherical surfaces with radii of 5.00 and 3.50 cm. The refractive index is 1.54. What is the focal length? What will be the image distance if an object is placed 18.0 cm away from the lens?

22. A narrow beam of light from a distant object is sent into a glass sphere of radius 6.00 cm and refractive index 1.54. Where will the beam of light be focused?

23. Suppose that you have a camera and take a picture of a 1.75 m tall friend standing upright 3.5 m away. The camera has an 85 mm lens (focal length). What is the distance between the lens and the image plane when the picture is taken? Are you able to fit the entire person within the image if the image is recorded on an old-fashioned film or a full-size CMOS 24×36 mm image sensor? How much of the person do you get in the picture if it is recorded with a CMOS photo chip of size 15.8×23.6 mm?

24. Repeat the previous task for the camera in a mobile phone. For example, an iPhone model has a true focal length of 3.99 mm, and the photo chip is about 3.99 × 7.21 mm (the numbers apply to the general purpose camera and not the telephoto variant of this camera).
25. When Mars is closest to the earth, the distance is about 5.58×10^7 km. The diameter of Mars is 6794 km. How large will be the image if we use a convex lens (or concave mirror) with focal length 1000 mm?
26. The old Yerkes telescope at the University of Chicago is the largest single-lens refracting telescope of the world. It has an objective that is 1.02 m in diameter and a f-number of f/19.0. How long is the focal length? Calculate the sizes of the images of Mars and the moon in the focal plane of this lens. (The angular diameter of the moon is about half a degree, and the angular diameter of Mars can be estimated from the information in the previous task.)
27. A telescope has a lens with a focal length of 820 mm and a diameter of 100 mm. The eyepiece has a focal length of 15 mm and a diameter of 6.0 mm. What magnification does the telescope have? How big is the image angle? Can we see the whole moon disc at once?
28. A slide projector (or data projector, for that matter) has a lens with a focal length of 12.0 cm. The slide is 36 mm high. How big will be its image on a screen 6.0 m from the projector (lens)? Is the image upright or inverted?
29. Suppose we have two glasses, one with optical strength +1.5 dioptres on both lenses and one with optical strength +2.5 dioptres on both glasses. We only find one of the glasses and would like to check if these are the stronger or weaker. Can you provide a procedure on how to determine the optical strength of the glasses we found?
30. (a) Where is the near point of an eye for which an optician prescribes a lens of 2.75 dioptres? (b) Where is the far point of an eye for which an optician prescribes a lens with lens strength −1.30 dioptres (when we look at things a long distance away)?
31. (a) What is the optical strength of spectacles needed by a patient who has a near point of 60 cm. (b) Find the optical strength of the spectacles for a patient who has a far point of 60 cm.
32. Determine the accommodation (in the sense of a change in optical strength) of a person who has a near point at 75 cm and a far point at 3.0 m.
33. In a simplified model of the eye, we see that the cornea, the fluid inside, the lens and the vitreous humour inside the eye have all a refractive index 1.4. The distance between the cornea and the retina is 2.60 cm. How big should the radius of curvature be for an object 40.0 cm from the eye to be focused on the retina?
34. A loupe has a focal length of 4.0 cm. What magnification will it give under "normal" use? Is it possible to get a magnification of 6.5 X using the loupe in a slightly different way than described as a standard (do not think about ocular projection)? If so, tell us where the object we consider must be placed, and say something about how we can now use the eye.

35. A spherical concave mirror will not collect all parallel rays at one point, because the effective focal length will depend on how close the optical axis of the beam hits the mirror.

(a) Try to set up a mathematical expression for effective focal length of a beam that comes in parallel with the optical axis a certain distance from the axis. The radius of curvature of the mirror is set equal to R. As a parameter, we can use the angle θ between the incoming beam and the line that runs between the centre of curvature of the mirror and the point where the beam hits the mirror surface.

(b) For which angle will effective focal length have changed with 2% relative to the focal length of the rays coming in very close to the optical axis?

(c) Can you explain why mirror scanners based on spherical mirrors often have high f-numbers (low brightness)?

36. Suppose you have a removable telephoto lens for a camera. The focal length is 300 mm. Suppose you also have a good quality 5X loupe. You want to make a telescope of these components and have a suitable tube in which the lenses can be held.

(a) What is the focal length of the loupe lens? What is the magnification of the telescope?

(b) State the distance(s) between the objective and loupe (used as the eyepiece) when the telescope is to be used to view objects from 10 m to infinitely far away?

(c) You would like to be able to use the telescope for a distance of 25 cm. Is it possible? (As usual: The answer must be supported by arguments.)

(d) The telephoto lens has a diameter of 60 mm. What is the f-number (corresponding to the largest aperture) this lens can have? (Simplify the discussion by regarding the lens as a simple thin lens.)

37. A lens telescope is to be used by an amateur astronomer. The focal length of the objective is 820 mm, and the diameter 10.0 cm. The objective is located at one end of the telescopic tube and the eyepiece holder at the opposite end. The eyepiece holder can be adjusted so that we get a clear picture of the starry sky and planets. In order to use slightly different magnification on different objects, the amateur astronomer has four different eyepieces with focal lengths 30, 15, 7.5 and 3.0 mm. The diameter of the lens in these eyepieces is 48, 20, 11 and 3.7 mm, respectively. We treat all lenses as if they were "thin".

(a) How long should the telescope tube be (the distance between the objective and ocular)?

(b) How much change in position must the eyepiece holder allow?

(c) How much longer should the eyepiece move if we want also to use the telescope as field glasses with the minimum object distance equal to 20 m?

(d) What is the f-number of the objective?

(e) What do we understand by the "magnification" of a telescope?

(f) Estimate how much magnification we get for the four different eyepieces.

(g) Estimate the approximate image angle we receive for the 30 and 3.0 mm eyepiece.

(h) Compare this with the image angle of the moon, which is about 0.5°.

(i) How big will Jupiter look under conditions best suited for observations, when we view it through our telescope with the 3.0 mm eyepiece? (Approximate radius of earth's orbit is 1.50×10^{11} m and of Jupiter's orbit 7.78×10^{11} m. Jupiter's diameter is about 1.38×10^{9} m.)

38. The telescope constructed by Galileo consisted of a convex lens and a concave eyepiece. Such a telescope is called today a Galilean telescope, and a principle sketch is shown in Fig. 12.29 for a case where the object is far away.

The image from the objective (red arrow in the figure) is for this configuration placed "behind" the eyepiece (the eyepiece is closer to the objective than the image formed by the objective). The Galilean telescope therefore becomes shorter than for a telescope where both objective and ocular were convex (positive focal width).

Let us analyse the light rays in the figure.

(a) We have assumed, for the sake of simplicity, that the objects we are looking at are "infinitely far away" and that the eyes focus as if the objects were placed infinitely far away. How are these assumptions manifested in the way we have drawn the rays in the figure?

(b) Explain in particular which construction rules for light ray optics we have used for the red and violet ray in the figure.

(c) Show that the (angular) magnification (numerical value) of the Galilean telescope is given by the relation $M = f_1/f_2$, where f_1 and f_2 are the numerical values of the focal lengths of the objective and eyepiece, respectively.

(d) (Somewhat difficult) Would it be possible to use eyepiece projection enabling this telescope to be used for recording pictures directly on an image sensor or film? Explain.

39. A laboratory microscope has an objective with focal length 8.0 mm, and an eyepiece with focal length 18 mm. The distance between the lens and the eyepiece is 19.7 cm. We use the microscope so that the eyes focus as if the object is placed infinitely far away. We treat the lenses as if they are "thin".

(a) What distance should there be between the object and the objective when using the microscope?

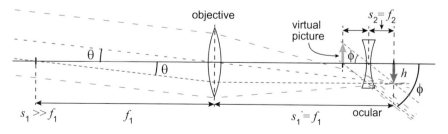

Fig. 12.29 Galilean telescope

(b) How large is the linear/real magnification provided by the objective (when used alone)?

(c) How large is the magnification provided the eyepiece alone?

(d) How is magnification defined for a microscope?

(e) What is the magnification of this microscope?

40. Show that when two thin lenses are in contact, the focal length f of the two lenses together will be given by:

$$1/f = 1/f_1 + 1/f_2$$

where f_1 and f_2 are the focal lengths of the individual lenses. We have a converging meniscus-shaped lens with refractive index 1.55 and radii of curvature 4.50 and 9.00 cm. The concave surface is turned vertically upwards, and we fill the "pit" with a liquid having a refractive index n = 1.46. What is the total focal length of lens plus liquid lens?

41. In this task, we will compare the cameras in an iPhone 5S and a Nikon D600 SLR camera with "normal lens". The following data are provided:

iPhone 5S: The lens has a focal length of 4.12 mm and aperture 2.2. The sensor chip has 3264 × 2448 pixels and is 4.54 × 3.42 mm in size. Nikon D600 with normal lens: The lens has a focal length of 50 mm and aperture 1.4. The sensor chip has 6016 × 4016 pixels and is 35.9 × 24 mm in size.

(a) Determine the effective diameter of the two lenses.

(b) Determine relative irradiance that falls on the photo chip in each camera. We ignore the influence of different image angles. (iPhone has a maximum image angle of 56° while the 50 mm lens on a Nikon D600 has a maximum image angle of 46°.)

(c) Fig. 12.30 shows identical snippets from photographs taken under approximately the same conditions with an iPhone device and a Nikon D600 device. Try to describe the difference in the quality of the pictures, and try to explain why there may be such a difference.

(d) Would the graininess in the iPhone image be larger or smaller if we increased the number of pixels in the iPhone to the same level as the D600?

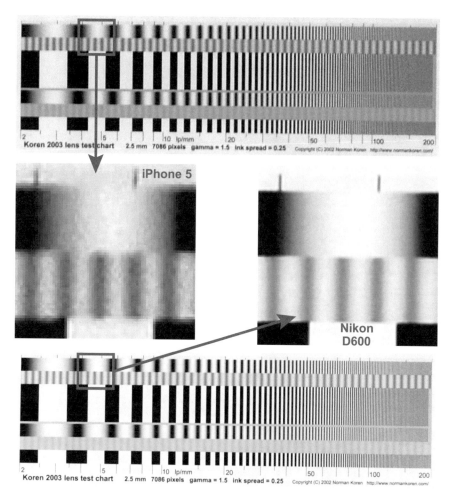

Fig. 12.30 Parts of photographs taken with the iPhone and Nikon D600. The length of Kohren's test chart was approximately 1/4 of the longer dimension of the picture in both cases. Notice not only how close the lines can lie and yet be distinct from each other, but also the graininess in the grey parts of the pictures. Details are likely to be better when the images are viewed on screen with some enlargement, than when they are viewed on A4-size paper. (Effective ISO values are not necessarily comparable.)

References

1. GodeNehler, https://commons.wikimedia.org/wiki/File:Canon_EF_400_DO_II.jpg. Accessed April 2018
2. Paul Chin (paul1513), https://en.wikipedia.org/wiki/History_of_photographic_lens_design. Accessed April 2018
3. BPK, https://commons.wikimedia.org/wiki/File:EIA_Resolution_Chart_1956.svg. Accessed April 2018

Chapter 13
Interference—Diffraction

Abstract This chapter is based on two extremely important properties of waves. One is diffraction, the property whereby a wave confined to a part of space gradually spreads out to nearby spatial regions if there are no barriers to prevent its spread; the other is interference, which arises from the fact that, in linear media, waves add to each other at amplitude level when they meet. Two waves of equal amplitude may at one extreme add so that they completely cancel each other or in the opposite extreme add constructively so that the resultant wave (with twice the amplitude) carries four times as much energy as does each individual wave. This chapter shows how these general principles can be applied when there are particular geometric constraints for wave motion, like a single slit, a double slit, a diffraction grating, a circular aperture or a spherical obstacle. Arago's spot is mentioned since it provides a historical perspective. Both analytical mathematics and numerical methods are used. Particular topics like Huygens's principle, Airy's disc, Rayleigh's resolution criterion, diffraction-limited optics and Babinet's principle are also presented.

13.1 The Nature of Waves—At Its Purest

In this chapter, we will describe some of the most *wave*-specific phenomena found in physics! These are phenomena that can be displayed for all waves in space, such as sound waves, waves in water and electromagnetic waves, including visible light. In many contexts, the experiments are simple and transparent, and the extension of the waves in time and space becomes a central and almost inevitable ingredient of any explanation model.

There are a number of phenomena that can be observed when two or more waves work together. Sometimes, the results are surprising—and often beautiful! In this chapter, we will primarily discuss interference and diffraction. Historically, we may say that the word "interference" was primarily used when two separate waves interacted, while the word "diffraction" was most commonly used when some parts of a wave interacted with other parts of the same wave. It is almost impossible to keep these two concepts apart in every situation, with the result that sometimes we are confronted with an illogical use of these words.

© Springer Nature Switzerland AG 2018
A. I. Vistnes, *Physics of Oscillations and Waves*, Undergraduate Texts in Physics,
https://doi.org/10.1007/978-3-319-72314-3_13

Whatever the names, diffraction and interference are, as already mentioned, some of the most wave-specific phenomena are known to us. Thomas Young's double-slit experiment is one of the most discussed topics in physics today, and interference is the main reason why one could not overlook the wave nature of light a hundred years ago when Einstein and others found support for the view that the light sometimes appears to behave like particles.

In this chapter, we will first and foremost illustrate interference and diffraction through phenomena related to light, but sometimes it is useful to resort to concrete water waves to better understand the mechanisms behind the phenomena.

When two or more waves work together, we need to know how to add or combine waves.

The basis for all interference and diffraction is the *superposition principle*:

The response to two or more concurrent stimuli s_i will at a given time and place be equal the sum of the response the system would have on each of the stimuli individually.

Superposition implies, in other words, additivity, which is expressed mathematically as:

$$F(s_1 + s_2 + \cdots + s_n) = F(s_1) + F(s_2) + \cdots + F(s_n) .$$

This means that F is a linear mapping. In other words, F must be a linear function!

In physics, we know that many phenomena behave approximately linearly. The most familiar examples are probably Ohm's law for resistors and Hooke's law for the extension/compression of a spring. As long as the "amplitudes" are small, an (approximately) linear relation applies. But we know that this law does not give a good description for larger "amplitudes". Then, the "higher-order terms" must be taken into account (the expression can be understood as referring to a Taylor expansion). We mention this to remind you that the superposition principle does not apply in every situation. In this chapter, however, we still limit almost exclusively to linear systems where superposition applies.

In this chapter, phenomena will sometimes be presented qualitatively, usually with a simple formula. In addition, we will provide a more "complete" mathematical description of three basic situations:

- Interference from a double slit,
- Interference from a grating (many parallel slits),
- Diffraction from a single slit.

The mathematical details of the actual derivation are of limited value (especially for gratings and single slit), but the main idea that lies at their root is of paramount importance, so be sure to get a firm hold on it!

Fig. 13.1 According to
Huygens's principle, we may
think of any point on a
wavefront as the source of
elementary waves

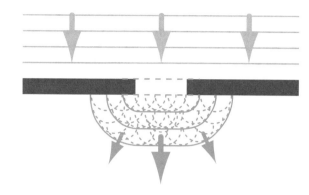

13.2   Huygens's Principle

Our description of interference and diffraction is based on Huygens's principle, which
states that:

> *Any point in a wave can be viewed as a source of a new wave, called the
> elementary wave, which expands in all directions. For following a wave motion,
> we can start from, for example, a wavefront and construct all conceivable
> elementary waves. If we go one wavelength along these elementary waves,
> their envelope curve will describe the next wavefront (Fig. 3.1).*

Fresnel modified the above view by saying that if we are to find the wave amplitude
somewhere in space (also well away from an original wavefront), we can sum up
all conceivable waves provided that we take into account both amplitude and phase
(and whether or not something obstructs the wave).

The Dutchman Christiaan Huygens[1] lived from 1629 to 1695 and the Frenchman
August-Jean Fresnel from 1788 to 1827, and we might wonder if such an old viewpoint has any relevance today, when we have Maxwell's equations, relativity theory
and quantum physics. Remarkably enough, the Huygens–Fresnel principle is still
applicable, and it is in a way a leading principle in quantum electrodynamics (QED),
the most accurate theory available today. True enough, we do not use the vocabu-
lary of Huygens and Fresnel for describing what we do in QED, but mathematically
speaking the main idea is quite equivalent. In quantum electrodynamics, it is said
that we must follow all possible ways that a wave can go from a source to the place
where the wave (or probability density) is to be evaluated. If a particle description
is used, the phase information lies at the bottom even in the quantum field. In other
words, the Huygens–Fresnel principle is hard-wearing (Fig. 13.1).

---

[1] Unusual pronunciation, see Wikipedia.

Throughout the chapter, we assume that the light is "sufficiently coherent". We will return to coherence in Chap. 15, and here we will only state that the light we start with (e.g. emerging from a slit) can be described mathematically as an almost perfect sinusoidal wave without any changes in amplitude or frequency as time flows. In other words, we assume complete predictability in the phase of the Huygens–Fresnel elementary waves in relation to the phase of the waves we start with.

13.3 Interference: Double-Slit Pattern

In 1801, when the Englishman Thomas Young (1773–1829) conducted his famous double-slit experiment, Newton's corpuscular (particle) model for light was the motivation. The corpuscular model seemed appropriate in so far as it accounted for the fact that light rays travel in straight lines and for the observed laws of reflection. And Newton's red, green and blue corpuscles provided an excellent starting point for explaining additive colour mixing.

If Newton's light particles pass through two narrow parallel slits, we would expect to see two strips on a screen placed behind a double slit. But what did Young observe? He saw *several* parallel strips! These strips are called interference fringes. This fringe pattern was almost impossible to explain on the basis of Newton's particle model. Young, and subsequently Fresnel and others, could easily explain this phenomenon, and we shall presently look at the mathematics (Fig. 13.2).

The two slits are assumed to be narrow (often 1–1000 times the wavelength), but "infinitely" long so that we can look at the whole problem as two-dimensional (in a plane perpendicular to the screens and slits).

We assume that light enters with a wavefront parallel to the slits so that the light starts with identical phase throughout the "exit plane" in both slits. We assume further

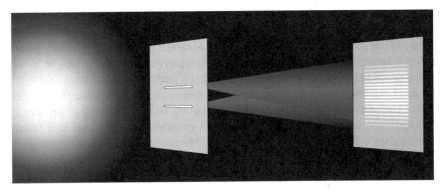

Fig. 13.2 Experimental set-up for Young's double-slit experiment. Slit sizes and strip patterns are greatly exaggerated compared to the distance between light source, slits and screen

Fig. 13.3 Schematic light
path from the double slit to a
given point on the screen at
the back. In reality, the
distance R from slits to the
screen is much greater than
the gap d_1 between the slits.
See text for details

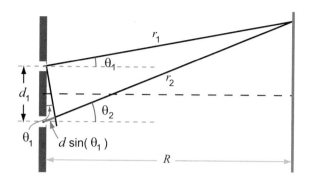

that each of the slits emits elementary waves, and for reasons just mentioned, these
waves will have a wavefront that is shaped as part of a cylindrical surface with the
slit as the cylinder axis. In a plane perpendicular to the columns, we will then get a
purely two-dimensional description (see Fig. 13.3).

We are dealing with light, that is, with an electromagnetic wave. The wave is
transverse and is described by an electric and a magnetic field, each of which has
a certain direction in space. We assume that we are considering the interference
phenomenon so far away from the slits that we can ignore the *difference in the
direction* in space for electrical fields originating from slit 1 compared to the field
originating from slit 2. It will be sufficient for us therefore to add the two electrical
fields as scalar quantities with the correct intensity and phase.

We want to find the electric field on a screen parallel to the plate with the slits,
in a direction θ relative to the normal vector between the slits (see Fig. 13.2). The
contributions from the two slits are then:

$$E_1(\theta_1) = E_{1,0}(r_1, \theta_1) \cos(kr_1 - \omega t - \phi)$$

$$E_2(\theta_2) = E_{2,0}(r_2, \theta_2) \cos(kr_2 - \omega t - \phi)$$

where ϕ is an arbitrary phase angle when space and time are given. Since the screen
with the slits and the screen where we capture the image are very far apart compared
to the gap between the slits, the angles θ_1 and θ_2 will be almost identical, and we
replace them both with θ:

$$\theta_1 \approx \theta_2 = \theta .$$

For the same reason, we will assume that the two amplitudes are identical, and write:

$$E_{1,0}(r_1, \theta_1) = E_{2,0}(r_2, \theta_2) = E_0(r, \theta) .$$

The total amplitude in the direction θ is therefore (according to the superposition
principle):

$$E_{tot}(\theta) = E_0(r, \theta)[\cos(kr_1 - \omega t - \phi) + \cos(kr_2 - \omega t - \phi)] \ .$$

Using the trigonometric identity

$$\cos a + \cos b = 2 \cos \left(\frac{a+b}{2} \right) \cos \left(\frac{a-b}{2} \right)$$

and get:

$$E_{tot}(\theta) = 2E_0(r, \theta) \ \cos \left(k \frac{r_1 + r_2}{2} - \omega t - \phi \right) \cos \left(k \frac{r_1 - r_2}{2} \right) \ .$$

Superposition always operates on amplitudes (i.e. to say, a real physical quantity, not an abstract quantity such as energy or intensity). Be that as it may, physical measurements are often based on intensity. When we view light on a screen with our eyes, the light intensity we sense is proportional to the intensity of the wave.

The intensity of a plane electromagnetic wave in the far-field zone is given by the Poynting vector, but the scalar value is given by:

$$I = cED = c\varepsilon E^2$$

where c is the velocity of light, E the electric field, D the electric flux density (electric displacement), and ε the electric permittivity. Hence:

$$I(\theta, t) = c\varepsilon E_{tot}^2(\theta, t) = 4c\varepsilon E_0^2(r, \theta) \ \cos^2 \left(k \frac{r_1 + r_2}{2} - \omega t - \phi \right) \cos^2 \left(k \frac{r_1 - r_2}{2} \right) \ .$$

This is the so-called instantaneous intensity that varies over time within a period. We are most interested in time-averaged intensity. The first $\cos^2$ term varies with time, and the time average of $\cos^2$ is 1/2. Accordingly:

$$I(\theta) = 2c\varepsilon E_0^2(r, \theta) \cos^2 \left(k \frac{r_1 - r_2}{2} \right) \ .$$

We define

$$r_1 - r_2 = \Delta r = d \sin \theta$$

where d is the distance between the slits. Furthermore, we bring in the wavelength through the relationship $k = 2\pi / \lambda$.

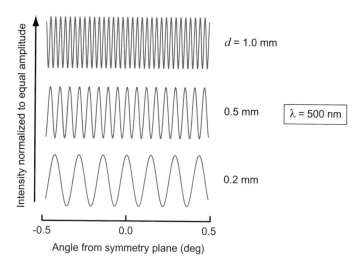

Fig. 13.4 Strip pattern on a screen behind the double slit. The distance between the slits is indicated

Whence follow the intensity distribution of the light that has passed a double slit (Fig. 13.4):

$$\overline{I}(\theta) = 2c\varepsilon E_0^2(r, \theta) \cos^2\left(\frac{d \sin\theta}{\lambda}\pi\right) . \qquad (13.1)$$

When $\theta = 0$, we get maximal intensity. The minima are obtained when the argument of the cosine function is an odd multiple of $\pi/2$:

$$\frac{d \sin\theta}{\lambda}\pi = (2n + 1)\frac{\pi}{2}, \qquad (n = 0, 1, 2, \ldots) .$$

The condition for a minimum is thus found to be:

$$\sin\theta = \frac{\lambda}{d}\left(n + \frac{1}{2}\right) .$$

The maxima occur when:

$$\sin\theta = \frac{n\lambda}{d} \quad \text{approximately} .$$

The word "approximately" has been added because the exact expression for the maxima depends also on how $E_0^2(r, \theta)$ varies with θ.

We should notice that usually, at least for light, the gap between the slits is large relative to the wavelength. That is, the angle between two minima (or between two

Fig. 13.5 Direction of the
interference lines can be
demonstrated by placing two
sets of concentric circles,
with the centre of each set in
the middle of a slit

maxima) is usually quite small. This means that we can in principle get an interference
pattern consisting of very many parallel bright strips on the screen with dark spaces
in between. Thus, we will not get just *two* strips, as a particle model of light would
predict.

How many strips do we really get? Well, it depends on $E_0^2(r, \theta)$. If we use Huygens's principle and only use one elementary wave, it should have the same intensity
in all directions (where the wave can expand). But the gap (between the slits) cannot be infinitesimally narrow, for in that case virtually no light would pass through.
When the slit has a finite width, we should actually let elementary waves start at any
point in the slit. These elementary waves will set up a total wave for slit 1 and a total
wave for slit 2, which will *not* have the same electric field in all directions θ. We will
address this problem below (diffraction from one slit).

Since $E_0^2(r, \theta)$ will be large only for a relatively narrow angular range, we get a
limited number of fringes on the screen when we collect the light from the double
slit. This will be treated later in this chapter.

In Fig. 13.5, we finally show a fairly common way of illustrating interference
by a double slit. With the centre in each of the two slits (and in a plane normal to
and in the middle of the slits), wavefronts are drawn, characterized by the property
that electric fields is, for example, maximum in a direction normal to the plane under
consideration. At all places where the crest (top, peak) of a wave from one slit overlaps
the crest of a wave from the other slit, there will be a constructive interference and
we will get maximum electric field. These are places where the circles cross each
other.

The places where a wave crest from a slit overlaps a wave trough (valley, minimum)
from the second slit (i.e. in the middle of two circles from this slit), there will be a
destructive interference and we will have almost a negligible electric field.

We can see from the figure that positions with constructive interference lie along
lines that radiate approximately midway between the two slits. It is in these directions

that we get the bright strips in the interference pattern from a double slit. In the midst of these, there is destructive interference and little or no light.

It is instructive to demonstrate how the angles between the directions of constructive interference change as we vary the distance between the centres in the circular patterns.

13.3.1 Interference Filters, Interference from a Thin Film

We have previously seen that when we send light towards a flat interface between air and glass, about 5% of the light is reflected at the surface (even more at larger angles of incidence). Such reflection deteriorates the contrast and image quality in general if lenses in a binocular or a camera are not given an anti-reflection treatment. But how can we deposit such a coating on a lens?

Figure 13.6 shows schematically how we can proceed. We put a thin layer of some transparent substance on the outside of the glass and choose a material that has a refractive index about halfway between the refractive indices of air and glass. We will then reflect about as much light from the air–coating interface as from coating–glass interface. If we ignore yet another reflection (in the return beam), we see that light reflected from the upper and lower layers will have the same direction when they return to the air. The two "rays" will superpose. If the two have opposite phase, they will extinguish each other. This means that the light *actually* reflected will (on the whole) be significantly less intense than if the coating was not present.

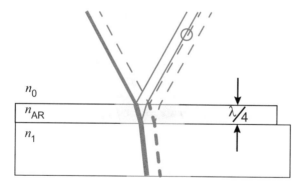

Fig. 13.6 An anti-reflection treatment of a lens or spectacle consists of a thin transparent layer with a refractive index roughly halfway between the refractive indices of air and glass. The layer must be about a quarter wavelength thick for the wavelengths where the filter has to give the best performance. A beam of light that is slightly inclined towards the surface is drawn to produce the sum of a part of the wave reflected on the surface of the anti-reflection layer (dashed) and a part of the wave reflected from the surface of the glass itself (solid line). The overlap between these is marked with a circle

Fig. 13.7 Play of colours in a cracked ice chunk

By carefully selecting all parameters, we can determine whether there will be destructive or constructive interference. In the first case, we get an anti-reflective layer as already shown. In the second case, we get more reflection. In this case, a coating consisting of several layers on top of each other is often used and the parameters are chosen so that light reflected everywhere comes in phase with other reflections and that the light transmitted from different layers is always out of phase with other transmission contributions. In this way, it is possible to make mirrors that can have more than 99.9% reflection for a particular wavelength and for a particular direction of a beam of light towards the mirror, while at other wavelengths we can look across the mirror! It is quite nice to experience such mirrors!

In nature and everyday life, thin films form spontaneously, for example in thin cracks or thin air layers between two glass plates. For example, if we put a "watch glass" (slightly curved glass for covering the dial of a pocket watch) on top of a flat glass surface, we get constructive and destructive interference between light reflected at the interfaces between air and the curved and flat glass surfaces. Since the effect is wavelength dependent, the circles are coloured and they are called Newton's rings.

In Fig. 13.7, one finds another example of the same effect. There is a chunk of ice in which a slight crack has occurred after a blow against the piece, and the play of colours is evident.

13.4 Many Parallel Slits (Grating)

If we have many parallel slits with the same mutual distance d, and if we collect the light on a screen far from the slits (compared to d), we get a situation that can be analysed in much the same way as the double slit. The difference is that we must sum up contributions from all N slits (see Fig. 13.8).

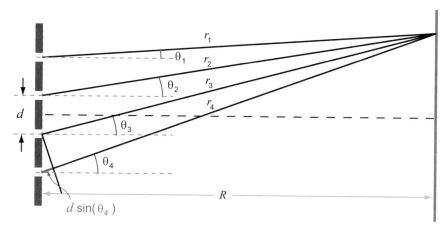

Fig. 13.8 Starting point for the mathematics for many slits is likewise that for the double slit. In practice, we often have several hundred slits per mm, illuminated by a light beam with a diameter of the order 1 mm. The screen is often 1 m away or more. Thus, all r_n are nearly equal and likewise for the θ_n. This simplifies the calculations

The resultant field will be:

$$E_{tot}(\theta) = E_1 + E_2 + \cdots + E_N$$

$$= E_0(r,\theta)\big[\cos(kr_1 - \omega t - \psi) + \cos(kr_2 - \omega t - \psi) + \cdots + \cos(kr_N - \omega t - \psi)\big].$$

In order to simplify the calculations further, we note that the absolute phase ψ relative to the selected position and time is uninteresting. When we only look at the time-averaged intensity, only phase differences due to different path lengths of the various elementary waves will count. For a given angle θ, the difference between two adjacent elementary waves will be given by $d \sin \theta$. This path difference represents a phase difference ϕ, and we have already shown above that this phase difference is given by $\phi = 2\pi d \sin \theta / \lambda$.

Notice:
The following page (slightly more) is a pure mathematical treatment and adds nothing to the physics. You may skip this and jump directly to the next figure and/or the grey marked text starting with "The intensity pattern ...".

Notice that if we start from one slit, the phase difference to the next will be ϕ, then 2ϕ, the next 3ϕ, etc. Then, we can write the resultant field in this simplified way:

$$E_{tot}(\theta) = E_0(r,\theta)\big\{\cos \omega t + \cos(\omega t + \phi) + \cos(\omega t + 2\phi) + \cdots + \cos[\omega t + (N-1)\phi]\big\},$$

$$E_{\text{tot}}(\theta) = E_0(r, \theta) \sum_{n=0}^{N-1} \cos(\omega t + n\phi) \ .$$

We use now Euler's formula $e^{i\theta} = \cos\theta + i\sin\theta$ and the symbol $\Re$ as before for taking the real part of a complex expression and get:

$$\sum_{n=0}^{N-1} \cos(\omega t + n\phi) = \Re \sum_{n=0}^{N-1} e^{i(\omega t + n\phi)} = \Re \left(e^{i\omega t} \sum_{n=0}^{N-1} e^{in\phi} \right) \ .$$

From the mathematics, we know that the sum of a geometric series with common ratio k can be written as:

$$1 + k + k^2 + \cdots + k^{N-1} = \sum_{n=0}^{N-1} k^n = \frac{k^N - 1}{k - 1} \ .$$

Applying this relation to the sum $\sum_{n=0}^{N-1} e^{in\phi}$ (k standing for $e^{i\phi}$), we get:

$$\sum_{n=0}^{N-1} \cos(\omega t + n\phi) = \Re \left(e^{i\omega t} \sum_{n=0}^{N-1} e^{in\phi} \right) = \Re \left(e^{i\omega t} \frac{e^{iN\phi} - 1}{e^{i\phi} - 1} \right)$$

$$= \Re \left(e^{i\omega t} \frac{e^{iN\phi/2}}{e^{i\phi/2}} \frac{e^{iN\phi/2} - e^{-iN\phi/2}}{e^{i\phi/2} - e^{-i\phi/2}} \right)$$

$$= \Re \left(e^{i\omega t} e^{iN\phi/2 - i\phi/2} \frac{2i\sin\frac{N\phi}{2}}{2i\sin\frac{\phi}{2}} \right)$$

$$= \Re \left(e^{i(\omega t + N\phi/2 - \phi/2)} \frac{\sin\frac{N\phi}{2}}{\sin\frac{\phi}{2}} \right)$$

$$= \cos(\omega t + N\phi/2 - \phi/2) \frac{\sin\frac{N\phi}{2}}{\sin\frac{\phi}{2}} \ .$$

Combining this with earlier expressions, the electric field in the direction of θ will be:

$$E_{\text{tot}}(\theta) = E_0(r, \theta) \cos\left(\omega t + \frac{N\phi}{2} - \frac{\phi}{2} \right) \frac{\sin\frac{N\phi}{2}}{\sin\frac{\phi}{2}} \ .$$

In the same way as for the double slit, we are interested in the intensity of the interference pattern we can observe. Again we have:

$$I(\theta, t) = c\varepsilon E_{\text{tot}}^2(\theta, t) \, .$$

When the time average is calculated, $\overline{\cos^2(\omega t + \frac{N\phi}{2} - \frac{\phi}{2})} = \frac{1}{2}$ as before. Accordingly:

$$I(\theta) = \frac{1}{2} c\varepsilon E_0^2(r, \theta) \left[\frac{\sin \frac{N\phi}{2}}{\sin \frac{\phi}{2}} \right]^2$$

$$I(\theta) = \frac{1}{2} c\varepsilon E_0^2(r, \theta) \left[\frac{\sin \frac{N\phi}{2}}{\sin \frac{\phi}{2}} \right]^2 . \tag{13.2}$$

The intensity pattern is then described by:

$$I(\theta) = I_0(r, \theta) \left[\frac{\sin \frac{N\phi}{2}}{\sin \frac{\phi}{2}} \right]^2 . \tag{13.3}$$

where $I_0(r, \theta)$ is the intensity contribution to the light passing *one* of the N slits and $\phi = 2\pi d \sin \theta / \lambda$ is the phase difference between two neighbour slits for the actual θ.

We can show (using L'Hôpital's rule) that when ϕ goes to zero, the expression inside the square brackets goes to N. That is, the intensity of the strip found at $\phi = 0$ becomes N^2 times the intensity we had from one slit only. The other maxima we find for $\sin \frac{\phi}{2} = 0$ (assuming we ignore the angular dependence of $E_0^2(r, \theta)$). It follows that maxima will occur when:

$$\sin(\pi d \sin \theta / \lambda) = 0$$

or, equivalently, when:

$$m\pi = \pi d \sin \theta / \lambda, \quad (m = \ldots, -2, -1, 0, 1, 2, \ldots)$$

$$\sin \theta = \frac{m\lambda}{d} . \tag{13.4}$$

These are the same directions as for interference maxima for a double slit.

We see that the positions of the intensity maxima are independent of the N, the number of slits.

Fig. 13.9 Intensity
distribution versus angle for
2, 8 and 32 slits

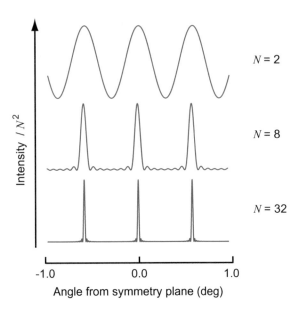

Figure 13.9 shows how the intensity distribution is for slightly different numbers of slits. We see that the most distinctive feature is that the peaks are becoming more pronounced when the number of slits increases.

Difference among the various interference lines

Equation (13.4) can be rewritten like this: $d \sin \theta = m\lambda$.

Thus, light from the same wavefront contribute equally with identical phase to the centre line ($m = 0$) of the diffraction pattern.

However, there is one wavelength difference between light passing one slit and light passing a neighbour slit, for light-contributions to the first line on each side of the central one ($m = \pm 1$) in the diffraction pattern. Thus, if 100 slits are illuminated, it will be 99 wavelengths difference between the contribution from slit 1 and slit 100.

For the second line on each side of the central one ($m = \pm 2$), there will be two wavelengths difference between contributions from one slit and the neighbour slit.

Thus, in order to have a result in agreement with our derivation in practice, it requires a very regular wave. We will discuss this a bit more when we talk about temporal coherence in Chap. 15.

It can be shown that the half-width of the peaks are given by:

$$\Delta\theta_{1/2} = \frac{1}{N\sqrt{\left(\dfrac{d}{\lambda}\right)^2 - m^2}} \tag{13.5}$$

where m gives, as before, the order. We see that the central line $m = 0$ has the smallest line width and that the line width increases when we examine lines further and further from the centre (question: Can $(d/\lambda)^2 - m^2$ be negative?).

In Fig. 13.10, we have drawn the same curves as in Fig. 13.9, but now with logarithmic y-axis. The purpose is to show details of the small peaks between the main peaks. We see that the small peak nearest to a main peak is about three log units (close to a factor 1000) less than the main peak. There are no dramatic deviations from this rule even though we change the number of slits significantly. However, we see that the width of each principal peak decreases with the number of slits, even if we include a few small peaks on each side of the principal peak. Furthermore, the logarithmic plot shows that the intensity of the main peaks relative to the minor peak approximately midway increases dramatically with the number of slits.

13.4.1 Examples of Interference from a Grating

We looked in Eq. (13.4) that the angle between the fringes in the interference pattern from a grating depends on the relationship between the wavelength and the slit separation in the grating. It is a blessedly simple relationship. An angle is easy to

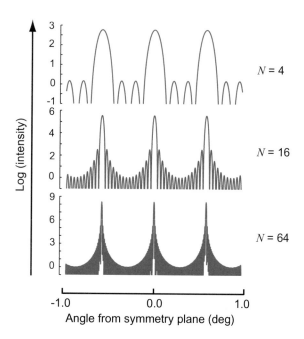

Fig. 13.10 Intensity distribution versus angle of 4, 16 and 64 slits but now drawn in logarithmic scale along the y-axis to study details near the zero line between the main peaks

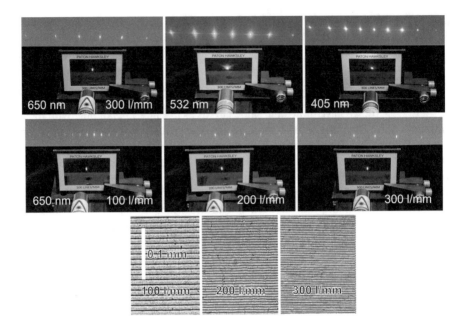

Fig. 13.11 Experimental images showing how the distance between lines (dots) in an interference pattern varies with wavelength and distance between the slits in the gratings (indicated as the number of lines per mm in the grating)

measure, and the distance between the slits in a grating is quite easy to measure (but the slits are tightly spaced, and we need a good microscope). And then, the wavelength of light is the last parameter.

In Fig. 13.11, we show examples of how the fringes look like (almost like dots, since we have used lasers with a rather narrow beam). The pictures in the top row show how the distance between the dots changes when the wavelength of the light changes. The gap between the slits in the grating is always the same (the "grating constant" is 300 lines per mm). We notice that red light gives the *largest* angles. This is in a way the opposite of what we saw in dispersion. Red light, on being dispersed by a glass prism, suffered the *smallest* deviation.

In the middle row of pictures in Fig. 13.11, we use red light all the time, but changed the distance between the slits in the grating. We see that the distance between the dots on the screen increases when the gap between the slits becomes smaller (when the number of lines per mm increases), completely in accord with Eq. (13.4). The bottom row shows photographs, taken through a microscope, of the three gratings used.

Experiments like these show that, in one way or another, wavelength must be a central part in the description of light and that wavelength must have a link with real distances in space since there are only distances in space in the gratings that vary in the experiments when we switch from one grating to another.

13.5   Diffraction from One Slit

> Suppose that we now have a single slit illuminated from one side with plane
> polarized waves with wavefront parallel to the "surface" of the slit. We can
> *model* the slit as a grating where the slits lie so close and are so wide that they
> completely overlap one another. If the single slit has an opening of width a,
> we can imagine that it consists of N parallel sub-slits with a centre-to-centre
> distance (from the centre of a sub-slit to the centre of the neighbouring sub-slit)
> $d = a/N$.

There are two different methods for calculating the light intensity on a screen
after the slit. The simplest method is based on an approach where the screen is
thought to be very far away from the slit, compared to both the width of the slit and
the wavelength. This case is called Fraunhofer diffraction and is characterized by
the supposition that the amplitude of the electric field from each of the sub-slits is
approximately identical on the screen and that the angle from a sub-slit to a given
position on the screen is approximately equal to the angle from another subdivision
to the same position.

If the distance between the slit and the screen is not very large relative to the slit
width and/or wavelength, we must use more accurate expressions for contributions to
the amplitude and angles. This case is called Fresnel diffraction and is more difficult
to handle than Fraunhofer diffraction. The difficulties can be surmounted by using
numerical methods, and we will return to this topic later in the chapter.

Let us now go back to the simple Fraunhofer diffraction, where we consider a slit
composed of N narrow parallel slots that lie edge to edge. We can now use a similar
expression as for the grating, Eqs. (13.2) and (13.3), if we replace d with a/N and
correct for amplitudes at the different slits. In the expression of the phase difference
ϕ, we now get the following relation:

$$\phi = 2\pi \frac{d \sin \theta}{\lambda} = 2\pi \frac{a \sin \theta}{N\lambda} = \frac{2\alpha}{N}$$

where

$$\alpha = \pi \frac{a \sin \theta}{\lambda} \ . \tag{13.6}$$

For the interference pattern for N equal slits in Eq. (13.3), we found that the
intensity peaks were N^2 times the intensity evolving from each slit. Since intensities
are proportional to electric field amplitudes squared, this corresponds to an efficient
amplitude of the sum signal that is N times the amplitude due to each slit alone.
This is the case since the contributions of light to a peak in the interference pattern
are exactly in phase with each other, whichever slit it went through (assuming high
coherence).

For the diffraction from one slit, *the contributions to the diffraction pattern from
all fictitious sub-slits will never add with the same phase.* We therefore have to assign

an effective amplitude E_{ss} for each sub-slit like

$$E_{ss} = E_{tot}/N \tag{13.7}$$

where E_{tot} is an "effective amplitude" attributed to the entire slit. This is by no means a strict mathematical description, but it provides a qualitative explanation why we need to use the $1/N$ factor for the amplitudes when we divide the slit into N sub-slits.

By applying the variables of Eqs. (13.6) and (13.7) in the expression for the total intensity distribution, according to Eq. (13.2), the result is:

$$I(r,\theta) = \frac{1}{2}c\varepsilon E_{ss}^2(r,\theta)\left[\frac{\sin\frac{N\phi}{2}}{\sin\frac{\phi}{2}}\right]^2 = \frac{1}{2}c\varepsilon\frac{E_{tot}^2(r,\theta)}{N^2}\left[\frac{\sin\alpha}{\sin\frac{\alpha}{N}}\right]^2 .$$

When N is chosen to be very large, the angle α/N will be so small that $\sin\frac{\alpha}{N} \approx \frac{\alpha}{N}$. The intensity distribution can then be written as:

$$I(r,\theta) = \frac{1}{2}c\varepsilon\frac{E_{tot}^2(r,\theta)}{N^2}\left[\frac{\sin\alpha}{\frac{\alpha}{N}}\right]^2 = \frac{1}{2}c\varepsilon E_{tot}^2(r,\theta)\left[\frac{\sin\alpha}{\alpha}\right]^2 .$$

When $\theta \to 0$, α also goes to zero, and $\sin\alpha/\alpha$ approaches unity, so that:

$$I(r,0) = \frac{1}{2}c\varepsilon E_{tot}^2(r,\theta) \equiv I_0(r) .$$

The intensity distribution in the single-slit diffraction pattern thus takes the form (see Fig. 13.12):

$$I(r,\theta) = I_0(r)\left[\frac{\sin\alpha}{\alpha}\right]^2 \tag{13.8}$$

where

$$\alpha = \frac{\pi a}{\lambda}\sin\theta .$$

The intensity vanishes when

$$\alpha = n\pi$$

where n is a positive or negative integer. This happens when

$$\sin\theta = n\frac{\lambda}{a} . \tag{13.9}$$

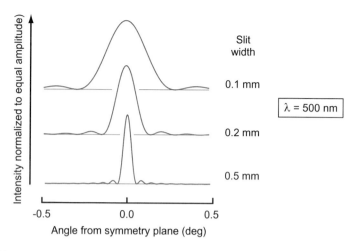

Fig. 13.12 Intensity distribution for strips after a single slit

It can be shown that the maxima lie approximately midway between the angles where the intensity is zero.

At first sight, Eq. (13.8) might look quite similar to the intensity distribution from a diffraction grating. However, it is a considerable difference.

The angle between the central top and the first minimum for a grating is given by:

$$\phi = 2\pi \frac{d \sin \theta}{\lambda} = \frac{2\pi}{N},$$

$$\sin \theta = \frac{\lambda}{Nd}.$$

Thus, the width of the central top is becoming narrower as the number of slits increases, and it does not depend on this approximation of the width of each slit.

For the single slit, however, we thought that the slit was divided into N close-lying sub-slits, the separation between which is a/N. However, the width does not depend on this fictitious division of sub-slits. The width depends on the width of the single slit only.

The angle that gives the first zero of the intensity distribution for a single slit can be determined through a different and simple argument. Figure 13.13 shows how we can think that a pair of fictitious sub-slits a distance $a/2$ apart from each other work together to get destructive interference for *all* the light passing through the slit.

We also see from the figure that the minimum for diffraction from one slit must always exist at a larger angle than that for diffraction from two or more separate slits (since the distance d between the slits must necessarily be greater than or equal to the slit width in a grating). In other words, the angular distance of the first minimum of

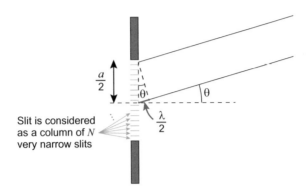

Fig. 13.13 Geometric conditions showing the direction for which the intensity of diffraction from a single slit will be zero. For any choice of a pair of fictitious sub-slits a distance $a/2$ apart, the difference in the light path will be equal to half a wavelength (which yields destructive interference)

a grating can easily be much less than for the angular distance to the first minimum in the diffraction pattern.

We can calculate the half-value for the intensity distribution from the single-slit pattern using Eq. (13.5) for a grating, but again replace the slit separation d with our fictitious gap a/N. Thus, we get:

$$\Delta\theta_{1/2} = \frac{1}{N\sqrt{\left(\dfrac{a}{N\lambda}\right)^2 - m^2}}$$

$$= \frac{1}{\sqrt{\left(\dfrac{a}{\lambda}\right)^2 - (Nm)^2}} \; .$$

The half-width for the central peak in the single-slit pattern comes out to be ($m = 0$):

$$\Delta\theta_{1/2} = \frac{\lambda}{a} \; .$$

We find, of course, that the expression does not depend on N.

A typical intensity distribution in the single-slit pattern looks approximately as shown in Figs. 13.12 and 13.14. There is a distinctive central peak with weak stripes on the side. It can easily be shown that we do not get more marked peaks than the central top (since the denominator never gets zero except for the central top).

Fig. 13.14 An example of the observed intensity distribution in a single-slit diffraction pattern. The central band is overexposed to make the side bands come out well

13.6 Combined Effect

In the development of the expression of the intensity distribution from a single slit, we did not pay particular attention to the fact that the strength of the electric field will vary with the angle θ. In treating the double slit and grating, we placed more emphasis on this. The reason is that it is actually the underlying diffraction from each slit that forms the envelope for $E_0^2(r, \theta)$! We do not get the clearest fringe pattern from a double slit or from a grating to extend beyond the central peak of the diffraction image from each single slit.

In practice, therefore, we will always have a combined effect of diffraction from a single slit and interference from two or more simultaneous slits. Figure 13.15 shows the combined effect of diffraction from each of the two parallel slits and interference due to the fact that we have two slits. The example is chosen to match an optimal double-slit experiment where there are a significant number of clearly visible fringes within the central diffraction peak.

13.7 Physical Mechanisms Behind Diffraction

So far, we have used the Huygens–Fresnel principle to calculate mathematically what intensity distributions we get from interference and diffraction. But what are the physical mechanisms behind diffraction? There are several different descriptions; among other things, it is popular to invoke Heisenberg's uncertainty relationship for this purpose. It is an "explanation" that does not really go back to physical

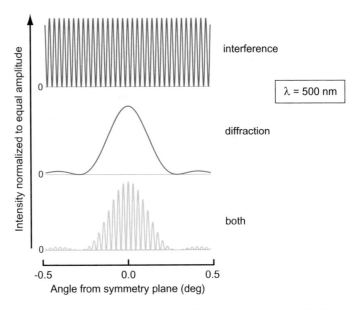

Fig. 13.15 Calculated intensity distribution for the fringe pattern from a double slit when each slit is 200 wavelengths wide and the gap between the slit centres is 2000 wavelengths

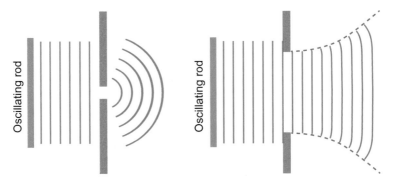

Fig. 13.16 Diffraction may appear for surface waves on water. Waves can be generated by a rod oscillating up and down the water surface. When waves are sent to a wall with a gap, the waves behind the wall get a form dictated by diffraction

mechanisms, and it is just a mathematical game. We will try to figure out more physical mechanisms for the phenomenon.

We choose to look at plane waves created by letting an oscillating rod go in and out of a water surface (see Fig. 13.16). The waves so generated move towards a vertical wall with a vertical opening (slit). The waves are parallel to the wall. The waves are only (approximately) plane over a limited length, but are at least so long that they cover the entire opening in the wall.

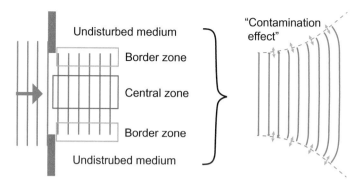

Fig. 13.17 With the starting point in the previous figure, we have tried to illustrate the situation when diffraction does not occur. See the text for further discussion of this hypothetical case

Diffraction causes the waves on the opposite side of the wall to adopt a fan shape if the slit is narrow (e.g. with a width approximately equal to the wavelength). If the gap is much wider, that is, a few wavelengths wide, the diffraction will lead to waves approximately as shown in the right part of the figure. The question is then: What are the mechanisms behind this diffraction?

In Fig. 13.17, we have shown in the left part how the waves would go after the wide gap if there is no diffraction. Then, the waves would continue as a train of waves having the same length as the width of the slit.

In the central part of the waves, the wave will initially continue as before. This is because the neighbouring area of every part of a wave in the central zone consists of waves that move alike. There are no possibilities for leakage sideways, and the wave continues as before.

In the border regions, the situation is very different. Try to visualize yourself a water wave that is sliced laterally and moves the water surface with a wave on one side and perfectly plane water surface on the other side of the partition. It would just not work out! Water from the wave edge would affect water that was originally thought to be outside the wave. There must be a continuous water surface also along the demarcation. This situation would give rise to a "contamination process" where energy is stolen from the peripheral areas of the waves and fed into the area where there would have been a flat, calm water surface without diffraction.

The contamination process will continue all the time, and the wave will therefore become wider and wider. The very same mechanisms lie behind the contamination process as those which propagate the wave and give it a definite speed. As a result, the diffraction pattern becomes almost the same for any diffraction situation as long as we scale the slit width with the wavelength. The waves will eventually be curved at the edges. Also, the region we called the central zone will eventually feel the influence of the edge, causing the wavefront to take the form of an almost perfect arc when it is far from the slit relative to the width of the slit. The radius of curvature will eventually equal the distance from the slit to the wavefront we consider (i.e. the waves far from the gap look as if they come from a point in the middle of the opening).

A physical explanation model completely analogous to that used for water waves can be applied to electromagnetic waves. It is impossible to have an electromagnetic field in a continuous medium (or vacuum) where there is a sharp separation between an area with a significant electromagnetic field and an adjacent area (all the way up to the previous) where there is no electromagnetic field. Maxwell's equations will ensure that the electromagnetic field will contaminate the area that without diffraction would be without fields and we have continuity requirements just like surface waves on water.

The key point is that a wave entails an energy transfer from a region in space to the neighbouring area, and such an energy transfer will always take place if there are physical differences between the regions, provided that there is actually a connection between the two areas.

Any situation where we create side effects between regions with waves and adjacent regions without waves (where the two are in contact with each other) is a source of contamination and thus diffraction. Contamination can propagate and appear even after the wave has moved far from the spatial constraints that created the unevenness in wave intensity.

13.8 Diffraction, Other Considerations

We have derived above the intensity distribution of the light falling on a screen after it has passed a narrow slit. The intensity distribution just after the slit can be considered a spatial square pulse. However, the intensity captured on a screen at a large distance shows an intense bell-shaped central peak with fainter lines on either side (see Fig. 13.12). The two closest side peaks have the intensity 4.72 and 1.65% of the intensity of the central maximum. Is there something magical about this change from a square to a bell-shaped intensity distribution? In a way, there *is*.

Figure 13.18 shows the square of the Fourier transform of the product of a sine curve and a square function. The Fourier transformed curve has the exact same shape as the intensity distribution we calculated for diffraction from a single slit. This is an example of a part of the optics called "Fourier optics".

If we multiply a sine function with a Gaussian curve instead of a square function, the square of the Fourier transformed becomes a pure Gaussian curve. If we start experimentally with a Gaussian intensity distribution in a beam, the beam can be made either narrower or wider, using lenses and diffraction, and still retain its Gaussian intensity distribution. In other words, diffraction will not cause any peaks beyond the centre line when the beam has a Gaussian intensity distribution.

It can be shown more generally that the intensity distribution for diffraction from a slit is closely related to the intensity distribution of the beam of light we start with. In other words, intensity distribution can be regarded as a form of "boundary conditions" when a wave spreads out after hitting materials that limit its motion.

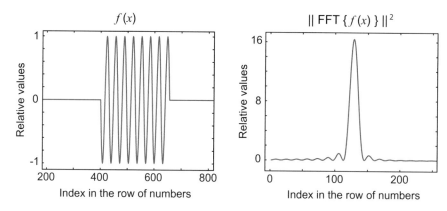

Fig. 13.18 If a sinusoidal signal is multiplied by a square pulse, we get a signal as shown in the left part of the figure (only the interesting area is included). Here, 4096 points are used in the description, the sine signal has 32 points per period, and the square pulse is chosen so that we get eight full periods within the square. If this signal is Fourier transformed and we calculate the square of the absolute value of the Fourier coefficients, we get the curve shown to the right of the figure (only the interesting area is included). The curve has the exact same shape as the curve we calculated for diffraction from a single slit

Modern optics often uses laser beams with Gaussian intensity distribution across the beam. Then, the beam shape will be retained even after the beam is subjected to diffraction.

A beautiful formalism based on matrices (called the ABCD method) has been developed that can be used to calculate how diffraction changes the size of a laser beam (assuming the intensity profile is Gaussian). In this formalism, first and foremost two quantities are included, which are of prime importance for the development of such a beam. One is the diameter of the beam (diameter between points where intensity has fallen to $1/e^2$ of the peak value). The second parameter is the radius of curvature of the wavefront as a function of position along the beam. The formalism is based on "small angles". This is for your orientation.

Test yourself:
The information given in the caption to Fig. 13.18 is associated with the figure itself. If you want to test how much you remember from Fourier transformation, try to answer the following questions:

1. Can you explain why the top of the right part of the figure ends where it is?
2. Is there any connection between the position where the square pulse occurred in the left part of the figure and the position/intensity in the right part of the figure? Explain as usually the answer!
3. If the left-hand square pulse was only half as wide as in our case, how would you expect the right figure to look like?

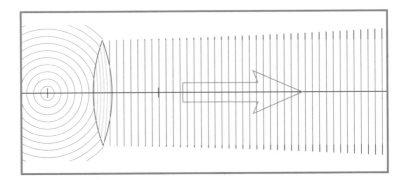

Fig. 13.19 When light from a light source is sent through a lens as shown in this wavefront diagram, diffraction will affect the edge of the light beam (highlighted in red). It is very easy to create this situation, while it is almost impossible to create the reverse process in practice (which corresponds to the reversal of the time flow)

13.8.1 The Arrow of Time

The laws of physics are often such that a process, in principle, works equally well both when time runs forwards and in the reverse direction. When we used light ray diagrams in the previous chapter, the diagram would have remained valid if we had switched the object and the image. The lens formula is also symmetrical in this sense.

The wavefront diagrams in the previous chapter will also be used (when ignoring some shadow effects) both forwards and backwards in time.

The conditions are different when we consider diffraction. In Fig. 13.19, we have examined how diffraction affects the beam shape after light has passed a lens. It is *in principle* possible to run the process backwards in time also in this case, but *in practice* it is impossible. It would require us to reproduce the wave conditions in the smallest detail.

Diffraction is therefore an example of physical processes in which the time arrow in practice cannot be reversed. Perhaps, the best-known example of a similar process is the diffusion of, e.g. molecules in a gas or liquid.

13.9 Numerical Calculation of Diffraction

The derivations we have carried out so far are based on analytical mathematics, which has given us closed-form expressions for intensity distributions in various diffraction patterns. These expressions, though absolutely priceless, are based on approaches that represent only some limiting cases of a far more complex reality.

We will now see how numerical methods can help us calculate diffraction patterns for a much larger range of variation in the parameters that enter a problem.

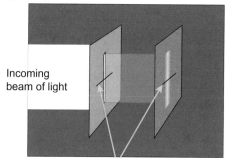

The "light source" is light   The diffraction pattern on a screen
passing through a slit   is the aim for our calculations

Incoming
beam of light

Along these two lines the light source
and the difraction picture is desribed

Fig. 13.20   Sketch showing where we describe the light source and the diffraction image by calculating diffraction at a slit

For simplicity, we consider the calculation of diffraction at a slit. Light is incident normally on a flat surface with a rectangular opening, a slit whose length is
much larger than the width. The light distribution after the slit has then approximately a cylindrical symmetry, and we therefore consider watching electric fields and
intensities along a one-dimensional line across the slit (see Fig. 13.20).

13.9.1   The Basic Model

The model for our numerical calculation is the same as that used for deriving the
analytical solutions, except that we do not have to make such drastic assumptions as
were introduced earlier. Figure 13.21 shows how we are going to proceed.

We will base our analysis on electromagnetic waves originating from N source
points along a line across the slit. The points have positions x_n ranging from $-a/2$
to $a/2$ since the width of the slit is a (see Fig. 13.21). The amplitude of the electric
field is A_n, so that the electromagnetic wave at the point x_n is

$$\vec{E}_n = A_n e^{i(kz - \omega t + \theta_n)} \vec{u}_n$$

where the symbols have their usual meanings, except $\vec{u}_n$, which is just a unit vector
that indicates the direction of the electric field (perpendicular to the direction of
motion of the wave at the specified location, assuming a plane polarized wave). θ_n
is an angle that gives relative phase from one point to another across the slit. If the
wavefront of the incoming light beam is parallel to the plane of the slit, all θ_n are
identical, and the parameter can then be dropped.

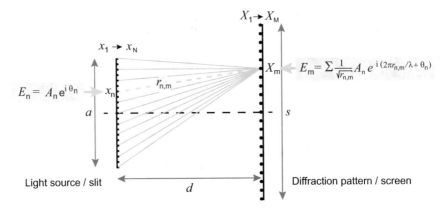

Fig. 13.21 Sketch that indicates how the Huygens–Fresnel principle is used in calculating diffraction from a slit

In our modelling of diffraction, we will take the starting point of electric fields at the same time in the entire slit. Then, $e^{-i\omega t}$ will be a constant phase factor that will disappear when intensities are to be calculated in the end. We therefore drop it at this point. Similarly, we will drop the corresponding factor when calculating the field on the screen where the diffraction image is captured.

If we put the slit in the xy-plane ($z = 0$), we end up with a simplified expression for the electric fields at different points *across the slit*:

$$\vec{E}_n = A_n e^{i\theta_n} \vec{u}_n \qquad (z = 0) . \tag{13.10}$$

Let us look at the diffraction pattern captured on a screen parallel to the slit, at a distance d from the slit. Numerically, we calculate the diffraction image in M points symmetrically positioned relative to the centre of the slit. The calculations span a width of s so that the position of the selected points X_m is from $-s/2$ to $s/2$. We must choose a suitable value for s to capture the interesting parts of the diffraction pattern (but not much more).

The electric field at any point X_m will be the sum of contributions from electromagnetic waves coming from all the points x_n in the slit. Since the distance $r_{n,m}$ between the relevant points changes as we pass all x_n, the contributions will have different phases at the screen. In addition, the distance differences make the amplitude of the electric field reduced. In total, we then get the following expression for summation of all contributions to the electric field at point X_m:

$$\vec{E}_m = \sum_n \frac{A_n}{\sqrt{r_{n,m}}} e^{i(2\pi r_{n,m}/\lambda + \theta_n)} \vec{u}_{n,m} \qquad (\text{since } kr = 2\pi r/\lambda) .$$

The expression is problematic, because there is no easy way to find the $\vec{u}_{n,m}$ on each electric field contribution (unless the light is polarized in a special way).

We are therefore more or less forced to process electric fields as scalar quantities in such formalism. As already mentioned earlier in this chapter, this is not a big problem when we consider the diffraction image far from the slit. However, very close to the slit, the scalar approach will be a clear source of error in our calculations.

The basic expression for numerical calculation of diffraction from a slit is then:

$$E_m = \sum_n \frac{A_n}{\sqrt{r_{n,m}}} e^{i(2\pi r_{n,m}/\lambda + \theta_n)} \tag{13.11}$$

where

$$r_{n,m} = \sqrt{d^2 + (X_m - x_n)^2} \, . \tag{13.12}$$

The intensity at any point is proportional to the square of the electric field.

Note that we have used the square root of the distance when calculating reduced electric field strength. This is because we have cylindrical symmetry. If we send out light along a line, the intensity through any cylindrical surface with the centre of the line will be the same. The area of the cylindrical surface is $2\pi r L$, where L is the length of the cylinder. Since the intensity is proportional to electric field strength squared, then the electric field itself must decrease as $1/\sqrt{r}$. Had we had spherical geometry, the intensity would have been distributed on spherical surfaces with an area of $4\pi r^2$, and the electric field would decrease as $1/r$.

13.9.2 Different Solutions

Calculations based on the expressions (13.11) and (13.12) may be demanding in some contexts, as calculations of sines, cosines, squares and square roots are included in each term. In addition, it needs $N \times M$ calculations. For modern computers, this is very affordable for straightforward calculations of diffraction. Nonetheless, if the diffraction calculations are included in more comprehensive calculations of image formation based on Fourier optics and more, the above expressions are in fact a bit too computer-intensive even today.

Historically, therefore, different simplifications have been made in relation to the above expressions in order to reduce the calculation time. In many current situations where we study diffraction images of light, $a \ll d$ and $s \ll d$ are in Fig. 13.21. We can then use a Taylor expansion in the expression of $r_{n,m}$ instead of Eq. (13.12). The result is (you may try to deduce the expression for yourself):

$$r_{n,m} = \sqrt{d^2 + (X_m - x_n)^2} \approx d \left(1 + \frac{1}{2} \frac{(X_m - x_n)^2}{d^2} - \frac{1}{8} \frac{(X_m - x_n)^4}{d^4} \right) . \tag{13.13}$$

In Eq. (13.11), the most important term, namely $r_{n,m}$, occurs in the factor $e^{i2\pi r_{n,m}/\lambda}$. If we substitute the approximate expression for $r_{n,m}$, we get:

$$e^{i2\pi r_{n,m}/\lambda} \approx e^{i2\pi d/\lambda}\; e^{i\pi \frac{(X_m - x_n)^2}{d}/\lambda}\; e^{-i\pi\frac{1}{4}\frac{(X_m - x_n)^4}{d^3}/\lambda}. \qquad (13.14)$$

or with a more readable way to write exponentials:

$$\exp\left[i2\pi r_{n,m}/\lambda\right] \approx \exp[i2\pi d/\lambda]\; \exp\left[i\pi\frac{(X_m - x_n)^2}{d}/\lambda\right]\; \exp\left[-i\pi\frac{1}{4}\frac{(X_m - x_n)^4}{d^3}/\lambda\right]$$

In different situations, some of these terms will be practically constant, and this is precisely the basis of some historical classifications of diffraction.

We will now try to provide an overview of different variants of computational accuracy:

1. **Less than a few wavelengths away from the edges of the slit**. Here, we must use Maxwell's equations and bring polarization and surface currents in the material surrounding the slit. "Evanescent waves" are part of the solution. (This is a complicated calculation!)
2. **For $d^3 \leq 2\pi a^4/\lambda$**. This is a problematic area where Maxwell's equations can be used for the smallest d, while the expressions (13.11) and (13.12) begin to work reasonably well for the largest d which satisfies the stated limit.
3. **For $d^3 \gg 2\pi a^4/\lambda$, we have Huygens–Fresnel diffraction**. The expressions (13.11) and (13.12) work. Even if we put $1/\sqrt{r_{n,m}} = 1/\sqrt{d}$ and we skip the last term of the Taylor expansion in Eq. (13.13), the result will be satisfactory.
4. **For $d \gg \pi a^2/\lambda$, we have Fraunhofer diffraction**. The expressions (13.11) and (13.12) work. Although we use the same approaches as for Huygens–Fresnel diffraction, and then $(X_m - x_n)^2 \approx X_m^2 + 2X_m x_n$ in the middle of the series of Eq. (13.13), the results will be satisfactory.

Figure 13.22 displays numeric calculations based on the expressions (13.11) and (13.12) directly. In the first case, we are relatively close to the slit (Huygens–Fresnel zone), while in the second case we are in the transition between the Huygens–Fresnel and Fraunhofer zones.

Note that when we are near the slit (Huygens–Fresnel zone), the diffraction image on the screen will have approximately the same size as the slit. However, some of the intensity at the edge of the slit leaks into the shadow section (marked with arrow in the figure), resulting in a continuous intensity distribution between shadow and full light intensity. We get characteristic fringe patterns in the image of the slit. There are larger "spatial wavelengths" on these fringes near the edge of the slit than towards the centre. There are only faint fringes in the shadow section on each side of the image of the slit.

Figure 13.23 shows a photograph of two diffraction patterns that have features similar to those used in the numerical calculation.

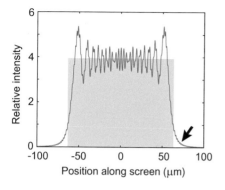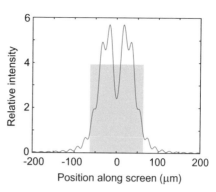

Fig. 13.22 Diffraction from a slit calculated from the Huygens–Fresnel principle. The left part of the figure corresponds to the screen being fairly close to the slit. The right part depicts the situation a little farther away from the slit, yet not quite as far as in Fraunhofer diffraction, which was treated analytically earlier in the chapter. The width of the slit is marked with a yellow rectangle

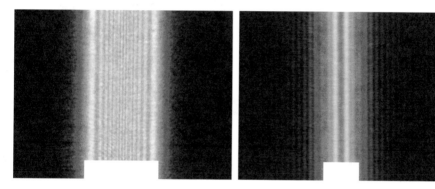

Fig. 13.23 Photograph of diffraction image of a slit with approximately the distances that were used in the calculations in Fig. 13.22 correspond to. The size of the slit is marked at the bottom

Similarly, we have shown calculations and an example of diffraction pattern in the border region between the Huygens–Fresnel and Fraunhofer zones in the right part of Figs. 13.22 and 13.23. We see some wavy features here both in the image of the slit and in the light falling on the shadow zone.

The Fraunhofer zone diffraction pattern is exactly the same as that derived analytically, and illustrative results are already given in Fig. 13.12 and a photograph in Fig. 13.14. In that case, we only have wavy features in the zone outside the central peak.

13.10 Diffraction from a Circular Hole

When a plane wave is sent to a circular hole, we also get diffraction (see Figs. 13.24 and 13.26), but it is more difficult to set up a mathematical analysis of that problem than for slits. As a result, the image that can be collected on a screen shows a distinctly

Fig. 13.24 Experimental set-up for observing diffraction from a circular hole

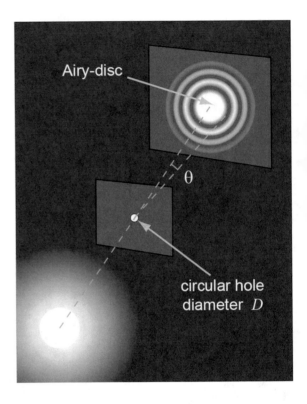

central bell-shaped peak, with weak circles. The central peak seems to form a circular disc, called the "Airy disc".

Mathematically, the intensity at an angular distance θ away from the centre line is given by:

$$I(\theta) = I_{\text{max}} \left[1 - J_0^2 (\tfrac{1}{2}kD \sin \theta) - J_1^2 (\tfrac{1}{2}kD \sin \theta) \right]$$

where J_n denotes the Bessel function of the first kind of order n, D is the diameter of the hole, and the k is the wavenumber. When the distance to the screen is much larger than the diameter of the hole, the intensity distribution becomes:

$$I(\theta) = I_{\text{max}} \left[\frac{2J_1(\tfrac{1}{2}kD \sin \theta)}{\tfrac{1}{2}kD \sin \theta} \right]^2$$

where the values of Bessel functions can easily be calculated numerically from the expression:

$$J_n(x) = \frac{1}{\pi} \int_0^\pi \cos(n\tau - x \sin \tau) d\tau \ .$$

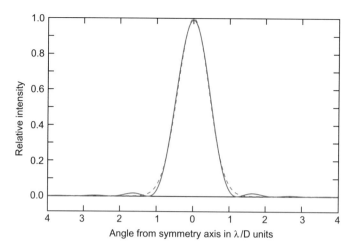

Fig. 13.25 Diffraction pattern for a circular hole far from the hole (*red*). The central peak that makes up the Airy disc has an intensity profile shape very close to a Gaussian (*blue dashed line*)

The angle for the first minimum is given by:

$$\sin \theta = \frac{1.22 \lambda}{D}$$

where D is, as stated above, the diameter of the hole. Since the angle is usually very small, we can use the approximation:

$$\theta = \frac{1.22 \lambda}{D} . \tag{13.15}$$

The next pair of dark rings have a radius of 2.232 λ/D and 3.238 λ/D, and the intensities of the first three rings are 1.75, 0.42 and 0.16% of the intensity at the centre of the central disc. See Figs. 13.25 and 13.26.

It is of interest to look into some details within the central peak both for a single slit and for an Airy disc. From Fig. 13.13, we can deduce that within the central peak of the single-slit diffraction pattern, the wavefront is very close to plane (given that the slit was illuminated by a plane wavefront). The deviation from the perfect plane wavefront is $\lambda/2$ or less. This is remarkable, since the central peak at a screen easily can be several thousand times as wide as the slit itself (for narrow slits).

Similarly, reasoning along the same lines, it can be shown that within the central peak (the Airy disc) of the diffraction pattern from a circular hole, the wavefront is very close to plane. The maximum deviation from the perfect plane wavefront is slightly larger than $\lambda/2$. Since the intensity of the rings around the Airy disc is much

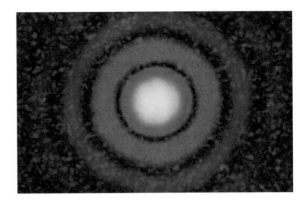

Fig. 13.26 Airy disc as it looks with some overexposure in the central portion to get the surrounding circles. Overexposure is difficult to avoid since the maximum intensity in the first ring is only 1.75% of the maximum intensity in the central disc. There are some "speckles" (bright spots), probably due to scattered laser light in the room

less than for the central peak, the diffraction leads to a kind of transformation from a narrow circular beam with constant intensity and flat wavefront throughout the cross section, to a much wider beam close to Gaussian intensity profile and a much larger cross section, but even so, with an almost flat wavefront.

The expression in Eq. (13.15) and the diffraction-of-light-through-a-circular-hole phenomenon has far-reaching consequences, and we shall mention some.

13.10.1 The Image of Stars in a Telescope

Light from a star comes towards a telescope. The light can be considered a plane wave when it reaches the objective, and the light is focused by a lens or a mirror. In geometric optics, we get the impression that we can collect all the light rays from a distant object at one point, the focal point, as indicated in the left part of Fig. 13.27. At the very least, it should be possible if the angular diameter of the object is very small, such as when we look at the stars in the sky. That is wrong!

The light beam from a star will follow a shape similar to that shown in the right part of Fig. 13.27. The light bundle has a minimum diameter of d which is significantly larger than what we would expect from the angular diameter of the object (the star). The reason is diffraction.

It is actually diffraction from a circular hole we witness. However, it is now a large hole with diameter equal to the diameter D of the objective of the telescope. According to Eq. (13.15), the angle to the first minimum will be very small. It will not be observable if we did not focus on the beam by the telescope objective.

We cannot use Eq. (13.15) directly to calculate the Airy disc size when we focus on the beam the way we do. We therefore use another kind of reasoning and refer to Fig. 13.28.

The light passes in the real life through the objective with diameter D, is focused and forms a diffraction picture where the Airy disc diameter is d in the waist and then continues to the right. However, waves can in principle move backwards as well

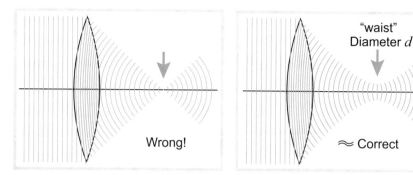

Fig. 13.27 According to ray optics, light from a tiny distant star should be focused on a point one focal length behind a telescope objective (*left*). However, due to diffraction the light beam will have a finite diameter d at its minimum before the beam size increases again (*right*). The part of the beam with minimum diameter is denoted the "waist". The Airy disc will in practice be *much* smaller than shown in the right part of this figure, but *in principle* it will have this pattern. Note also the difference in wavefronts in the two descriptions

Fig. 13.28 To get an estimate of the size of the Airy disc in the case where light passing through the telescope objective is focused, we can imagine that the waves are moving backwards *from* the Airy disc in the waist *towards the objective*. See text for details

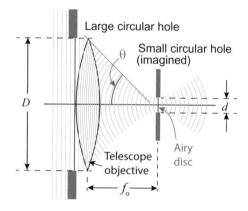

as forwards and follow the same pattern. Thus, we may imagine that we actually start with a reasonably flat wavefront passing through a small circular hole of the same size as the Airy disc, and let the wave move *to the left*. We should then have a situation not very differently from the starting point which lead to Eq. (13.15). The diameter of the beam should increase in size so that the diameter is roughly equal to D when the beam reaches the objective's position.

Based on the wave-moving-backwards argument, diffraction from the tiny Airy disc at the waist will cause the light beam to spread out at an angle θ (relative to the centre line of the beam)

$$\theta \approx \frac{1.22\,\lambda}{d}$$

This diverging conical light bundle due to diffraction must match in its extension the converging conical light beam from the convex lens. Thus:

$$\frac{1.22\,\lambda}{d} = \theta \approx \sin\theta \approx \tan\theta = \frac{D/2}{f_o}$$

where f_o is the focal length of the objective.

The radius of an Airy disc in the focal plane will then be:

$$\frac{d}{2} = \frac{1.22\,\lambda f_o}{D}\,.$$

The Airy disc of a star will have this extension, even though the angular extension in the sky is vanishing small. All stars will make equally sized luminous discs in the focal plane, but the intensity of the disc will reflect the brightness of the star under observation.

You will certainly have noticed that our argument with backward waves has obvious difficulties. We mix perfectly plane wavefronts with a wavefront not exactly flat, and we neglect the difference between a constant intensity across a hole and a more Gaussian intensity profile. We also neglect the rings around the Airy disc. Even so, a more rigorous treatment leads to roughly the same conclusion as we have arrived here.

The diffraction has important consequences. Two stars close to each other in the sky will form partially overlapping discs in the focal plane. If the overlap is very large, we will fail to notice that there are two discs, and will consider them as one. If the overlap is small, we will conclude that there are two discs, representing two stars.

Lord Rayleigh addressed this problem in the following manner:

When two objects (or details in objects) are viewed in a telescope, the ability to separate the two objects will reach its limit when the central maximum in the Airy disc of one object coincides with the first diffraction minimum of the other object. This description is known as **Rayleigh's Resolution Criterion**.

The minimum angle ψ where we can see that there are two Airy discs is then given by:

$$\psi \approx \frac{d/2}{f_o} = \frac{1.22\,\lambda}{D}\;. \tag{13.16}$$

In other words, with a lens of diameter D and focal length f_o, we can distinguish two stars (or other point-like objects) from each other if the angular distance between the stars is at least ψ.

Examples:

As we have just seen, we are not able to distinguish detail that subtends an angle of less than $1.22\lambda/D$, no matter how much we enlarge the image. For a prism binocular with an objective of about 5 cm diameter, the smallest angular distance we can resolve with 500 nm light becomes

$$\frac{1.22 \times 500 \times 10^{-9}}{0.05}$$

which corresponds to $0.00069°$. For the Mount Palomar telescope, with a mirror of 5 m diameter, the best resolution is 1/100 of this angle. The Mount Palomar telescope can resolve details that are approximately 50 m apart from each other on the moon, while a prism binocular will only be able to resolve details located 5 km from each other.

The diameter of the pupil in our eye is about 5–10 mm in the dark. This means that without the help of aids we can only distinguish details on the moon which is at least 25–50 km apart (the moon's diameter is 3474 km).

In a prism binocular, the magnification is almost always so small that we cannot see the Airy disc. In a telescope where we can change eyepieces and the magnification can be quite large, it is common to see the Airy discs. A star does not look like a point when viewed with a large magnification through a telescope. The star looks exactly like the diffraction image from a small circular opening on a screen, with a central disc (Airy disc) surrounded by weak rings. The rings are often so faint that it is hard to spot them.

The optical quality of many binoculars and telescopes are so poor, that e.g. spherical aberration, chromatic aberration or other imperfections so that we do not get a nice Airy disc if we enlarge the image of a star. Instead, we get a more or less irregularly illuminated surface that covers an even greater angular range than the Airy disc would have done. For such telescopes, we fail to resolve the fine details that the Rayleigh criterion indicates.

A telescope so perfect that its resolution is limited by the Airy disc is said to have *diffraction-limited optics*. This is a mark of excellence!

Today, it is possible to use numerical image processing in a smart way so that we can reduce the effect of diffraction. We theoretically know what intensity distribution we will get when light from a fictitious point source goes through the optical system we use (telescope or microscope). By an extensive iterative method, one can then slowly but surely generate an image with more details than the original. The image so can get close to represent what we would observe in the absence of diffraction. In this way, today, in favourable situations, we can attain about ten times better resolution in the images than can be achieved without the extensive digital image processing.

13.10.2 Divergence of a Light Beam

At the Alomar Observatory on Andøya, an ozone detector has been installed where a laser beam is sent 8–90 km up in the atmosphere to observe the composition and movements of molecules up there. The beam of light should be as narrow as possible far up there, and we can wonder how this may be achieved.

The first choice might be to apply a narrow laser beam directly from a laser. The beam is typically 1–2 mm in diameter. How wide would this beam be, for example, at a height of 30 km?

We use the relationship of diffraction from a circular hole and find the divergence angle θ:

$$\sin \theta = \frac{1.22 \times \lambda}{D} .$$

For light with wavelength 500 nm and an initial beam diameter of 2.0 mm, at the start, we get:

$$\sin \theta = \frac{1.22 \times 500 \times 10^{-9}}{0.002} = 3.05 \times 10^{-4} .$$

The angle is small, and if the radius of the beam at 30 km height is called $D_{30\,km}$, we find:

$$\frac{D_{30\,km}/2}{30\,km} = \tan \theta \approx \sin \theta = 3.05 \times 10^{-4}$$

$$D_{30\,km} = 18.3\,m .$$

In other words, the laser beam that was 2 mm in diameter at the ground has grown to 18 m in diameter at 30 km altitude!

An alternative is to expand the laser beam so that it starts out much wider than the 2 mm. Suppose we expand the beam so that it is actually $D = 50$ cm in diameter at the ground. Suppose the wavefront is flat at the ground so that the beam at the beginning is parallel (so-called waist) and eventually diverges.

How big will the diameter be at $R = 30$ km height?

We must be meticulous in stating the divergence angle:

$$\frac{D_{30\,km}/2 - D/2}{R} \approx \tan \theta \approx \sin \theta = \frac{1.22 \times \lambda}{D} .$$

On solving this equation for $D_{30\,km}$, we get 57.3 cm. In other words, a beam that starts out as 50 cm wide only becomes 57.3 cm wide at 30 km height! This is significantly better than if we start with a 2 mm thin beam.

We can, however, make things *even* better! We can choose not to place the laser (light source) exactly at the focal point of the 50 cm mirror we used in town (as a part of making the beam wide). If we place the laser slightly beyond the focal point, the beam will actually converge before it reaches the "waist" (corresponding to the Airy disc) and then diverges again. See Fig. 13.27. How small can we make the waist (Airy disc) at 30 km altitude?

We can then work *backwards* and consider the "waist" at 30 km height to be the *source* of a diverging beam (on both sides of the waist, since we have symmetry here). In that case, the beam will on its way from the waist to the mirror have diverged to D equal to 50 cm at the location of the mirror (imagining that the beam goes backwards). The calculator will look like this:

$$\frac{D/2 - D_{30\,km}/2}{R} \approx \tan\theta \approx \sin\theta = \frac{1.22 \times \lambda}{D}$$

$$D_{30\,km} = 42.7\,\text{cm} .$$

In other words, we can even get a smaller beam than the one we started with.

Conclusion: A laser beam that has a 2 mm diameter at the ground becomes 18 m in diameter at 30 km altitude. However, if we start with a beam of 50 cm diameter and focus it so that the waist will be found 30 km above the ground, the beam is "only" 43 cm in diameter at this height. The energy density in the cross section is then over 400 times as great as in the first case.

13.10.3   Other Examples *

1. Laser beams have often a "Gaussian intensity profile". It may be shown that if you send such a beam through mirrors and lenses, the Gaussian shape will be preserved ("beam optics"), even though the width is changed. We do not get any diffraction rings around the central beam of a Gaussian beam.
2. Diffraction takes place even in our eyes. The pupil's opening is typically 6 mm or less during daily tasks. Thus, diffraction sets the lower limit on the angular distance between two details in our visual field that we can distinguish. Another limitation for the resolving ability of the eye is the size of the light-sensitive cells in the retina. Evolution seems to have chosen an optimal solution since the size of the Airy discs is roughly the same as the effective area of our light-sensitive cells (rods and cones).
3. A camera is not necessarily well adapted. If we choose a sensor chip that gives many pixels per image, it does not necessarily mean that we can *exploit* this resolution. If the Airy disc for the selected lens and aperture (see Chap. 12) is larger than the size of a pixel in the sensor chip, the effective resolution in the image is not as good as the number of pixels indicates. You may test this on your own camera!

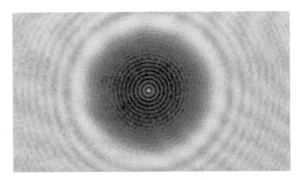

Fig. 13.29 Photograph of Arago's spot in the shadow image of a small ball. The ball was held in place by gluing it up with a small (!) drop of glue on a thin piece of microscope cover glass. In addition to Arago's spot, we see a number of details due to diffraction, both in the shadow section and in the illuminated party. Note that there is no clear boundary between shadow and light

4. As we have seen, the width of the central peak in the diffraction image from a single slit is given by:

$$\Delta\theta_{1/2} = \frac{\lambda}{a} . \tag{13.17}$$

In quantum physics, this result has occasionally been taken as a manifestation of Heisenberg's uncertainty relationship. Rightly enough, the expression in E-q. (13.17) may support this point of view if we threat the phenomena superficially, but there are so many other details in our descriptions of diffraction that the Heisenberg's uncertainty relationship cannot give us.

Also, in other parts of this book we have had relationships that are reminiscent of Heisenberg's uncertainty relationship. In all these situations, there are fundamental wave features that lie behind.

It is therefore no wonder that many today perceive Heisenberg's uncertainty relationship as a natural consequence of the wave nature of light and matter and that it has only a secondary link with the uncertainty of measurement.

5. Diffraction has played an important role in our perception of light. At the beginning of the nineteenth century, Poisson showed that if light had a wave nature and behaved according to Huygens's principle, we would expect to see a bright spot in the shadow image of an opaque sphere (or circular disc). Arago conducted the experiment and found that there was indeed a bright spot in the middle (see Fig. 13.29). The phenomenon now goes under the name of Arago's spot (or the Poisson–Arago spot).

Fig. 13.30 Photograph from a military grave field in San Diego, where the graves are placed very regularly. In some directions, we see many gravestones in line. If these tombs emit elemental waves, we would get interference patterns similar to those we have discussed for gratings in this chapter. We would get a series of interference fringe pattern, but the centre of each set would correspond to the different directions to the lines we see in the image. The distance between the lines in each set would depend on the distance between the source points along the direction we consider

13.10.4 Diffraction in Two and Three Dimensions

In our treatment of interference and diffraction, we have so far only considered the summation of waves from elementary wave sources located along a straight line. It is a normal situation for interference and diffraction of light.

For other types of waves, we can find diffracting centres that form two- or three-dimensional patterns. Most well known is perhaps X-ray diffraction. When X-rays are transmitted to a crystal of some substance, single atoms will spread the X-rays so that the elementary waves come from each individual atom in the area of the X-ray.

The atoms in a crystal are in a regular pattern. If we pick out atoms that are on a line in a plane, the elementary waves from these atoms will provide interference lines or interference points that can be calculated with similar equations as those we have been through in this chapter.

Both physics and chemistry provide so-called X-ray diffraction information that can be used to determine the structure of the crystals under investigation. It is this type of research that lies behind almost all the available detailed information about positions of the atoms in relation to each other in different substances.

Figure 13.30 illustrates that points that are regular to each other form lines that can cause interference/diffraction in many different directions.

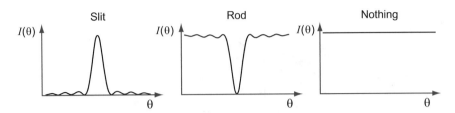

Fig. 13.31 Intensity distribution from a slit and a stick is complementary (really only when we operate at amplitude level and not at intensity level as here). In this case, "Nothing" means light with even intensity everywhere within the given θ interval

13.11 Babinet's Principle

The superposition principle can be used in a very special way where we utilize symmetries.

We have learned the form of the intensity distribution of light diffracted by a narrow slit. How would the interference pattern look at the complementary structure, which is a stick of exactly the same size as the slit? Babinet's principle tells us something about this:

> *Suppose that a wave is sent to a body* **A** *that leads to diffraction (e.g. a long slit a light-tight screen). We send the same wave to a body* **A'** *complementary to the first one (e.g. a long stick with the same position and width as the slit on the screen of* **A**), *and we see a different diffraction. If we overlap the diffracted wave in the first case with the diffracted wave in the other, we get a waveform that is identical to the one we would have had if neither* **A** *nor* **A'** *existed.*

Figure 13.31 shows the principle. The figure is a simplification, since we specify intensities, but *summation of waves occurs always at amplitude level.*

If we send a relatively narrow laser beam towards a gap and then to a thread of the same thickness as the slit, we can use the principle of Babinet to find out how the two diffraction patterns relate to each other. However, the relationships are quite different from the very wide light-beam/plane-wave situation we show in Fig. 13.31. Outside the narrow laser beam, the intensity is virtually zero when the gap or thread is not found in the light path. But with a slit or thread inside the narrow laser beam, we get diffraction patterns also in the area where there would otherwise be no light. This can be understood by the fact that the superposition principle can only be applied at amplitude level, not at intensity level. A wave $E = E\cos(kz - \omega t)$ will have the very same nonzero intensity as a wave $E = -E\cos(kz - \omega t)$, but the sum of these two waves is zero.

Fig. 13.32 Diffraction of a laser beam from a human hair

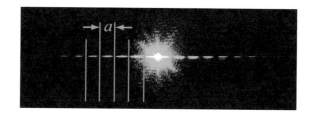

Babinet's principle is handy because we can use the theory of diffraction from a slit also by analysing the diffraction image from a thread. Figure 13.32 shows the diffraction image from a single hair placed in the beam from a laser pen. With very simple means, by measuring the distance between the minimum points between the light spots we can determine the thickness of the hair, provided we know the wavelength of the laser. A task at the end of this chapter provides a concrete example of how a boiled measurement may fall.

13.12 Matlab Code for Diverse Cases of Diffraction

Here is a Matlab program that can be used to see examples of how diffraction is affected by different initial light intensity patterns. The program is especially interesting for calculating diffraction when it is not a very long distance between, for example, a single (or double) slit and the screen where we capture the diffraction image. Then, there is a wealth of details that cannot be calculated analytically, but that matches what we can observe in experiments.

For all excitations, we assume that we have coherent light with wavefront in the excitation plane and that we have a form of cylindrical symmetry for each strip in the excitation plane.

The program must be used with a code (a number) as a parameter, such as:

```
diffraction(4)
```

if the intensity distribution on a screen after a double slit is to be calculated. Widths of gap, distance between slots and screens, etc., must be manually entered in the program (in function parameters). A bit of trial and error is required for the calculation area to cover the entire diffraction image we are interested in (but also not much more). Good luck!

Main Program

The code is available at the "supplementary material" Web page for this book at http://www.physics.uio.no/pow.

```
function diffraction(code)

% This program calculates and plots intensity patterns for a
% variety of diffraction and/or interpherence phenomena with
% cylindrical symmetry.
% Functionalities: code = 1: One slit, 2: Gaussian intensity
% profile, 3: Straight edge, 4: Double slit, 5: Read excitation
% data from file (amplitude + phase)
% Program is written by AIV. Version 15. October 2017

% Establishes essential parameters for the calculations.
% Results depend critically on these. See the code for this
% function.

[lambda,a,b,nWavel,N,twopi,Nhalf] = parameters;

% Allocates arrays for the calculations:
[x,x2,x0,x1,sc,r] = allocateArrays(nWavel,N);

% Generate or read in excitation data:
[x0] = generateExcitation(code,lambda,a,N,Nhalf,twopi,x0);

% Calculates sines, cosines, distances and relative phase
% differences for vectors between the plane of excitation and
% the screen for the final pattern:
[sc,r] = generateRelPositionData(N,b,lambda,twopi);

% Sum all contributions to every point on the screen (main
% loop):
[x1] = summation(N,x0,r);

% Plots intensities for diffraction pattern along with a
% marking of the excitation
plotDiffraction(x,x0,x1);

% Calculates and write out linewidths in case the excitation
% was a single slit or Gaussian profile, and write to
% screen the actual linewidth of the intensity profile.
if (code==1) || (code==2)
    linewidth(N,lambda,x1);
end;

% Plots expected theoretical intensity profile for a single
% slit:
if code==1
```

```
        plotTheoreticalSingleSlit(N,a,b,twopi,x,x1);
end;


% Option: Save data to a file (as a string of floating point
% numbers):
%writeToFile(x1);


% Removes all plots when we leave the program (cleans up):
input('Close all figures');
close all
```

Choose Parameters for the Calculations

```
function [lambda,a,b,nWavel,N,twopi,Nhalf] = parameters

% Choose parameters for the calculation.
% Written by AIV. Version 15. October 2017

% Choose resolution, distance to screen, and the width of the
% area on screen the calculation should include. Some constants
% are defined. Results depend critically on the parameters
% set by this function.
% Whether the result will be mainly a Fresnel- og Franuhofer
% diffraction depend on the b parameter. nWavel must be
% increased if b is large to include the full diffraction
% pattern within the calculated area on screen.
% The parameters given in this particular code is suitable
% for a double slit in the Fresnel regime (quite complicated
% pattern).

lambda = 4;          % Four points per wavelength resolution in
% excitation points
a = 20;              % Width of single slit, given in
% # wavelengths
b = 4000 * lambda;   % Distance to screen is b wavelengths
nWavel = 1024*3/2;   % # wavelengths along the screen (an
% integer!)
N = nWavel*lambda;   % Width of excitation area as well as
% screen in # wavelengths
twopi = 2.0*pi;      % Somewhat innecessary, but speeds up
% a bit...
Nhalf = N/2;
return;
```

Allocate Arrays We Need

```
function [x,x2,x0,x1,sc,r] = allocateArrays(nWavel,N);
% Allocates space for various arrays
% Function is written by AIV. Version 15. October 2017


x = linspace(-nWavel/2, nWavel/2, N); % A relative position
% array for plot
x2 = linspace(-N,N,2*N+1);   % Simil, but for plot/test of
% hjelp functions
x0 = zeros(N,2);             % Excitation data, amplitudes
% and phases
x1 = zeros(N,2);             % Amplitudes at screen, amplitudes
% and phases
sc = zeros(2*N + 1,2);       % Store sin/cos for component
% calculations
r = zeros(2*N + 1,2);        % Distance-table: reduction
% factor and phase-correction
                             % based on path length
return;
```

Generates the Various "Excitations" (Single or Double Slit, etc.)

```
function [x0] = generateExcitation(code,lambda,a,N,Nhalf, ...
twopi,x0)
% Generate or read in excitation data. NOTE: There are
% specific requirements for the various excitations that
% can only be changed in the code below.
% Function is written by AIV. Version 15. October 2017


switch code
   case (1)
      disp('Single slit')
      m = a * lambda / 2;   % Slit is a wavelengths wide
      x0(Nhalf-m:Nhalf+m-1,1) = 1.0;
      %x0(:,2)= [1:N].*0.05; % Phases are modifies so that
                             % it mimics a ray is not coming
                             % perpendicular towards the slit.
   case 2
      disp('Gaussian excitation')
      % Intensity
      width = 200*lambda/2.0;
      dummy = ([1:N]-Nhalf)./width;
      dummy = (dummy.*dummy);
```

```
        x0(:,1) = exp(-(dummy));
        % Phase
        R = 1000; % Radius of curvature in # wavelengths
        y = [-Nhalf:Nhalf-1];
        R2 = R*R*lambda*lambda*1.0;
        dist = sqrt((y.*y) + R2);
        fs = mod(dist,lambda);
        x0(:,2) = fs.*(twopi/lambda);
        %figure;   % Plot if wanted
        %plot(x,x0(:,2),'-r');

    case 3
        disp('Straight edge')
        % Excitation is a straight edge, illuminated part: 3/4
        x0(N/4:N) = 1.0;

    case 4
        disp('Double slit')
        % For the double slit, use sufficient large b in
        % 'parameters' in order to get the well known result
  x0 = zeros(N,2);
  a = 20*4;
  d = 200*4;
  kx = d/2 + a/2;
  ki = d/2 - a/2;
  x0(Nhalf-kx+1:Nhalf-kx+a,1) = 1.0;
  x0(Nhalf+ki:Nhalf+ki+a-1,1) = 1.0;

    case 5
        disp('Reads excitation data from file')
        % (often earlier calculated results.)
        filename = input('Give name on file with excitation ...
            data: ', 's');
        fid = fopen(filename,'r');
        x0(:,1) = fread(fid,N,'double');   % Need to know #
         % elements
        x0(:,2) = -fread(fid,N,'double');
        status = fclose(fid);
        % figure;   % Testplot to check if data was read properly
        % plot(x,xx0(:,1),'-g');
        % figure;
        % plot(x,xx0(:,2),'-r');
        % aa= xx0(Nhalf);
        % aa   % Test print for one single chosen point
```

```
    otherwise
        disp('Use code 1-5, please.')
end;
return;
```

Calculate Relative Position Data (from Excitation to Screen)

```
function [sc,r] = generateRelPositionData(N,b,lambda,twopi);
% Establish sine and cosine values for vectors from one
% position in x0 to all positions in x1, and find distances
% and relative phase differences between the points.
% Function is written by AIV. Version 15. October 2017

y = [-N:N];
b2 = b*b*1.0;
y2p = (y.*y) + b2;
rnn = sqrt(y2p);
sc(:,1) = b./rnn;
sc(:,2) = y./rnn;
r(:,1) = 1./sqrt(rnn);
fs = mod(rnn,lambda);
r(:,2) = fs.*(twopi/lambda);
% mx = max(r(:,1));  % For testing if field reduction vs
 % distance is correct
% r(:,1) = mx;
%  plot(x2,r(:,2),'-k'); % Test plot of these variables
%  figure;
return;
```

Summation of all Contributions

```
{\footnotesize
\begin{verbatim}
function [x1] = summation(N,x0,r)
% Runs through x1 (screen) from start to end and sum all
% contributions from x0 (the excitation line) with proper
% amplitude and phase.
% Function is written by AIV. Version 15. October 2017

for n = 1:N
    relPos1 = N+2-n;
    relPos2 = relPos1+N-1;
    amplitude = x0(:,1).*r(relPos1:relPos2,1);
    fase = x0(:,2) - r(relPos1:relPos2,2);
```

```
       fasor(:,1) = amplitude .* cos(fase);
       fasor(:,2) = amplitude .* sin(fase);
       fasorx = sum(fasor(:,1));
       fasory = sum(fasor(:,2));
       x1(n,1) = sqrt(fasorx*fasorx + fasory*fasory);
       x1(n,2) = atan2(fasory, fasorx);
    end;
    return;
```

Plot the Diffraction Pattern

```
    function plotDiffraction(x,x0,x1);
    % Plots intensities for diffraction picture along with a
    % marking of the excitation. Some extra possibilities are
    % given, for testing or special purposes.
    % Function is written by AIV. Version 15. October 2017

    %plot(x,x1(:,1),'-r');    % Plots amplitudes (red) (can
     % often be skipped)
    figure;
    x12 = x1(:,1).*x1(:,1);   % Calculation of intensities
    hold on;
    scaling = (max(x12)/8.0);
    plot(x,x0(:,1).*scaling,'-r');    % Plot initial excitaion
    plot(x,x12(:,1),'-b');   % Plot relative intensities (blue)
    xlabel('Position on screen (given as # wavelengths)');
    ylabel('Relative intensities in the diffraction pattern');

    % figure;
    % plot(x,x1(:,2),'-k');    % Plot phases (black) (most
      % often skipped)
    return;
```

Calculates Linewidths (FWHM) for Single-Slit and Gaussian Intensity Profile

```
    function  linewidth(N,lambda,x1);
    % Calculates linewidths (FWHM) for single slit and Gaussian
    % intensity profile.
    % Function is written by AIV. Version 15. October 2017

    x12 = x1(:,1).*x1(:,1);   % Calculation of intensities

    mx2 = max(x12(:,1))/2.0;
    lower = 1;
```

```
upper = 1;
for k = 1:N-1
        if ((x12(k,1)<=mx2) && (x12(k+1,1)>=mx2))
            lower = k;
        end;
        if ((x12(k,1)>=mx2) && (x12(k+1,1)<=mx2))
            upper = k;
        end;
end;
disp('FWHM: ')
(upper-lower)*1.0/lambda
return;
```

Plot Theoretical Single-Slit Pattern

```
function plotTheoreticalSingleSlit(N,a,b,twopi,x,x1);
% Plots the theoretical intenstity pattern for our single slit.
% Function is written by AIV. Version 15. October 2017

%figure;
theta = atan2(([1:N]-(N/2)),b);
betah = (twopi*a/2).*sin(theta);
sinbetah = sin(betah);
theoretical = (sinbetah./betah).*(sinbetah./betah);
x12 = x1(:,1).*x1(:,1);  % Calculate intensities
scaling = max(x12);
plot(x,theoretical.*scaling,'-g');
return;
```

Write Data to File (for Other Purposes Later)

```
function writeToFile(x1);
% Write data to file (as a string of floating point numbers)
% Function is written by AIV. Version 15. October 2017

filename = input('Give the name of new file for storing ...
results: ', 's');

fid = fopen(filename,'w');
fwrite(fid,x1(:,1),'double');
fwrite(fid,x1(:,2),'double');
status = fclose(fid);
return;
```

13.13 Learning Objectives

After working through this chapter, you should be able to:

- Explain the principle of Huygens–Fresnel.
- Derive the condition of constructive interference from a double slit (when the slits are assumed to be very narrow).
- Describe the interference pattern from a double slit, and indicate why the attempt by Thomas Young had a great historical significance.
- Give the main idea of a regular anti-reflection treatment of optics.
- Specify how the interference image changes qualitatively when using more than two parallel identical slits.
- Explain the qualitative intensity distribution in a diffraction pattern for a narrow single slit when we consider the pattern far from the slit.
- Calculate using numerical methods interference pattern also for Fresnel diffraction.
- Specify how the diffraction pattern looks like for light passing through a circular hole.
- Explain how diffraction sets limits on how close two stars can be on heaven before we can no longer distinguish them when we view them through a telescope.
- Calculate the maximum achievable angular resolution for lenses in many different contexts (eye, camera objective, telescope, etc.).
- Know Babinet's principle.
- Know the so-called Arago spot (also called Poisson's spot) and why this phenomenon has a historical significance.

13.14 Exercises

Suggested concepts for student active learning activities: Superposition, Huygens–Fresnel principle, wavefront, coherent, double slit, single slit, optical grating, grating constant, interference pattern, half-width of peaks, interference filter, thin film, diffraction, border region, Huygens–Fresnel diffraction, Fraunhofer diffraction, Airy disc, beam optics, beam waist, Rayleigh's resolution criterion, Arago's spot, X-ray diffraction, Babinet's principle, amplitude level summation.

Comprehension/discussion questions

1. Is it possible to conduct Young's double-slit experiment with sound? Discuss a possible experimental set-up and whether there is a difference between longitudinal and transverse waves in this context.

2. We use the superposition principle "on amplitude level" instead of "on intensity level". Explain why.

3. We have a telescope and want to check if an object we observe is a double star or not. In other words, we need *slightly* greater resolution, and we assume that the telescope has so-called diffraction-limited optics. What do we mean by this expression? Can we increase the resolution by "shuttering down" so we only use a central part of the lens? Or can we increase the resolution by inserting a filter that transmits light either in the blue area or in the red area?

4. In a diffraction experiment with a single slit and light with wavelength λ, there is no intensity minimum. What can we say about the width of the slit?

5. A regular rainbow we get when the drops are above a certain size. For very small drops, the rainbow becomes almost white. How small do you think the drops must be for it to happen?

6. Good speakers (in stereo systems) are often composed of at least one bass speaker (woofer, low frequencies) and a treble speaker (tweeter, high frequencies). The former often has a relatively large diameter, while the latter is usually only a few inches in diameter. Try to give one explanation of this choice based on what you know about diffraction. Also, come with an explanation that is based on a physical mechanism different from diffraction.

7. Why is a diffraction grating (with many slits) better than a double slit if it is to be used in a spectrometer by means of which we can measure wavelengths?

8. Diffraction from a single slit also affects the interference pattern from a grating. Explain the relation.

9. Try to describe the essence of Fig. 13.27. Pay particular attention to the similarities and inequalities between the left and right parts of the figure.

10. Will the interference intensity pattern depend on the diameter of the laser beam when you send a laser beam through an diffraction grating? Explain.

Problems

11. Two coherent sources (always the same phase) for radio waves are located 5.00 m apart, and the waves have a wavelength of 3.00 m. Find points on a line passing through the two sources where we have constructive and destructive interference (if such points exist).

12. Two slits with a mutual distance of 0.450 mm are placed 7.5 m from a screen and illuminated with coherent light with wavelength 500 nm. What distance is there between the second and third dark lines in the interference strips on the screen?

13. An anti-reflection coating on a lens has the refractive index $n = 1.42$ (and that for the glass is 1.52). What is the minimum thickness the coating must have for red light with a wavelength of 650 nm to have minimal reflection?

14. In a Young double-slit experiment, place a piece of glass with refractive index n and thickness L in front of one of the slots. Describe qualitatively what happens to the interference pattern.

15. We use a 10 cm diameter biconvex lens with focal length 50 cm for focusing the sunlight as a "burning glass". The light does not accumulate at one point, but in a disc with a diameter of d. There are two contributions to the size of the disc,

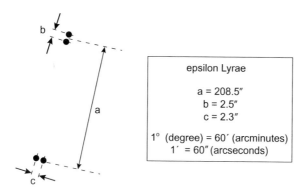

Fig. 13.33 Angular distances between the four stars we perceive with the naked eye as one star, namely Epsilon Lyrae

namely that the sun is imaged by the lens and that the lens causes diffraction. Determine the two contributors to see which one is more important in this case.

16. The "star" Epsilon Lyrae is a double star in the constellation of Lyra, where each of the components is again a double star. The angular distance between the stars is as shown in Fig. 13.33. What demands must we impose on a telescope so that it enables us to distinguish the first pair (observing "two stars"). What requirements must be met to observe all four stars on an evening of calm and clear air?

17. A digital SLR camera has a CMOS chip that is 15.8×23.6 mm in size and has 2592×3872 pixels. A 35 mm focal length lens is used with f-number 3.3/22 (min/max). What size is the largest and smallest Airy disc from the lens? Enter the answer in both absolute measure and relative to pixel size.

18. We consider the diffraction pattern from a human hair held in the beam of a green laser pen of wavelength 532 nm. It is 16.2 cm between two minimum points with 11 light areas between when the laser pen (hair) is 185 cm from the screen where the measurements were made. How big diameter does the hair have? Is the value you arrive at reasonable based on available information about diameters for human hair?

19. A diffraction grating has its third-order light band at the angle 78.4° for light with wavelength 681 nm. Determine how many lines the grating has per centimetre. Also, determine the angles of the first- and second-order bands. Is there a fourth order of bands?

20. We light with a standard He–Ne laser wavelength 632.8 nm perpendicular to a CD. The "grooves" in a CD are 1.60 μm apart. What are the angles of reflections from the CD?

21. The Hubble Space Telescope has an aperture (opening) of 2.4 m and is used for visible light (400–700 nm). The Arecibo Radio Telescope in Puerto Rico is 305 m in diameter (built in a valley) and is used for radio waves of wavelength 75 cm.

(a) What is the smallest crater size on the moon that can be separated from a neighbouring crater with the two telescopes? (Distance to the moon is about ten times the perimeter of the earth, more specifically 3.84×10^8 m.)

(b) Suppose we want to turn Hubble into a spy satellite that goes into a new orbit around the earth. If we were able to read the number plates for cars with the telescope, what height would the new path to Hubble be?

22. Observe the moon with only the eyes. Try to notice the smallest structure you can distinguish. Find a picture of the moon, and find the structure there. Determine the distance across the structure, and compare it with what you would expect from Rayleigh's resolution criterion.

23. Take a photograph of a distant bright spot with your camera. Analyse the image to see if you can detect the Airy disc. This would require blowing up the image you took until you can see single pixels in the image. Attempt to calculate how large the Airy disc is expected to be.

24. Start with Fig. 13.30. Assume that the distance between the gravestones sideways is a and that they are a distance b behind each other. Determine the angle between each row of gravestones that are behind each other as we see in the photograph. Also, determine the distance between adjacent gravestones along the lines we see (the distance that will match the slit separation distance in a diffraction grating).

25. From a hotel window, interference-like patterns were observed when small light spots were viewed through curtains (made of netting and partially see-through, see Fig. 13.34). Examples of the light phenomenon we observed at night through the curtains are shown in Fig. 13.35). The image does not change if we move closer to or farther away from the curtain while viewing the light coming from outside.

Fig. 13.34 Picture of a light curtain which it was possible to look through. Details show how the fibres in the curtain were in relation to each other. The bar in the middle part is originally 2.0 mm long

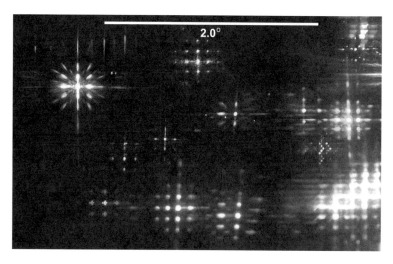

Fig. 13.35 Picture of distant light points observed through the curtain in the previous figure. The bar indicates an angle of 2.0°

(a) Describe which details in the observed light pattern which indicate that diffraction/interference is responsible for what we see.

(b) Carry out calculations that can support such a conclusion (there is probably an estimated 20% uncertainty in the measurements indicated in the figures).

Chapter 14
Wavelet Transform

Abstract The aim of this chapter is to present a time-resolved frequency analysis of a signal. This is more demanding than one might be inclined to think, due to the time-bandwidth product (a classical analogue of Heisenberg's uncertainty relationship). We have chosen a so-called continuous wavelet transform with Morlet wavelets, since it offers an extremely useful rationale for optimization. Morlet wavelets are presented and a brute force method of analysis is outlined, followed by a more elegant transform utilizing FFT repeatedly in a kind of frequency selection procedure. A computer program in Matlab or Python is given. The user must specify frequency range as well as frequency resolution (through a K parameter). We discuss in detail how to optimize frequency resolution vs time resolution in the analysis.

14.1 Time-Resolved Frequency Analysis

Fourier transformation, as described in Chap. 5, is well suited for "stationary" signals (whose character does not change appreciably over time). For signals that change over time, the temporal information is distributed over all frequency components, and it is a hopeless task to determine how the frequency spectrum varies over time. Therefore, the FFT of a signal that varies widely from one time interval to another within the same data string is of very little value indeed.

There are several methods to explore how the frequency spectrum changes over time. In frequency analysers and in analyses of long-lasting signals (very large data sets), a so-called short-time Fourier transformation or piecewise transformation is used (see Fig. 14.1). "Short-time FT" or "short-term FT" are shortened to STFT, and mathematically STFT can be stated as follows:

$$\text{STFT}\{x(t)\}(\tau, \omega) \equiv \mathscr{X}(\tau, \omega) = \int_{-\infty}^{\infty} x(t)w(t-\tau)e^{-i\omega t}\,dt \ .$$

© Springer Nature Switzerland AG 2018

A. I. Vistnes, *Physics of Oscillations and Waves*, Undergraduate Texts in Physics,
https://doi.org/10.1007/978-3-319-72314-3_14

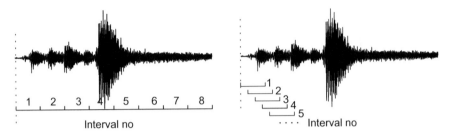

Fig. 14.1 To get temporal information by analysing a long string of time series data, we can break up the total observation into several intervals and perform Fourier transformation interval by interval. The intervals can be chosen so that they do not overlap each other (left part) or overlap each other (right part)

Here, $w(t - \tau)$ is a so-called window function that has a bell-like shape. In practice, the so-called Hanning or Gaussian forms are used, which have a significant value only for a limited period of time around the reference time τ and fall to zero outside the limited time period.

In practice, a fast Fourier transform (FFT) is used for analysing each time window. By choosing a narrow window function w, we get a high time resolution and a poorer time resolution with a wide window function.

In the analysis, we let τ slide through the entire data string to be analysed. We can often choose whether the next window function should overlap the previous one or not, and if so, how large the overlap should be. The result is often shown in a diagram where the intensity of Fourier components in each time window is plotted as a function of time. Intensity is usually indicated by colour coding. The result is usually called a "spectrogram" (see left part of Fig. 14.2).

The advantage of this method is that, if we wish, we can analyse continuous signals for weeks on end. There is no limitation on the length of the data string, since, in practice, we only pick out a limited segment for each round of analysis.

The downside is that we get a frequency resolution that is inversely proportional to the time analysed in each window (i.e. how long the time window lasts). This means that we get the exact same frequency resolution (and thus also time resolution) whether we analyse low-frequency or high-frequency signals. We must select the width of the window function for some typical frequency in the signal. However, if there are widely different frequencies at the same time in the signal, it is impossible to find an optimal window width suitable for all circumstances.

This is an important detail for STFT and for all time-resolved frequency analysers. Due to the time-bandwidth product with which we became acquainted when we studied Fourier transformation, it is impossible to get very precise information about time and frequency at the same time. If we use a window w that extends over a long period of time, we can get fairly accurate frequency information. However, we cannot get precise information about changes over time. It is the classic analogy to Heisenberg's uncertainty relationship that surfaces again.

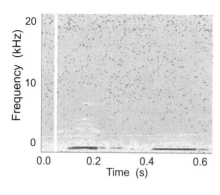
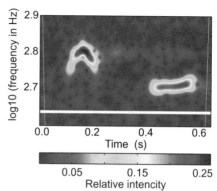

Fig. 14.2 Spectrograms of the sound of a cuckoo bird. To the left is a result of Matlab's built-in STFT function, based on FFT of 300 points at a time, and a so-called Hamming window with 400 points. The overlap from one analysis to the next is 200 points. The vertical white stripe indicates that this spectrum is built up line to line with vertical lines. To the right, a spectrogram is calculated using wavelet transformation with Morlet wavelets, with the computer program given later in this chapter. The horizontal white stripe indicates that such a chart is constructed line by line using horizontal lines. See text for other details

By a "sliding filter" method, as in STFT, where the window changes a data point at a time before new analysis, we avoid jumps in the results arising from randomness in how the intervals are chosen. The problem is, however, that we must carry out some apparently unnecessary calculations. This can be avoided by making a jump in the position of the window function from one analysis to another, but the appropriate jump length will depend on the frequency being analysed. With the STFT function in Matlab, we can choose both the width and the degree of overlap for the window function w, but it is difficult, on account of several reasons, to optimize the choice using STFT.

In this chapter, we will consider another method that can provide time-resolved frequency information, the so-called continuous wavelet analysis with Morlet wavelets. We end up with a chart showing how the frequency picture changes over time, just as we often do with STFT, but there are major differences in how the analysis is performed mathematically [see right part of Fig. 14.2)].

In STFT, we use FFT to analyse each segment of the time signal (picked out using the w window) and then switch to the next segment of the time signal. We therefore build up the "Fourier coefficient versus frequency and time" diagram ("STFT spectrogram") stripe by stripe, using *vertical* stripes. We get all frequencies from zero to half the sampling frequency, whether or not we are interested in the entire range.

In wavelet analysis, we get a "wavelet spectrogram", which may look quite similar to an STFT spectrogram in certain contexts. In wavelet analysis, we also build the spectrogram line by line, but now with *horizontal* lines (see white lines in Fig. 14.2). Therefore, we must choose the entire time interval we want to analyse before the analysis starts, which in some contexts is a disadvantage. The advantage is, however, that we can choose which frequency range will be studied (see Fig. 14.2). We can further select a logarithmic frequency axis if we wish to fully exploit the fact that the *relative* frequency resolution of this method is the same for all frequencies.

"Continuous wavelet analysis" resembles a sliding, short-time Fourier transform (STFT), but the wavelet analysis with Morlet wavelets gives the same relative frequency resolution for all frequencies. The secret is to use different lengths of time, depending on the frequency to be analysed.

It may be mentioned that there are also many variants of wavelet transformation wherein we make as few transformations as possible with only a fairly limited loss of information. Such a transformation is much more efficient than the continuous variant and is used in technological contexts where speed is of paramount importance. The disadvantage of such a wavelet transformation is that the transformed signal is far more difficult to understand than the usual Fourier spectrum. This is the main reason why we do not go into that method here.

Wavelet analysis is a wide-ranging field of mathematics/informatics, and courses are offered on the subject at many universities. We will not go into details concerning the strictly mathematical or computational aspects of the subject. The purpose of including wavelets in this book is to point out that Fourier transformation is often unsuitable for nonstationary signals and to draw attention at the same time to a method of analysis that is preferable in such circumstances. In addition, work with wavelets can contribute to a deeper understanding of time-limited phenomena in general and the corresponding frequencies. Among other things, there are close analogues between Heisenberg's uncertainty relationship and wavelet analysis.

Some of you will probably use wavelet analysis in the master thesis or in a PhD project (and later employment). For this reason, we place emphasis on showing when wavelet analysis is useful and when the method does not have much to offer. Wavelets are used for analysing solar spot activity (and changes in the spot cycle over time), El Niño Southern Oscillations, glacial cycles, roughness, grain size analysers, analysis of, e.g. cancer cells vs. normal cells and much more.

Technologically, there is an extensive use of wavelets in, among other applications, JPEG compression of image files and in MP3 compression of sound.

14.2 Historical Glimpse

Let us recapitulate the story of Fourier transformation: the French mathematician Joseph Fourier (1768–1830) "discovered" Fourier transformation almost 200 years ago. (Fourier also worked with heat flow, and was probably the first to discover the greenhouse effect.)

Fourier transformation is largely used in analytical mathematics. In addition, the transformation gained enormous currency in the data world after J. W. Cooley and J. W. Tukey discovered in 1965 the so-called fast Fourier transform (FFT) that makes it possible to perform Fourier transformation much faster than before. In FFT, symmetries in the sine and cosine functions are used to reduce the number of multiplications in the calculation, but to get the most effective transformation, the number of data points must be an integer power of 2, i.e. $N = 2^n$.

It has been said that the Cooley–Tukey fast Fourier transform was actually discovered by Carl Friedrich Gauss around 1805, but forgotten and partially reinvented several times before 1965. The success of Cooley and Tukey's rediscovery is due to the emergence of computers at about the same time.

Wavelet analysis is of much later origin. Admittedly, wavelets were introduced already around 1909, but the method was first taken seriously around 1980. There is far greater scope for special variants of wavelet analysis than in Fourier transformation. It is both an advantage and a disadvantage. We can by far tailor-make a wavelet analysis to suit the data we wish to analyse. The downside is that the wide variety of possibilities causes us to use our head a little more in wavelet analysis than in Fourier transformation, both when the transformation is to be carried out and when we interpret the results. But the results are often the more interesting!

14.3 Brief Remark on Mathematical Underpinnings

14.3.1 Refresher on Fourier Transformation

We have gone through Fourier transformation in Chap. 5, but let us recall the mathematical expressions here too.

Let $x(t)$ be an integrable function of time. We can then calculate a new function $X(\omega)$, where ω denotes the frequency, in the following manner:

$$X(\omega) = \frac{1}{2\pi} \int_{-\infty}^{\infty} x(t) e^{-i\omega t} \, dt . \tag{14.1}$$

The interesting feature about this function is that we can make a corresponding inverse transformation:

$$x(t) = \int_{-\infty}^{\infty} X(\omega) e^{i\omega t} \, d\omega \tag{14.2}$$

and recover the original function. Note the change in the sign of the exponent in the exponential function.

We see from Eqs. (14.1) and (14.2) that when $x(t)$ is real, $X(\omega)$ will be complex. This is necessary in order that $X(\omega)$ should be able to indicate both how large the oscillations are at different frequencies, *and* the mutual phase of the different frequency components. (A symmetry in X ensures that x, given by the inverse transformation, will be real, as it was originally.)

It should also be mentioned that x and X (called conjugate variables) generally do not have to be time and frequency. In Chap. 8, we also used FT to analyse a spatial description of a wave. The result was a description in "spatial frequencies" which we may equally well have called "wavenumbers". Thus, position and wavenumbers are also conjugate variables. Even more conjugates variables are in use in physics.

The above expressions are used for analytical calculations. When we use a computer, we do not fully know how $x(t)$ varies in time. We only know x_n, the value of x at discrete times t_n, where n is an index that varies from 1 to N, where N is the number of measurements of x that have been made. We assume that the measurement times are equally spaced. The total time over which x is measured is then $T = N\delta t$ where δt is the interval between two successive times of measurement (details discussed in a previous chapter).

When Fourier transformation is performed on discrete data, a discrete transformation is used. This can be stated as follows:

$$X_k = \frac{1}{N} \sum_{n=1}^{N} x_n e^{-i2\pi f_k t_n} = \frac{1}{N} \sum_{n=1}^{N} x_n \exp\left[-i2\pi f_k t_n\right] \qquad (14.3)$$

where we have used the two common ways to write an exponential function, and $k = 0, 1, 2, \ldots, N-1$. Further, $f_k = 0, f_s/N, 2f_s/N, \ldots, f_s(N-1)/N$ where f_s is the sampling frequency. Finally, $t_n = 0, T/N, 2T/N, \ldots, T(N-1)/N$ with $N/T = f_s$.

It may not be easy to grasp the expression, but what we really do to determine the Fourier transform at a frequency f_k is to multiply (term by term) the digitized function x_n with a cosine function of frequency f_k and sum up all the terms that appear. (For the imaginary part of the Fourier transform, we multiply with a sine function of frequency f_k.)

The corresponding "inverse" transformation is given by:

$$x_n = \sum_{k=1}^{N} X_k e^{i2\pi f_k t_n} \qquad (14.4)$$

for $n = 1, 2, 3, \ldots, N$.

14.3.2   Formalism of Wavelet Transformation

Wavelet transformation can be stated in an apparently similar manner to a Fourier transformation:

> Let $x(t)$ be an integrable function of time. We can then calculate a new function $\gamma_K(\omega_a, t)$ which provides information about frequency and time simultaneously.
>
> ω_a can be termed "analysis angular frequency". K is a "sharpness" parameter (also known as "wavenumber") related to whether we want high precision in time (K small) or high precision in frequency (K large).
>
>   Individual values for the new wavelet transformed function can be found in the following way:
>
> $$\gamma_K(\omega_a, t) = \int_{-\infty}^{\infty} x(t + \tau)\Psi_{\omega_a, K}^*(\tau)\mathrm{d}\tau \ . \tag{14.5}$$
>
> Here, $\Psi_{\omega_a, K}(\tau)$ is the wavelet itself, and the asterisk denotes complex conjugation. For the Morlet wavelet used here, $\Psi_{\omega_a, K}^*(\tau) = \Psi_{\omega_a, K}(-\tau)$ (see below).

The special thing about wavelet analysis is that we can choose from almost infinitely many different wavelets depending on what we want to get from the analysis. In our context, we only use Morlet wavelets, whose real part can be expressed as an *cosine* function (plus a small amount of constant correction) multiplied by a Gaussian envelope. The imaginary part is a *sine* function multiplied with the same Gaussian envelope as before (see Fig. 14.3).

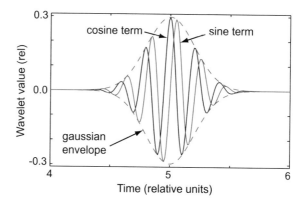

Fig. 14.3 Example of a Morlet wavelet for $K = 6$. Both the real part (the cosine term) and the imaginary part (sine term) are displayed

A Morlet wavelet can be described as:

$$\Psi_{\omega_a, K}(\tau) = C\{\exp(-i\omega_a \tau) - \exp(-K^2)\} \exp\left[-\omega_a^2 \tau^2/(2K)^2\right] \quad (14.6)$$

where C is a "normalization constant". When we describe $\Psi_{\omega_a, K}(\tau)$ numerically, it is advantageous to use the following expression for C:

$$C = \frac{0.798\,\omega_a}{f_s K} \quad (14.7)$$

where f_s is the sampling frequency.

Some remarks:
There is still no uniform description of wavelets. Different sources indicate formalism in different ways; including expressions such as "scaling parameter", "mother" and "daughter wavelets" are key concepts. We have chosen to use a presentation that is close to an article by Najmi and Sadowsky (see the bibliography) because its similarity to other formalism in this book. The "constant" C, I found through trial and error, after imposing the criterion that a wavelet transform of a pure sinusoidal signal should be close to the amplitude of the sinusoid regardless of the parameters ω_a, f_s and K (with a deviation usually not exceeding 1%). The actual expression for a Morlet wavelet we will not be used in practice, except as an illustrative example. For efficient wavelet transformation, we will be using the Fourier transform of the wavelet directly. Details are given in the text which follows.

In the wavelet analysis, we check if the signal we study contains different frequencies at different times. The angular frequency used for the analysis is ω_a. The τ parameter specifies the time at which a specific wavelet has a maximum and corresponds to the centre for the small time interval we investigate.

The parameter K, a real constant, can be called the "width" of the wavelet. Some call it "wavenumber" because it specifies the approximate number of waves under the Gaussian envelope for the wavelet, given by the last factor on the right-hand side of Eq. (14.6). It is recommended that K is 6.0 or larger.

Because of the second factor on the right-hand side of Eq. (14.6), we see that the wavelet Ψ is complex.

Figure 14.3 shows an example of a Morlet wavelet. We see that it bears the correct name, because "wavelet" means a "small wave". Be sure that you thoroughly understand how the wavelet is formed, namely, as the product of a complex harmonic function and a Gaussian envelope centred around time τ.

Note that the expression in Eq. (14.6) is a general description. When it is implemented in a computer program for analysing a specific signal, we must know the sampling frequency used. This enters the normalization constant C. If the specific signal is described in N equidistant points in time, then the total time for sampling equals $T = N/f_s$. We then choose to describe any Morlet wavelet used in the analysis by an array with the same sampling frequency and the same length as the specific signal to be analysed.

If we compare Eq. (14.1) with Eq. (14.5), we see that the expressions look similar to each other. We integrate the product of a function x and a wave. Both are thus linked to an "inner product" within mathematics, but as already said, we would not go into mathematical details here.

However, there are more differences than we might think at first. A significant difference is that the wavelet transformation leads to a three-dimensional description (value of γ as a function of both ω_a and t), while a description based on Fourier transformation is only two dimensional (value of X as function of frequency).

It is possible to make an "inverse" wavelet transformation similarly to an inverse Fourier transform. This is essential when wavelets are used in JPEG image compression and MP3 music file compression. However, we do not include details regarding this formalism in our context. (Those interested may consult the last reference in the list of wavelet resources at the end of the chapter.)

14.3.3 *"Discrete Continuous" Wavelet Transformation*

First, a few words about the use of the words "discrete" and "continuous". A digitized signal will be called discrete because we only have a finite number of measurement results (equidistant in time). However, we will designate the particular wavelet transformation described in this chapter as "continuous", meaning that the "sliding filter" (just the right part of Fig. 14.1) moves by one point across the digitized signal each time a new calculation is performed. An alternative would be to shift the wavelet by, for example, half the wavelet width.

Wavelet transformation is used almost exclusively on discrete signals, since the calculations are so extensive that they are virtually impossible to perform analytically (except in very simple model descriptions).

For digitized signals (discrete signals), the Morlet wavelet itself can be expressed as:

$$\Psi_{\omega_a, K, t_k}(t_n) = C\{\exp(-i\omega_a(t_n - t_k)) - \exp(-K^2)\} \exp\left[-\omega_a^2(t_n - t_k)^2/(2K)^2\right].$$
$$(14.8)$$

Here, it is assumed that the signal to be analysed is described in equidistant points using the number string x_n for $n = 1, 2, \ldots, N$. The time t_k indicates *the centre of the wavelet (!)*.

The wavelet transformation for one particular frequency and one particular instant will be:

$$\gamma_K(\omega_a, t_k) = \sum_{n=1}^{N} x_n \Psi_{\omega_a, K, t_k}^*(t_n).$$
$$(14.9)$$

Fig. 14.4 A sinusoidal
signal (in the middle) along
with a (Morlet) wavelet with
the same period of time (*top*)
and a wavelet with a shorter
period of time (*bottom*)

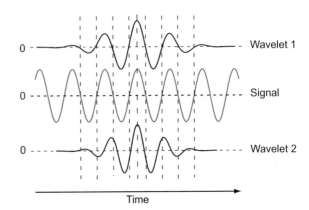

Let us visualize the process in order to acquire a better understanding of what it entails. In Fig. 14.4, we show a section of a time string along with two different choices of wavelets. Wavelet transformation consists in point-by-point multiplication of the signal with the analysing wavelet, and calculating the sum of all the products. The result is the wavelet transform of the signal for exactly the frequency the wavelet represents and for the exact point in the signal where the wavelet has its maximum value.

For wavelet 1, we see that the signal changes signs approximately at the same instant at which the wavelet changes its sign. That is, the product at any point becomes positive, and the sum of products is therefore quite large (since $\int \cos^2 \omega t \, dt$ is positive). For wavelet 2, the signal and the wavelet do not change sign at the same instant. Some of the products are therefore positive and some negative. The sum of products is significantly lower than for the first case (since $\int \cos \omega_1 t \cos \omega_2 t \, dt$ is often close to zero when $\omega_1 \neq \omega_2$).

We have thus attempted to show that the wavelet transformation of a regular sinusoidal wave will have a maximum when the "periodicity" (or frequency) of the analysing wavelet corresponds to the "periodicity" of the signal in the time interval where we perform the analysis.

To analyse the signal x_n for other periodicities, we need to change the wavelet, and this is done by using, for example, the ω_a parameter.

14.3.4 A Far More Efficient Algorithm

Wavelet transformation defined in Eq. (14.9) is, what is called, a convolution of the time signal $x(t)$ with the wavelet $\Psi^*_{\omega_a, K}$. The appearance of the convolution integral is of interest to us because it is easy to show that the Fourier transform of a convolution is similar to the inner product (pointwise product of two functions) of the Fourier transforms of each of the two functions that are included.

Let us denote the Fourier transform of x and the Fourier transform of the wavelet Ψ with, respectively, $\mathcal{F}(x)$ and $\mathcal{F}(\Psi)$. Let us also denote the Fourier transformed of $x * \Psi$ by $\mathcal{F}(x * \Psi)$. Then, the convolution statement states that

$$\mathcal{F}(x * \Psi) = \mathcal{F}(x)\,\mathcal{F}(\Psi) \tag{14.10}$$

where right-hand side is pointwise multiplication of the two Fourier transforms. But then we can make an inverse Fourier transform $\mathcal{F}^{-1}$ of the right- and left-hand side of this equation and get:

$$\mathcal{F}^{-1}\left(\mathcal{F}(x * \Psi)\right) = x * \Psi = \mathcal{F}^{-1}\left(\mathcal{F}(x)\,\mathcal{F}(\Psi)\right) . \tag{14.11}$$

The Fourier transform of the signal, $\mathcal{F}(x)$, can be calculated easily and can be used unchanged for the rest of the wavelet analysis. Fourier transformation of the wavelet Ψ itself must, in principle, be repeated every time we change the analysis frequency or wavenumber. However, we have an analytical expression of the Fourier transform of a Morlet wavelet (see below), which makes the calculation significantly faster.

When we then take an inverse Fourier transform of the product $\mathcal{F}(x)\,\mathcal{F}(\Psi)$, we get the entire $x * \Psi$ envelope in a jiffy. That is, we get the entire time variation in the convoluted signal for the selected analysis frequency (and K value) at one time.

To get a full wavelet analysis, we then have to carry out the procedure for an array of analysing frequencies (which we basically choose for ourselves). For example, if we choose to do the analysis at 1000 frequencies, it means that the calculations take about 1000 times longer than a simple Fourier transform. So, although the method based on the convolution theory is very effective compared to the brute force method, the calculation of continuous wavelet transformation with Morlet Wavelets takes a long computer time.

Now let us look at the Fourier transform of a Morlet wavelet defined in Eq. (14.6). The Morlet wavelet is complex, and it is very satisfying to find that when we calculate the FFT of this complex function, we get a purely real result; moreover, there is no mirroring in the spectrum!

The Fourier transform of a Morlet wavelet (Eq. 14.8) can be stated as follows:

$$\mathcal{F}(\Psi) \equiv \hat{\Psi}_{\omega_a, K}(\omega) = 2\{\exp\left(-[K(\omega - \omega_a)/\omega_a]^2\right) - \exp(-K^2)\exp\left(-[K\omega/\omega_a]^2\right)\} . \tag{14.12}$$

We see that this is a bell-shaped (Gaussian) feature (apart from a rather insignificant correction term for most selections of K). The peak of the Gaussian function is at the analysis frequency.

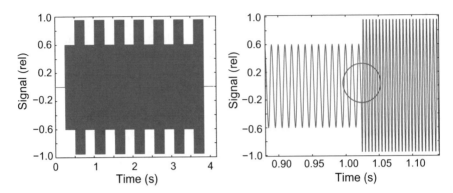

Fig. 14.5 Generated is used in our example. The left part shows the entire signal, while a detail of this is shown in the right part. The amplitude of the 100 Hz signal is 0.6, while the amplitude of the 200 Hz signal is 1.0. The signal is continuous everywhere, even at the instances where the frequency changes abruptly (*detail to the right*)

14.4   Example

We will now show in practice an example of wavelet transformation and start by generating a signal as a function of time.

The signal we generate changes between 100 and 200 Hz at fixed intervals (see Fig. 14.5). The outermost parts of the signal are set equal to zero. Note that when we generate a variable frequency signal during the time the signal exists, we will *insist*, for reasons that need not be spelled out here, that there is no discontinuity in the signal at the instant when the frequency changes. We will be able to meet this demand if we keep an eye on the *phase* of the signal throughout and upgrade the phase at each new time step. This feature of the signal is demonstrated in the expanded plot on the right half of Fig. 14.5.

We then implement the wavelet transform directly from Eqs. (14.9) and (14.8), or we can use the more efficient method described by Eqs. (14.11) and (14.12). The frequency of the analysing wavelet was selected as $\omega_a = 2\pi \times 100$ (which equals 100 Hz in the signal itself).

If we use Eqs. (14.9) and (14.8), we will shift the peak of the wavelet (along the time axis), for each new point in the wavelet transform calculation, from being completely at the outer left edge to being completely at the outer right edge. The result is shown in Fig. 14.6. We see that we get a value of about 0.6 for the times when the original signal had a frequency equal to the analysis frequency.

Figure 14.7 illustrates the more efficient method described by Eqs. (14.11) and (14.12). The Fourier transform of the signal itself is multiplied point by point with the Fourier transform of the wavelet (wavelet with analysing frequency 100 Hz and the given K value). The Fourier transform of the wavelet is a bell-shaped function (almost Gaussian form) with a position of 100 Hz in our case and has no "folded" component. The result of the pointwise multiplication is that only one of the four

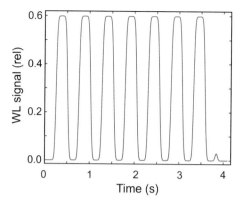

Fig. 14.6 Wavelet transform of the time signal in Fig. 14.5 for an analysis frequency of 100 Hz. The K parameter was 12, which means that the wavelet was roughly about $12 \times (1/100)$ s $= 0.12$ s long. The width of the wavelet leads to rounding of sharp corners in the diagram

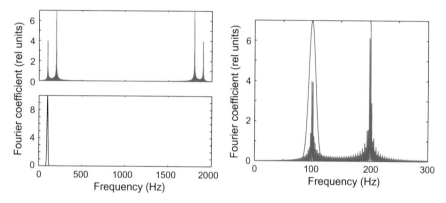

Fig. 14.7 Fourier transform of our time signal at the top left, together with the Fourier transform of the wavelet at the bottom left. The details on the right show that the Gaussian curve in a way functions as a filter and will pick out only parts of the frequency spectrum of the signal. See the text for details

"peaks" in the frequency range of the signal survives. An inverse Fourier transform of this signal is then calculated, and absolute values are taken. Plotting the result, we get exactly the same curve as shown in Fig. 14.6.

The wavelet diagram does not show any sign of the 200 Hz signal, but that is because we have only analysed the 100 Hz signal. To get a more complete wavelet diagram, one has to repeat the procedure for a whole set of frequencies. Then, the wavelet diagram becomes three dimensional with time along the x-axis, frequency along the y-axis and the intensity as a function of time and frequency indicated by colour.

A detail is worth noting here already: The curve in Fig. 14.6 has rounded corners. This is due to the fact that the wavelet has a definite extent in time and therefore will "detect" a 100 Hz sequence even before the wavelet peak is within the 100 Hz range. Similarly, the wavelet will "discover" areas with no 100 Hz, even when the peak of the wavelet is within the 100 Hz range. We come back to this effect in great detail since.

It seems appropriate to list the steps involved in the calculation of a wavelet diagram (using the most effective method):

- Calculate the Fourier transformed of the time signal we are going to analyse.
- Calculate directly the Fourier transformed of a Morlet wavelet with the analysis frequency ω_a and wavenumber K of interest.
- Multiply these with each other, point by point. (Note that there must be consistency between the frequencies f_k in the Fourier transforms of the signal and of the analysing wavelet.)
- Perform an inverse Fourier transform.
- The absolute value of this will then provide information about the time at which the original signal contained frequencies equal to the analysis frequency.
- By changing the Morlet wavelet to the next analysis frequency, we gradually build new horizontal lines in the wavelet diagram until we have covered as many analytical frequencies as we want.

Since fast Fourier transform is such an efficient operation, the method we just outlined is sufficiently fast to be useful.

In a program code a little later in this chapter, there is an example of how wavelet transformation with the effective algorithm can be implemented.

14.5   Important Details

14.5.1   Phase Information and Scaling of Amplitude

In the usual Fourier transformation, we effect in principle two transformations simultaneously, one of the type

$$X(\omega) = \int_{-\infty}^{\infty} x(t) \cos \omega t \, dt \qquad (14.13)$$

and the other of the type

$$X(\omega) = -\mathrm{i} \int_{-\infty}^{\infty} x(t) \sin \omega t \, \mathrm{d}t \, . \tag{14.14}$$

The reason is that we have both sine and cosine terms and that we must be able to capture, for example, a sinus signal in x no matter what phase it has.

In the usual Fourier transformation, we have *one* starting point for the analysis. This means that it is easy to find the relative phases of the different frequency components.

In continuous wavelet analysis, we have different starting points and lengths of the analysis window along the way in the calculations. That makes it much more difficult to keep track of phases. This is one of the reasons why we almost exclusively take the absolute value of the wavelet transformation in one or the other variant when the output of wavelet analysis is presented. (However, if we were to do an inverse wavelet transformation afterwards, we would have to take care of the phase information.)

There are several ways of specifying signal strength in a wavelet diagram. Often, the *square* of absolute values is used, which gives the *energy* of the signal.

Based on experience, I do not like to use the square of absolute value because the difference between the powerful and weak parts is often so great that we lose information about the weak parts. Then, it is often better to use absolute value directly ("amplitude level").

However, I often prefer to use *square root* of the absolute value. Then, the weak parts show up even better than when the absolute value is plotted.

We are free to choose how the results of the wavelet transformation are plotted, but we must not lose sight of our choice when we extract quantitative values from the diagrams.

14.5.2   *Frequency Resolution Versus Time Resolution*

We figured out from Fig. 14.4 that, when the wavelength of the signal x is exactly equal to the wavelength within the wavelet, wavelet transformation will give maximum value. If we make a *small* change in the wavelet by changing the analysis frequency, the transformation will give a lower value, but not zero value. In other words, a wavelet analysis will give rise not only to a different frequency from that corresponding to the signal but also to nearby frequencies.

The theme of this section is to know how far this "adulteration effect" goes.

Let us assume that a wavelet transformation involves a "digital filtering" of a signal, as illustrated in Fig. 14.7. The sharpness of this filtering is determined by the width of the Gaussian function used in the filtering. *We need to find the relation between the width of the frequency picture and the width of the wavelet in the time picture.*

In Fig. 14.8, on the left, there are three different choices of wavelets [(calculated from Eq. (14.8)] and to the right is the Fourier transform of the wavelet [calculated from Eq. (14.12)].

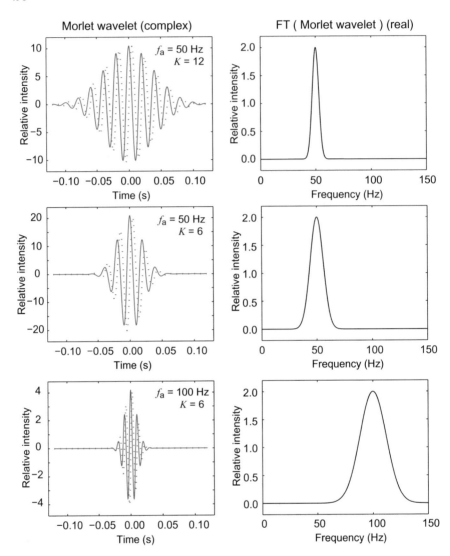

Fig. 14.8 Three different wavelets that indicate how the parameters ω_a and K (analysis frequency and "wavenumber", respectively) affect the wavelet. A wavelet has a limited extent in time (left part). We can specify a width for the envelope curve, e.g. by using the $f1/e$ criterion. If we make an inverse Fourier transform of this wavelet, we get the frequency responses shown to the right. Notice both the position in the frequency spectrum and the width of the Gaussian-shaped curves. There is a relationship between the widths in the time domain and the frequency domain. If we increase one the other will decrease, and vice versa

We know from before that the frequency spectrum of the Fourier transform of the product of a sine signal and a Gaussian curve is itself a curve with a Gaussian envelope. Again, this is get confirmed by Fig. 14.8.

The temporal width of the wavelet can be determined by starting from the envelope curve [from Eq. (14.8)]. If we define the width as the time difference between the peak value and a point where the amplitude of the envelope curve has dropped to $1/e$ of the maximum value, the half-width is:

$$\Delta t_{1/e} = 2K/\omega_a \, .$$

The corresponding width of the Fourier transform of the wavelet is quite close to [(as follows from Eq. (14.12)]

$$\Delta f_{1/e} = f_a/K = \omega_a/(2\pi K) \, . \tag{14.15}$$

It is interesting to note that

$$\Delta t_{1/e}\Delta f_{1/e} = (2K/\omega_a) \, (\omega_a/(2\pi K)) = 1/\pi \, .$$

We can calculate the "standard deviation" for time and frequency by using statistical formulas:

$$\sigma_t^2 = \frac{\int t^2\Psi^2(t)\mathrm{d}t}{\int \Psi^2(t)\mathrm{d}t}$$

and

$$\sigma_f^2 = \frac{\int f^2\hat{\Psi}^2(f)\mathrm{d}f}{\int \hat{\Psi}^2(f)\mathrm{d}f} \, ,$$

and it can be shown that

$$\sigma_t^2\sigma_f^2 = \frac{1}{2\pi} \, . \tag{14.16}$$

This relation is analogous to the Heisenberg uncertainty relation. Examples conforming to this relation are shown in Fig. 14.8.

The relationship is very important for wavelet analysis. If we allow a wavelet to extend for a long time, the width in the frequency domain will be small and vice versa. In other words: *We cannot get accurate temporal details of a signal at the same time as we get an accurate frequency description.*

An interesting consequence of Eq. (14.15) is that

$$\Delta f_{1/e}/f_a = 1/K \, .$$

In other words, in a wavelet analysis, we usually keep K constant throughout the analysis. Then, the relative uncertainty in the frequency values is constant throughout the diagram.

It is then natural to choose a logarithmic frequency axis, in the sense that the analysing frequencies we choose are related to each other as

$$(f_a)_{k+1} = (f_a)_k \, f_{\text{factor}} \, .$$

We have chosen logarithmic axes for the selected analysis frequencies in all examples in this chapter, but it is of course possible to choose the analysis frequencies on a linear scale, at least if the difference between the smallest and the largest analysis frequency is small (e.g. factor two or less).

Comparing wavelets with piecewise FT
If we use piecewise FT, there will within each 'piece' (window) be room for only a few (or less) time periods for a low-frequency signal but many time periods for a high-frequency signal. This means we would get a poor frequency resolution for the lowest frequencies (measured as relative frequency), but a far better frequency resolution for the higher frequencies. That means we would end up with an analysis that would not be optimal.

The procedure used in wavelet analysis provides an optimum time resolution for *all* frequencies. But we *can* nevertheless choose to *somewhat* emphasize time resolution at the expense of frequency resolution and vice versa, depending on what we want to study. This makes the method a very powerful aid in many contexts.

14.5.3   Border Distortion

When we calculate a wavelet transform, we basically multiply a signal, point by point, with a wavelet and sum all the products. We then move the wavelet and go through the same steps. This is repeated over and over again, beginning with the situation in which the centre of the wavelet lies completely at one end of the signal and finishing when the centre of the wavelet is located at the other end of the signal.

We then change the frequency of the analysing wavelet and go through the same routine.

Here, however, we meet a problem. As long as the wavelet is not complete within the data range, we would expect a different result than if the entire wavelet was used in the calculations. This is illustrated in Fig. 14.9. For the position the wavelet has in relation to the data in this figure, only about half of the wavelet will be used in practice. This means that the sum of the products is expected to be much lower (about a half) than what it would be if we had complete overlap.

It is therefore common to mark the wavelet spectrogram with what is called a "cone of influence" in order to indicate the region where the analysis is susceptible to border distortion (the name used for problems caused by incomplete overlap at the edges).

In Fig. 14.10, an example of a wavelet spectrogram which shows an analysis of temperature oscillations in the South Pacific. Figures (numbers) and colours are

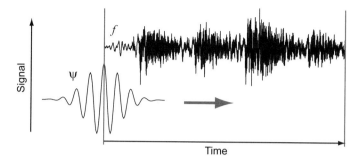

Fig. 14.9 It is not possible to get a correct wavelet result for times and frequencies where the entire wavelet does not come within the data range during the calculations

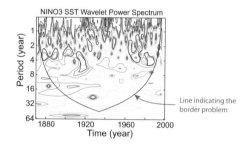

Fig. 14.10 Example of a wavelet spectrogram for temperature oscillations in the South Pacific. The figure was produced with data and software provided by C. Torrence and G. Compo made available at http://atoc.colorado.edu/research/wavelets/ and retrieved April 2016 [1]

used to display "energy" in different forms of oscillation (periodicity) as they have evolved over the last one hundred years.

In this diagram, a V-shaped curve is also drawn, the abscissa of whose cusp is at the middle of the scale and whose arms rise symmetrically on both sides, at first slowly and then steeply near the edges. This V curve is the above-mentioned cone of influence (COI), and it marks the area where most of the wavelets are complete within the data string: everything above the COI represents reliable data, but the results below the curve are suspect.

In the program examples given in the rest of this chapter, we have chosen to place a mark to indicate where the outer part of the wavelet with a value less than $1/e$ of the maximum is outside the diagram. We cover such a small frequency range that in our own examples so that we do not get the entire V curve, but only a small near-vertical part of the total V curve (except for Fig. 14.11). All parts of the wavelet spectrogram that lie between these marks have insignificant errors attributable to border distortion. We provide details below on how to set the selections.

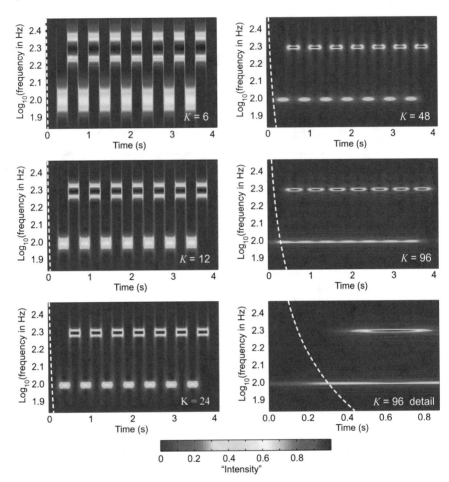

Fig. 14.11 Wavelet diagrams for the time signal in Fig. 14.5 for six different "wavenumbers" K. See the text for details

14.6 Optimization

Wavelet analysis is more demanding than normal Fourier transformation. One must choose what kind of wavelet we are going to use. Even though *we* stick to the Morlet wavelets, we do have to decide what "wavenumber" to use.

We have previously seen that by increasing the "wavenumber" K, the wavelet will have a significant value over a longer period than at low "wavenumber" (at the same analysis frequency). Further, we have seen that when the "width" of the wavelet in the time domain is large (that is, large K value), the "width" on the Fourier transform of the wavelet will be small. The product of the width of the wavelet in the time domain and the width of the wavelet in the frequency domain is constant.

The consequence is that there are no panaceas. If we want to get an accurate indication of the time course, small K values will be preferred. If we want to get as accurate frequency indications as possible, the K value should be large. In principle, we want as good time resolution and frequency resolution as possible, but always have to settle for a compromise.

The optimum result is often achieved if we keep an eye on the signal itself. The signal has often inbuilt uncertainty in time and/or frequency. We can never get a better resolution in time by wavelet analysis than the resolution in the signal itself, and likewise for frequency analysis.

Spelled out even more clearly: when the signal itself in the time domain has passages where oscillations are relatively constant in frequency and amplitude for M oscillation periods, we often get best results in a wavelet transform if the K-value is roughly equal to M (or slightly higher).

Figure 14.11 shows wavelet spectrograms of the signal (described above) that alternated between 100 and 200 Hz. Five different wavenumbers K are used. We see that for low K values the time resolution is very precise, but the frequency determination is poor. For high K values, the opposite holds: the frequency resolution is good, but the time resolution is poor.

In this case, there is really not much more to get in frequency resolution when we go from $K = 48$ to $K = 96$. This is because the signal itself has a "uncertainty in frequency" since the *duration* of each period of "constant frequency" is limited. In this case, there are 25 periods of oscillation within each 100 Hz interval and 50 periods of oscillation within each 200 Hz interval.

In Fig. 14.11, we have enhanced the marking of the left border distortion area. We see that the distortion increases with increasing K value. It may be interesting to note that the border distortion marking changes with the analysis frequency. Furthermore, it is useful to observe that the distance from the side edge to the border distortion mark also indicates smearing in the precision of time in the wavelet diagram. All time information in the analysis is smeared out to an extent that exactly corresponds to the distance from the edge to the border distortion mark.

What are the "best choices" of all the analyses presented in Fig. 14.11? Well, it depends on what we want to get out of the analysis. The diagram for $K = 6$ demonstrates that the change from 100 to 200 Hz (and vice versa) takes place very sharply in time. The $K = 96$ chart shows that the frequency is as uniform as it can be within each of the time intervals. If we need an overall optimization, perhaps $K = 48$ or so would be a good choice.

A standard Fourier transformation of the signal would yield two peaks, one for 100 Hz and one for 200 Hz. Had we taken the absolute value of the frequency spectrum we would not have seen any trace that could show that the signal varied between 100 and 200 Hz in time.

14.6.1 Optimization of Frequency Resolution (Programming Techniques)

Another form of optimization lies in the choice of frequency range for the analysis. In a digital fast Fourier analysis, we automatically get "all" frequencies between zero and the sampling frequency (but only half is useful due to folding). For a continuous wavelet analysis, we usually choose to narrow down the frequency range to the area where the frequency content is of interest.

In Fig. 14.11, we only chose to include frequencies between 70 and 300 Hz in the analysis. The reason is that we knew that the signal contained only frequencies close to 100 and 200 Hz. It may often be an advantage to start with a regular Fourier transform to ensure that we choose a frequency range that is suitable.

However, it is important to consider how many intermediate frequencies we will include in the analysis. In this context, we have to go back to the "width" of the wavelet in the frequency domain. This width, as we have seen before, was:

$$\Delta f = f_a / K .$$

This "width" was determined by the Gaussian frequency curve having fallen to $1/e$ of the maximum value. We do not want to take such great steps in frequency from one analysis frequency to the next, but maybe just a fraction of this.

Practical testing shows that an optimal choice of the difference between one analysis frequency and the next is then about

$$f_{a,\text{next}} - f_{a,\text{now}} = f_{a,\text{now}} / 8K . \qquad (14.17)$$

If we are going to cover a frequency range $[f_{start}, f_{end}]$, then we can easily show that we should use M analysis frequencies in a logarithmic order where

$$f_{\text{end}} = \left(1 + \frac{1}{8K} \right)^{M-1} f_{\text{start}} .$$

The number of analysis frequencies is thus:

$$M = 1 + \log(f_{\text{end}} / f_{\text{start}}) / \log \left(1 + \frac{1}{8K} \right) . \qquad (14.18)$$

14.6.2 Optimization of Time Resolution (Programming Techniques)

A continuous wavelet diagram may sometimes consist of very many points. For example, if we start with sound digitized at a sampling rate of 44.1 kHz, and we study sound with frequencies in the range of 10^2–10^4 Hz, we can in practice pick out only every fourth point in time from the diagram without losing significant information. Once we know that the wavelet has a width of about K times the period of the analysis frequency, we realize that we can remove even more points in time from the diagram without being detected in a wavelet diagram.

It may sometimes be of interest to optimize a wavelet diagram with the time indication. Not least, this is important to get plot files that are so small that they are easily incorporated into reports and the like.

In practice, one finds that it is enough to give each Pth item in the time dimension in a wavelet diagram (without loss of information) when P is given by:

$$P = \text{Integer-Value-Of}\left(\frac{K}{24}\frac{f_s}{f_{a,\max}}\right). \qquad (14.19)$$

In the computer program, the "floor" function is used to get the integer value.

14.7 Examples of Wavelet Transformation

14.7.1 Cuckoo's "coo-coo"

Figure 14.12 shows an example of optimized wavelet analysis. The signal is a CD quality audio file that gives the sound of a cuckoo singing its "coo-coo". The signal is given in three variants, namely as a pure time signal, as a frequency spectrum after a regular FT and finally as wavelet diagram.

In the total work plan, the first step is to select a suitable section from the audio file. This is done by selecting the starting point and total number of points to be retrieved from the available data file. Next, a Fourier analysis is performed. From the Fourier spectrum, we see that the sound usually contains only frequencies between 450 and 750 Hz. Accordingly, the wavelet analysis is limited to this frequency range.

Finally, we must try different K values and choose the "best" compromise between good time description and frequency description at the same time. We need to decide whether we want to prioritize time resolution (by having a small K), but at the expense of a fairly wide frequency response, or accept a slightly poorer resolution (by choosing a larger K) to get a slightly better frequency resolution. What is

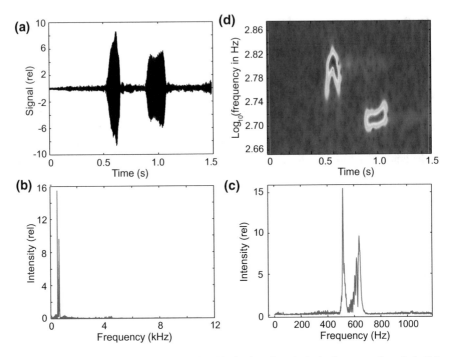

Fig. 14.12 A cuckoo bird's "coo-coo" analysed in the time domain, in the frequency domain (within a window) and in the combination frequency and time in the form of a wavelet diagram

optimal depends on the signal we analyse and will depend on what is important to the individual analysing the data. In our example, $K = 40$ was used.

Note the beautiful details that appear in the wavelet analysis. Have you been aware, for example, that the sound in the first "coo" actually changes significantly within the short period during which the sound lasts? Details in the wavelet analysis of birdsongs allow ornithologists to recognize birds individually. The details are finer than what human auditory apparatus is capable of perceiving.

It should be obvious that, for a sound of such type, wavelet transformation provides far more interesting data than a standard Fourier transform.

14.7.2 Chaffinch's Song

We include two further examples of wavelet analysis. The first (Fig. 14.13) is similar to the one we had for the cuckoo. We have chosen the chirping of a chaffinch, which dominates the birdsong in April. Chaffinch's song is characterized in several different ways. I even like the characteristic "*tit tit tit tit tit …I-love-you*". The joy of wavelet analysis is that the sound image is far more complicated than what we perceive. There is a very fast variation in frequency within each "tit", which we do not perceive. The K value used in the analysis was 48.0.

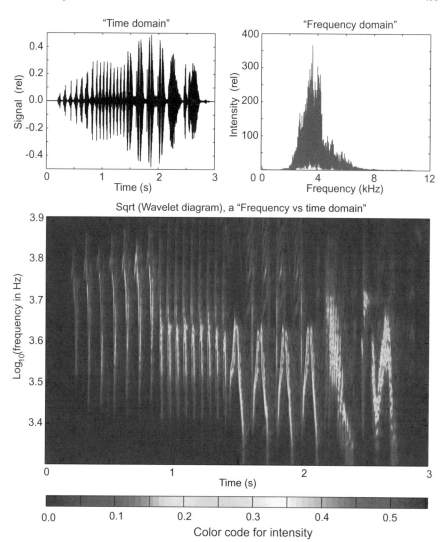

Fig. 14.13 Chaffinch's "tit tit tit tit tit…I-love-you" analysed in the time domain, in frequency domain and after wavelet transformation

14.7.3 Trumpet Sound, Harmonic in Logarithmic Scale

The last example is a wavelet analysis of a trumpet sound (Fig. 14.14). We have chosen a slice in time when the trumpet holds the same tone, and the intensity of the sound we experience is quite constant. The time-domain picture of the sound shows more variation in intensity than we perceive. However, the frequency spectrum (frequency domain) is the way we expected it. It consists of a series of sharp lines that show the fundamental tone and its harmonics.

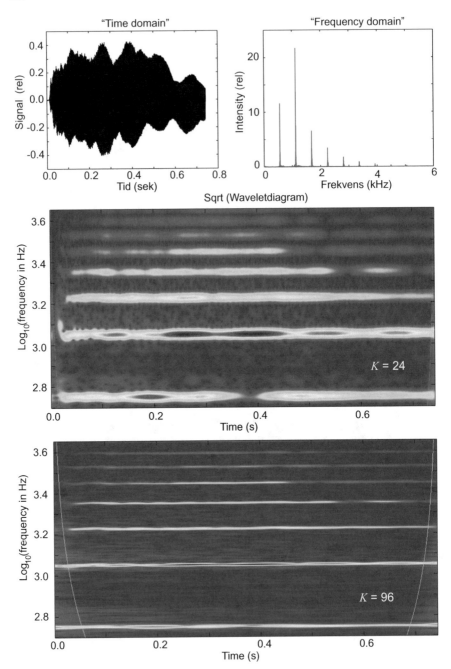

Fig. 14.14 A pure trumpet sound analysed in the time domain, in the frequency domain and in a wavelet transform. Two different K values are used in the wavelet analysis

The frequency axis is linear, and thus, the distance between two adjacent harmonics is constant, more particularly the frequency of the fundamental tone.

A wavelet analysis of such a signal fails to match the sharpness of the frequency spectrum; if we are primarily interested in the frequency of the fundamental tone and the harmonics *for a sustained tone*, the Fourier analysis method is to be chosen.

However, if we are interested in variations in the sound over time, Fourier analysis is not suitable. Then comes the wavelet analysis. We have included two different variants of analysis based on the wavenumbers $K = 24$ and $K = 96$. In the first case, the frequency resolution is rather poor, but the time resolution fits the signal. In the latter case, the frequency resolution is good, but the time resolution is poor.

Are we getting something more out of wavelet analysis than out of Fourier analysis? Yes, as a matter of fact. We see that the strength of the fundamental tone and harmonics varies slightly in time. We also see that there is a certain exchange between the intensities of the fundamental tone and the first harmonic: when one is powerful, the other is weak, and vice versa. This gives life to the sound and shows an example that it is difficult to replace real sound with synthetic sound.

Additionally, note that the distance between the harmonics is not constant in a normal wavelet diagram since we usually have a logarithmic frequency axis.

The frequencies f_n of the harmonics are, we recall, $f_n = nf_1$ where f_1 is the fundamental frequency. We have

$$\log(f_n) = \log(nf_1) = \log(f_1) + \log(n)$$

for $n = 1, 2, 3, \ldots$. It follows that:

$$\log(f_1) = \log(f_1)$$
$$\log(f_2) = \log(f_1) + \log(2) = \log(f_1) + 0.301$$
$$\log(f_3) = \log(f_1) + \log(3) = \log(f_1) + 0.477 = \log(f_2) + 0.176$$
$$\log(f_4) = \log(f_1) + \log(4) = \log(f_1) + 0.602 = \log(f_3) + 0.125$$
$$\log(f_5) = \log(f_1) + \log(5) = \log(f_1) + 0.699 = \log(f_4) + 0.097$$

The distance between successive harmonics on a harmonic scale is seen to be:

$$0.301, \ 0.176, \ 0.125, \ 0.097, \ \ldots$$

regardless of the frequency of the fundamental tone (see Fig. 14.14), in sharp contrast with the usual Fourier spectrum.

It is sometimes very handy to use these distances both for recognizing harmonics and for making sure that you have chosen a frequency range for the analysis that includes the fundamental tone (where important).

14.8   Matlab Code for Wavelet Transformation

We have included a Matlab program that can be used to analyse bits of audio files. The code is available at a "Supplementary material" Web page at http://www.physics.uio. no/pow, so you will not need to re-type it.

The code is divided into four different functions: a "main program", a function that reads wav format audio files and plots the time frame of the signal, a function that performs a Fourier transform and plots the result, and a function that performs a wavelet analysis of the signal (given a set of parameters that we choose ourselves). The "main program" uses the three other features in turn.

Some remarks:
The program is self-written and bears imprints which reveal that I learned programming in an earlier era. Feel free to write your own program version more in line with modern trends and requirements. However, if you are short of time, you may want to use our Matlab or Python program directly to get what you need to do, and in the longer term, improve the program on your own.

Main Program

The code is available at the "Supplementary material" Web page for this book at http://www.physics.uio.no/pow.

```
function WLofWAV

% ''Main program'' that reads a wav-file, plots the signal
% in time and frequency domain as well as the wavelet
% trasformed signal. NOTE: All parameters in this program
% has to be chosen carefully according to your signal!
% The program is written by AIV. Version 16. October 2017


% Reads a wav-file
c = 'gjok.wav';   % Name of file
N = 1024*32;      % Number points to be analyzed
nstart = 31000;   % First point in file to be read
[fs,h] = readWavFile(c,nstart,N);


% Fourier transformation
[FTsignal] = fftAndPlot(h,N,fs);


% Wavelet-analysis
fmin = 400.0;   % Use FFT and preferences to choose this!
fmax = 800.0;   % (as above)
K = 32;         % Must be optimized for every signal
[msg] = wltransf(FTsignal,fmin,fmax,K,N,fs); % Wavelet
                                             % transform
```

Reading Wav-File and Plotting the Results

```
function [fs,h] = readWavFile(c,nstart,N)

% This function reads a wav-fil with the name c.
% Reading starts at point number ''nstart'' and N points are
%  read. The sound is played, and the signal is plotted.
% The function returns the sampling frequency and
% one channel of the stereo signal in the wav file.
% The program is written by AIV. Version 16. October 2017.

nend = nstart+N-1;
[y, fs] = audioread(c, [nstart nend]); % Read array y(N,2)
% from file.
% 'fs' is usually 44100 (sampling frequency at CD quality)
h = zeros(N,1);   % Picks only one channel from stereo signal
h = y(:,1);
sound(h,fs);      % Play the sound track read from file
T = N/fs;         % Total time for the sound track read

% Plot the signal in time domain
t = linspace(0,T*(N-1)/N,N);
plot(t,h,'-k');
title('Wav-file signal');
xlabel('Time (sec)');
ylabel('Signal (rel units)');
return;
```

Calculating FFT and Plotting the Results

```
function [FTsignal] = fftAndPlot(h,N,fs)

% This function perform a FFT of a signal h described in
% N points.
% Sampling frequency is fs. The Fourier transformed signal
% (absolute values) is plotted, but the full complex
% Fourier transform is returned to the calling function.
% The function is written by AIV. Version 16. October 2017.

% Calculate FFT of the time descripion of signal: h
FTsignal = fft(h);   % This is what is returned at exit.

% Plot the frequency spectrum (absolute values only)
```

```
f = linspace(0,fs*(N-1)/N, N);
nmax = floor(N/2);   % Plot only lower half (due to aliasing)
figure;
plot(f(1:nmax),abs(FTsignal(1:nmax)));
xlabel('Freqency (Hz)');
ylabel('Relative intensities');
title('Frequency spectrum of the signal');
return;
```

Calculating the Wavelet Transform and Plotting the Results

```
function [msg] = wltransf(FTsignal,fmin,fmax,K,N,fs)

% This function carries out a wavelet transform using Morlet
% wavelets.
% Input is a full FFT of the signal that should be analyzed,
% as well as min and max frequencies for the wavelet analysis.
% K and N are ''wavenumber'' in the Morlet and number of points
% and sampling frequency for the input signal, respectively.
% fs is the sampling frequency.
% The function is optimized so that it chooses both the
% resolution of frequencies and time in the final diagram.
% Two different choises of intensity scaling are possible:
% intScale 1 and 2 correspond to a high or low ''dynamical
% range'' in the wavelet plot (Use 1 to make weak signal
% details visible). At the end the wavelet transformed signal
% is plotted.
% The function is written by AIV. Version 16. October 2017.

% Calculate # frequencies for analysis, write to screen,
% make list of frequencies ready for plot

intScale = 1; % 1 makes weak signal details visible

% Make sure that the FT signal is a column array
SZ = size(FTsignal);
if SZ(1) > SZ(2)
    FTsignal = transpose(FTsignal);
end;

% Calculate/define parameters what will be used later
M = floor(log(fmax/fmin) / log(1+(1/(8*K)))) + 1;
NumberFrequenciesInAnalysis = M
fstep = (fmax/fmin)^(1/(M-1));
f_analysis = fmin;
```

```
T = N/fs;                  % Total time for the sound track chosen
t = linspace(0,T*(N-1)/N,N);
f = linspace(0,fs*(N-1)/N, N);


% Allocate array for the wavelet diagram and array for
% storing frequencies
WLdiagram = zeros(M,N);
fused = zeros(1,M);


% Loop over all frequencies that will be used in the analysis
for jj = 1:M
    % Calcualate the FT for the wavelet directly
    factor = (K/f_analysis)*(K/f_analysis);
    FTwl = exp(-factor*(f-f_analysis).*(f-f_analysis));
    FTwl = FTwl - exp(-K*K)*exp(-factor*(f.*f));   % Minor
                                                   % correction term
    FTwl = 2.0*FTwl;   % Factor (different choices possible!)
    % Calculate a full line in the wavelet diagram in one
    % operation! (Inverse of the convolution, see textbook.)
    if intScale == 1
        WLdiagram(jj,:) = sqrt(abs(ifft(FTwl.*FTsignal)));
    else
        WLdiagram(jj,:) = abs(ifft(FTwl.*FTsignal));
    end;
    fused(jj) = f_analysis;  % Store frequencies actually used
    f_analysis = f_analysis*fstep;  % Calculate next frequency
end;
% The main loop finished! The wavelet diagram is complete!


% Reduce file size of the wavelet diagram by removing a lot
% of redundant information in time. The purpose is just
% to make the plotting more managable.
P = floor((K*fs)/(24 * fmax));   % The number 24 may be changed
                                 % if wanted
UseOnlyEveryXInTime = P     % Write to screen (monitoring)
NP = floor(N/P);
NumberPointsInTime = NP     % Write to screen (monitoring)
for jj = 1:M
    for ii = 1:NP
        WLdiagram2(jj,ii) = WLdiagram(jj,ii*P);
        tP(ii) = t(ii*P);
    end;
end;


% Make a marking in the plot to visualaize border of
```

```
% distortion
maxvalue = max(WLdiagram2);
mxv = max(maxvalue);
for jj = 1:M
    m = floor(K*fs/(P*pi*fused(jj)));
    WLdiagram2(jj,m) = mxv/2;
    WLdiagram2(jj,NP-m) = mxv/2;
end;

% Plot wavelet diagram
figure;
imagesc(tP,log10(fused),WLdiagram2);
set(gca,'YDir','normal');
xlabel('Time (sec)');
ylabel('Log10(frequency in Hz)');
if intScale == 1
    title('Sqrt(Wavelet Power Spectrum)');
else
    title('Wavelet Power Spectrum');
end;
colorbar('location','southoutside');

msg = 'Done!';
return;
```

It should be noted that once in a while we may want to use an 'intensity versus time plot' instead of a full wavelet diagram, to demonstrate particular details (as we did in Fig. 14.6). It is easy to pick a particular horizontal (or vertical) line in the WLdiagram2(i, j) array and make a normal 2D plot of the result. However, care has to be taken to get the right frequency since we use a logarithmic frequency axis in our wavelet diagram.

14.9   Wavelet Resources on the Internet

1. A.-H. Najmi and J. Sadowsky: "The continuous wavelet transform and variable resolution time-frequency analyses." *Johns Hopkins APL Technical Digest*, vol 18 (1997) 134–140. Available on http://www.jhuapl.edu/techdigest/TD/td1801/najmi.pdf accessed May 2018.
2. http://www.cs.unm.edu/ williams/cs530/arfgtw.pdf "A really friendly guide to wavelets", C. Valens and others. Accessed May 2018.
3. http://tftb.nongnu.org/, "Time-frequency toolbox". Accessed May 2018.
4. http://dsp.rice.edu/software/, "Rice Wavelet Toolbox." Accessed May 2018.
5. http://www.cosy.sbg.ac.at/ uhl/wav.html, Several wavelet links. Accessed May 2018.

6. A 72-page booklet by Liu Chuan-Lin: "A tutorial of the wavelet transform" (dated 23 February 2010) is available on http://disp.ee.ntu.edu.tw/tutorial/WaveletTutorial.pdf Not available May 21th 2018 (Will be available through the web-pages for this book if it is a continuous problem with access.). The booklet also deals with the use of wavelets in image processing.

14.10 Learning Objectives

After working through this chapter, you should be able to:

- Describe similarities and differences between Fourier transformation and wavelet transformation.
- Describe for which signals Fourier transformation is preferred and for which signals wavelet transformation is preferred. Explain why.
- Explain what we can deduce from a given wavelet diagram.
- Explain how we can adjust a wavelet transformation to accentuate temporal details, or details in frequency.
- Explain qualitative analogues between wavelet transformation and Heisenberg's uncertainty relationship.
- Use a wavelet analysis program and optimize the analysis.

14.11 Exercises

Suggested concepts for student active learning activities: Short-time Fourier transform, time domain, frequency domain, fast Fourier transform, Morlet wavelet, wavenumber K for wavelets, "discrete continuous", optimization, classical analogy to Heisenberg's uncertainty relation, frequency resolution, time resolution, absolute value of the transform, cone of influence.

Comprehension/discussion questions

1. What is the most important difference between Fourier transformation and wavelet transformation?
2. In what situations does Fourier transformation provide a rather useless result?
3. What are the disadvantages of wavelet transformation compared to Fourier transformation?
4. When was Fourier transformation implemented on a large scale (FFT) and when did wavelet transformation come into vogue?

5. Wavelet transformation is affected by "border distortion". What is meant by this? How big is the border zone?

6. Can you outline how wavelet transformation might be used to generate notes directly from a sound recording? What problems do you think may occur?

Problems

7. (a) Calculate a Morlet wavelet (in time domain) for analysis frequencies 250 and 750 Hz when the sampling rate is 5000 Hz and the K parameter is 16. Plot the result with correct time on the x-axis (a figure similar to Figure 14.3).

(b) Calculate the Fourier transform of each of the two wavelets. Use both an FFT directly on the Morlet wavelet described in the time domain and by calculating the Fourier transform directly using Eq. (14.12). Plot the results with correct indications of frequency on the x-axis.

(c) Make sure that the peak occurs at the place you would expect. Do you see mirroring?

(d) Repeat points a–c also when $K = 50$.

8. In this task, the underlying theme is the analogy with Heisenberg's uncertainty relation.

(a) Generate a numeric data string representing the signal

$$f(t) = c_1 \sin(2\pi f_1 t) + c_2 \cos(2\pi f_2 t) \,.$$

Use 10 kHz sampling rate and $N = 8192$ points, $f_1 = 1000$ Hz, $f_2 = 1600$ Hz, $c_1 = 1.0, c_2 = 1.7$. The signal must last throughout the period under consideration. Plot an appropriate section of the signal in the "time domain" (amplitude as a function of time) so that the details will become noticeable. Be sure to provide correct numbers as well as text along the axes, preferably also a heading.

(b) Calculate the Fourier transform of the signal. Plot an appropriate section of the signal in the " frequency domain" (choose absolute values of Fourier coefficients as a function of frequency), with numbers and text along the axes as above.

(c) Calculate the wavelet transform of the signal (may well be based on the programs given in this chapter, or you can write the program more or less from scratch yourself). Use Morlet wavelets, and let the analysis frequency go, for example, from 800 to 2000 Hz (logarithmically scaled as usual within wavelet transformation). Then, plot the result for the wavenumber K equal to 24 and 200. Comment on the result.

(d) Let the signal be a harmonic signal as before, but now multiplied with a Gaussian function so that we get two "wave packets":

$$f(t) = c_1 \sin(2\pi f_1 t) \exp\left(-[(t - t_1)/\sigma_1]^2\right) + c_2 \cos(2\pi f_2 t) \exp\left(-[(t - t_2)/\sigma_2]^2\right)$$

where $t_1 = 0.15$ s, $t_2 = 0.5$ s, $\sigma_1 = 0.01$ s and $\sigma_2 = 0.10$ s. Calculate the Fourier transform of the signal and also the wavelet transform of the signal. Plot the signal in the time domain, the frequency domain (match the section) and the wavelet transform of the signal for $K = 24$ and 100 (please test more values!), and other parameters as given in point (c) above. Comment on the results!

9. Analyse the song of a blackbird (svarttrost in Norwegian) using wavelet trans-formation. An audio file is available from the "Supplementary material" Web page. Use the program in this chapter (or own version) and analyse a time string of 1.4 s. Parameters of analysis: Filename: 'Svarttrost2.wav', Nstart = 17000, data string length "64 k" $= 2^{16}$, frequency range 1500–8000 Hz, wavenumber K equal 12 and 96 (and preferably some values in between as well). The signal consists of five different audio groups. We are primarily interested in the fourth of these!

Plot the signal in the time domain, in the frequency domain, and the signal anal-ysed by wavelet transformation for this fourth audio group. Be sure to include some sections of the original plots to get details. This applies in particular to the time of the original sound! Hopefully you will then recognize a signal we have encountered at least twice in previous chapters. You should recognize how we can make such a signal mathematically.

Careful analysis of the fourth bit of the sound signal in the (1) time domain and (2) wavelet diagram makes it possible to see how a close analogy to Heisenberg's uncertainty relationship comes into play. To get the full benefit, you should extract time differences and frequency differences in the charts and compare these with the wavelet evolution in the time and frequency domain for the two selected K values.

If you are a student and have offer for help from teachers, we strongly recom-mend that you discuss the relevant details with the teacher until you get a firm grasp of the relationships we wish to bring to the fore. There is a lot of valuable knowledge to extract from this problem, knowledge that can be valuable also in many other parts of physics!

10. Make a wavelet analysis of chaffinch sound (the audio file "bokfink" on the "Supplementary material" Web page). Try a K factor twice as large as in the example in Fig. 14.13. Also, try an analysis for half the K value used in the above figure. Describe the differences you see.

11. In this problem, the theme is to explore various ways to display a wavelet trans-formed signal so that you get as much information from the diagram as possible. The "dynamic rage" in a colour coded diagram varies with the different choices. Repeat wavelet analysis of chaffinch sound again for $K = 48$. Choose succes-sively "power spectrum" (absolute value after inverse Fourier transformation

squared), wavelet analysis "on amplitude level" (absolute value after inverse Fourier transformation directly) and wavelet analysis with the square root of the wavelet analysis (square root of absolute value after inverse Fourier transformation). Make your assessments of which of these methods you like best for this particular signal. Perhaps you would prefer one of the other viewing modes (squares or square roots) for another signal?

12. Use the knowledge from Chaps. 2 and 3 to calculate the timing of a spring pendulum after it is set in motion by a harmonic force with a frequency equal to the resonance frequency. Follow the oscillations also some time after the harmonic force is removed. Then, perform a wavelet analysis of the oscillation process. Attempt to optimize the analysis with respect to the K value. Do you find an apparent correlation between the Q value for the pendulum oscillation and the K value that gives the optimal wavelet diagram?

13. Use the various wav files available at the "Supplementary material" Web page in order to get used to wavelet transform, how to use it in practice, choice of parameters and analysis of the results. Everyone need practice in order to utilize new tool as well as possible.

14. Select yourself an audio file that you can transform into a .wav file and select a slice that you may want to analyse. Optimize the analysis and tell what information you get from the chart.

16. Find data online that show a time sequence you think might be interesting to study. It may be weather data, solar spots, power consumption or what you need to find. Analyse the data set both in traditional Fourier transformation and wavelet analysis. Which method do you think is best suited to the data you selected? (Should have data with some form of loose periodicity with at least 20–30 periods within the data you have available.)

17. Compare the voices of Maria Callas and Edith Piaf (sound files available at the "Supplementary material" Web page at http://www.physics.uio.no/pow.). Is the vibrato an oscillation in frequency and/or in intensity? Which one of the two artists has the highest number of harmonics? Could you guess this just by listening to their voices?

Reference

1. C. Torrence, G. Compo, http://atoc.colorado.edu/research/wavelets/. Accessed 20 April 2016

Chapter 15
Coherence, Dipole Radiation and Laser

Abstract This chapter is focused mainly on coherence, a vital concept if one wants to go beyond a rudimentary understanding of waves; the notion is rooted in the statistical properties of a wave. We also elucidate how mixing of real physical waves leads to unexpected relationships: for example, it would be disastrous if the members of a choir managed to sing in perfect tune. This is related to the difference between light from a lamp compared to laser light. A computer program is provided for numerical explorations of both temporal and spatial coherence. The discussion of coherence is followed by a conceptual explanation of how electric charges in motion may lead to radiation of electromagnetic waves, and the radiation diagram for a dipole antenna is presented. The chapter concludes with a brief description of the basic principles for generating laser light.

15.1 Coherence, a Qualitative Approach

"Coherence" is a very useful concept when we want to describe how regular a wave is. In modern physics, it is clear that waves may interact very differently in various systems, depending on the statistical features of the waves. It is insufficient to characterize a wave only with amplitude, frequency, wavelength and the spatial volume where the wave is found. Figure 15.1 tries to illustrate this point using pictures of various surface waves on water.

The dictionary meaning of the word *coherence*, "the quality of being logically consistent", may be inferred from its etymology (*co* together + *hæreo* to stick). In physics, however, coherence is defined differently, and they are comparatively recent concepts. They describe important statistical properties. Wikipedia states that "two wave sources are perfectly coherent if they have *a constant phase difference* and *the same frequency*, and *the same waveform*". Additionally, it says: "Coherence describes all properties of the *correlation* between physical quantities *of a single wave, or between several waves or wave packets*" (emphasis added).

© Springer Nature Switzerland AG 2018
A. I. Vistnes, *Physics of Oscillations and Waves*, Undergraduate Texts in Physics, https://doi.org/10.1007/978-3-319-72314-3_15

Fig. 15.1 Three examples of surface waves on water, illustrating the need for a statistical description of waves in addition to the parameters we have used so far

Thus, coherence is a term related to waves and other time-varying signals. When we say that waves at two points A and B in space are coherent, we mean that there is a certain relationship between the waves passing at points A and B at any time.

We distinguish between temporal and spatial coherence. For **temporal coherence** (or longitudinal coherence), we consider the wave at two points A and B which lie along the direction of the wave motion (see Fig. 15.2). We then check if there is a sustained definite relationship between the wave at point A and the wave at another point B which the same part of the wave passed a little earlier.

For **spatial coherence**, we compare the wave at a location A with the wave at location B, the two points being adjacent to each other, but the direction from A to B is perpendicular to the direction of the wave motion.

For real waves, there is often a high degree of correlation between the waves at A and B if the two points are very close to each other (less than the smallest wavelength found in the wave). On the other hand, it is always true for real waves that the correlation between the waves at A and B becomes exceedingly poor if the distance between A and B is made big enough.

If the distance between A and B must be less than a few wavelengths to get a high degree of correlation, we say that the wave is *incoherent*. If, on the other hand, we can find a high degree of *sustained definite relationship* between the wave at point A and the wave at point B even when the distance between A and B is very many

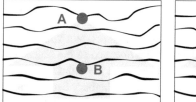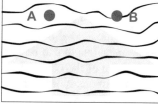

Fig. 15.2 Temporal (also longitudinal) and spatial (also transverse) coherence tell us something about the regularity in waves in the direction along which the wave moves or in the perpendicular direction, respectively. The wave is believed to be two dimensional in this case (e.g. surface waves on water). The black stripes indicate wave peaks, and their thickness indicates the amplitude of the wave at the current location. The wave in this figure is rather irregular

wavelengths, we call the wave *coherent*. However, there is a continuous transition between incoherence and coherence.

The term *coherence length* will be used for the largest distance between A and B for which significant correlation can still be found. We can specify both a temporal and a spatial coherence length. For temporal coherence, we can also operate with *coherence time* which is the time the wave uses in traversing the temporal coherence length.

Since there is always a certain degree of variability and unpredictability in real waves, and randomness can be described by statistics, the degree of coherence may be quantified statistically.

15.1.1 When Is Coherence Important?

Coherence is always important when two or more waves superimpose on each other. Thus, coherence is an important condition for interference. In deriving the intensity distribution at a double slit, we assumed that the wave has the same amplitude at all times in the two slit openings. It is equivalent to saying that the spatial coherence length must be at least as large as the distance between the two slits. In order to have several interference fringes outside the central one, it is also necessary that the temporal coherence length is at least several wavelengths since we add, in that case, one wave with a time-shifted part of the same wave.

An implicit consequence of a long temporal coherence length is that the wave must last for at least as long as the coherence time. This implies that there is a close relationship between temporal coherence length and width of frequency (or wavelength) distributions. The relationship is given by the time-bandwidth product discussed in Chaps. 3 and 5. Thus, narrow linewidth light emitted from atoms will have a long temporal coherence length, while light with a broad frequency distribution will have a short temporal coherence length. Light from the sun has a temporal coherence length of only a few wavelengths. In comparison, a laser may have a temporal coherence length of the order a million wavelengths.

If a source of waves is very small ("point-like", of the order one wavelength or less), the source will radiate waves with circular wavefronts. If the medium is isotropic, there will be more or less perfect circular wavefronts, at least within a sector of radiation. For such a system, the spatial coherence length can be very large, even if the temporal coherence length is short.

Even for a very extended object with many independent sources of radiation, the spatial coherence length can be large if the distance to the object is very large compared to the size of the object. We will return to the point later in this chapter when we discuss a famous experiment performed by Hanbury Brown and Twiss more than 60 years ago. We also often use a so-called pinhole in optics in order to increase spatial coherence length of light, as will be discussed later in this chapter.

15.1.2   Mathematical/Statistical Treatment of Coherence

Waves can be irregular in many different ways. Several methods are worked out to characterize the irregularities. We will limit ourselves to one of the simplest methods based on the calculation of a first-order correlation function.

If we want to characterize a wave, we can either record the amplitude at one point in space as a function of time, or we can record the amplitude at one instant as a function of position in space. In both cases, we acquire real measurements as a row of numbers, an array. If we make measurements at two points, as is indicated in Fig. 15.2, we end up with two sets of numbers, two signals.

There are many ways to compare two such arrays, but for analysing coherence physicists chose many years ago a strategy similar to Fourier transformation. In Fourier transformation, we actually do a correlation analysis between the signal we are studying and a perfect mathematical harmonic function with a given frequency (and we change the frequency to get the entire frequency range). Fourier transformation thus involves calculating the inner product between the signal we analyse and a perfect harmonic function.

When analysing correlation between signals (waves), we also calculate the inner product, but now between the two signals we compare! In a manner of speaking, we use one of the signals as a reference to check how much it resembles the other signal.

If the signal recorded at point A is called $f(t)$ and the signal at B is called $g(t)$, our predecessors have chosen to calculate the correlation between f and g as follows:

$$C = \frac{\int f(t)g(t)\,dt}{\sqrt{\int f^2(t)\,dt\ \int g^2(t)\,dt}} \ . \tag{15.1}$$

The integrations extend over an arbitrary interval of time. If the signals are stationary (in the sense that their statistical character does not change over time) and also ergodic (in which case the statistical information derived by analysing many independent signals will be equivalent to the statistical information derived by following only one sufficiently long-lasting signal), C will approach a well-defined value when the integration time increases.

Defined this way, correlation is simply a number. There will be no function before we find some means of systematically changing how f and g are measured or generated.

15.1.2.1 Autocorrelation

In the left part of Fig. 15.2, we chose to compare waves at two points A and B situated along the direction in which the wave moves. Practical measurements could be carried out if we chose to analyse sound, because very small microphones are available that perturb the wave so little that the signal at A would not be affected by the presence of the microphone at point B.

In other contexts, it is not feasible to place a sensor at B without interfering with the wave that reaches point A.

This is one of the reasons why we often choose, when we analyse temporal coherence of a wave, to use only one detector, for example at point B, and no detector at point A. We assume that the shape of the overall wave pattern (as given by the black lines in Fig. 15.2) does not change much during the time it takes for the wave to move from B to A; the signal in point A will be approximately the same as at point B, only time shifted (because of the wave pattern moving at a given speed).

In such cases, we will have

$$f(t) \approx g(t + \tau)$$

where τ is the time used by the wave (wave pattern) to move from B to A.

The temporal correlation between the wave at A and at B is then given by:

$$C = \frac{\int g(t)g(t + \tau)\,dt}{\sqrt{\int g^2(t)\,dt \ \int g^2(t + \tau)\,dt}} . \tag{15.2}$$

If the statistical properties do not change over time, we say that the wave is *stationary*. For such waves

$$\int g^2(t)\,dt \approx \int g^2(t + \tau)\,dt .$$

With this viewpoint, we are able to calculate correlations for many different distances between points A and B. In practice, we do this by changing the time shift τ in Eq. (15.2) above. If we let τ vary continuously from zero onwards, the correlations $C(\tau)$ will become what we call *autocorrelation function* for the signal.

Thus, the expression for the autocorrelation function for a signal g becomes:

$$C(\tau) = \frac{\int g(t)g(t+\tau)\, dt}{\int g^2(t)\, dt}. \tag{15.3}$$

For a digitized signal $g_i \equiv g(t_i)$ for $i = 1, 2, \ldots, N$ (discrete instants), the corresponding expression is:

$$C(j+1) = \frac{\sum_{i=1}^{M} g_i g_{i+j}}{\sum_{i=1}^{M} g_i g_i} \tag{15.4}$$

for $j = 0, \ldots, N - M$.

Note that since we shift the selection of points used for describing the signals at points A and B from the same data string, $M < N$. The largest j value we can calculate for a signal described by N points is $j = N/2$.

The left part of Fig. 15.3 shows an example of an autocorrelation function for a wave. Along the x-axis, we have the time difference as $f(t) = g(t + \tau)$ is offset relative to $g(t)$. If the sampling rate is F_s, the relation between the index j and the time delay τ will be:

$$j = \text{round}(F_s \tau)$$

where "round$(\cdots)$" means the integer nearest to the numerical value of the expression enclosed in the parentheses.

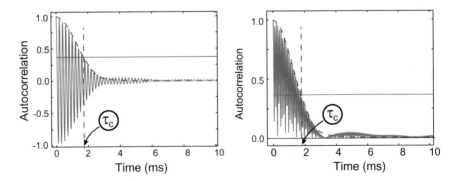

Fig. 15.3 An example of how the autocorrelation function of a wave might look like. To the left, the usual autocorrelation function is plotted, while in the right part the absolute value is plotted. The time shift τ_c that causes the correlation to decrease to $1/e$ of the maximum value is called the coherence time of this wave. (Note: The curve in the right part of this figure seems to never touch the x-axis. The reason for this is that the sampling frequency is not much larger than the signal frequency. We just do not happen to measure values very close to zero.)

We see that the autocorrelation varies from $+1$ to near -1 and oscillates up and down while the amplitude of the variation decreases to zero. Correlation equal 1 (for the standard formula we used here) corresponds to the fact that f and g are identical. They are always at the first point (no shift). Correlation -1 would mean that $f = -g$. For a wave, just say that when f has a wave peak, g will have a valley, and vice versa. In that case, there is still perfect correlation, but the sign has reversed.

When the distance between points A and B in Fig.:15.2 increases, in Fig. 15.3, there is a gradual transition from high correlation (dashed upper envelope curve has a value close to 1) to a smaller and smaller correlation (envelope near zero). After about 1.8 ms, the number has fallen to $1/e$ of max. We say that *the temporal coherence time* τ_c for this signal is 1.8 ms. The corresponding temporal coherence length is $\tau_c v$ where v is the wave phase velocity.

In the right part of Fig. 15.3, we have plotted the absolute value of the autocorrelation function as a function of the displacement τ. We have seen that correlation $+1$ and -1 both correspond to a perfect correlation, only with a change of sign in the amplitude in the latter case. It is therefore best to draw a hypothetical envelope curve touching the peaks in the above plot, when we want to determine the value of τ at which the correlation decreases from 1 to $1/e$.

It should be noted that in practice, the autocorrelation will never vanish when the distance between A and B increases. However, if the average value of g is zero and if the total observation time is much longer than τ_c', the asymptotic value of the autocorrelation ratio will come so close to zero that the remaining variation does not affect the determination of coherence time.

It turns out that the variation in the first part of the autocorrelation function is quite stable if we make more subsequent recording of the signal $g(t)$, but the oscillations around zero when we have passed at least twice the coherence time will change from one data record to the next.

If we take the average of many runs and add the autocorrelations, the first part of the correlations will be added constructively while what we like to call the "noise" around zero will eventually be considerably reduced. The autocorrelation function can then often be written almost:

$$C(\tau) = \cos(\bar{\omega}\tau)\exp[-(\tau/\tau_c)^2] \tag{15.5}$$

where $\bar{\omega}$ corresponds to the mean of the angular frequencies in the original signal. τ_c is the correlation time and corresponds (with some reservations) to what we also call the "coherence time" for our wave/oscillation. The coherence time is then defined by Eq. (15.5).

If g was a perfect sinusoid, a shift of $\tau = 2\pi n$ (n an integer) would cause $g(t)g(t+\tau)$ in Eq. (15.3) in practice to be a $\sin^2$ function. The autocorrelation function C will then simply be the mean of $\sin^2$ divided by the same value, which is 1.

Similarly, we can show that when τ equals $(2n+1)\pi$ in the perfect sinus signal g, the calculation will mean the mean of $-\sin^2$ and the answer would be -1. For a quarter wavelength offset relative to the cases we have already mentioned, C will be the mean of a $\sin \times \cos$ expression, which is equal to zero.

The autocorrelation function for a perfect sine will therefore be a periodic function that varies from $+1$ to -1. The autocorrelation function simply gets a cosine shape and will never get an amplitude that decreases to zero as shown in Fig. 15.3. Coherence time would then become infinite.

No real waves have infinite coherence time, but if the time frame we have available for analysis is no more than twice the coherence time long, we cannot determine the coherence time.

15.1.3 Real Physical Signals

We will now take a few real physical signals to show important features when we mix real waves. We have at our disposal three different analytical methods to study the time signal, namely Fourier transformation, wavelet analysis and calculation of the autocorrelation function.

The signals are microphone signals after sampling sound waves from a person who sings "eeeeee" throughout the sampling process. Eight separate audio recordings of the same person have been made. From these recordings, we have constructed three: signal 1 is simply one of the sound recordings. Signal 2 is the sum of two recordings, and signal 3 is the sum of all eight recordings. Signal 2 simulates a data recording of two people who sing "eeeeee" simultaneously, and likewise for signal 3. The signals are shown in Fig. 15.4 together with the analyses we have performed.

Signal from one source
We see in the left-hand column in Fig. 15.4 that the amplitude of the signal (and thus the sound intensity) remains quite similar throughout the period. The frequency range has a width of about 15 Hz. The wavelet diagram shows how the frequency and the amplitude changed during data recording. At the bottom of the figure, the first part of the autocorrelation function is given, and we can estimate the coherence time of the signal to about 0.18 s.

The sound speed in air is about 340 m/s. This means that we can predict correlation in the phase of the signal in hand within a range of about $\Delta L = 340 \times 0.18$ m, that is, about 60 m. This quantity we call the *coherence length* for our wave (more precisely the *temporal coherence length*).

There seems to be nothing special about data recording for one source. Everything is as we expected, but we realize that the instability in frequency (pitch) about the middle of the data recording probably affects the calculated coherence time. We also notice that the autocorrelation function does not settle down to the zero baseline after we have passed the coherence time, which is as expected.

Signal from several sources
We see from the middle column in Fig. 15.4 that the amplitude in the sum of two almost similar waves varies drastically during data recording, although the amplitudes of the individual contributions remained rather stable. The reason for this is

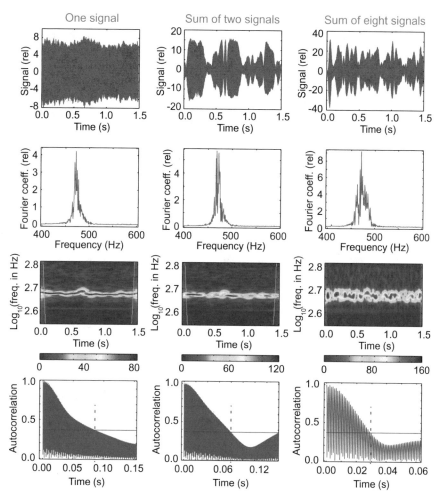

Fig. 15.4   Three examples of real sampled audio signals, with left to right one, two and eight voices at the same time. Frequency analysis is also shown (a little second harmonic component is not included), as well as wavelet analysis and the first part of the autocorrelation function. For the wavelet analysis, $K = 100$ was applied to the first two signals and $K = 32$ to the last

that the sound waves from singer 1 and singer 2 add constructively for some periods, so that the amplitude becomes about twice as large as that for each single wave. At the same time, the sound waves in other periods happen to add destructively, and the total amplitude of the sound waves falls almost to zero.

All waves seem in general to have the following properties:

The amplitude at an arbitrary place in the wave can have many contributions. However, the amplitude has only one value within a sufficiently small volume (lengths in each direction much smaller than the shortest wavelength in the waves that contribute). For example, the sound pressure will only have one value in small volumes when a sound wave passes, the height of the water surface has only one value in any place, and the electric field in the sum of all electromagnetic waves has only one value in each small volume even though many electromagnetic waves contribute.

We can only recognize different contributions to the waves and their origins by examining the pattern in the sum of all the waves and seeing how this pattern evolves in time. Contribution from circular ripples on a water surface impacted by a stone can only be recognized by looking at the rings in the region surrounding the small volume under consideration.

It is the *summed wave* which evolves in time, not every single contribution separately. However, when the physical system behaves linear, we can still describe the evolution of a wave *mathematically* as a sum of several contributions that individually conform to the wave equation. That we can use mathematics in this way should, however, be considered more a happy exception than the rule (because when we deal with nonlinear processes, it does not apply).

When we add more independent waves, there will always be some periods of constructive interference and some periods of destructive interference. The duration of the periods of constructive interference depends very much on the frequency variation in the signals that are added (which is related to the temporal coherence time). We will show more examples of this a little later in the chapter. The effect is manifested particularly well in continuous wavelet analysis with Morlet wavelets.

Besides displaying the characteristic fluctuation in amplitude due to the summation, Fig. 15.4 shows that the frequency spectrum becomes larger as more signals contribute. Each signal has its centre frequency and variation, and the sum of signals therefore gets a wider width than each individual contribution. In our case, the centre frequencies of the eight contributions do not differ by more than 1 Hz (tested separately, data not shown).

An increase in the width of frequency distribution affects also the coherence time. With more contributions, we get a shorter coherence time than with individual signals. However, it is impossible to draw conclusions about relationships between, for example, the width of the frequency spectrum and coherence time, on the basis of these files alone. The statistics are too poor for the task. We will come back to the issue about a little.

The wavelet diagrams in columns 2 and 3 in Fig. 15.4 show essentially the same characteristic features as the amplitude variation in the time domain. However, the frequency is so well defined that sometimes we can see small changes in the dominant frequency in the sum signal as time passes.

Amplitude and spatial distribution of the summed signal

It is worth noting that the average amplitude of the sum of eight similar signals is about 20 on a scale where the average amplitude of one of the signals is about 7. It can easily be shown that $20 \approx 7 \times \sqrt{8}$. With the addition of independent waves of only approximately the same amplitude, frequency and degree of variation over time, the sum of amplitudes is not proportional to the number of signals that contribute, but only roughly proportional to the square root of the number of signals. The intensity (proportional to the amplitude squared) is proportional to the number of signals.

However, if the contributions were close to perfect harmonic signals with the same frequency, amplitude and phase, the sum would have got an amplitude proportional to the number of signals we add, and the intensity proportional to the *square* of the number of signals added. That is what makes the intensity of a laser beam so impressive even if its power is only a few milliwatts.

Let us go back to our singers who sing "eeeeee"s. If the singers had sung exactly and constantly in phase, eight singers would give an intensity equal to 64 times the intensity of each individual. This is quite different from the eight times intensity we found in practice with our real signals. Is this something we can utilize?

Unfortunately, *"There ain't no such thing as a free lunch"*. We get nothing for nothing. It may appear that eight hypothetical singers who sing in unison would give *eight times greater* sound intensity than eight singers who do not sing coherently. How, then, can the requirement for energy conservation be satisfied?

The energy ledger will stand up if *spatial* relationships are also taken into account. Eight hypothetical singers who sing coherently will not give eight times larger sound energy everywhere in space, only in those places where the signals from all eight are in phase with one another. At other places in space, where the signals are out of phase, the sound energy could drop to almost zero. That is not what happens with the real singers. In no region of space will there be permanent constructive or destructive interference for these singers. We hear the sound of the real singers everywhere.

If we integrate the sound energy over all space, it will be about the same regardless of whether the singers are singing coherently or incoherently.

These spatial considerations are analogous to the intensity distribution in the interference fringes from many slits. When the number of slits increases, the fringes become narrower and narrower, and the bright streaks become more intense even though the total luminous flux out of the slits is unchanged.

15.2 Finer Details of Coherence

As stated above, there is a close relationship between temporal coherence length and width of frequency (or wavelength) distributions. We also claimed that the spatial coherence length may depend on the size and other features of the source of the waves.

In this sub-chapter, we will explore finer details of coherence which in fact can be very useful to know—useful for experimental physics and for better understanding of some exciting phenomena in physics.

We are now going to analyse signals where we can choose between different widths in the frequency (or wavelength) distributions of the wave. We could avail ourselves of audible noise found in nature or other types of waves with similar characteristics. However, we choose to generate the signals numerically that makes it easy to change the characteristics.

15.2.1 Numerical Model Used

We will make a kind of noise signal, a signal similar to the sound of a large waterfall. The sound may be composed of many sources that act independently of each other (the sound of small and large masses of water hitting rocks and water surface at the bottom of the fall). Since the sound is created in a variety of unrelated processes, we call it "random" or "stochastic". The sound has many frequency components that cover an entire frequency band.

We choose an approach in which we create the frequency spectrum of the signal we wish to work with, and then, we use an inverse Fourier transform to generate the signal in the time domain.

We choose that the frequency spectrum (frequency image) should have many frequency contributions with an optional centre frequency and a Gaussian distribution of nearby frequency components. The width of the frequency distribution must be optional. In order to get a large variation each time we generate a signal, we allow each frequency component to have an arbitrary value between zero and the variance of the Gauss distribution. In addition, we allow the phase of each frequency component to be arbitrary, lying between 0 and 2π.

A Matlab function that generates such arbitrary signal with given centre frequency and given full spread in the frequency distribution is given at the end of this chapter.

In Fig. 15.5, one sees three examples of arbitrary signals generated in this way, along with the analysis of the signals by the same three methods of analysis as before. The same centre frequency (5000 Hz) has been selected for all three signals, but three different full widths (down to $1/e$ of max), namely 50, 500 and 5000 Hz. 2^{16} points are used, and the sampling rate is 44,100 Hz (same as audio on CDs). There are a number of interesting results.

15.2.2 Variegated Wavelet Diagram

It is particularly interesting to see the wavelet diagram for these arbitrary signals. Already in Fig. 15.4 we saw that the sum of several signals led to the amplitude in some time intervals being large, but small in other periods. We got a "clumping" of the signal in time. However, in Fig. 15.4 there was little difference in frequency.

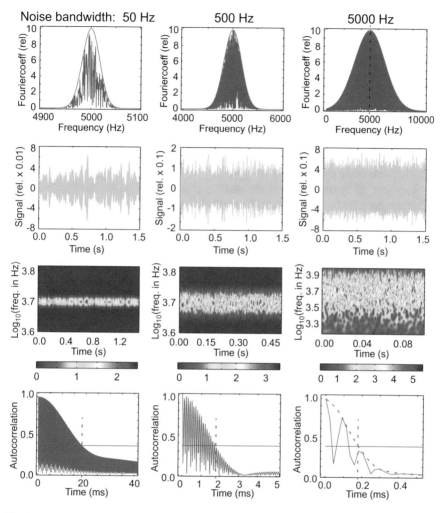

Fig. 15.5 Three examples of synthetic chaotic fluctuations, with a width in the frequency range of, from left to right, 50, 500 and 5000 Hz. The centre frequency is 5000 Hz. Frequency analysis, wavelet analysis and the first part of the autocorrelation function are also shown. For the wavelet analysis, $K = 120$, 36 and 12 were used on the analyses from left to right (almost optimal values). Note the differences in time scales for the three signals in the wavelet diagrams and autocorrelation graphs

We see this much better in Fig. 15.5. The clumping of the signal occurs in a rather chaotic manner both in frequency and time when the width of the frequency distribution is large enough. There are up to thousands of contributions (frequency components) to the final signal (all frequency components within the frequency distribution), and the result is a rather chaotic variegated pattern in the wavelet diagram.

If we scale frequencies and sampling rates in our calculations to visible light, the last column in Fig. 15.5 would be comparable to electromagnetic waves from the sun. In that case, it is tempting to make believe that one is dealing with photons. Each red spot in the wavelet diagram could then be associated with a photon coming at a given time and having a certain frequency. But we know from the way we have generated this signal that the effect is due to a straightforward summation of many random waves with a width in frequency distribution. This is a fingerprint of the sum of many simultaneous independent wave contributions which is quite natural since the light emitted from some parts of the sun arises in a chaotic manner and there is no correlation between light coming from one part of the sun surface with what is coming from other parts.

Note that the time excerpt of the wavelet diagrams is ten times smaller for the right one compared to the left one. It shows that the duration of each red spot (periods with a significant amplitude for the frequency to which the spot corresponds) becomes shorter when the width of the frequency distribution increases. It is difficult to estimate some sort of average duration for the red spots, and the result is partly also dependent on the choice of K value for the wavelet analysis. Nevertheless, we can give a (very rough) estimate of the duration of the spots as follows:

Frequency width (Hz)	Duration of red spots (ms)
50	20–50
500	6–10
5000	1–2

We see that the duration of the spots decreases as the width of the frequency distribution increases. We further note that for the smallest width of the frequency distribution it is not possible to detect that more frequencies occur simultaneously, but when the width of the frequency distribution increases, there are many examples that more than one frequency can be significantly present at the same time.

15.2.2.1 Width of Frequency Distribution Versus Coherence Length

We find the following relation between the width of frequency distribution and coherence length.

Frequency width (Hz)	Coherence time (ms)	Product-of-these
50	18	0.9
500	1.8	0.9
5000	0.18	0.9

The interesting point is that if the width of the frequency distribution is small, it takes a relatively long time between the occurrence of constructive and destructive interference (for fictitious sub-signals with slightly different frequencies). Each time

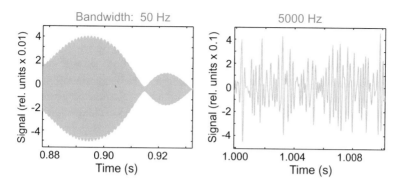

Fig. 15.6 Sections from the signal in the time domain to show that each "bubble" with relatively pure signal is much larger when the frequency width is small than when it is large. Note the difference in time intervals displayed

the signal gets an increased amplitude, it will take a time equal to the inverse of the width in the frequency distribution before the amplitude again decreases towards zero.

Figure 15.6 shows a detail in the time-domain description of a signal generated with widths (in the frequency distribution) of 50 Hz (left) and 5000 Hz (right). The centre frequency is still 5000 Hz. Note that for the smallest frequency band, each bubble takes in the time domain 15–40 ms (more than 100 time periods, individual oscillations do not appear in the plot). With a hundred times greater width in frequency distribution, the width of each bubble (to the extent that it is possible to define such) is only about 0.3–0.5 ms (about two periods). With goodwill, we can say that an increase in the width of the frequency distribution by a factor of hundred led to a 100-fold reduction in the duration of each bubble. This accords with the relationship between width of the frequency distribution and coherence time of the signal.

The result is related to the so-called *Wiener–Khintchine theorem* which states that the Fourier transform of the autocorrelation function of a function is equal to the power spectrum of the function (also called the spectral power density). We do not go into details about this last relationship.

15.2.3 Sum of Several Random Signals; Spatial Coherence *

So far, when we have discussed random signals (chaotic signals), we have only studied each signal in itself. In calculating coherence time, however, we have in a way compared a chaotic signal with itself, but slightly shifted in time. We saw that coherence length was very small when the width of the frequency distribution was about as large as the centre frequency.

We shall now study *spatial* coherence; that is, we will investigate the correlation between the wave passing point A with the wave passing point B in the right part of

Fig. 15.2. A key for understanding spatial coherence is a spatially distributed source of waves.

This subsection deals with the experiments carried out by Hanbury Brown and Twiss more than 60 years ago. Our treatment explains how we can generate light with considerable spatial coherence by sending a light beam (initially with much less spatial coherence) through a "pinhole" with a few micrometre diameter.

The description is, however, somewhat demanding and is not expected to be treated at bachelor level in physics. Jump to Sect. 15.3 if you want to skip these finer details.

A hypothetical "point source"

Suppose we have "point source" of a wave with an extent less than a wavelength. Assume further that the wave exits from this source with an almost spherical symmetry (at least for the part we are interested in). Assume further that A and B are equally distant from the source. In that case, the wave at A will be equal at each instant to the wave at B. A wave peak will pass A and B at the same time. We can say that A and B are on the same well-defined wavefront, which is part of a spherical surface with the source at the centre. In that case, the spatial coherence is as long as the extent of the part of the sphere where this relationship holds.

This means that we could insert a double slit perpendicular to the wave motion direction and see interference on a screen behind the double slit. If the coherence time is more than a few time periods, we will be able to get more fringes in the interference pattern.

Similarly, we could put a spherical obstacle and demonstrate the existence of Arago's spot (bright point) in the centre of the shadow (image) of the obstacle.

A more realistic source of light

However, for light it is difficult to create a light source with an extent less than the wavelength. It is possible in the so-called quantum dots, but when a filament lamp emits light, we can look at the filament as a mass of independent light sources, each emitting chaotic light signals. If we now study light waves passing two points A and B at the same distance from the incandescent lamp, there will no longer be a good correlation between the waves at the two points. This is because there is a certain distribution of distances between the different parts of the filament and the point A, and a *different* distribution of distances between the same parts of the filament and the point B.

The same reasoning can also be used for light from the surface of a star. Hanbury Brown and Twiss developed an elegant method in 1954 and 1956 that can be used to measure the extent of a star using the properties of chaotic waves (see references at the end of this chapter).

A simple model to point out the essence we want to discuss

In Fig. 15.7, the principle of the Hanbury Brown and Twiss effect is shown. For the sake of simplicity, we have only included two independent sources with the same type of chaotic signal, near each other, and two detectors A and B. The signal into A consists of the sum of waves from sources 1 and 2 travelled equal path lengths.

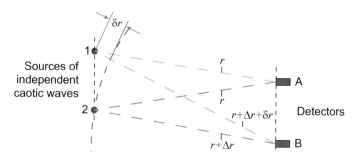

Fig. 15.7 Principle sketch to show the basis of the so-called Hanbury Brown and Twiss effect. See the text for details

The signal into B is also a sum of the signals from the sources 1 and 2, but this time the path length is slightly different. The signal from source 2 is delayed relative to the signal from source 1, which is different to the situation at A.

The signals from the sources 1 and 2 are both chaotic, and their sum is also chaotic. The spots in the wavelet diagram will have short duration and will be distributed chaotically in time and frequency (assuming a large width in the frequency spectrum).

Since the path length are equal between A and source 1 and 2, and equal to the path length between B and source 2, but not between B and source 1, the wavelet diagrams of A and B may be quite different. This means that there may be a bad correlation between the signals into detectors A and B. However, if we placed A and B in the same location, the correlation would be maximum.

Although the distribution of distances is always different at points A and B, if these points do not coincide, the difference may not be demonstrated in practice. If the differences in the distribution of distances of A and B differ by significantly smaller than a wavelength, we cannot expect to see any difference in the waves at points A and B. In that case, there will be a high degree of correlation between the waves in these points. If we make the distance between A and B larger, we will sooner or later get a difference of more than one wavelength in distributions of distances from different parts of the light source to point A and corresponding to B. In that case, we would expect the correlation between the waves at A and B to decrease.

The largest distance between A and B for which we can still get a significant correlation will be called the spatial coherence of the wave at the place under consideration.

If the light source, which consists of independent chaotic parts, has an extent given by an angular diameter θ judged from observer, and the average wavelength is λ, the spatial coherence length a will be in magnitude

$$a = \lambda/\theta$$

Hanbury Brown and Twiss used this relationship in 1956 to calculate the size of the star Sirius which is 8.6 light years away from us. The spatial coherence length of the light from Sirius was about 8 metres here on earth. The angular diameter was estimated at 0.0068 arc seconds which correspond to 3.3×10^{-8} rad.

If we use this relationship and consider a halogen bulb (with a filament of 1 cm extension) as light source and considering the light from this 100 m away, we will be able to find correlation in the signals at points A and B, which are up to about 6 mm apart. In other words, the spatial coherence of this lamp at 100 m distance would be about 6 mm.

For random (chaotic, stochastic) light, we have so far only considered how the spots in a wavelet diagram change, for example, when two chaotic signals are added with and without a time offset. Basing our discussion on the signals with the given frequency and the width of the frequency distribution, we have calculated temporal correlations in the amplitude (first-order correlations). This approach works well when the frequency is less than a few GHz, but we have no detectors that can follow the time variation of the signal for visible light (6×10^{14} at 500 nm). Accordingly, we cannot sample and calculate the autocorrelation function for light waves.

The detectors for light are so-called square law detectors that provide a response proportional to the *square* of the amplitude of the light coming in. The detectors cannot follow the instantaneous intensity, which varies as fast as the underlying sinusoid itself, but provides an integrated intensity over a significantly longer period. This means that light detectors can only detect time variations in intensity in a frequency range below 1 GHz. When Hanbury Brown and Twiss performed their famous experiment about Sirius in 1956, the available bandwidth of detectors and amplifiers was only 38 MHz. How could they follow the much faster changes in the light signals themselves?

The clue is that when we sum up two frequencies and squares the sum, we get the following:

$$(\cos \omega_1 t + \cos \omega_2 t)^2 = 1 + \frac{1}{2}\cos(2\omega_1 t) + \frac{1}{2}\cos(2\omega_2 t) + \cos\left[(\omega_1 + \omega_2)t\right] + \cos\left[(\omega_1 - \omega_2)t\right].$$

In addition to the constant term, the frequency of three terms is about twice the original, and for the detection of light, they are completely beyond the possibility of detection. The term $\cos\left[(\omega_1 - \omega_2)t\right]$ is, however, a kind of "beat frequency term". For continuous frequency distributions we have worked with, this term will provide contributions from the frequency zero to a frequency that corresponds to the width of the frequency distribution.

Because of this "beat frequency term", Hanbury Brown and Twiss (and everyone else for that matter) could transform the variation in the visible light frequency range to a much lower frequency range. The signal that forms within the "beat frequency range" can in our modelling case also be analysed by wavelet analysis, and we will have similar chaotic patterns there as well. This illustrates that the correlation of signals as shown in Fig. 15.7 can also be studied in cases where detection occurs with "square law detectors" in a completely different frequency range than the original waves.

Spatial coherence in the light from the sun

If we use the same relation for the light from the sun seen from the earth, we find that the spatial coherence length is only about 60 μm. This means that it is impossible to use sunlight directly for double-slit experiments and to detect Arago's spot. What

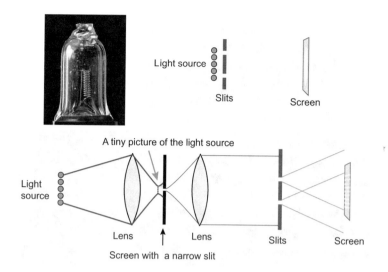

Fig. 15.8 We do not get interference fringes when light from a halogen bulb is sent directly through a double slit. However, if we first send the light from the bulb through a very narrow slit, we can get sufficiently large spatial coherence that interference fringes can be detected. To get a sizable amount of light through the slit and enough light to be able to see the fringes on a screen or on a photographic plate, it is advantageous to use convex lenses (cylindrical) both before and after the first slit

did Young do in 1801 and Arago in about 1820? The secret lies in making the angle of the light source we actually use small enough. We can achieve that by sending light through a so-called pinhole. We can make a pinhole by sticking a very thin needle through an aluminium foil, or we can buy foils with a well-defined pinhole in a holder for about a hundred US dollars. We have a wide choice of the hole sizes, and if we choose, for example, the diameter to be $10\,\mu$m, the spatial coherence length of the light passing through the hole will be about 5 cm after the light has travelled 1.0 m from the hole (at 550 nm).

However, the cross-sectional area of a hole with a diameter of $10\,\mu$m is exceedingly small. If we want to experiment with a filament lamp or sun as a light source (incoherent light source), we can increase the intensity of the light that passes through the hole by passing the light through a convex lens and positioning the hole exactly where the image of the sun (or filament) is formed. In this case, one captures only a small part of the light from the source, but much more than if one did not use a lens.

It may also be advantageous to use a lens after the hole to prevent the light from spreading too much. This last lens should then be positioned so that the hole is at the focal point of the lens (see Fig. 15.8).

Fig. 15.9 Surface waves on water at some point. Within small patches on the surface, we have a rather "pure" wave. See the text for further discussion

15.3   Demonstration of Coherence

It is no easy matter to get a good understanding of coherence. We therefore choose to include a photograph of surface waves on water to illustrate coherence in an altogether different way.

Within small patches on the surface we have a rather "pure" wave (see Fig. 15.9). Within these patches, it is possible to predict with fair confidence mutual phase relationships in the direction along which the wave is moving (red lines). Within the patches, the phase and amplitude of the wave are approximately constant in a direction that is normal to the propagation direction (yellow lines). The patches are very different in size. In the direction of propagation, the patches vary between two and twelve wavelengths. This means that the temporal coherence length is of the order of 5–7 wavelengths, but this is hardly a sound estimate of size. In a direction normally to the direction of propagation, the yellow lines in this case are on average about as long as the average red line. This means that the spatial coherence length is about as long as the temporal in this case. Perspective conditions, however, make it difficult to specify the width of the patches in terms of uniform waves.

If we consider waves in three dimensions, the "patches" where one sees moderately well-defined waves will be replaced with small volumes where there are fairly well-defined (and almost flat) waves.

However, the patches or volumes with fairly well-defined waves will change in time, which aggravates the complexity even more. One readily appreciates what an

enormous statistical challenge it is to describe this dynamic situation, which one
often comes across in practice when waves propagate in space.

A small detail in Fig. 15.9 may be worth reminding. At any point, the water surface
at a certain moment has a fairly well-defined height. Put another way: The height
of the water surface does not have multiple values at the same time! We are so used
to thinking that "there are several waves at the same time", but at one and the same
place, the local air pressure has only one value at a given moment for sound waves
in air, and at one place, the electric field has only one value and only one direction
for the sum of all contemporaneous electromagnetic waves at this location.

This is a property well worth pondering over!

15.4 Measurement of Coherence Length for Light

Visible light has such a high frequency that we cannot detect the sinusoidal vibration
of the electric field as the wave passes. We cannot use the mathematics mentioned
above directly.

However, we can perform an *analogue* calculation of a quantity closely related
to the autocorrelation function. This is done by splitting a light beam into two sub-
beams by a so-called beam splitter. The two sub-beams are then reunited, but only
after one has been made to travel a longer path than the other. When the sub-beams
are brought together, their electric fields are added, and so are the magnetic fields,
and we deal with the intensity of the sum. We simply consider:

$$
\begin{aligned}
G(\tau) &= 1/T \int_0^T [f(t) + f(t+\tau)]^2 dt \\
&= 1/T \int_0^T \left[f^2(t) + 2f(t)f(t+\tau) + f^2(t+\tau) \right] dt \qquad (15.6) \\
&= 1 + 2/T \int_0^T f(t)f(t+\tau) dt \ .
\end{aligned}
$$

Here is $f(t)$ to be considered as, for example, the electric field in the beam after
it is divided into two and the amplitude is normalized to 1.

This means that we can simply change the path of one sub-beam compared to the
other before they combine, and then, we get the autocorrelation function just like
the curve in the left part of Fig. 15.3, except that the entire curve is offset by $+1$ so
the minimum is zero (intensity cannot be negative). Figure 15.10 shows the principle
of a so-called Michelson interferometer commonly used in such measurements. The
path difference between the two sub-beams is $\Delta L = 2L1 - 2L2$.

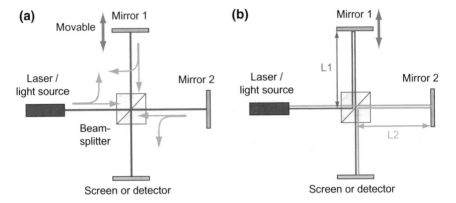

Fig. 15.10 In a Michelson interferometer, a beam of light is brought to a beam splitter. Half of the beam goes to a fixed mirror and is reflected from here, while the other half goes to a movable mirror. Half of the light reflected from the mirrors is sent to a screen or detector where electric fields from the two contributions are added. In the right part of the figure, the light path for the two sub-beams is marked schematically

Light from thermal light sources, such as incandescent lamps, may have a temporal coherence length of only a few wavelengths (that is, just a few microns). Light with such small coherence length is called "incoherent". Light from a good laser can have a coherence length of up to several hundred metres. A laser that costs a few thousand dollars typically has a coherence length of a few centimetres (i.e. in the order of 100,000 wavelengths). Light with long coherence length is called "coherent". There is no sharp boundary between incoherent and coherent light.

Albert Abraham Michelson (1852–1931)
was an eminent experimental physicist. He is perhaps best known for the Michelson–Morley experiment in 1887. Michelson and Morley concluded that their experiment and showed no evidence for the relative motion of the earth and ether. Michelson measured the speed of light with great precision. Furthermore, he developed stellar interferometers, thus measuring the diameter of distant stars, and measuring the distance between star pairs ("binary stars" in English). He received the Nobel Prize in Physics in 1907, the first American to win this distinction.

15.5 Radiation from an Electric Charge

Coherence is linked to the mechanisms of how waves are generated. We have previously discussed mechanisms for producing waves on a string, sound waves and surface waves on water. For electromagnetic waves, we have so far shown that, for

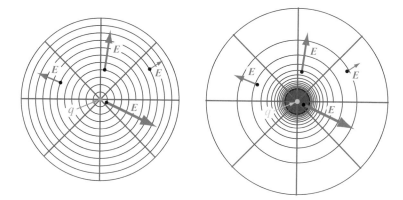

Fig. 15.11 A charge at rest has spherical equipotential surfaces around it (black circles) and radial electric field lines (red lines). The electric field is drawn at four points in the plane. The field is strong close to the charge and decreases with increasing distance. In the left part, the distance between the equipotential surfaces does not reflect the differences in potential between them. In the right part, the potential difference is equal for every set of neighbour surfaces, so that lines close to each other reflect a space with higher electric field than a space where the lines are further apart from each other. However, for point sources the equipotential curves get so close to each other near the charge that the lines overlap. Equipotential lines are therefore often drawn with a somewhat arbitrary selection of electric potentials (as in the left part)

example, plane waves are solutions of Maxwell's equations in the remote field zone in vacuum (at least in the absence of free charges). But what is usually the source or the mechanism behind the generation of electromagnetic waves? We will barely touch this vast field of physics. First, we will see how charge in motion can produce waves, and then, we will look at some of the main features behind the laser.

We can through calculations show that we can make electromagnetic waves in the radio frequency range by sending an alternating current to an antenna. In this case, we have free charges and free currents in action, and Maxwell's equations give us an inhomogeneous second order partial differential equation for the electric field $\vec{E}$ and a corresponding equation for the magnetic field $\vec{H}$. Calculations of this type can be done with finite element methods as mentioned earlier. We do not go into details here.

We choose a "picture and words" presentation instead of a rigorous mathematical treatment, but hope that it will be sufficient to throw light on the key features.

Figure 15.11 shows schematically that a charge q at rest has electric field lines that point radially outward (if q is positive). Equipotential surfaces are spherical shells centred at the charge.

If the charge is moving at a constant speed (left part of Fig. 15.12), the equipotential surfaces according to the theory of relativity will become "squeezed", that is, slightly discus-shaped with the shorter axis in the direction of motion. In a system where the charge is at rest, there is only an electric field. In a system where the charge is in motion, there will be both an electric and a magnetic field. However, when we talk

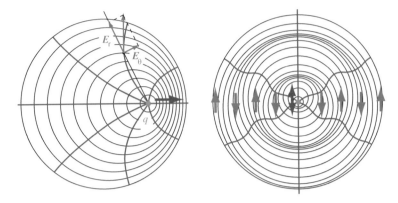

Fig. 15.12 *Left part*: A charge in steady rectilinear motion has spherical equipotential surfaces around it, but only in the sense that the equipotential surface at a certain distance is centred at the charge position at the time d/c earlier, where d is the distance from the equipotential plane and the charge at the earlier time. On account of the relative displacement of the equipotential surfaces, all the electric field lines, except those moving in the same (and opposite) direction as the charge, do not remain purely radial and acquire a tangential component. *Right part*: An oscillatory charge (black arrow) will give equipotential surfaces that are slightly offset relative to each other, as suggested. This causes electrical field lines in tangential direction (*blue arrows*), and these change in principle similarly as in an electromagnetic wave

about generating waves, we must include the so-called *retarded potentials*. We will not go into a more advanced treatment of this topic, but only look at some superficial features.

We assume that *changes* in electrical and magnetic fields move in space with the velocity of light. We do not see a supernova when it happens, but only after light has travelled the enormous distance from the nova to us. What we see today is the supernova as it was for the exact time d/c since, where d is the distance between us and the supernova and c is the velocity of light.

This is also true when we move a charge in space. The field somewhere in space has a distribution corresponding to the location of the charge at the time

$$t' = t - d/c$$

where t is the present time and d is the distance from the charge to the point where the field is being measured at the instant t'.

If we draw equipotential surfaces from a charge moving with constant speed, the planes will have relative positions as indicated in the left part of Fig. 15.12. The effect is greatly exaggerated as the charge would actually have a velocity above half the light velocity as the figure is now drawn.

The electric field is usually given as the gradient of the electrical potential, and we apply this rule also when we use retarded potentials. Then, we get electric field lines that are curved, as shown in the figure.

Suppose that we are at rest and a charged particle is going past us at constant speed. We will first experience an electric (and magnetic) field at our location that has a time development in which the electric field has the same direction as that of the moving charge (for positive charge), and then, we will experience a much stronger field perpendicular to this direction as the charge passes, and end up with a weak field in the opposite direction. This is a "pulse" of electrical (and magnetic) field, and not a wave in the usual sense.

An observer who happens to be following a charge moving at constant speed will describe the electric field as static, and it would appear to him to conform to Fig. 15.11. Such a situation does not qualify for the radiation of energy. In our own reference system, where the charge is in motion, the electric fields will be built up at one place in space, while a completely equivalent depletion of fields takes place somewhere else in space. To be sure, the region that has the highest electromagnetic field energy density will move, in the same way as the charge, but this displacement is of local character, and does not represent energy flowing out of the region around the charge.

To get a wave that extends beyond the vicinity of the charge, we must strive for a situation similar to that of an electromagnetic wave in Chap. 9. Electrical (and magnetic) fields must oscillate and have a direction perpendicular to the direction of wave motion. To get this with our charge in motion, we must have a charge that is subjected to an *acceleration*. For example, the charge can oscillate back and forth in space, preferably in a harmonic motion. The electric field a little away will then oscillate as outlined in the right part of Fig. 15.12. This change in electrical field will have both a radial and a tangential component relative to the radius vector from the charge to the point we consider.

The component in the *radial* direction (when we are at some distance from the charge compared to the amplitude of charge oscillation) will (almost) not change over time. Therefore, this component will (almost) not give rise to any wave that will propagate.

However, the component *perpendicular to the radial direction* (in the plane perpendicular to the charge oscillation direction) will oscillate (almost) as a sinusoid over time. This component could give rise to an electromagnetic wave that spreads into space.

The curvature of the electric field increases with the speed of charge while it oscillates. The time derivative of this again determines how large $\partial E/\partial t$ becomes. These two factors together cause the radiated energy to be proportional to the square of the frequency of oscillation. Therefore, we often say that the radiation is proportional to the *acceleration* of the charge.

Some side remarks:
It may be tempting to think that the electric field from a charge "radiates" outward all the time. Such thoughts could be nourished by the notion of a retarded potential where we think the field in one place is due to the charge where it was a while ago. However, a constantly radiating electric field would soon contradict energy conservation, etc. Fortunately, there is no need to think along these lines. There are *changes* in electrical and magnetic fields that propagate with the light velocity. Before a particular change has spread and reached a given location, it is the field distribution that is rooted in the relationship *before* the change that applies. The electric field from a charge at rest is in equilibrium with itself. It is a solution of Maxwell's equations, and there are no changes in fields and no transport of energy. As soon as movement and particularly acceleration enter, things become different.

15.5.1 Dipole Radiation

An alternative way of generating electromagnetic waves is to use an electrical (or magnetic) dipole that varies in time. This is a very effective way to make waves. We can understand this by considering electrical field distribution from a permanent electrical dipole (see left part of Fig. 15.13). The electric field is directed perpendicularly to the radial direction in the plane normal to the dipole direction.

If we change the polarity of the dipole in a harmonic way, we get an electric field in this equatorial plane that will vary just the way we want it for generating an electromagnetic wave that can propagate in space (electric field perpendicular to the direction of motion). In the direction of the dipole itself (and in the opposite direction), the electric field from the dipole is nearly radially directed and has negligible component across the radial direction. In these two directions, virtually no waves are transmitted.

In the right part of Fig. 15.13 is shown a diagram of electric field distribution near a dipole antenna at a given time. The electric field is the strongest where the field lines are closest. The entire pattern moves outwards with the velocity of light, and new loops form near the antenna twice for each period (direction of the field changes direction in the two systems of loops that form each period in the dipole variation). An animation of the time course (and much other information) is available on Wikipedia under the heading "dipole radiation".

[A remark: The right part of Figs. 15.12 and 15.13 has a certain relation to each other, but is still different. Try to point out differences and similarities.]

It is common to draw *direction diagram* for antennas. A direction diagram indicates the relative temporal intensity of the transmitted waves for different directions

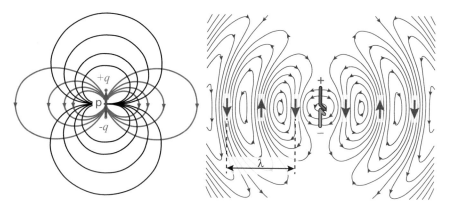

Fig. 15.13 *Left part*: A static electric dipole consists of two identical charges, but with opposite signs, slightly spaced apart. Equipotential surfaces (black, with lobes pointing upward and downward) are drawn as well as electric field lines (red, with belly outwards to the sides). The physical extent of the dipole is greatly exaggerated in relation to the field line pattern. *Right part*: An oscillating electric dipole will create an electric field in the surrounding space as indicated. The field pattern moves outwards at the speed of light

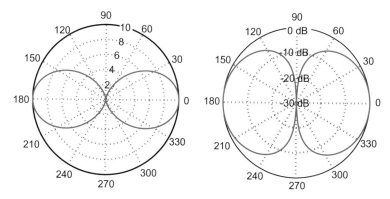

Fig. 15.14 Directional diagram for a single vertical dipole antenna (also called "antenna diagram" or "radiation diagram"). The diagram only applies to a distance from the antenna which is large in relation to the length of the antenna. Linear scale for the intensity in the radial direction is used in the left part of the figure and a logarithmic scale in the right part. Intensities are all relative to the maximum value. The diagram is read as follows (*left part*): the relative intensity at 0° is set to "10". Then, the relative intensity at 30° is about 7.3 and at 60° about 2.8. For the *right part*, see the text

in space, in a so-called polar diagram. Figure 15.14 shows the direction diagram in a vertical plane passing through a single vertical dipole antenna.

The intensity (time-averaged Poynting vector) as a function of angle θ deviation from the meridian plane of the dipole is given by:

$$I(\theta, r) = constant \times f^4 \cos^2(\theta)/r^2$$

where f is the frequency of the alternating current and r the distance from the dipole (has to be many wavelength away to avoid near-field conditions).

A direction diagram can be given with a linear scale in the radial direction (left part of the figure), but most commonly used is a logarithmic scale in the radial direction (right part of the figure).

A remark:
Suppose we create a polar graph with logarithmic scaling of intensity in the radial direction. Since the logarithm of 0 does not exist, a cut-off intensity needs to be chosen when the graph is drawn. In our case, a relative intensity of 1.0 corresponds to the outer radius in the chart, while a relative intensity of 0.001 is chosen at the centre of the diagram. Relative intensities less than 0.001 would be negative on a logarithmic scale and would appear on the opposite side of the chart. To avoid misunderstandings, we remove negative values before plotting.
The Matlab program used to create Fig. 15.14 is given below:

```
function antennaDiagram3
N = 1024;
theta = linspace(-pi/2.0,3.0*pi/2.0,N);      % Angles
costheta = cos(theta);
intensities = costheta.*costheta;
intensities = log10(intensities*1000.0);
for i = 1:N
    if(intensities(i)<0)
        intensities(i)=0;
    end;
end;
polar(theta,intensities);
```

A dipole antenna can in some ways be viewed as a single slit of very small width. The radiation becomes identical in all directions perpendicular to the direction of the dipole. However, if we insert two dipoles next to each other and feed both antennas with identical signal, the radiation diagram will look like a double-slit pattern. By placing many identical antennas in sequence, we get a radiation diagram similar to a single slit (in a "meridian plane" including the complete line of antennas). By inserting reflectors and directors, we can further influence the radiation diagram, and an example is given in Fig. 15.15. There is an antenna diagram for an antenna that is widely used in base stations for mobile telephony. Note that the diagram is radial with decibels in radial direction.

The thinking that lies behind the antenna pattern diagrams is much the same as for the intensity distribution of light after passing one or more slits. The principal motif in the calculations is interference between sufficiently coherent waves. Thus guided, we use the idea of differences in the path lengths and add various contributions with the correct mutual phase.

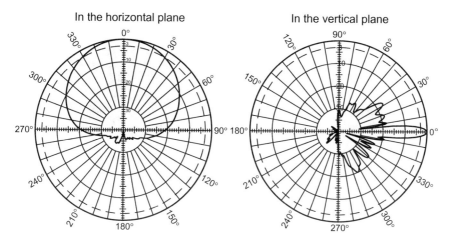

Fig. 15.15 Directional diagrams for a commonly used base station antenna for GSM 900 mobile telephone (Kathrein 80010621). One diagram gives angular distribution in the vertical direction and the other for horizontal direction

15.6 Lasers

One of the most important light sources in science nowadays, lasers, is used in everyday technological appliances, such as CD, DVD players and laser printers. Lasers are also used for cutting metals and other materials, and in medicine, for example, by reshaping the cornea in our eyes, and other operations. Even my dentist has switched to using lasers for "drilling" in the teeth. Some car manufacturers now use lasers as headlights. The range of laser applications is impressive and still increasing!

The word laser is an acronym for Light Amplification by Stimulated Emission of Radiation. Theodore Maiman at Hughes Research Laboratories managed to make the world's first laser (see Fig. 15.16). This happened on 16 May 1960. The laser celebrated its 50th anniversary in 2010. However, there are many physicists who have been involved in the development and utilization of the laser, and it was Charles H. Townes who in 1964 received the Nobel Prize in Physics "for the development of laser principles". Also, other Nobel prizes in physics are fairly closely linked to the laser in one way or another. It is therefore natural that we devote some time to the concepts that lie behind a laser. However, only the main principles are mentioned, and here too we will use a "picture and words" approach.

A laser is based on the so-called *stimulated emission*. Einstein had already shown in 1917 that we could get stimulated emissions from, for example, atoms. By that we mean that we do not have to excite an atom and wait until it finds it convenient to send out light and fall back to its ground state. By shining some light on an excited atom, we can actually *trigger/stimulate* it to return to the ground state. The laser requires a medium that is amenable to stimulation and an arrangement where light produced by

Fig. 15.16 Photograph of the first laser. A flash tube encircles a ruby rod, which is coated with an almost 100% reflective mirror at one end and approximately 95% reflective mirror at the other end. The laser emitted pulsed coherent light. Image courtesy of HRL Laboratories—Malibu California

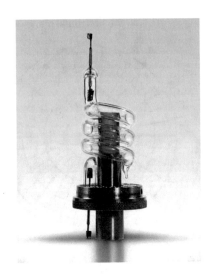

Fig. 15.17 Main constituents of a traditional laser (schematic)

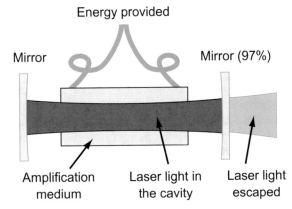

Energy provided

Mirror

Mirror (97%)

Amplification medium

Laser light in the cavity

Laser light escaped

stimulation may lead to the stimulation of even more light from the medium. In this way, we get a positive feedback in the process that makes it almost self-sustained. However, since the energy that is consumed in bringing about the emission of laser light must be compensated for, an external source of energy is necessary if the process is to be sustained.

Figure 15.17 shows the main ingredients in a traditional laser. It contains an amplification medium to which energy can be supplied from an external source. The medium is in an optical cavity ("cavity" or "box") limited by two mirrors. Light that is produced in the medium will initially spread in all directions, but light that hits one mirror tends to be reflected back through the media, hit the mirror on the other side, is reflected once more and sets up a standing wave of electric and magnetic fields in the cavity. The light that is reflected back and forth many, many times can stimulate the medium to give even more light.

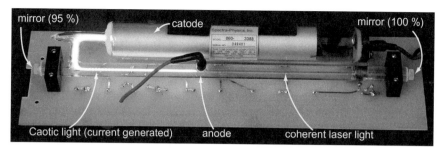

Fig. 15.18 A photograph showing the innards of a standard HeNe type laboratory laser. An electric current is sent through a low-pressure mixture of helium and neon and leads to the emission of chaotic, incoherent light emitted in all directions. Part of this light energy is (at proper conditions) building up in a cavity between two mirrors. The result is a strong, coherent beam of light between the two mirrors. We do not see this beam from the side because the light beam in the cavity is almost perfectly aligned along the axis between the mirrors, and does not exit from the sides. A tiny fraction of this light beam is transmitted through the mirror to the left in this figure

With ordinary sources, the light comes from many atoms or molecules that appear quite independent of each other. Then, we get a state that corresponds to our real singers and our chaotic waves mentioned earlier in this chapter. The light intensity from the light source increases approximately proportionally with the number of atoms/molecules emitting light and the light goes in "all" directions.

The conditions are different in lasers, which will be exemplified through a description of a HeNe laser (see Fig. 15.18). A low-pressure mixture of helium and neon (in a ratio of approximately 10:1) is held in a glass receptacle. Between a cathode and an anode a high voltage is applied, which generates an electric current through the intervening gas mixture. Electric current through the gas volume between the cathode and anode leads to the emission of chaotic, incoherent light emitted in all directions. This process is similar to what is found in a standard fluorescent lamp used in coloured advertisement lighting at night ("neon lights").

Energy of helium atoms (the majority species) excited by the electric current is transferred to the neon atoms during collisions. The excited neon atoms emit light initially through the so-called spontaneous emission in the same manner as the helium atoms.

Part of the light emitting atoms is located in a cavity between two mirrors. Some light will be reflected back and forth between the mirrors as described above, and standing electromagnetic waves may form. The presence of the electromagnetic field with correct wavelength leads to an enhanced probability for an excited atom to emit light, and the process is called "stimulated emission". Light coming from the different atoms through stimulated emission has nearly the same phase and is directed in the same directions as the standing waves between the mirrors. Then, the amplitudes of electrical and magnetic fields will be added directly and the intensity of the light within the beam will be proportional to the *square* of the number of emitting atoms. This leads to increased stimulated emission and thus a positive feedback. The light intensity within the cavity builds up in time, but after few seconds the intensity will reach a plateau which depends among others on the efficiencies to excite He atoms and transfer of energy to Ne atoms.

One of the two mirrors reflects only 95–99% of the light. About 1–5% of the light intensity will be transmitted and is responsible for the laser beam available for use in the laboratory.

It should be mentioned that the energy levels for helium and neon allows us to make lasers with several different wavelengths. Usually, the wavelength of light from a HeNe laser is about 633 nm. Other wavelengths would lead to less efficiency and even light that is not visible for the human eye. Special tricks are used in order to avoid building up standing waves with wavelengths different from about 633 nm.

Since the light waves of a laser are created in a "cavity" (with mirrors at both ends, see Fig. 15.18), the light will form standing waves as mentioned. Then, the frequency will be very precise, in a similar way as the sound of a guitar string attached to both ends is pretty precise. If the distance between mirrors is 30 cm for a HeNe laser wavelength of about 633 nm, there will be about 473,940 wavelengths between the mirrors. The actual line width of the energy transition we use in the neon gas is so wide that sometimes there may be more simultaneous wavelengths in the cavity (the line width is broadened due to collisions with other atoms). With 473,940 wavelengths between the mirrors, the wavelength will be 632.9915 nm, but with one wavelength more or less than this, the wavelength will be 632.9902 and 632.9929 nm, respectively. We are talking about "modes" for the laser light. Mechanical heating of the laser cavity will cause small changes in the distance between the mirrors. In that case, the wavelengths will also change, which will lead to what is called "mode hopping".

In some contexts, lasers are constructed to play an active part in the various modes the laser can operate in. We can then achieve many wavelengths whose mutual difference is almost constant, and the phenomenon is called a "frequency comb". Theodor Hänsch received the Nobel Prize in 2005 for creating a "frequency comb synthesizer" that made it possible for the first time to measure the oscillations in light with extreme precision. The method forms the basis for our most modern atomic clock.

The laser light that escapes from the cavity through the 95–99% reflecting mirror is quite different from light emitted by, for example, an incandescent lamp. If we compare the phase of the electromagnetic wave on a plane normal to the beam, the phase throughout the plane will be almost identical. As stated earlier in this chapter, this is equivalent with a high degree of spatial coherence. This is unlike the light that originates from many atoms which have no definite phase relationship with each other, for example light from a filament lamp. Such light has little spatial coherence, and the wavefront is very uneven.

The stimulated emission in the cavity is fairly stable, and the frequency is so well-defined all the time that we can predict the phase of the laser light beam many wavelengths in the future. Thus, even the temporal coherence is high in laser light.

Since the wavefront is very well defined across the beam, while at the same time we can predict the phase of the laser light long distances along the beam itself and a laser is a far superior light source for interference and diffraction experiments compared to so-called thermal light ("incoherent" light). It also means that a laser beam will maintain a very well-defined shape where diffraction is kept to a minimum. The light in a laser beam is one of the closest we can come to a mathematically idealized wave description in practice. Laser light is therefore sometimes called "classic light", but such a term confuses more than it instructs.

15.6.1 Population Inversion

While we are explaining lasing action, we cannot completely leave out a detail called population conversion. We will not go into detail since this theme is not so important in our context. Nevertheless, a brief review is given below.

An atom may be in one of the several different energy states. We often draw the energy states schematically as in the left part of Fig. 15.19. The ground state is usually labelled E_0 and the first excited state as E_1. An atom can be exited from the ground to the first excited state by, among other ways, placing it in an electromagnetic field with the frequency $v = (E_1 - E_0)/h$ where h is Planck's constant. An atom in an excited state can fall back to the basic state entirely on its own. The transition may then be accompanied by the emission of light (called spontaneous emission which is a radiative process), or it may be through a so-called nonradiative processes. We can also stimulate the transition with an electromagnetic field with the same frequency as indicated for absorption (a radiative process called stimulated emission).

For the simple system in the left part of Fig. 15.19, energy is stolen from a beam of light to excite the atom at an absorption, while in stimulated emission, energy is released by the atom. There is the same probability of one transition as the other per atom, assuming that the pertinent initial state is occupied. In order to release more light than we insert (as required by a laser), there must be

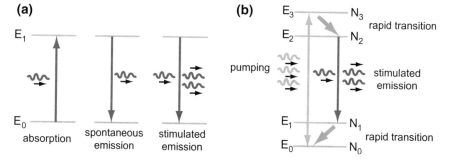

Fig. 15.19 *Left part*: Two energy states in one atom, and schematic transitions between these (very simplified). *Right part*: Population inversion can be achieved by pumping between energy levels other than those involving the emission of laser light. See text for details

Fig. 15.20 Energy levels which are involved in a common HeNe laser. XuPanda, Wikipedia Commons, CC BY-SA 4.0. Modified from [1]

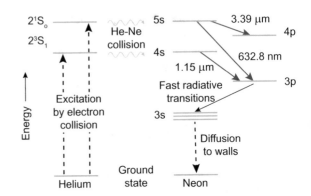

more atoms in the excited state than in the state to which the atom falls back after emission. The population of energy levels usually follows Boltzmann statistics. Then, there are more atoms in a low energy state than in a higher one. That is, it is usually impossible to form a laser from atoms in a state of thermal equilibrium and few energy transition possibilities available.

The right part of Fig. 15.19 shows a way to get to a higher population in an energy state than in a lower state. The principle is used in neodymium YAG lasers and is based on four energy levels. The atoms are excited, by the use of strong light from some light source, from the ground to the fourth energy level (E_3). The atom then spontaneously falls rapidly to the energy state E_2 through a nonradiative process, but stays here (in a "metastable state"). There is also a fast spontaneous transition from E_1 down to the ground state. However, the transition from E_2 to E_1 is not fast, and after some pumping, there are more atoms in E_2 than in E_1. We have got a population version!

If we now send a (weak) light with the frequency $\nu = (E_2 - E_1)/h$, we will get more emitted light than absorbed light from the atoms, and we may form a laser. However, the intensity of the laser will be limited by how fast we can pump atoms from the initial state to E_3.

For the helium–neon laser, the energy levels involved are a bit more complicated. A sketch (for orientation) is shown in Fig. 15.20. In this case, helium atoms are excited by sending an electrical current through a gas mixture of helium and neon. Electrons with significant speed supply the excitation energy for helium atoms. Helium has two excited levels that are "metastable" so that helium can be in these states quite long before they fall into lower energies. If such an excited helium atom collides with a neon atom, the excited energy can be transferred from helium to neon. The neon atom can then be further deexcited through a transition that gives light at 632.8 nm. It is this red light we recognize from a HeNe laser. The transition from the 3p to the ground state is rapid and involves both radiative and nonradiative processes.

Today, there are many different ways to make a laser. Most people have more and more lasers in their homes, as CD and DVD players use lasers. In addition, many also have a laser pointer. In all these examples, semiconductor laser diodes are used.

The light from such laser diodes is continuous in time. There are also lasers that only give off, in part, very short light pulses. The pulse length can be completely down in the so-called femtosecond area, even down to the so-called attosecond area (uses this term for pulses a bit shorter than 1 fs). The wavelength is not well defined for such short laser pulses!

Relevance for us?

When we went through the generation of electromagnetic waves using an oscillatory charge or oscillating dipole, we based the treatment on Maxwell's equations. The process was described as continuous, and we obtained an electromagnetic wave that lasted as long as the oscillation continued.

When we explained the laser, we used energy levels and jump from one energy state to another. Such an image is based on quantum physics, but only a quantum physics based on energy states where we use perturbation theory to look at probabilities for transitions. How should such energy diagrams as shown in Fig. 15.19 be understood? When a transition first takes place, does it take place immediately or does the transition take some time? A good answer to this question is difficult to get, since there are no consensus on this matter!

We like to draw "photons" like small wave packages (wavy lines) in diagrams like the ones in Fig. 15.19, which implies that there is a small wave when a photon is released from an atom. But how long is this wave? Can we have waves that come out having a phase memory (coherence length) that corresponds to several hundred thousand wavelengths, but which itself has almost no extent?

And how come that electrons at lower frequencies provide a continuous, sustained wave in Maxwell's formalism, but all of a sudden electromagnetic waves with frequencies corresponding to light cannot be described as continuous waves (but as photon particles)?

There are similarities in how quantum mechanics and classic electromagnetism describe the emission of electromagnetic wave/light, but the interpretation of the formalism is quite different.

In spite of the fact that physicists today often refer to light as "photons" which have a slightly fuzzy particle nature, the vast majority of phenomena involving light may be explained by the wave model of light. There are very few phenomena where we must use a particle model.

But what is meant by waves and what is meant by particles after all? Could it be the wave–particle duality and the apparent paradoxes that follow from such an opinion may disappear if we try to get a little more precise in our description?

The problems regarding the wave–particle dualism is intimately connected with philosophy. If we take Niels Bohr's view that the purpose of physics is not to tell how the world *is*, but just to find relations between properties we can measure, it is no problem to switch between a particle and wave description of light. However, if we have a philosophical stance close to realism, the huge differences between a physics behind particle and wave phenomena in nature make the wave–particle dualism unacceptable.

It is now 100 years since the last time physicists shifted from one paradigm about light to another paradigm. Perhaps, the time is now ripe for a new paradigm shift?

15.7 A Matlab Program for Generating Noise in a Gaussian Frequency Band

Actually, the full program code for the Hanbury Brown and Twiss model, is available from this Web page (Matlab version only).

```
function  [xx] = whiteNoiseGauss(Fs,N,fcenter,fullFwidth)

% Parameters: Fs : Sampling frequency, N : # data points
% fcenter, fullFwidth : The frequency spectrum has a gaussian
% distribution with center frequency fcenter and full width
% (1/e) in frequency spectrum equal to fullFwidth

% In the calling program, use for example:
% Fs = 44100;
% N = 2^16;
% fcenter = 400.0;
% fullFwidth = 100.0;

fsigma = fullFwidth/2.0;
y = zeros(N,1);
xy = zeros(N,2);
T = N/Fs;
t = linspace(0,T*(N-1)/N,N);
f = linspace(0,Fs*(N-1)/N,N);
ncenter = floor(N*fcenter/(Fs*(N-1)/N));
nsigma = floor(N*fsigma/(Fs*(N-1)/N));
gauss = exp(-(f-fcenter).*(f-fcenter)/(fsigma*fsigma));
ampl = rand(N,1);
ampl = ampl.*transpose(gauss);
phases = rand(N,1);
phases = phases*2*pi;
y = ampl.*(cos(phases) + i*sin(phases));

% Mirror of lower half (Nhalf+1) to make upper half correct
Nhalf = round(N/2);
for k = 1:Nhalf -1
    y(N-k+1) = conj(y(k+1));
end;
y(Nhalf+1) = real(y(Nhalf+1));
y(1) = 0.0;
% plot(f,abs(y),'-g');  % Plotting as a check if desired
% figure;

xy = ifft(y);
xr = real(xy*400);
xx = xr;

plot(t,xr,'-b'); % Plotting if wanted
hold on;
plot(t,imag(xy),'-r');
xlabel('Time (s)');
ylabel('Sound signal (rel units)');
sound(xr,Fs); % Playing the sound for proper frequencies if wanted
```

15.8 Original and New Work, Hanbury Brown and Twiss

R. Hanbury Brown and R.Q. Twiss. A test of a new type of stellar interferometer on Sirius. Nature 178, 1046 (1956).

R. Hanbury Brown, R.C. Jennison, and M.K.D. Gupta. Apparent angular sizes of discrete radio sources: Observations at Jodrell Bank Observatory, Manchester. Nature 170, 1061 (1952).

R. Hanbury Brown and R.Q. Twiss. Correlation between photons in two coherent beams of light. Nature 177, 27 (1956).

A relatively new article 6:37980—https://doi.org/10.1038/srep37980 "The colored Hanbury BrownTwiss effect" based on a quantum approach by Silva et al. is available for free at Nature.

15.9 Learning Objectives

After working through this chapter, you should be able to:

- Explain what distinguishes a real wave from an idealized simple mathematical description of a wave.
- Explain what is meant by temporal coherence and coherence length, and how they can be determined.
- Explain what is meant by spatial coherence.
- Explain why coherence plays a role in experiments involving interference.
- Explain qualitatively why the line width in a frequency spectrum is related to coherence length.
- Explain why summation of N perfectly coherent waves leads to an intensity of N^2 times that of each wave, while the intensity is only N times that of each wave for the summation of N incoherent waves.
- Explain how energy conservation is fulfilled in the two cases in the previous learning objective.
- Explain why we claim that it would have been disastrous if all members in a choir sung perfectly in tune with each other.
- Explain how we can measure coherence lengths with a Michelson interferometer.
- Explain qualitatively why an oscillating charge leads to the emission of electromagnetic waves.
- Provide a qualitative correlation between a composite radio frequency antenna and the diffraction pattern of light from two or more slits.
- Explain why a laser basically gets a (temporal and longitudinal) coherence length that is far greater than does thermal light from, for example, an incandescent lamp.
- Explain qualitatively why population inversion is important for making a laser work.

15.10 Exercises

Suggested concepts for student active learning activities: Temporal/longitudinal coherence, spatial coherence, autocorrelation function, correlation function, coherence time, coherence length, chaotic signals, intensities after summation of coherent/noncoherent waves, frequency width, pinhole, interferometer, retarded potential, dipole radiation, dipole antenna, antenna diagram, laser, population inversion, stimulated emission, spontaneous emission, cavity, standing waves.

Comprehension/discussion questions

1. Attempt to explain why the signal in the middle of Fig. 15.4 varies so much in amplitude even though the signals we started with had a more even distribution of amplitudes.
2. If you have sung in choirs, you have probably noticed that the volume reached with more and more singers does not increase as much as we might think. A soloist with a good voice is not outsung by a choir of maybe 30 people. Why is not that one voice totally drowned by that of the 30?
3. Will a singer who sings a "dead" tone be expected to have a greater or less coherence time for his/her voice than a singer who has significant "vibrato" in his song?
4. Indicate the advantages and disadvantages of people who sing in a choir having relatively short coherent times for the voice (the sound).
5. In Fig. 15.21, red lines mark the portion of the wave pattern that will pass a double slit as waves are passing by. Do you think you would be able to detect interference fringes for the case on the left and/or the case on the right? Explain what would happen in the two cases.
6. Some believe that when we talk about coherent and incoherent waves, there are two well-defined types of waves. In reality, there is a continuous range all the way from "incoherent" to "coherent". Explain.
7. There are advantages and disadvantages associated with both coherent light and incoherent light. In which contexts would you prefer one and when would you prefer the other? As usual, explain your answers.

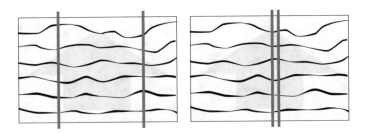

Fig. 15.21 Marking the position of the two apertures in a double slit relative to the wave pattern. The gap between the slits is much larger in the example to the left than in that to the right

8. What will it take to obtain at least five interference fringes after sending light through a double slit when the light we use comes from the sun?

9. Try to describe in your own words what we mean by coherence and coherence time for a wave. What is the difference between spatial coherence and temporal coherence?

10. What is the main idea behind the method of Hanbury Brown and Twiss portrayed in a simplified form in Fig. 15.7?

11. What is the main idea behind "retarded potential"?

12. What is most important for creating an electromagnetic wave from an oscillatory charge or oscillating dipole: the radial component of the electrical field or the tangential component? Explain!

13. Antennas are often designed as electrical dipoles as indicated in Fig. 15.13. Occasionally, however, magnetic dipoles are used as antennas (a simple circular current loop). Sketch the magnetic field around a magnetic dipole, and use this as a basis to argue that the antenna can be expected to be quite effective.

14. In a previous chapter, we found intensity distribution of light diffracted at a single slit. How would intensity as a function of angle from the symmetry axis be expressed in a polar graph (qualitative)? Would the plot have some resemblance to what we find in one of the antenna diagrams in this chapter?

15. Explain how population inversion can be achieved in a four-level system.

Problems

16. Use the program code earlier in the chapter to generate a chaotic signal with centre frequency 5000 Hz and a 3000 Hz width in the frequency distribution. Create a function that can calculate the autocorrelation function. Make a plot. Is the coherence time you arrive at close to that expected? Comment on the results.

17. Use the program code given earlier in the chapter to generate a chaotic signal with centre frequency 5000 Hz and width in the frequency distribution of 3000 Hz. Make a Fourier and wavelet analysis of the signal. Then, calculate the square of the original signal (elementwise squaring). Make a Fourier and wavelet analysis of the squared signal (be sure to include a large enough frequency range in the wavelet analysis). Comment on the results.

18. Determine the coherence time for your own voice. Specifically, the task entails the following steps:

(a) Create a computer program where you can digitize audio, calculate the autocorrelation function, and plot a selected part (see program snippets in Chap. 5). Use the plot to estimate the approximate coherence time of the signal. If you have created a program for digitizing audio when working with Chap. 5, you may save a lot of time using it here as well.

(b) In particular, explain how you chose to utilize the data string you received when digitizing in the analysis. Specifically: How did you choose to let the i and j in Eq. (15.4) run in relation to the total data string?

(c) Determine the approximate coherence time of your own voice when you sing

Fig. 15.22 Waves on water surface at some instant

"eeeee" as evenly as you can. Perform this for 2–3 different pitches. Does the correlation time seem to change much with the pitch?

(d) Perform a wavelet analysis of the signal. Comment on the results.

(e) Digitize another sound and determine the coherence time also for this (Suggested sound: Own voice, same pitch as you used in point c, except that you are now singing "oooooo" instead of "eeeeee". Alternatively: Audio from a piano, guitar or some other musical instrument.). Do you find any interesting differences or similarities compared to what you found in point c?

19. Try to mark, in Fig. 15.22, regions where the waves are relatively well-defined. How big are these regions approximate? And how much of the entire surface have you taken into account in the selection of these areas? Enter lengths in numbers of "approximate wavelengths". [It is helpful to draw a line indicating the approximate direction of waves moving in. To get a correct picture of coherence lengths, the *entire* surface must be included in the statistical processing.]

20. Can you suggest a way to make an interferometer for sound that corresponds to a Michelson light interferometer? (Must be used to measure coherence lengths for sound from, for example, different musical instruments.)

21. Let us look at the antenna diagrams in Figs. 15.15 and 15.23. The radiation pattern in the horizontal plane is approximately what one expects for a single dipole, except that the radiation backward is damped heavily. The radiation diagram in the vertical plane is different and has something in common with the diffraction pattern for a single slit, but with a great deal of deviations. Figure 15.23 shows a photograph of a regular base station for mobile telephony and indicates how the antenna is built up of multiple dipole antennas that have identical radiance.

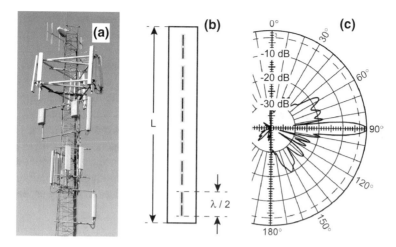

Fig. 15.23 a Base station with many different antennas for mobile telephony. **b** Each antenna is assembled by pair of dipole antennas opposite each other (here are six identical antennas drawn). Each dipole antenna is a half wavelength long. **c** Radiation diagram in the vertical plane of the antenna in Fig. 15.15 (Kathrein 80010621). For this antenna $L = 1.4$m, and it is intended for signals at approximately 2 GHz

This basic construction shows that the comparison with diffraction from a slit is not completely groundless. Note, however, that "slit" may be completely confusing since we use the *length* of the antenna as *slit width*.

(a) Use the radiation diagram in Fig. 15.23c as well as the formula of the angle to the first minimum by diffraction from a single slit:

$$\sin \theta = \lambda/a$$

and the information that the antenna is used for approximately 2 GHz signals, to calculate a, the "effective slit width" based on the given formula.

(b) Compare a with the antenna's outer vertical length. By the way, approximately how many dipole antennas (half a wavelength long) would there be room for within L?

(c) There is a mismatch between L and a even if they are of the same order of magnitude. Do you have any ideas about what the disagreement can bump into? Could you test your ideas using numerical calculation?

22. (a) In Figs. 15.15 and 15.23 are given antenna diagrams for a much used base station antenna. In a specific case, such an antenna is on a mast 22 m above the ground. Determine the intensity at ground level at a distance of 30 m from the mast (measured along the ground) relative to the intensity 500 m from the antenna in that direction in the horizontal plane where the intensity is greatest. [Hint 1: The radial direction in the antenna diagram indicates relative intensity in the number of dB for waves that go out in different directions (in a vertical

plane) from the antenna. Hint 2: We can only specify *relative* intensities. In the right part of the figure, the origin indicates that the intensity is 40 dB lower than in the maximum direction (0°).]

(b) Perform the same calculation if instead a single dipole antenna was used (use Fig. 15.14).

(c) Is it favourable that the base station antenna has the intensity profile it has, or would it be advantageous if a single dipole antenna was used instead?

23. (a) Use the "bandwidth theorem" (the classic counterpart of Heisenberg's uncertainty relationship) to determine the width of the frequency spectrum of a laser pulse using 40 fs (femtoseconds) to pass a point in space (image from the laser is shown on the first page of this chapter). The wavelength corresponding to the centre of the frequency spectrum is 810 nm. [The pulses have a near Gaussian envelope curve in both time and frequency.]

(b) How many wavelengths are there in a pulse?

(c) How "long" is the light pulse (measured along the direction of propagation)? [It will get rather messy if you mix the theory of relativity into such calculations, so we recommend that you stick to a nonrelativistic description.]

(d) Repeat the same calculations as above for a 7.7 fs pulse. [This is a laser currently used in Munich in their attempt to create attosecond laser. Interested is referring to a notice on 7 May 2015 in http://www.Photonics.com headed: "Laser Design Brings Attosecond Spectroscopy Closer" (http://www.photonics.com/Article.aspx?AID=57412 accessed 10 May 2015.)

It is interesting to see what the researchers behind this work write:

"This field of ultrafast physics focuses on phenomena such as electron motions in molecules and atoms, which can take place on attosecond time scales. The ability to generate attosecond laser pulses would effectively permit electron motions to be 'photographed'." As you can see, exciting development is taking place in physics today!

24. Search the Internet to get an impression of the status of lasers in the X-ray region. How far has the development come? Which applications will an X-ray laser find?

Reference

1. XuPanda, https://en.wikipedia.org/wiki/Helium-neon_laser#/media/File:HeNe_Laser_Levels.png. Accessed April 2018

Chapter 16
Skin Depth and Waveguides

Abstract The last chapter begins by asking how far electromagnetic waves penetrate into a metal and introducing the concept of skin depth. It is pointed out that this behaviour will depend both on the frequency of the waves, and whether we are considering a near-field or a far-field situation. This is followed by a treatment of waveguides and how these may be used for transporting well defined ("single mode") electromagnetic waves in the microwave and optical region (single-mode optical fibres). The concept of a "cut-off frequency" is introduced.

16.1   Do You Remember …?

We have previously pointed out in the book that the solution of a wave equation largely depends on the boundary conditions. In Chap. 9, we echoed the same remark in the context of electromagnetic waves. The well-known plane electromagnetic waves are found far from the source and far from structures that can perturb the electrical and/or magnetic field. Plane waves are just one solution of Maxwell's equations, a solution that is only valid in media without free charges, in the remote zone.

What happens if an electromagnetic wave is an incident on a flat metal plate or some other material containing free charges? The charges will be influenced by electromagnetic forces, including the Lorentz force, and will move. The movement will set up a secondary field that will tend to counteract the original field. The free electrons will be able to move over distances amounting to several atomic radii. During their movement, these electrons will collide with atoms and some of their energy will be converted to heat. It is then natural to expect that the electromagnetic field will decrease as it penetrates the material deeper and deeper. The term "skin depth" quantifies this effect and tells us how far into the metal the waves penetrate.

In other situations where the geometry is different, there may sometimes be solutions of the wave equation (or Maxwell's equations) completely different from planar waves. This opens up the possibility to transport waves without significant loss over long distances, and the waves are then transmitted through so-called waveguides. This chapter will deal with skin depths and waveguides.

© Springer Nature Switzerland AG 2018

A. I. Vistnes, *Physics of Oscillations and Waves*, Undergraduate Texts in Physics,

https://doi.org/10.1007/978-3-319-72314-3_16

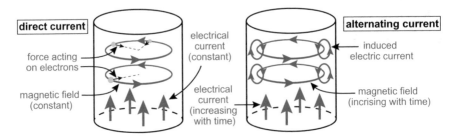

Fig. 16.1 Electric and magnetic fields inside a cylindrical metal conductor with an electrical current. To the left, we have a constant direct current and to the right an alternating current. The directions of the induced currents pertain to the period during which the current grows over time. In a period during which the current decreases, the local induced current loops go in the opposite direction

16.2   Skin Depth

When electromagnetic waves are incident normally on a metal surface, they will be damped as they propagate inside the metal. However, we start with a simpler picture to get the underlying mechanisms.

When we send an alternating electric current through a conductor, the current will not spread evenly over the entire cross section. The current tends to be greatest in the outer parts (or the "skin") of the conductor. The thickness of the layer where the current density is greatest, we call the skin depth.

When we send an alternating current through a cylindrical metal conductor, it is relatively easy to explain the most important mechanism responsible for the skin effect.

A snapshot of the resulting current and fields that this generates is shown in Fig. 16.1. The electrical current will generate circularly oriented magnetic fields perpendicular to and centred in the axis of the conductor. If direct current is flowing through the lead, the electrons will be affected by a force that pulls them towards the centre of the conductor. Called "Hall effect", this phenomenon gives rise to a small potential difference between the outer part of the conductor and the axis of the conductor. The potential difference quickly leads to an electric field that precisely counteracts the transport of electrons towards the centre of the conductor. Aside from this "once and for all" effect that comes into play when power is turned on, the current will be distributed relatively evenly across the cross section with direct current.

With an alternating current, the situation is different. In addition to the effects we have for direct current, *change* in current with time will lead to local current loops that will try to counteract the magnetic field increase ("Lenz's law"). The local current loops cause the current density in the central parts of the conductor to be counteracted while the current density in the outer part of the conductor increases (see Fig. 16.1). However, the local current loops are phase shifted in relation to how current changes over time. Therefore, the overall picture becomes rather complex when we take into account phase shifts, the sum of more contributions to the electron motion, and geometry. As a result, we get a skin effect that causes the alternating current to be greater in the outer parts of the conductor than in the central ones. Therefore, the alternating current does not utilize the entire cross section of the conductor equally efficiently. This means that the resistance of the conductor for AC is different from that for DC.

The induced current loops influence the local current density more and more effectively as the frequency increases. As a result, the layer where the current is flowing becomes thinner with increasing frequency. Skin depth is frequency dependent.

We will shortly derive an expression for skin depth, but can already mention that for aluminium, which is often used in power lines, the skin depth is 11–12 mm at 50 Hz. This means that for thick power lines with a diameter of about 3 cm, most of the flow will involve an outer layer about 1 cm thick and to a lesser degree the central parts of the wire. Occasionally, such power lines are made hollow because the central part does not contribute significantly to the overall conductivity anyway. On other occasion, a steel wire is used as the central core with an aluminium sleeve around it. The steel core provides increased strength to the lead, and the poorer conductivity of steel compared to aluminium plays little role since the current density in the centre is still quite modest.

Instead of one wire that is extra thick when transferring large amounts of power (high current), one sometimes chooses to add two ("duplex") or three ("triplex") lines within each of the three phases of one power line. The two or three wires are then kept at a constant mutual distance of 10–20 cm for, among other reasons to reduce the overall skin depth effect.

16.2.1 Electromagnetic Waves Incident on a Metal Surface

What will happen if an electromagnetic wave in the radio frequency range falls normally onto a metal surface?

In Chap. 9, we showed how Maxwell's equations lead under certain conditions to the following wave equation:

$$\frac{\partial^2 \vec{E}}{\partial t^2} = c^2 \frac{\partial^2 \vec{E}}{\partial z^2} \tag{16.1}$$

where

$$c = \frac{1}{\sqrt{\varepsilon_r \varepsilon_0 \mu_r \mu_0}} \equiv \frac{1}{\sqrt{\varepsilon \mu}}. \tag{16.2}$$

The reader is supposed to be familiar with the symbols.

When the wave is perpendicular to a medium where the conductivity $\sigma \neq 0$ (e.g. a metal), the current density is also different from zero. It can be shown that the wave equation under these conditions gets the form:

$$\frac{\partial^2 \vec{E}}{\partial z^2} = \mu\sigma \frac{\partial \vec{E}}{\partial t} + \mu\varepsilon \frac{\partial^2 \vec{E}}{\partial t^2} . \tag{16.3}$$

We can guess a solution in which the fields decrease exponentially in the metal:

$$E = E_0 e^{i(kz-\omega t)} \tag{16.4}$$

where k now can be complex.

If we substitute this trial solution in Eq. (16.3), we get:

$$k = \sqrt{\mu\omega}\sqrt{i\sigma + \varepsilon\omega} .$$

We see that the wavenumber k in this expression is a complex quantity. It is in line with the fact that the exponent on the right-hand side of Eq. (16.4) has an exponentially decreasing term, as expected.

If the conductivity is large, or more precisely: if $\sigma \gg \varepsilon\omega$, the k expression can be simplified to:

$$k = \sqrt{i}\sqrt{\mu\sigma\omega} .$$

Since $(1 + i)^2 = 1 + 2i - 1$, it follows that

$$\sqrt{i} = \frac{1}{\sqrt{2}}(1 + i) .$$

Consequently k can be expressed as:

$$k = \sqrt{\frac{\mu\sigma\omega}{2}}(1 + i) \equiv \frac{1}{\delta}(1 + i)$$

where δ is the skin depth. By inserting this expression in Eq. (16.4), we obtain:

$$E = E_0 \, e^{i(z/\delta - \omega t)} \, e^{-z/\delta} .$$

The physical solution is the real value of the expression, which is:

$$E(z, t) = E_0 \cos\left(\frac{z}{\delta} - \omega t\right) e^{-z/\delta} . \tag{16.5}$$

The question is, however, whether this is too simple a solution. We assumed above $\sigma \gg \varepsilon\omega$. If we set the current sizes for copper, we will:

$$\frac{\sigma}{\varepsilon\omega} = \frac{6.4 \times 10^{18}}{\omega} \, \text{F}^{-1} \, \Omega^{-1} \, .$$

It turns out that the approximation we made holds for all electromagnetic waves from about the X-ray region and longer wavelengths. However, the formula is only valid for frequencies that are far from significant atomic or molecular resonance frequencies, and also from the normal collision frequency of electrons in their migration through the metal under consideration. For nonmetals, a somewhat more complicated correlation between skin depth and electromagnetic properties is derived from the material, but we do not deal with these details here.

Equation (16.5) seems to be adequate for the chosen geometry. The equation shows that the electromagnetic wave continues inside the metal, but its amplitude decreases exponentially, the attenuation factor for each distance δ (the skin depth) being $1/e$. We put in the data for copper in the expression of the skin depth:

$$\delta = \sqrt{\frac{2}{\mu\sigma\omega}} \qquad (16.6)$$

we find the skin depth to be
- 9 mm at 50 Hz
- 66 μm at 1 MHz
- 100 nm at 30 GHz (radar)

This means that the waves at radio frequencies and higher are severely attenuated in the outer part of a metal. For low frequencies, the damping is far less pronounced.

Figure 16.2 shows the relationship between the skin depth δ and the frequency f for five different metals or alloys in a log–log plot.

From the figure, we see that at 1 MHz the skin depth of aluminium is 90 μm, and for 0.9–1.8 GHz mobile phone frequencies the aluminium skin depth has decreased to about 3 μm! This means that, so far as resistance is concerned, at such high frequencies there is little to be gained by making the wires much thicker. A large surface is more important than total cross section. The word "skin depth" seems to be a good choice!

Skin depth lies also at the basis of induction cookers. The commonly used frequency here is around 24 kHz. Using steel pots, in which the conductivity is not particularly high and the relative magnetic permeability is close to 1 (nonmagnetic material), the skin depth becomes so large that large portions of the electromagnetic field from the stove passes straight through the bottom of the pots. Only when we have materials that have a high relative magnetic permeability (containing magnetizable iron), almost all energy in the fields from the oven will be deposited as heat in the bottom of the pan.

Fig. 16.2 Skin depth as a
function of frequency for
different metals (idealized).
A log–log plot is chosen to
cover many decades. Zereks,
Wikipedia Commons, CC0
1.0, Modified from original
[1]

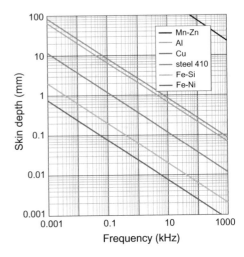

In pots and pans intended for induction cookers, magnetic steel, such as carbon
steel 1010 or stainless steel 432, is used, both of which have a relative magnetic
permeability of about 200. From Eq. (16.6), we see that the skin depth then drops
considerably compared to nonmagnetic material. The skin depth at 24 kHz will only be 0.1–0.2 mm, and accordingly virtually all the energy from the stove will be
deposited as heat in the bottom of the pot.

Comments
The derivation of the expression for the skin depth must be put in perspective. We
have shown that Eq. (16.5) is one possible solution of Maxwell's equations. It has
not been said that the solution in a concrete case actually *is* this solution! Far from
that! We pretended that the solution could be written as a plane wave meaning that
the solution does not depend on x and y. To be applicable, the physics must be such
that there are no boundary conditions that affect the wave in the x- and y-direction
near the place we consider.

This means that Eq. (16.5) must be used with great caution. Geometry in specific
situations is often much more important than skin depths calculated blindly from
Eq. (16.5).

16.2.2 Skin Depth at Near Field

The mathematical derivation in the previous section was based on electromagnetic
waves in the remote zone. That is a situation involving basically electrodynamic
conditions in which time variation in electric field creates a magnetic field, and time
variation in magnetic field in turn again creates an electric field.

When we are dealing with near field, the situation is different. For example, we
might have a power line with strong electric fields without any particular magnetic

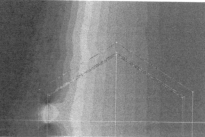

Fig. 16.3 Electric field from a power line (50 Hz) is damped strongly by new woodwork (*left part*: conductivity $\sigma = 1 \times 10^{-6}\,\Omega^{-1}\,m^{-1}$) while old crushed wood does not dampen the electric field (*right part*: conductivity $\sigma = 1 \times 10^{-9}\,\Omega^{-1}\,m^{-1}$). The 50 Hz magnetic field went through the walls without noticeable cushioning in both cases. The figure is derived from a master thesis in physics at University of Oslo: Ellen Røhne: Electrical Fields in Houses Near Power Lines—Measurements and Element Method Calculations, 1997

field, and the converse. In such situations, it is often meaningless to talk about skin depth.

A static magnetic field is not noticeably damped, for example, by an aluminium plate, even if it is thick. At 50 Hz, the induced currents will be so small that the induced magnetic field only causes moderate attenuation of an outer magnetic field. The effect is also highly dependent on geometry. If aluminium plates are used to dampen a 50 Hz magnetic field, they must be fully welded so that the induced currents should flow as freely as possible. At the outer edge of the area the aluminium plates cover, the magnetic field is often stronger than if there were no plates there.

It is completely different with electric fields. Static electric field is shielded very efficiently by having a conductive screen connected to earth. Then charges will be drawn to the screen and will neutralize the field on the opposite side of the source. Even at 50 Hz it is easy to remove, for example, electrical fields in homes near power lines. Even a chicken wire under the roof and connected to the ground provides a very effective damping. In fact, even the small electrical conductivity found in relatively new wood is often sufficient to provide a good damping of electric fields from a power line inside a wooden house close to a power line (see Fig. 16.3). The time period of a 50 Hz period is so long (10 ms for each half-period) that there is sufficient time to draw enough charges through the wood to get a good neutralization of the outer electric field. For old, very dry wooden houses, however, the conductivity of the wood is not good enough to provide a good damping.

Summary: In houses near power lines, the magnetic field from the power lines suffers little damping as it permeates through the walls, while the electric field often becomes quite efficiently damped. This is the reason that in the 1980s and 1990s, there was much focus on magnetic fields from power lines and possible health damage, while the electric fields did not attract similar attention. This difference between electric and magnetic fields shows that skin depth is often an inappropriate term in cases where near fields dominate.

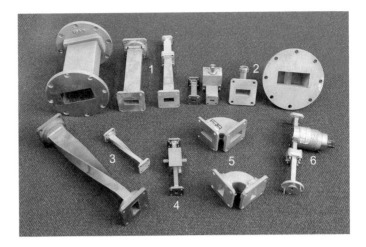

Fig. 16.4 Photograph of some microwave components where waveguides are involved. **1** Straight waveguides, **2** Waveguides with a semiconductor diode working as a detector, **3** Twisted straight waveguides to rotate the polarization 90°, **4** A "phase shifter" for microwaves, **5** 90° bend, one conserving the direction of E, the other conserving the direction of H, **6** A "wave metre", a resonance cavity, to determine the microwave frequency accurately

16.3 Waveguides

A waveguide is a mechanical structure that directs waves from one place to another. In old boats, there was usually a metal pipe from the wheelhouse to the engine room. Someone talking into one end of the pipe could be heard by those at the opposite end several metres away.

An even more well-known waveguide is the doctor's stethoscope. Sound from the heart and lungs is caught in a small funnel held against the skin, and the sound is directed to the ears of the doctor. There is more physics involved in a stethoscope than many are aware of!

In our context, we will concentrate on waveguides for electromagnetic waves. At the bottom lie Maxwell's equations and the wave equation derived in Chap. 9, but now the differential equations must be solved with a set of boundary conditions completely different from what we had in the far field and representing plane electromagnetic waves.

Waveguides for electromagnetic waves are common in the microwave range, that is, frequencies between 2 and 40 GHz (wavelengths from 15 to 0.67 cm). [The range is actually even wider.] The most commonly used forms are hollow rectangular metal pipes, like those shown in Fig. 16.4.

When Maxwell's equations are to be solved for such geometry, the boundary conditions are as follows:

- Electromagnetic waves do not pass through the metal, but are reflected.
- Any electrical field that meets a metal surface must be (approximately) perpendicular to this surface.
- Any magnetic field that meets a metal surface must be (approximately) parallel to the surface.

The electric and magnetic field can of course have other directions towards the metal than those we just listed. However, the above listed boundary conditions are chosen to find a solution of Maxwell's equations that cause as small currents as possible in the metal. It is necessary that the wave does not lose too much energy per unit length as it moves through the waveguide.

There are generally a number of different solutions of Maxwell's equations for a waveguide with a rectangular cross section. Electric and magnetic fields have very different distributions and direction in space compared with the planar wave solution in the remote field zone discussed in Chap. 9.

However, for a given frequency there are only a finite number of possible solutions, and if the wider dimension of the waveguide cavity is less than half the wavelength, it is actually no solution. When the wider dimension in the cavity is between a half and an entire wavelength, and the shortest dimension is only half the longest, there is only one possible solution of Maxwell's equation that corresponds to a wave. The wave pattern we obtain in the waveguide is uniquely determined. We say we have *single-mode transmission*. The lowest frequency that can be sent through a waveguide is called "cut-off frequency".

If we increase the frequency of the electromagnetic waves so that the wider dimension in the cavity of the waveguide is larger than a wavelength, there are at least two different solutions of Maxwell's equations. Then the wave can go through the waveguide in (at least) two different ways. We get a multimode propagation.

In a rectangular waveguide, the smallest dimension is usually half the size of the wider dimension. This ensures that the polarization of the electromagnetic waves can be one way only.

Referring to Fig. 16.5, the following list state the names of frequency band, approximate dimensions for waveguides, cut-off frequencies and optimal frequency ranges:

Frequency band	Wider dimension (mm)	Cut-off frequency (GHz)	Optimal frequency range (GHz)
G	58	2.6	3.2–4.9
X	27	5.6	6.9–10.5
Ka	8.3	19	22–34
Q	5.9	25	32–48

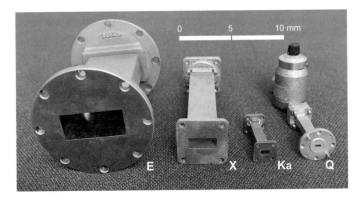

Fig. 16.5 Photograph of waveguides for four different frequency bands. The dimension varies with the frequency so that only one mode will be present for the signal of interest

16.3.1 Wave Patterns in a Rectangular Waveguide

Figure 16.6 shows a schematic representation for the field distribution in a so-called TE10 waveguide. TE stands for "transverse electric". The electric field is perpendicular to the wider surface of the waveguide with rectangular cross section. The field distribution is not the same as a plane electromagnetic wave. Where does the difference lie?

Imagine a plane electromagnetic wave as we discussed it in Chap. 9. If we had such a field distribution within the rectangular waveguide, the electric field would be parallel to two side edges. Such a field would cause large currents of electrons in the metal wall of the waveguide, and thereby a large loss.

> In a waveguide, initial conditions and boundary conditions force a solution of Maxwell's equations that can be at least as "beautiful" as the planar wave solution. The field distribution in a TE10 waveguide is such that the electric field is always perpendicular to the larger internal surface, but the field decreases towards zero as we approach the side surfaces. As a result, there will be far weaker electrical currents in the side surfaces than would be with a plane wave.

Occasionally, it is said that the wave pattern of a waveguide corresponds to a planar wave being reflected back and forth between the walls of the waveguide. This is a misleading description. The waves are solutions of Maxwell's equations under the given boundary conditions and are a distinctive solution. However, when the waveguide dimension becomes large relative to the wavelength, there are many different solutions of Maxwell's equations. In such cases, it makes sense to compare solutions with reflected plane waves through the waveguide.

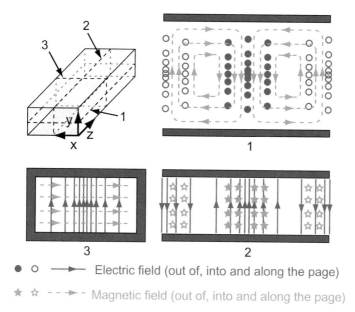

● ○ ——▶— Electric field (out of, into and along the page)

★ ☆ – – ▶ – Magnetic field (out of, into and along the page)

Fig. 16.6 Field distribution of a TE10 mode for the electric field inside a rectangular waveguide. To fit the dimensions of the waveguide relative to the wavelength, only the TE10 mode survives. The wave moves at close to the speed of light in vacuum in the z-direction (right in sections 1 and 2)

However, the electrical field lines across the waveguide start and end in electrical charges on the surface inside the waveguide. Since the wave moves along the waveguide, these charges must also move. This causes induced currents in the inner surface of the waveguide. This is unfortunately not shown in our figure. The inner surface of waveguides is usually coated with silver or gold in order to make the conductivity as large as possible. Then the loss will be minimal. The silver or gold need only be a few microns thick since the skin depth at these frequencies happens to be so small.

Electromagnetic waves with frequencies in the range of 2–60 GHz have traditionally been used for radar, but now these frequencies are also used for mobile telephony and data transmission. Particularly for radar purposes, large powers on the signal transmitted from a transmitter to the radar antenna are often used. It is problematic to send such signals through common wires and coaxial cables—waveguides can often withstand higher powers in the transmission. The microwaves then follow the tube system up to several metres from the generator (preferably so-called klystron) to the antenna where the microwaves are transmitted.

The waveguides are usually made as tubes with rectangular cross sections and flanges to unscrew different pieces. Some pieces can turn the field 90°, and other pieces can make a 90° break on the waveguide itself (see Fig. 16.4).

One attempts to avoid cracks in the waveguides, which prevents currents in the surface. The currents go along the wide walls. To avoid interrupting these currents, we can only make long slots *along* waveguide if the slot is made on the wide side.

By placing two waveguides on top of each other and making a common hole through the walls (on the broad side), some of the waves from one waveguide can be allowed to leak into the other. That way one can make wave dividers and wave combiners.

If a semiconductor diode is placed across the waveguide (and one end is directed as a separate wire), we get a detector that gives a signal proportional to the intensity of the waves passing (an example is given to the far right in Fig. 16.4).

If the wavelength is less than the wider dimension of the waveguide, the electric field may form several different patterns/distributions (multiple "modes") in rectangular (and circular) waveguides. Waves that have different motion patterns, or modes, go at slightly different speeds through the waveguide. For certain layouts, this is unfortunate. A "single-mode" solutions are preferable. We do not enter any mode other than TE10 in this round.

It is an interesting challenge to use Maxwell's equations to determine the direction of a TE10 wave when we have a drawing of the field distribution in a waveguide (see problems at the end).

16.4   Single-Mode Optical Fibre

A single-mode optical fibre consists of a very thin cylindrical core made of very pure silica or fused quartz and diameter only a few microns, surrounded by a layer of another type of glass with a slightly different refractive index than the core. It is usually surrounded with plastic sleeves of different types.

The core may also be made of plastic, but the attenuation is then larger than for glass. Plastic core fibres can only be used for communications over short distances.

It is common to hear that in an optical fibre the light stays in the fibre because of total reflection (based on Snel's refraction law). We have partly done the same earlier in the book.

For large-diameter optical fibres in relation to the wavelength, it is perfectly appropriate to use such an explanatory model. In that case, the interface between the core and the casing satisfies the preconditions we made when we derived reflection laws based on Maxwell's equations.

When the diameter of the core of the optical fibre is shrunk to about six times the wavelength, it will be different. Then we can no longer consider the light as plane waves, because plane waves will not survive in such a fibre.

Then there are other solutions of Maxwell's equations that force themselves forward. In Fig. 16.7, is shown the cross section of several possible solutions of the wave equation for this type of geometry and wavelength. We show different patterns that show where the electromagnetic field is greatest (red and blue only indicate that if the electric field across the fibre in a red area has a maximum value, the field in a

$l = 0, m = 1$	$l = 1, m = 1$	$l = 2, m = 1$	$l = 3, m = 1$
$l = 0, m = 2$	$l = 1, m = 2$	$l = 2, m = 2$	$l = 4, m = 1$

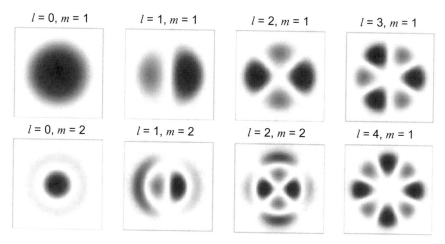

Fig. 16.7 Distribution of electric field across an optical fibre for eight different "modes". Only the simplest survives in a "single-mode fibre" ("single-mode fibre"). Red and blue signify different directions of the electric field in the two areas. The modes are classified using two numbers that give the symmetry properties of the mode. Try to find out what the two parameters really tell us. Figure generated with the software RP Fibre Power. R. Paschotta, Modified from original [2]. Reproduced with permission from the author and publisher

blue area is negative along the current direction). We say that the field has different modes to organize itself within an optical fibre. A complete description of the modes would require a three-dimensional sketch, but we do not go into detail here.

The point is that when the diameter of the fibre is made smaller and smaller, the higher modes will not be able to propagate along the fibre. For an appropriate diameter, only the simplest mode will survive. If we shrink the diameter still further, even this mode will not survive over long distances.

An optical "single-mode fibre" is therefore characterized by a "cut-off wavelength" and can be used to "clean up" laser light that does not have a perfect Gaussian intensity profile.

When light in the infrared region is sent through a single-mode fibre that has the right dimensions and has ultrapure glass in the core, the loss is incredibly low! Furthermore, as seen from Fig. 8.6 in Chap. 8 the index of refraction n is very close to constant over a wide wavelength range in the infrared region (e.g. FK51A glass). Thus, dispersion is very low indeed.

Since both the loss and dispersion are incredibly low, the IR light can move in a very carefully defined manner, and short pulses can be sent many, many kilometres

Fig. 16.8 To connect laser light from a laboratory laser into a single-mode optical fibre, one uses a microscope objective mounted on a three-dimensional stage with precision screws. The Airy disc at the waist of the laser beam after passing the microscope objective should be of the same size as the effective fibre diameter. Thus, diffraction comes into play even in this part of physics

before the pulse shape needs to be cleaned before the signal is forwarded. The result is a very high pulse rate and flow of information.

These are the optical fibres that ensure our impressive Internet. In other words: a special solution of Maxwell's equations, where initial conditions and boundary conditions are alpha and omega, in concert with development of materials with low dispersion, is what keeps the Internet going! Plane waves are not involved!

One disadvantage of using single-mode fibre is that the diameter of the core is so small that it is a challenge to get sufficient light from an laser beam in open air into the fibre. In our laboratory, we often use single-mode fibres, for example, to clean up laser light with a wavelength of 405 nm. The inner part of the fibre (where the light is going to go) is then only 2.7 μm in diameter. Around this core, a "cladding" zone with a lower refractive index extends to 125 μm and a "coating" zone is added to a diameter of 245 μm. Outside this comes a protective layer made of plastic.

If we start with a laboratory laser that normally sends the beam with a diameter of at least 1 mm into the open air, the beam must be focused strongly. It is done with a microscope objective (see Fig. 16.8). The end of the fibre must then be placed just in the focal plane of the focused beam, and the fibre must have a direction that completely coincides with the optical axis of the beam. It is a great patience test to get as much of the light into the fibre as possible! Also when the light is released by a single mode fibre, we often need to use a microscope objective to prevent the laser beam from diverging too much (freshly from Chap. 13 how is light going through round holes with very small diameter!).

For telecommunications, special adapters have been developed that make the connection far easier. In such systems, laser beams in air are not used at all.

16.5 Learning Objectives

After working through this chapter, you should be able to:

- Explain the term skin depth when an alternating current passes through a metal wire.
- Explain the concept of skin depth when electromagnetic waves meet a metal surface.
- Know what parameters affect the size of the skin depth and know about skin depths for a few frequencies and metals.
- Explain that a simple analysis of skin depth may have significant weaknesses.
- Explain the distribution of electrical and magnetic fields and electrical currents in the walls inside a TE10 rectangular waveguide if you are given a figure like Fig. 16.6.
- Explain why Snel's refraction law is not relevant for explaining how a single-mode optical fibre works.
- Indicate why single-mode fibres are attractive in research and technology.
- Explain why it is a challenge to connect light from an open laboratory laser into a single-mode optical fibre, as well as coupling from such fibre back to a free laser beam in air.

16.6 Exercises

Suggested concepts for student active learning activities: Skin depth, waveguides, single mode, multimode, field distribution, wave pattern, boundary conditions, rectangular waveguide, circular waveguide, cut-off wavelength/frequency, optical fibre.

Comprehension/discussion questions

1. Why does an old-fashioned aluminium casserole not work on an induction cooker?
2. What is the big difference between the physics involved when sending electromagnetic waves against a piece of glass and a corresponding piece of metal?
3. Why do we need to change the dimensions of a rectangular waveguide when we switch the frequency of microwaves to be transmitted through the waveguide?
4. The cross section of a conductor used in power lines can sometimes look as shown in Fig. 16.9. Try to explain why the conductor is built in this special way.

Problems

5. (a) Can you tell from the field distribution shown in Fig. 16.10 the direction in which the microwaves propagate in the rectangular waveguide?
   (b) Point out where there must be charges on the inner surface of a waveguide, and how these charges must move as the microwaves move through the waveguide.

Fig. 16.9 Cross section of a type of conductor used in power lines hanging between large masts across large parts of the country. Clark Mills, Wikimedia Commons, CC BY-SA 3.0, [3]

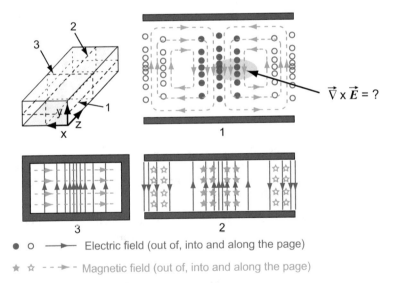

● ○ ———▶ Electric field (out of, into and along the page)

★ ☆ ---▶- Magnetic field (out of, into and along the page)

Fig. 16.10 Consider the field distribution in the shaded area and calculate the specified size. With the help of Maxwell's equations, you should be able to predict the time development

(c) Occasionally, for various reasons, we want to make narrow slits across a wall of a waveguide. In which direction should the gap be made to disturb as little as possible the propagation of the wave ? Justify as always the answer.

6. (a) A single-mode fibre designed for light of wavelengths between 450 and 600 nm has a core diameter of about 3.5 μm. Calculate about how many per cent of the energy flux in a laser beam we had received into the fibre if such a fibre was directly inserted into the beam from a regular laboratory laser without using a microscope objective to focus the beam onto the fibre. The beam diameter for many laboratory lasers is about 1.5 mm.

(b) When the light returns from the fibre to air at the other end, we get diffraction. Calculate the beam diameter 1 m after the light went out of the fibre.

(c) Approximately how much focal length should be on a microscope objective that we can place just after the light emanates from the fibre to create a laser beam about 1.5 mm in diameter?

7. Go to the web pages of, for example, ThorLabs (www.thorlabs.de) and search for "single-mode optical fibre" to find the mode field diameter (the diameter of the core) of three different single-mode fibres (calculated for different wavelengths). (Do not get confused by the cladding and coating diameters. The cladding is glass

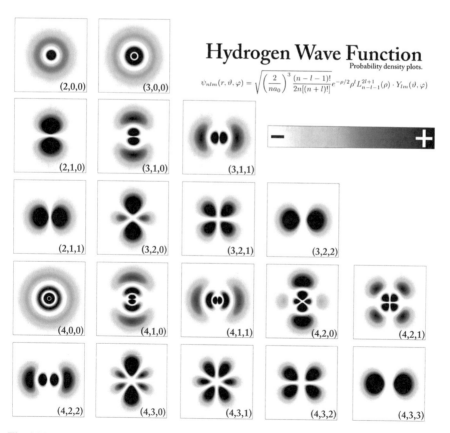

Fig. 16.11 Electron orbitals for the hydrogen atom in a quantum mechanical description. Read the assignment text for details. PoorLeno, Public Domain, Modified from original [4]

with a different index of refraction than the core, and the coating is often plastic to protect the tiny glass string). You may want to check data for "Single-Mode FC/PC Fibre Optic Patch Cables" to find fibres for widely different wavelengths. Do you find some regularity as to how the diameter of the core varies according to the wavelength?

8. Fig. 16.7 shows the patterns in the different modes of how electromagnetic waves (light) can organize themselves as they pass through an optical fibre. In Fig. 16.11, there is an overview of some of the common electron orbital of the hydrogen atom. The orbits show an average of how the quantum mechanical wave function differs when we cross the atom. The distribution in Fig. 16.7 is based on classical physics, yet there are certain similarities between the two figures. We need to be somewhat careful about the comparison since the fibre is a two-dimensional problem while the electrons are linked to a three-dimensional problem. In spite of this: Can you understand that there is a kind of "quantization" both in the classical system and in the quantum mechanics? What is the underlying reason that we get "quantization" in these cases.

References

1. Zureks, https://commons.wikimedia.org/wiki/File:Skin_depth_by_Zureks.png. Accessed April 2018
2. R. Paschotta, *Encyclopedia of Laser Physics and Technology*, vol. II. (Copyright Wiley-VCH Verlag GmbH & Co. KGaA, 2008), p. 805
3. Clark   Mills.   https://commons.wikimedia.org/wiki/File:Sample_cross-section_of_high_ tension_power_(pylon)_line.jpg. Accessed April 2018
4. PoorLeno,   https://en.wikipedia.org/wiki/Atomic_orbital#/media/File:Hydrogen_Density_ Plots.png. Accessed April 2018

Appendix A
Front Figure Details

The figure on the front of this book is the diffraction pattern of He–Ne laser light after passing through at 10 μm diameter circular hole. The central peak is far more intense than the surrounding circular rings with increasing diameters. Thus, the central peak and the first ring are heavily overexposed in order to see the ring structure outside this peak. The size of the Airy disc (central peak) as well as the width of the first ring is considerably larger than it would have been without the overexposure.

© Springer Nature Switzerland AG 2018
A. I. Vistnes, *Physics of Oscillations and Waves*, Undergraduate Texts in Physics,
https://doi.org/10.1007/978-3-319-72314-3

Index

© Springer Nature Switzerland AG 2018
A. I. Vistnes, *Physics of Oscillations and Waves*, Undergraduate Texts in Physics,
https://doi.org/10.1007/978-3-319-72314-3

Printed in the United States
By Bookmasters